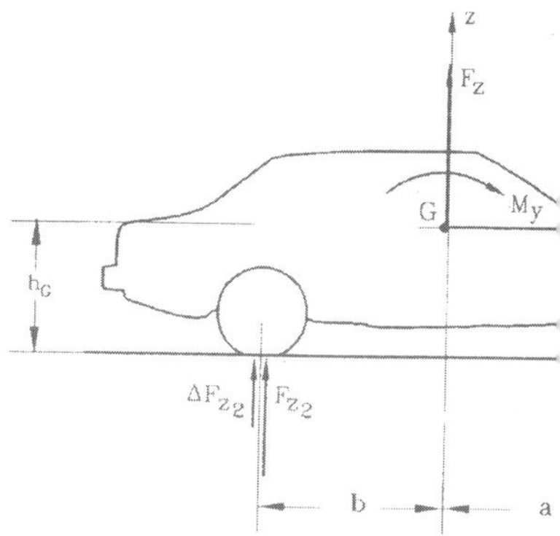

MOTOR VEHICLE DYNAMICS

Modeling and Simulation

Series on Advances in Mathematics for Applied Sciences – Vol. 43

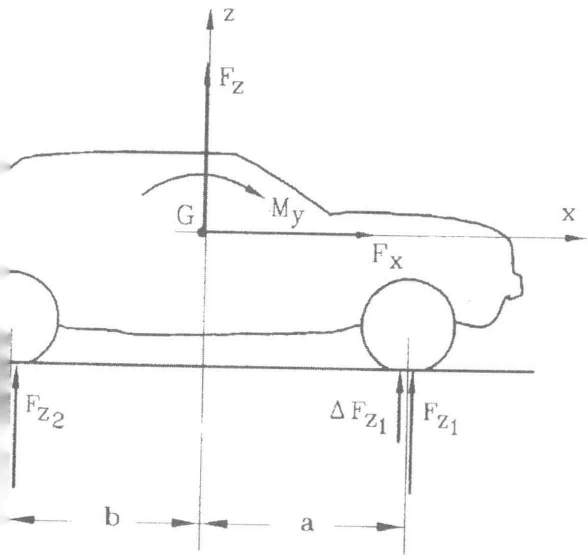

Giancarlo Genta
Dipartimento di Meccanica
Politecnico di Torino, Italia

MOTOR VEHICLE DYNAMICS

Modeling and Simulation

World Scientific
Singapore • New Jersey • London • Hong Kong

Published by

World Scientific Publishing Co. Pte. Ltd.
5 Toh Tuck Link, Singapore 596224
USA office: Suite 202, 1060 Main Street, River Edge, NJ 07661
UK office: 57 Shelton Street, Covent Garden, London WC2H 9HE

Library of Congress Cataloging-in-Publication Data
Genta, G. (Giancarlo)
Motor vehicle dynamics : modeling and simulation / G. Genta.
p. cm. -- (Series on advances in mathematics for applied sciences : vol. 43)
Includes bibliographical references and index.
ISBN 9810229119 (alk. paper)
1. Motor vehicles -- Dynamics -- Mathematical models. 2. Motor vehicles -- Dynamics -- Computer simulation. I. Title. II. Series.
TL243.G46 1997
629.2'31--dc21 96-51020
CIP

British Library Cataloguing-in-Publication Data
A catalogue record for this book is available from the British Library.

First published 1997
Reprinted 1999, 2003 (in bigger trim size)

Printed in Singapore by World Scientific Printers (S) Pte Ltd

To Franca and Alessandro

Preface

A motor vehicle is generally intended as a vehicle with mechanical propulsion which can move freely on the surface of the Earth (or of any planet with solid surface, so as to include in the definition the vehicle used by the Astronauts in the Apollo missions and all future vehicles for planetary exploration), without being mechanically constrained to follow a predetermined trajectory. A modern motor vehicle is usually supported on wheels with pneumatic tires and, with some exceptions as trolleybuses, has an autonomous energy storage system which provides the energy for propulsion.

The diffusion of motor vehicles is at present so large that modern civilization owes to it many of its characteristics. Their role in providing personal mobility and transportation of goods makes them an essential necessity in everyday life: the lives of a great number of men are influenced by motor vehicles, at least in developed countries, while in developing ones is often their insufficient availability which influences, sometimes with tragic issues, human lives.

Their widespread, and sometimes excessive, use is however not free from drawbacks, which in some instances can be very notable and can overshadow their benefits. A great number of books, papers, lectures and discussions has been devoted to this subject, in which experts of different disciplines did express opinions of any possible type. From time to time, waves of public opinion aiming to limit or to liberalize the use of motor vehicles, particularly in city centres, were registered.

The decisions concerning the best use of motor vehicles, and particularly the relationships between public and individual transportation or between the use of motor vehicles and other means, are mainly political and economical concerns. It is however a duty of automotive engineers the study and the implementation of all those devices which are able to boost their useful characteristics or to reduce their disadvantages.

Among the technical measures which were originated by new exigences and concerns of the public opinion, those having the aims of enhancing road safety, reducing pollution and energy consumption due to motor vehicles can be remembered. In the sixties the problem of road safety started to be felt in a new, or at least stronger, extent. Many studies aimed to enhance the safety of road vehicles were started, and many *ESV* (Experimental Safety Vehicles) were built. Before those studies yielded their fruits, a new consciousness of "ecological" problems caused by the high concentration of motor vehicles in urban areas, and particularly of air pollution, focused the interest on this problem. Again, the standards regarding pollution had not yet been met that the soaring cost of energy caused a new redirection of the priorities and aims of technical studies and law measures towards energy conservation.

As a result of the research and development work which was dedicated to these three aspects of vehicle construction, notable improvements which involved the structure of vehicles, their power plant and even some details which can appear to have more an aesthetical function, did take place; the importance of these changes is often far deeper than the user can realize. As there is always a certain time delay between the innovation and its actual implementation, particularly in mass production industries in which any change requires huge capital investments, many of the results of these studies will show out in the future, causing the motor vehicle of the eighties and nineties, which is the result of this research effort, to be deeply different from that of only twenty years before. New technological advances, particularly those linked to a larger use of electronics in motor vehicle industry, are now under development.

Motor vehicle mechanics, intended as the branch of applied mechanics studying the behaviour of motor vehicles and their performances, did play an important role in the recent progress, undergoing deep modifications.

Actually, together with the new needs expressed by the public opinion and those linked with energy conservation, a fast and unprecedented progress in the field of computation machines allowed to undertake the mathematical solution of problems which only few years before could be tackled only through long and painstaking experimentation.

It is to this combination of new problems and of new means to solve them that the recent progress of motor vehicle mechanics can be ascribed.

The present text is a revised and expanded edition of a collections of notes of the lectures of Motor Vehicle Mechanics given by the author to the students of Mechanical Engineering and of the School of Motor Vehicles at the Politecnico (Technical University) of Torino, Italy. The aim was that of collecting in a single text the foundations of the discipline and those more recent developments which, being published mainly on journals and proceedings, are not readily available to students and technicians working in industry.

The text was later expanded in order to deal with the discipline in a more systematic way, although the author is well aware of its lack of completeness.

The actual treating of the subject is preceded by some short notes illustrating the historical development which eventually lead to modern motor vehicles, with the assumption that the knowledge of the historical development of concepts and practical solutions is of great use to those who must deal with them in their everyday work.

In the first two chapters of the text the forces acting on motor vehicles are studied. The behaviour of pneumatical tires and the aerodynamics of vehicles are treated in detail but limiting the study to those concepts which are directly linked with the behaviour of the vehicle, without entering into those details usually found in specialized texts.

The following two chapters deal with the study of the dynamics of vehicles modeled as rigid bodies. Such models allow to predict most of the performances of road vehicles, particularly for what motion on straight road is concerned.

The study of the dynamic behaviour of motor vehicles in normal conditions is then continued taking into account the presence of suspensions and, even if briefly, the deformation of some parts of the structure itself.

A last part of the text deals with the dynamics of road accidents, and in general with the motion of the vehicles in abnormal conditions.

An effort was made in order to describe the whole subject in a way readily usable for the use of modern numerical computers. All systems of reference are unified and formulae are written in such a way that can be implemented without having to decide each time about sign conventions. This is done even in those cases in which it forced to go against a consolidated practice.

All formulae were written using consistent units, and where numerical examples or experimental values are given, reference to the International System is done.

The author is grateful to the colleagues and the students of the Mechanics Department of Politecnico di Torino for their suggestions, criticism and general exchange of ideas. The contributions due to the students who have performed their thesis work in the field of vehicle dynamics, sometimes in connection with car or motorcycle racing teams, are gratefully acknowledged, as is the thrust to keep the material updated due to endless discussions with my son, Alessandro, engineering student and sport cars enthusiast.

Particular thanks are once more due to my wife, Franca, for her encouragement and for having done as usual the tedious work of revising the manuscript.

Giancarlo Genta
Torino, January 1997

Contents

List of symbols

a	acceleration; distance between centre of mass and front axle
b	distance between centre of mass and rear axle
c	viscous damping coefficient
c_{cr}	critical damping
c_{opt}	optimal damping
d	distance between tractor and trailer; distance
e^*	restitution coefficient
f	friction coefficient; rolling coefficient
f_0	rolling coefficient at zero speed
f_r	rollover factor
f_s	sliding factor
g	acceleration of gravity
h	height of the road surface on a reference level
h_G	height of the centre of mass on the road
i	grade of the road; imaginary unit ($i = \sqrt{-1}$)
i_t	transversal grade of the road
k	stiffness
l	wheelbase; reference length
m	mass
m_e	equivalent mass
m_s	sprung mass
m_u	unsprung mass
$\vec{n}$	normal unit vector
p	pressure
p, q, r	components of the angular velocity along x, y and z axes

q	specific consumption of the engine
s	crush
t	temperature; time; track
$\vec{t}$	surface force per unit area
u, v, w	components of the velocity along x, y and z axes
v	slip velocity
v_a	ambient wind velocity
x_N	distance between centre of mass and neutral/steer point
z	height on sea level
$\{z\}$	state vector
$[A]$	dynamic matrix
$[B]$	input gain matrix
C	cornering stiffness
$[C]$	damping matrix
C_i	$i = f, x, y, z, M_x, M_y, M_z$: aerodynamic coefficients
C_γ	camber stiffness
E	energy; Young's modulus of the material
E_n	energy needed for motion
F	force
$\mathcal{F}$	Rayleigh dissipation function
F_c	centrifugal force
G	centre of mass; shear modulus of the material
H	thermal value of fuel
$H(\lambda)$	frequency response
I	area moment of inertia, impulse
$[I]$	identity matrix
J	mass moment of inertia
$[J]$	inertia tensor
J_e	equivalent moment of inertia
J_w	moment of inertia of the wheel
K	coefficient of the term in V^2 of rolling resistance; stability factor; crushing modulus
$[K]$	stiffness matrix
K^*	understeer gradient
K_r^*	impact resistance modulus
M	moment; torque
$[M]$	mass matrix
M_f	braking moment
M_s	steering moment
$\mathcal{M}_s$	static margin

M_z	aligning moment
$Oxyz$	vehicle reference frame
$Ox''y''z''$	"wind" reference frame
$OXYZ$	earth-fixed (inertial) reference frame
P	power; stiffness of the tire
P_a	available power
P_e	engine power
P_n	power needed for motion
Q	fuel consumption per distance travelled
Q_i	i-th generalized force
R	radius of the wheel (unloaded); radius of the trajectory
R_c	radius of the trajectory (low speed conditions)
$\mathcal{R}_e$	Reynolds number
R_e	effective rolling radius of a tire
R_l	radius of a tire under load
$\mathcal{S}$	reference surface
$S(\lambda)$	power spectral density
$\mathcal{T}$	kinetic energy
T_e	engine torque
$\mathcal{U}$	potential energy
V	vehicle speed
V_r	relative velocity
V_R	relative velocity
X_r	rolling resistance
W	weight
α	sideslip of the tire; aerodynamic incidence; grade angle of the road
α_t	transversal grade angle of the road
β	sideslip angle of the vehicle
β_a	aerodynamic sideslip angle of the vehicle
β_0	structural index
γ	camber angle
δ	steering angle
δ_c	steering angle (low speed steering)
$\delta\mathcal{L}$	virtual work
δx	virtual displacement
ϵ_f	efficiency of the brake
ζ	damping ratio ($\zeta = c/c_{cr}$)

η	efficiency
η_e	engine efficiency
η_f	braking efficiency
θ	tractor-trailer angle; pitch angle
λ	frequency; circular frequency; friction coefficient on the contact surface
λ_n	natural frequency
μ	dynamic viscosity; traction coefficient
μ_x	longitudinal force coefficient
μ_{x_p}	longitudinal traction coefficient
μ_{x_s}	sliding longitudinal traction coefficient
μ_y	cornering force coefficient
μ_{y_p}	lateral traction coefficient
μ_{y_s}	sliding lateral traction coefficient
ν	kinematic viscosity
ρ	density of air
ρ_f	density of fuel
ρ_c	constant of proportioning valve
σ	normal pressure; longitudinal slip
τ	shear stress; transmission ratio; time delay; nondimensional time
τ_s	steering transmission ratio
ϕ	roll angle
ϕ_t	phase delay
χ	torsional stiffness
ψ	yaw angle
ω	angular velocity
ω_e	engine speed
ΔF_z	load shift
Γ	damping coefficient for rotations
Π	torsional stiffness of the tires of an axle
$\Im$	imaginary part
$\Re$	real part
$_{,x}$	differentiation with respect to variable x

Chapter 1

Short historical notes on motor vehicles

1.1 Ground vehicles with mechanical propulsion

Ground vehicles with mechanical propulsion appeared only quite recently in the history of human civilization. They can be said to be one of the main consequences of the industrial revolution.

To allow their construction a prime mover able to move itself, together with the vehicle structure and the payload, was needed. Remembering that the power needed to move the mass m at the speed V on a level surface with coefficient of friction (sliding or rolling) f is equal to $P = mgfV$, it is easy to conclude that the minimum value of the power/mass ratio of a prime mover able to move itself is

$$\frac{P}{m} = \frac{gfV}{\eta\alpha} , \tag{1.1}$$

where α is the ratio between the mass of the engine and the total mass of the vehicle and η is the total efficiency of the mechanism which transfers the power and propels the vehicle.

Prime movers with an adequate power/mass ratio were practically not available until the XIX century, and this consideration is sufficient to explain the above mentioned delay.

The development of a suitable prime mover is however not sufficient for the construction of a successful automotive vehicle. The problems related to the construction of suitable transmission, propulsion, control and guidance systems must be successfully solved as well. The development of automotive vehicles is based consequently on that of vehicles propelled by animals and on the developments of the wheel, in its two aspects of supporting and propelling device.

1.2 Vehicles on wheels from prehistory to the end of the Roman Empire

1.2.1 *The supporting wheel*

For hundreds of thousands of years man did live without using any particular means of transportation. When he had to move an object, he simply lifted and carried it, if he was strong enough. If the object was too heavy, he arranged to drag it along. It is well possible that occasionally branches or other round objects were slipped under the load to reduce friction, but no evidence of this practice remains.

With the neolithic revolution the need for transportation greatly increased and, on the other hand, the practice of taming animals opened new perspectives. The development of agriculture caused the need to transport seeds to the field and crops back to the homestead and the number of objects which were considered important and which men needed to carry with them increased as a result of the new exigencies of village life.

It is given for sure that sleighs were used in northern Europe before 5000 B.C., and their use at that time in other places can be inferred. Sleighs and drags can actually be used for transportation not only on snow and ice, but also on grassland (American Indians used the travois well in the XIX century) and even in deserts and sometimes on rocks.

It is impossible to state when a drag was for the first time mounted on a pair of wheels or who operated this technical revolution. Ancient wheels were made mainly of wood, and little direct archeological evidence could remain.

About 3500 B.C. the potter's wheel was introduced in order to produce pots with axial symmetry. The use of the potter's wheel can be inferred from the marks left on pots made with it. The supporting wheel for vehicles is thought to have been originated about the same time.

The more ancient evidence of a wheeled vehicle is from a pictogram on a tablet from Inanna temple in Erech, Mesopotamia. Such document dates back to slightly later than 3500 B.C., and includes a small sketch of a cart with four wheels, together with that of a drag (Fig. 1.1a).

The vehicle shown in Fig. 1.1b has two features which were typical of all vehicles of its times for more than a thousand years: The wheels are discs made from three planks of wood and the animals are harnessed to a central shaft. This uniformity of types of wheels and of driving systems, particularly if compared with the great variety of vehicle structures, has

Fig. 1.1 (a) Pictograms representing a sleigh and a wagon. From a tablet from about 3500 B.C. found at Erech, Mesopotamia. (b) Copper model of a war chariot with 4 onagers harnessed. It was found in a tomb at Tell Agrab in Mesopotamia (third millenium B.C.).

led to the opinion that the wheel was "invented", or better, developed, in a certain place and then started a slow diffusion in all the ancient world. In various places where it was introduced, the local type of sleigh was adapted to the new vehicle, by using the standard wheels and harness.

The place where the wheel was first developed is not known, but it is possible to infer that it was in Southern Mesopotamia, where the wheel was for sure used about 3500 B.C. The diffusion of the wheel was quite slow. Evidence of its use dates from 3000 B.C. in Elam and Assyria, 2500 B.C. in Central Asia and Indus Valley, 2250 B.C. in northern Mesopotamia, 2000 B.C. in southern Russia and Crete, 1800 B.C. in Anatolia, 1600 B.C. in Egypt and Palestine, 1500 B.C. in Greece and Georgia, 1300 B.C. in China and about 1000 B.C. in northern Italy. Some centuries later it reached northern Europe.

It is not possible to understand from ancient pictures whether the axle did turn together with the wheels or was stationary. The fact that the central hole of the wheel disc was round has little meaning as a circular hole can also be explained by the ease of construction. It is likely that both solutions were used, as there are still people who use those primitive technologies nowadays.

It is however likely that the wheel did not derive from the roller: The types of wheels used would rule that out, and it is likely that, in the mind of the ancient wheelmaker, the wheel and the roller had little in common.

The need to build lighter wheels for war chariots probably led to the

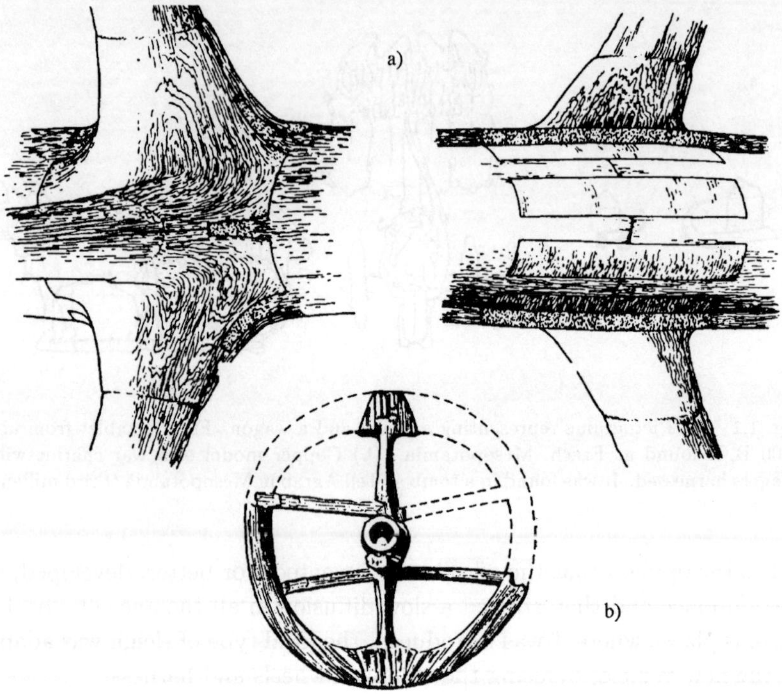

Fig. 1.2 (a) View and cross section of the central part of a wheel of an Egyptian war chariot (1350 B.C.), found in a tomb near Thebes. (b) Wheel found in Mercurago, northern Italy (about 1000 B.C.).

development of the spoked wheel, which is much more efficient for such use. The wheel with spokes was first used probably at about 2000 B.C. and by 1600 B.C. it reached its fully developed form, particularly in Egypt. The central part of a wheel of that type (with 8 spokes) is shown in Fig. 1.2a. It is a part of a chariot dating back to 1350 B.C., found in a tomb near Thebes. The spokes are fitted in the hub; the felloe is usually built in various parts but some examples of felloes in one piece, bent in a circular shape, were found as well.

A wheel which seems to be a stage in the evolution between the disc and the spoked wheel is shown in Fig. 1.2b. The time this wheel was built has however led one to think that it was simply an attempt to copy a spoked wheel by a wheelwright used to disc wheels; but wheels of the same type are represented on more ancient Greek paintings.

In many cases, even in very ancient times, wheels had a hoop or tire or at least some device to strengthen the rim. Some disc wheels have a wooden rim, in one or more pieces. Sometimes the rim of the wheel is inlaid with copper nails, to reduce wear or perhaps to keep in position a leather tire. Certainly many Egyptian war chariots had wheels covered with leather. In some pictures, even very ancient ones, something which looks like a metal tire can be seen. The evidence of such practice is however much more recent, dating back to about 1000 B.C. These metal tires were built in various parts, welded together and then shrink-fit to the wheel.

1.2.2 *Animal traction*

As it was already stated, only after animals were tamed could wheeled vehicles be propelled in a proper way. In Mesopotamia both transportation vehicles and war chariots were pulled by onagers. Also oxen were without any doubt used for transportation.

The spoked wheel, which appeared about 2000 B.C., was accompanied by the use of the horse for the traction of war chariots. It is not known where horses were first tamed and used for that purpose, but the scarce archeological evidence indicates that it should have happened in north-east Persia, and that from that region the use of horses spread to the whole ancient world, from China to Egypt and Europe.

The lack of a proper knowledge of the anatomy of animals led to harnessing onagers and horses in the same way as oxen. A collar was added to the yoke, as horses could not exert much force directly with the shoulders, like oxen. The collar however did press on the windpipe of the animal causing a decrease of the physical efficiency when a strong tractive force was needed. Horses could not exert on the harness all their force and this situation did not change substantially until the end of the Middle Ages, even if some progress was slowly obtained. In roman times horses are reputed to have been able to provide a maximum tractive force of about 62 kg. The failure to understand this simple fact had severe consequences on the development of ground transportation, and in general on all mechanism driven by animals.

Ancient people were not used to shoe horses. Horses, camels and mules were fitted with provisional shoes of metal, leather or straw when they had to walk over slippery or very hard surfaces. The use of adequate horse shoes is thought to have been introduced in the Roman Empire from the East about the II century A.D., but it became widespread only in the VIII century.

1.2.3 *Vehicles*

The possibility that carts originated as *sleighs on wheels* has already been mentioned. Ancient evidence of both two-wheel carts and four-wheel wagons is available. It is likely that the axle-load had to be limited to a very low value, due to the lack of prepared roads and to the small width of the wheels. It was then necessary to use wagons with four wheels for transportation. War chariots, which had to be light and easily manoeuvrable, always had two wheels.

The front axles of all wagons were not articulated on the body of the vehicle, at least until the beginning of the Christian era, and no suspensions were used. Wagons were consequently scarcely manoeuvrable and could be forced on a curved trajectory only due to the very low "cornering power" of the wheels. At any rate, to change the direction of motion of a wagon on soft soil must require a considerable muscular strain.

The lack of suspensions made the wagon a statically undetermined structure, and consequently when travelling on rough road only three wheels could remain in contact with the ground, unless the structure of the vehicle was quite compliant.

Vehicles were in general built using wood, but often a body made of a wickerwork chest was added on the wooden frame, showing that five thousands years ago the importance of lightweight structures in transportation had already been understood.

In many cases vehicles were built in such a way that they could be easily dismantled, a characteristic which should have been important in all vehicles operating on distances greater than those separating the fields from the village. Vehicles had to be carried by men or animals whenever the conditions of the ground made it necessary, in the same way as boats and rafts had to be carried when the river had waterfalls or rapids.

War chariots had more strict requirements about lightweight construction and manoeuvrability. In the third millennium B.C. war chariots were widely used and represented a decisive weapon in the wars which opposed the various states of Mesopotamia. The military power of a nation was measured by its number of war chariots, like in modern days by the number of nuclear warheads, and Egypt and the states of Mesopotamia were engaged in a race to build a great number of better and lighter chariots.

Chariots became a status symbol for the mighty ones, who used to be

Fig. 1.3 Egyptian war chariot, from a tomb near Thebes (XV century B.C.).

buried with their vehicle. This habit provided us of some war chariots very well preserved, while no wagon used for transportation remained.

Apart from throwing arrows from the chariot, thing which should have been at any rate quite difficult, vehicles were not used for actual fighting but for transporting the warrior in the battlefield and giving him a greater mobility. The crew of a chariot was of two men: A warrior and a charioteer. This situation is still that described in the Iliad, and it is easy to understand that the availability of a good number of chariots could give military superiority in an age when all soldiers were infantrymen, as the art of riding was still unknown.

As only powerful states could build and maintain a good number of chariots with the related animals, the superiority of these states over groups of plunderers, rebels and small tribes was granted. It is likely that chariotry represented also a powerful tool for social control.

The structure of an Egyptian chariot of the XV century B.C. is shown in Fig. 1.3. It represents without doubt the best of the state of the art of its times, and remained unchanged for centuries.

The progress from the Sumerian vehicle shown in Fig. 1.1 is great, and if the greater power of the two horses compared with that of the onagers is considered, it is easy to understand why some historians ascribed to the use of this weapon the expansion of the Hittites in Anatolia, the Achaei in Greece and of the Hyksos who in the XVIII Century B.C. invaded Egypt, teaching the new technology to the Egyptians.

Chariotry became obsolete when the knowledge of riding became widespread. Donkeys were used as pack animals and for transportation of per-

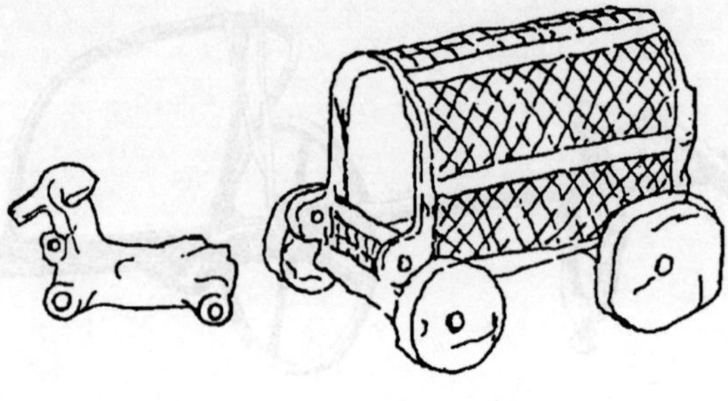

Fig. 1.4 Model of a wagon from Tepe Gawra, Assyria (III millenium B.C.).

sons in the third millennium B.C. and surely also horses were occasionally used in the same way. The lack of those devices intended to give stability to the rider as the saddle and the stirrups limited the use of riding on horseback for centuries. In the first millennium B.C. cavalry substituted gradually war chariots in the Middle East, North Africa and Europe. Likely the use of cavalry was one of the factors which allowed Alexander the Great to conquer his Empire.

In Europe only the Celtic tribes continued to use war chariots, which at the beginning were carried north of the Alps by Etruscans. Celtic wheelwrights learned the art of building wheeled vehicles and made significant progress.

When the military pressure was over, the progress of vehicles on wheels slowed down. Greeks and Romans used chariots only for ceremonies or races. In the latter case the crew was reduced to a single charioteer.

The transportation vehicle of roman times, both in the form of cart or wagon, is no more advanced than that of more ancient times, and above all wagons had neither steering nor suspensions. The wagon shown in Fig. 1.4, from Assyria and dating back to the third millennium B.C., remained unchanged for thousands of years. It would be at any rate interesting to know whether the vehicle whose model is shown in Fig. 1.4 was drawn by a single animal: In such case the harness would have been based on two poles or straps and not on the usual central shaft and consequently would have been more "modern" than many more recent types.

Roman vehicles spanned from the *birota* (two wheeled cart), with a payload of about 60 kg, to the heavy wagons carrying about 500 kg. For heavier loads it was necessary to arrange particular solutions when needed; at any rate water transportation was preferred.

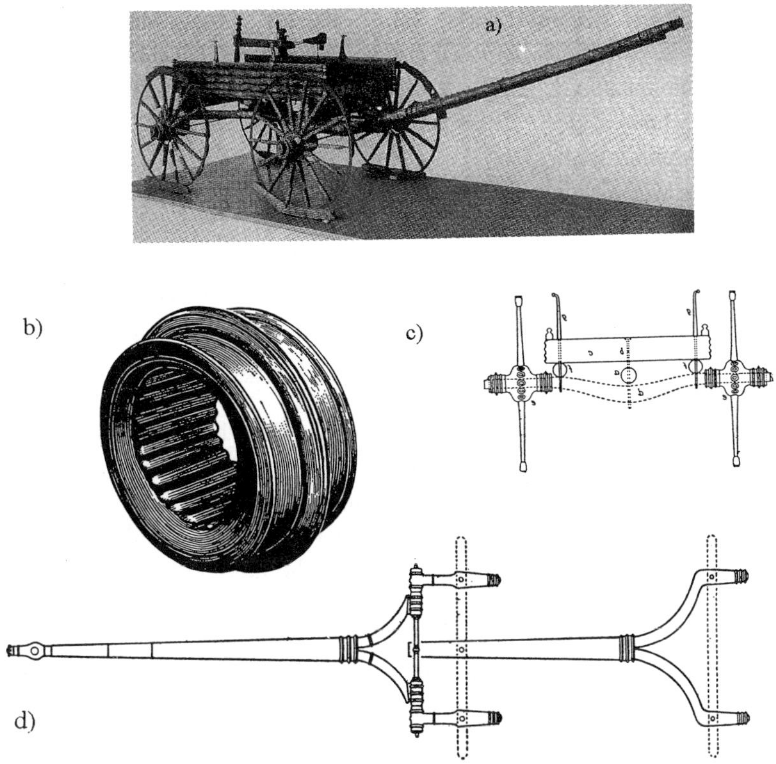

Fig. 1.5 Picture (a), plan (d) and cross section (c) of a Celtic wagon found near Dejbjerg. (b) Roller bearing of wood from the hub of a wheel of the same wagon. (I century B.C.)

As already said, Celtic tribes continued the progress of chariots. The remains of the wagon found at Dejbjerg are shown in Fig. 1.5. It is the first example of a wagon with steering on the front axle, but it can be considered an articulated vehicle made by two chariots. It is however unlikely that the solution was actually used for transportation; it looks more an insulated solution, for ceremonial (burial) purposes. At any rate it incorporated other interesting features, like the wooden roller bearings in the hubs.

1.2.4 *Road transportation*

Many roads were used in the ancient world, but for thousands of years they were simply footpaths, unsuitable for vehicles on wheels. Wagons could be used only for local transportation, or in particular geographical areas.

The only circumstance which caused the use of vehicles on wheels over large distances was war, and in these cases the armies were required to build roads when needed. Tiglath-Pileser I (about 1100 B.C.) wrote that during a campaign among the mountains of Elam his army had to open a road with bronze pickaxes for his war chariot, but that in many instances he had to proceed on foot.

Only in roman times a road system extending to a large area could be made practicable by vehicles. Well paved roads were on the contrary built in the vicinity of important cities for ceremonial and image purpose. The most ancient sacred road known was built near Hattusas, the capital of Hyttite kingdom, about 1200 B.C.

Often paved roads were provided of rails for the wheels of wagons. Such rails, cut by purpose, were likely to be needed to guide the wheels of wagons which, being without steering, behaved more like railway cars than road vehicles. Also in the most difficult sections of roads such rails could keep vehicles in the track, avoiding side slipping. The more ancient remains are from Malta, from neolithic age, but they were common in Greece, Sicily, northern Africa and other places of the eastern Mediterranean Sea. The gauge of such rails is quite uniform, from 1350 mm in Malta to 1500 mm. In Greece were even found switches, to allow the crossing of vehicles (Fig. 1.6).

The roads had often natural paving, but there are examples of roads paved with stones or wood. Roman roads were generally well paved and drained.

The crossing of rivers was usually performed at wading points. The bridge of Babylonia was very well known, but remained unique for centuries. In roman times many bridges and some tunnels were built.

Urban streets had usually natural paving, even if eventually stone paving was slowly adopted: It was only in roman times that the diffusion of artificial paving of the streets of the most important cities took place. A good example are the streets of Pompei, paved with stones, with elevated crossings for pedestrians. On both sides of those crossings there were rails, likely to guide the wheels of vehicles between the stones.

The streets were usually very narrow, as it was a common opinion that narrow streets were useful against wet and unhealthy winds. Vehicular traffic was forbidden in towns during the day, except for religious feasts or for games. Transportation was done by night, but it should have been quite difficult as the owners of houses on both sides of streets used to close them

Fig. 1.6 Greek road with stone rails. A "switch" for vehicle crossings is shown.

with chains, perhaps for fighting the noise. This problem was without any doubt a very important issue, as wheels with iron tire started to be widely used and the stone paving was quite rough. Iron tires caused a strong wear of the roads, but the alternative of inlaying iron nails or studs in wooden rims was not better.

1.2.5 *War engines on wheels*

Apart from war chariots, ancient military technology used many war engines which had to be moved, at least for short distances. Often catapults, siege towers, battering rams and other devices of this type had consequently wheels. Their diffusion on a large scale is ascribed to Assyrians, from 1100 B.C. An Assyrian siege tower used at about 870 B.C. and often compared to a modern armored vehicle, is shown in Fig. 1.7. Such machines had the main problem of the vulnerability of the drawing animals, if they were harnessed in the usual way. Catapults, which could operate from a greater

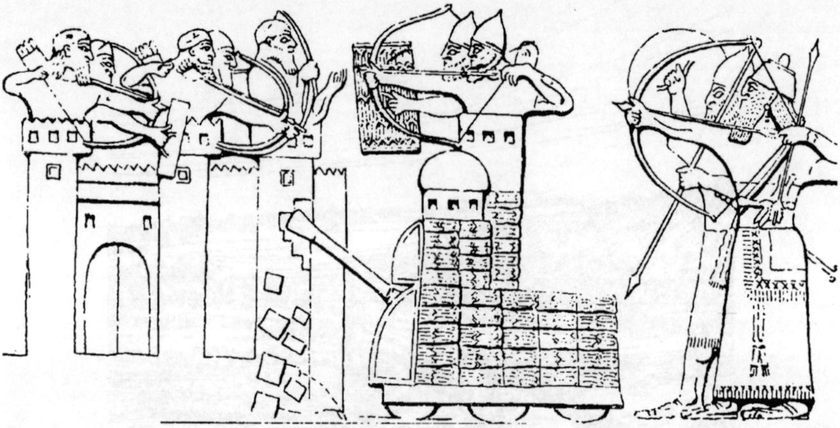

Fig. 1.7 Battering ram with siege tower used by Assyrians in 870 B.C. against a sieged city.

distance and were seldom moved under enemy fire, could be pulled as carts, like field artillery until recent times, but for siege towers and rams a different layout had to be used. A group of animals or even men could push the device remaining behind it, but they were at any rate in a dangerous position. The only safe solution was that of pushing the engine remaining inside it, under its armor Some pictures show this arrangement, but there is no evidence that propulsion was provided by using a device which moved the wheels. Wheels maintained their pure supporting function and men or animals pulled the engine, directly reacting on the ground.

Vitruvius writes of movable towers on wheels at least 56 m high, with a base with sides of 10 m; their weight is estimated in at least 100 t. The weight of such machines compelled to use a large number of wheels, which were much larger than those of wagons.

1.2.6 *Prime movers*

In the ancient world the only source of power available to men was their own muscular power and that of tamed animals; wind power was exploited only for sailing. As already stated, the harnessing system used for horses prevented full exploitation of their power and the advantage of using horses instead of men was marginal. A horse developed about the same power of four slaves, but needed a correspondingly greater amount of food, so the trade-off was influenced by the availability and other occasional circumstances.

Mechanical prime movers were introduced at the beginning of Christian Era, in the form of the water wheel. It could not be used directly for transportation and there is no evidence of its application to pull loads with ropes (like cable cars). Windmills were introduced after the beginning of the Middle Ages, and there is no evidence of possible attempts to build sail vehicles on wheels.

After the fall of Roman Empire communications in Europe experienced increasing difficulties, even if commercial traffic was more intense during the Middle Ages than what is commonly thought. Technology continued to be upgraded.

The water wheel started its diffusion in the first century of the Christian Era. Its use did certainly stimulate a general mechanical progress, which was useful also for the construction of vehicles, but particularly led to the development of those elements (bearings, gear wheels, transmission shafts, etc.) which were later essential for the construction of automotive vehicles.

Another type of prime mover started its diffusion in Europe from the XI century A.D.: The windmill. At least from a theoretical point of view, the windmill can be used for vehicle traction. Military engineers of XIV century were well aware of this fact, and they tried to use it for the construction of vehicles which could overcome the above mentioned disadvantages of animal traction.

The first drawing of a wind wagon we are aware of was produced by Guido da Vigevano, and is found on a treatise on war engines written in 1328 for Philip V of Valois. The vehicle is operated by a windmill through a chain of gear wheels, and is also provided of a hand-crank, possibly to be used when wind failed (Fig. 1.8a). In the same book other vehicles operated through gear wheels are shown.

It is clearly a machine which was never built, or at least, which could not work as expected, but it is quite interesting as it contains the first description of a driving wheel.

The same idea of a wind wagon was later developed by Roberto Valturio (Fig. 1.8b) and by Mariano Jacopo, said Taccola (Fig. 1.8c), in the first half of the XV century. It is to be noted that such wind wagons had no steering system, and that in the last two examples the driving wheels act also as gear wheels.

In spite of the fact that an ancient wagon with steering front wheels was found (see Section 1.2.3), it looks like such device had no application until

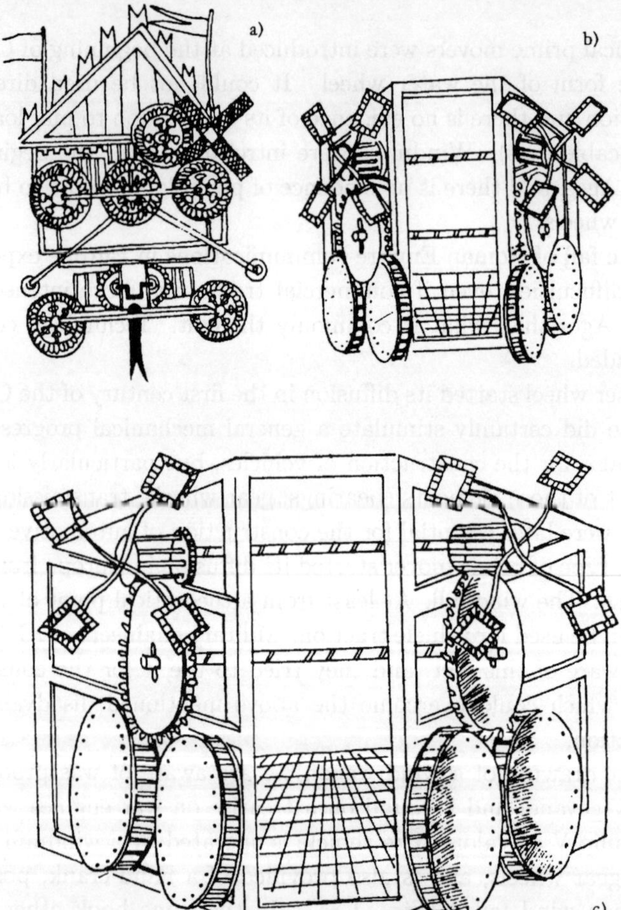

Fig. 1.8 Evolution of the concept of the wind wagon. (a) Guido da Vigevano, (b) Valturio, (c) Taccola.

the XV century, and even at that time, it did not have a fast diffusion.

A fully developed steering mechanism can be found in some drawings by Francesco di Giorgio Martini, who worked in the second half of XV century. The vehicle shown in Fig. 1.9 has four driving and steering wheels. The power is supplied to the wheels by four devices similar to the ones used in capstans. Similar devices operate also the steering mechanisms.

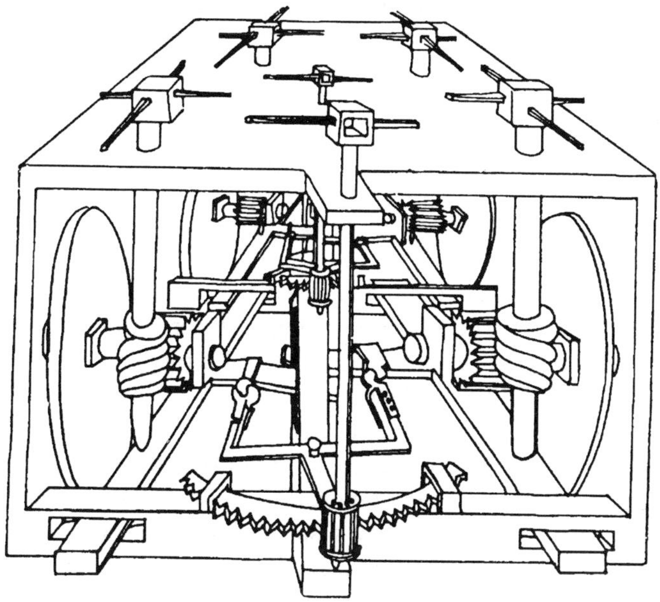

Fig. 1.9 Four wheel drive wagon drawn by Francesco di Giorgio Martini.

The drawing is well detailed, but the articulations of the steering are not shown. As it is drawn, the mechanism cannot work, and it seems that the designer failed to solve the difficult problem to power the steering wheels. This wagon, which is at any rate powered by men, or animals, if it is big enough, is another imaginary machine.

Apart from these new ideas, which were clearly too advanced for the technical possibilities of their time, a rapid progress in exploiting the possibilities of animals took place starting from the X century. A new form of collar exerting its pressure on the shoulders instead of on the neck, was introduced. Also the practice of fitting permanent iron shoes to horses became widespread, and consequently the available traction for road vehicles increased a great deal.

From the first half of XIV century the diffusion of firearms started. It caused the need of moving rapidly very heavy objects like guns on little prepared soil, and this led to the development of stronger wheels and vehicles. The supplies and equipment of armies became more numerous and heavy, and better and faster means of transportation were needed.

1.3 From the Renaissance to the industrial revolution

1.3.1 *Vehicles*

The increase of commerce and the new needs of personal mobility which took place in the last centuries of the Middle Ages continued at a greater pace in the following decades. A new confidence in progress and in the possibilities of technology led to try new experiments and to refine existing machines.

As a demonstration of the confidence in technological progress which was widespread in the XVII century, a writing of Roger Bacon can be mentioned in which he forecasts the construction of self-propelled ships, automotive vehicles and even flying machines. At any rate, in spite of all the progress in the means of transportation, journeys on land were still made on foot or on horseback (more often on the back of donkeys or mules). Still in 1550, in Paris only five coaches were reported, including that used by the queen.

The first coach with the body suspended on belts or chains dates from the beginning of the XV century, but the innovation did not spread out fast. A wagon with the body suspended on springs is reported by Fausto Venanzio in his treatise *Machinae Novae*, in 1595. From the XVII century the use of suspensions for coaches became common, at least for the vehicles of rich people.

In 1665 steel springs were introduced for the suspension of vehicles, but only at the beginning of XIX century this practice became general and it was possible to suspend the heavier vehicles on steel springs.

In the second half of XVII century the first coaches for public service were introduced, at first in London (1624), then in Paris (1630) and in many other cities.

The conditions of roads were at any rate quite poor, and remained so for a very long time. The increase of the traffic and the lack of road maintenance often made communications difficult. In order to improve the situation the design of vehicles was improved and larger wheels were introduced.

Wheels with iron tires, which were already in use for centuries, became larger in order to decrease the pressure on the ground. Conical wheels were also introduced (Fig. 1.10a). In spite of some improvements in the construction of roads, the situation did not change considerably until the XIX century, when more satisfactory and long lasting road surfaces were developed. A large wagon with conical wheels of the beginning of the XIX

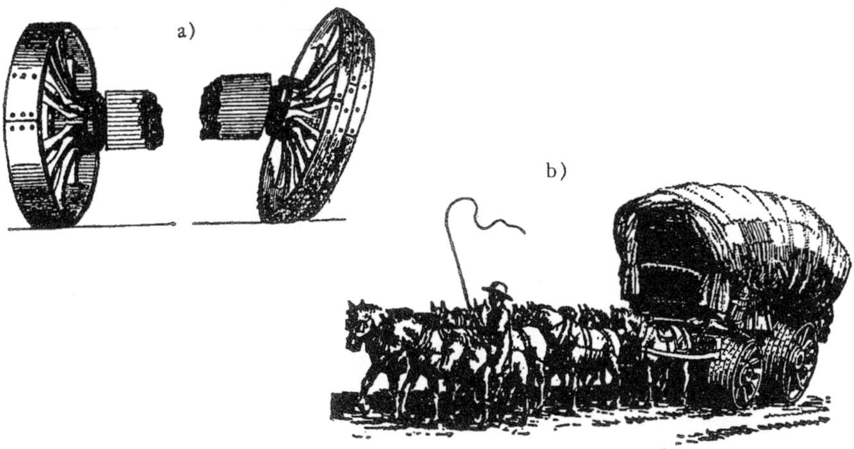

Fig. 1.10 (a) Wheels with a large iron tire. (b) Large wagon with conical wheels (beginning of the XIX century).

century is shown in Fig. 1.10b.

Transportation of goods was increasingly performed using water ways, particularly owing to the construction of a good network of canals, with bridges, tunnels, locks and devices to lift boats between canals at different levels.

1.3.2 *Prime movers and automotive vehicles*

Two prime movers were available at the end of the Middle Ages: The water-wheel and the windmill. They had already gone through a long evolution, and were at that time used for many different applications, apart from the original one of grinding cereals.

The power supplied by those machines and their power/weight ratio were at any rate quite low. It is difficult today to realize how scarce was the available power for all human activities, not only in the ancient world but even well into the industrial revolution.

As an example it is possible to remember that the power of the *Machine of Marly* (Fig. 1.11), the largest group of water wheels built in 1682 by Rennequin of Liege by order of Louis XIV to supply water for the fountains of Versailles, was theoretically of only 90 kW. In actual operation it supplied about 56 kW, and in 1796, owing to poor maintenance, it was derated to

Fig. 1.11 The machine for pumping water for the fountains of the Royal Palace of Versailles at Marly. It had 225 pumps worked by 14 water wheels.

less than 2 kW.

The power of water wheels was in most cases of about 3-4 kW, seldom being more than 7 kW. Rankine and D'Aubuisson, in the XIX century, report the values of the power of some prime movers listed in Table 1.1.

Table 1.1 Power developped by ancient prime movers following Rankine and D'Aubuisson.

man turning a handcrank	0.03 - 0.06 kW
man pushing the bar of a winch	0.035 kW
horse pulling a winch	0.3 - 0.4 kW
water wheel, about 5.5 m diameter	1.5 - 3.5 kW
pole wind mill	1.5 - 6 kW
tower wind mill	4 - 10 kW

Taking into account these data, and remembering that the efficiency of transmission mechanisms was certainly very low (the efficiency of the transmission of a Dutch wind mill used in 1648 to pump water was evaluated at about 39%, and this value can be considered as typical), it is clear why wind wagons did not enter the stage of practical realization.

The use of sails could yield better results, and in 1599 Simon Stevin built a sail wagon for Maurice of Orange. It could carry 28 persons, and is said to have reached a speed of 12 km/h. It did not have however practical applications.

Some vehicles with 3 or 4 wheels powered by men from the inside were built. Those by Richard (1630), Stephan Farffler (1685), Jacques de Vaucauson (1748) and John Vevers (1769) can be mentioned, among those which are known. It is likely that no knowledge remained about many other attempts.

The wagon built by the German clock-maker Hans Hautsch can also be remembered. It was powered by steel springs, like in clockwork mechanisms. It was said to be able to travel at about 1.5 km/h, but its range is not known.

In order to build a truly successful automotive vehicle it was necessary to wait until a viable thermal engine was available. The *Aeolipile* built by Heron is usually considered as a forerunner of the steam engine, or better of the steam turbine; the idea remained however for centuries at the stage of a technological toy, as it had only the capability of fractional horsepower.

Heron is also said to have built a simple machine using the expansion of hot air, which can consequently be considered as a forerunner of gas thermal engines. Simple wheels with blades operated by hot air or smoke were repeatedly suggested and it is likely that they were even used, but always to produce very small power. An example are the roasting-jacks worked by smoke described by many authors.

Leonardo da Vinci worked on machines of this type, but in spite of drawings which seem to suggest that he designed a workable steam engine, it is a common opinion that he did not make significant advancements in this field. Giovanni Branca in 1629 described a steam turbine, which however had no issue (Fig. 1.12).

Steam was at first used to make vacuum, or at least to raise water, by condensation, as Giovanni Battista della Porta wrote in 1606. The idea of building an atmospherical engine, operated by the pressure of the atmosphere on a piston moving in a cylinder in which vacuum was obtained by steam condensation, was a result of the discovery of atmospheric pressure.

In 1681 a steam vehicle (likely a reduced scale model) is said to have been built in China by the missionary F. Verbist. The engine is simply said to have been an Aeolipile.

In a sketch in a work of Isaac Newton a vehicle propelled by the reaction of a steam jet is shown (Fig. 1.13). It is quite likely that the drawing is

Fig. 1.12 Steam turbine drawn by Giovanni Branca in 1629.

only a sketch to show the reaction principle and that Newton never thought seriously to that application, but it could have inspired practical attempts.

In these times steam engines, all atmospherical engines, were used to pump water. When a mechanical power output was needed, the engine was used to pump water to an elevated site and then the potential energy of water was used via a water wheel.

The engines built by Savery and later those built by Newcamen were all of this type. Their efficiency was very low, in the range of 0.5%. The introduction of a condenser separated from the cylinder by Watt allowed the efficiency to increase to about 3% and, at the end of XVIII century, to 4.5%.

In 1784 Watt built a steam engine in which the oscillatory motion of the beam was converted in rotational motion by a planetary gear mechanism. Watt could not use the simpler crank mechanism, which had been patented

Fig. 1.13 Sketch showing a reaction steam vehicle by Isaac Newton.

by one of his employees. In the same year a patent regarding a steam vehicle was granted to him, but he never actually built it.

For the construction of steam engines with a power/weight ratio adequate to vehicular use it was necessary to use steam at pressures higher than those used by Watt in his engines. Trevethick experimented with such machines in the last years of the XVIII century, and when in 1800 the patents owned by Watt expired, he obtained a good success with them.

Nevertheless, some steam vehicles powered by the engines of XVIII century were, at least from a technical point of view, a success. This statement must be intended however in the sense that they succeeded to move under their own power.

The first and better known of these vehicles is the three-wheeled wagon built by Capt. Nicolas Joseph Cugnot (Fig. 1.14). It was meant for artillery use, and was completed and tested in 1769. It was powered by a steam engine with the boiler in front of the front wheel, which had both driving and steering functions. It is possible that this solution was introduced to enhance traction, but it is reasonable to doubt about the ability of the machine to drag heavy guns.

This vehicle had a weight of about 5 tons, and was at any rate able to move in a controlled manner. It is said to have reached the speed of 5 km/h. The tests were discontinued after the vehicle overturned as a consequence of striking a wall. The whole project was abandoned, and it is no longer an issue.

Fig. 1.14 Model of the steam wagon built in 1769 by N.J. Cugnot.

Also the steam coach by William Smyghton (1786) and that by Robert Furness (1788) did not have better success.

In the XVIII century a patent was granted to Philip Waugham for metal ball bearings for vehicular use, and some attempts were carried out for the construction of internal combustion engines using gunpowder burning within a cylinder.

1.4 Motor vehicles in XIX century

1.4.1 *The age of steam coaches*

As already said, the XVIII century steam engine was essentially an atmospherical engine, and even the engines built by Watt, in which the expansion of steam was used to produce power, did operate at very low pressure. Such design choice was mainly due to the impossibility of machining cylinders with the more strict tolerances needed for avoiding strong leakage at high pressures.

At the beginning of the XIX century, the improvements of machine tools allowed William Trevethick to build engines working at higher pressures. He worked more on stationary engines, but did also build steam vehicles with three wheels for transportation of goods and passengers. The vehicle he built in 1802 (Fig. 1.15) opened the era of steam coaches.

That particular coach had little commercial success, but others followed the path he opened and in 1822 the first commercial service of steam coaches

Fig. 1.15 Steam coach by Trevethick.

was started. In a few years this form of transportation became common and a commercial war started between water transport, steam coaches and the newly developed railway. At least in England, this commercial war was particular violent, and was not only fought with commercial and legal battles. There were even physical aggressions and other acts of violence. Some accidents, like the one in which a steam coach overturned after striking a pile of stone, perhaps purposely set on the road, and the subsequent explosion of the boiler killed several persons, caused the parliament to vote laws limiting the use of automotive road vehicles.

A first law limited their speed to 16 km/h, and then in 1865 the limit was lowered to 6.4 km/h and the presence of a man walking in front of the vehicle with a red flag was prescribed. These measures ended the short age of steam coaches, which could not compete with railways.

Actually steam coaches were uncomfortable and possibly dangerous and even without those limitations their future was quite uncertain. The conditions of roads were still poor, even if it did improve in the first years of the century, and above all the design of wheels was inadequate to the new vehicle. Wooden wheels with steel tire were used, and this gave steam coaches a questionable stability and traction.

Fig. 1.16 Steam coach built by Church. It was used for passenger service between London and Birmingham.

After 1860 steam coaches disappeared in England even if some attempts continued in France, where in the following years steam buses able to reach a speed of 40 km/h were built and used, and in other countries (Fig. 1.17).

1.4.2 *The motor car*

In the second half of the XIX century two new inventions allowed the development of motor vehicles as we now know them: The internal combustion engine and the rubber tire.

In 1846 Thomas Hancock built solid rubber tires to be added to the steel tires of the wheels of coaches. They had a thickness of about 30 mm and a width of 40 mm and proved to be very useful in reducing noise and improving comfort. They were used on coaches and particularly on the newly developed bicycles.

The year before Robert William Thompson built a tire made of a rubber inflated tube covered with leather. It was however unpractical and was not used.

In 1888 John Boyd Dunlop, a veterinary surgeon of Belfast, built a pneumatic rubber tire for the tricycle of his 3 years old son (Fig. 1.18). It proved practical, particularly after the tire-wheel attachment system was simplified in 1890.

Fig. 1.17 Steam coach built by gen. Virgilio Bordino at the military factory of Torino, Italy, in 1854.

From these beginnings the pneumatic tire was continuously improved, but many years had to pass before its use became general. Many motor cars built at the end of the XIX century and at the beginning of the XX century had solid rubber tires.

The problem of improving the comfort on uneven roads (tarmac or concrete roads had still a very small diffusion at the end of XIX century) led to the introduction of "elastic" wheels with the outer rim, covered with a solid rubber tire, connected to the hub by a system of springs (Fig. 1.19). These complex structures were without doubts quite costly and had little diffusion; they were at any rate abandoned in favor of pneumatic tires. The need of lighter wheels, particularly for bicycles, led to the introduction of steel spokes, but the older wheel with wooden spokes was used well into the XX century.

The development of automotive vehicles was marked in the second half of XIX century by the introduction of a new thermal engine: The internal combustion engine. As already stated the idea of a thermal engine in which combustion takes place in the cylinder is old. Huyghens described a machine powered by a small quantity of gunpowder burning in a cylinder in 1778,

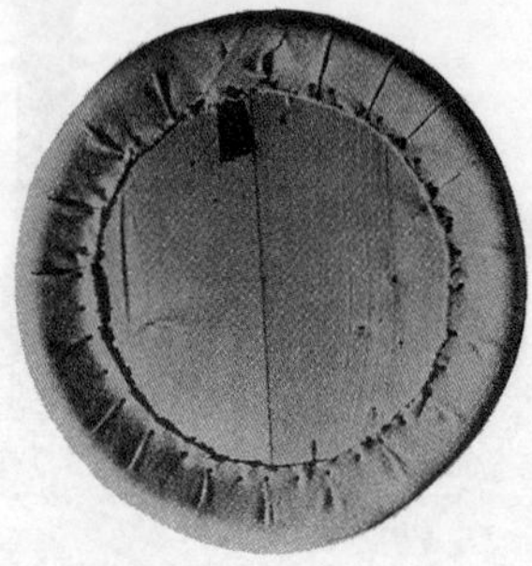

Fig. 1.18 The first experimental tire by Dunlop, on a wooden wheel.

but it did not succeed in obtaining good results with it.

In 1856 Barsanti and Matteucci built in Italy an internal combustion engine operated by a mixture of air and gas compressed before ignition. In 1860 Lenoir built a similar engine in France. He is said to have used his engine on a road vehicle and to have traveled several times from Paris to Jonville-le Pont (a journey of 20 km) with it, but there are strong doubts on this issue. The construction of the first road vehicle powered by an internal combustion engine in 1864 is so ascribed to Siegfried Marcus.

The engines built by Lenoir were quite heavy and slow and their fuel efficiency was poor. They were not superior to steam engines for traction purposes. Beau de Rochas patented a four stroke engine in which the air-gas mixture was compressed in the upstroke before ignition and N.A. Otto built such engine in 1878. It was a great commercial success, and in few years more than 35000 engines were sold. It was however a stationary horizontal engine, working on gas.

The tricycle built by Karl Benz in 1885 (Fig. 1.20) was the first vehicle powered by an internal combustion engine which could compete with the steam road vehicles which were still occasionally built. It was provided of single cylinder engine placed with vertical axis and could travel at 13 km/h.

The following year, again in Germany, Gottlieb Daimler built an engine which for the first time could operate at higher speed. It was applied in 1886 to a motor bicycle and the following year to a four-wheeled vehicle.

In 1889 Daimler built a twin cylinder V engine, which was used by others for vehicle traction, motor boats and fixed plants.

The vehicle built by Benz and the engine by Daimler, both working on

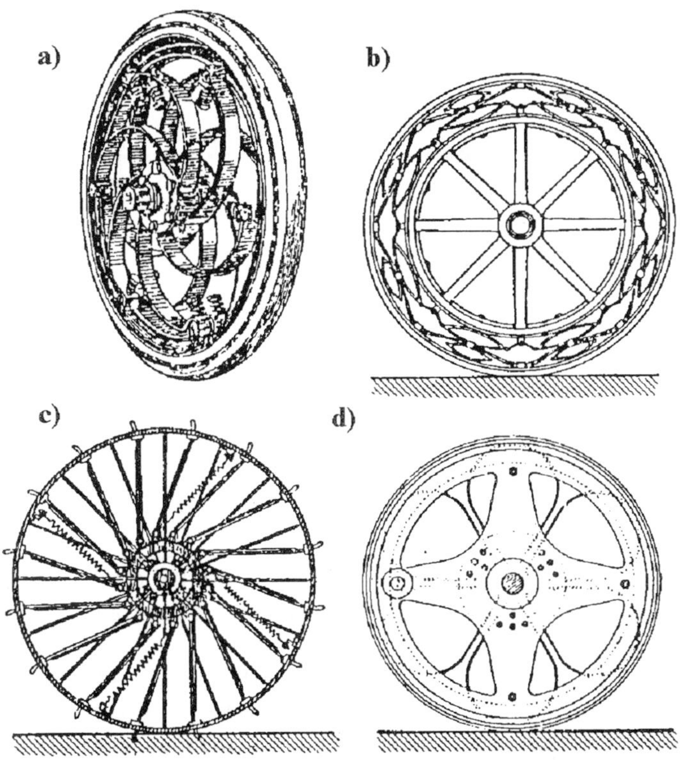

Fig. 1.19 Elastic wheels of the end of the XIX century. (a) Cardigan; (b) Gigli; (c) Groseclove; (d) Mac Laren and Morris.

gasoline, were sold in many units and were commercially successful. The last years of XIX saw a rapid progress which caused the newly developed motor vehicle with internal combustion engine to become a relatively safe and reliable means of transportation and that put the premises for its great diffusion of the following years.

While the motor car powered by the internal combustion engine was in its initial developing stage, some steam vehicles were still built, and particularly many efforts were devoted to overcome the limitations coming from the weight and bulk of the earlier steam coaches. The steam tricycles built in 1887 by De Dion and, in the following year, by Leon Serpollet were

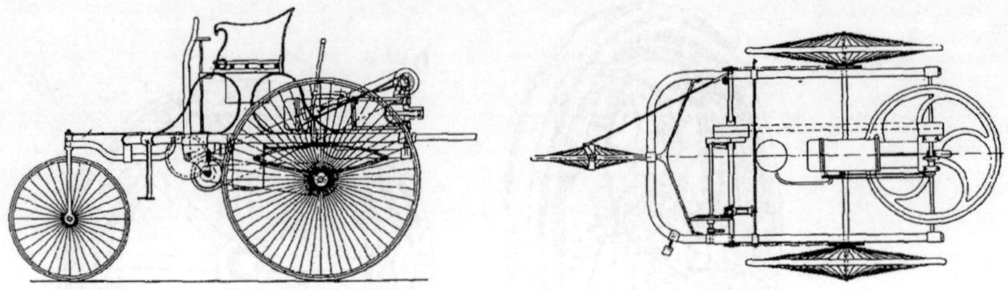

Fig. 1.20 The first really successful vehicle with internal combustion engine: The tricycle by Benz of 1885.

comparable with similar vehicles operated by internal combustion engines. Some of them were produced in many units with good commercial success, and some are still working.

Also the electric vehicle was developed in the second half of XIX century. The earlier attempts performed before 1881, when the lead-acid accumulator of Planté and Faure was improved and made of practical use by Joseph Swan, were based on primary batteries and had little success. The newly improved accumulator produced many new attempts, and after the vehicles by Radcliff Ward (1886) and Magnus Volk (1887), which had technical success but were commercial failures, Pouchain built in 1893 an electric car which could transport 6 persons and had some diffusion.

The cleanliness and low noise of electric vehicles were a key factor in its success; its performances were also not lower than those of internal combustion engine vehicles, at least as far as the speed and reliability are concerned.

In 1897 a public service operated by electrical coaches was started in London with vehicles (Fig. 1.21) having a range of 80 km with each charge. The used batteries were simply substituted by fresh ones and charged separately. The service had however no commercial success, and in 1899 36 complete cars and 41 incomplete ones were sold. In the same year the vehicle *Jamais Contente* (Fig. 3.18) operating on lead-acid batteries, built by Camille Jenatzy established a world speed record with 105.9 km/h.

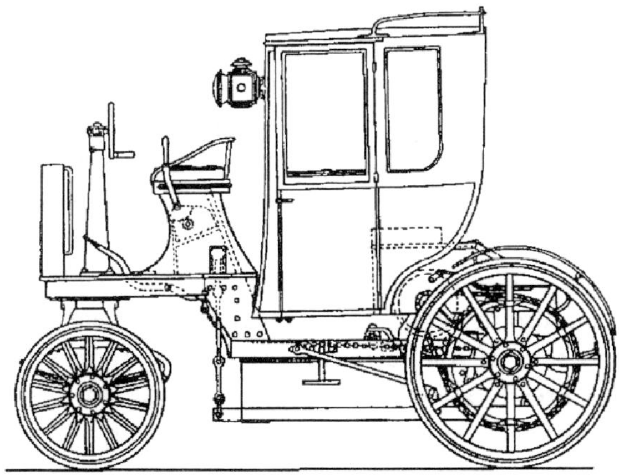

Fig. 1.21 Electric car built by Bersey in 1897, used for public service in London.

1.5 The age of motor cars

At the turn of the century the new road automotive vehicle was still not much reliable and diffused, but had great potentialities of technical improvement and commercial diffusion.

It is impossible to describe in a few words the development of motor vehicles in the first half of the XX century, as a great number of innovations and changes of the structure, engine and other mechanical elements took place at an increased pace. Above all, refinements of the production techniques allowed motor vehicles to be produced at decreasing costs, causing an unprecedented diffusion.

The result of such innovations is the modern motor vehicle, with its advantages and disadvantages, but with the unquestionable merit of giving man a mobility which has no precedent in history.

As it is always the case with technologies reaching their maturity, a few designs emerged from a multitude of attempts, even if some distinctive features, based on tradition, image and intended market section remained among the various firms.

Among the most important individual steps forward of the technology, it is possible to mention the use of steel for structural parts and now its substitution (still very limited) with reinforced plastics, the use of hydraulic and pneumatic (on industrial vehicles) braking systems, the braking on all wheels, the use of hydraulic shock absorbers, and monocoque structures. The pneumatic tire, which became common in the first years of the century, continued its evolution until modern radial tires.

In the seventies a new impulse towards the evolution of what seemed to be a mature product took place. At present (1997) the introduction of the computer and related electronic equipment in all human activities is producing and will produce even more in the future a revolution not only in industry but also in the way of living of millions of people. It will produce a true revolution also in automotive industry, under many aspects.

The introduction of computer assisted design, from structural analysis to drawing, already allows one to rationally design many parts which had to be dimensioned by experience and by testing with trial and error procedures. Proceeding along this way will without doubt lead to deep innovations.

On the other hand, computer aided manufacturing will produce even greater consequences on production techniques, and consequently on the product itself.

But the introduction of electronic devices on board of vehicles will carry on a revolution involving the concept of motor vehicle itself. Often it is a commonplace that the innovation caused by the introduction of electronics concerns not so much on "how" a certain product is designed or built, but more on "which" product to design or build.

If the introduction of electronics on board will be limited to some gadgets, it will have no deep impact on vehicles, but the number of functions which are already performed by electronic devices on a modern car is quite large. It is clear that before vital functions of vehicles can be switched to electronic devices, many problems concerning reliability and maintainability similar to those encountered and solved in the aerospace field, must be solved at costs which are compatible with automotive constraints. But once such problems can be solved, designers will gain new and unprecedented degrees of freedom and deep changes will be at hands.

To illustrate this consideration it is sufficient to think what the introduction of a system which controls the brakes of each wheel can mean for the architecture of vehicles, giving optimal braking performance to a vehicle whose mass distribution is far from being the optimum one or even is completely wrong for today standards. Or the consequences of a system which can control the slip angle of each wheel, making the definitions of understeering and oversteering meaningless, and giving the vehicle an artificial stability.

Some inventive minds did even anticipate vehicles in which advanced control systems will make not necessary the use of wheels, which require a prepared road to roll on, being able to remain in equilibrium in their run on a certain number of articulated legs.

It is obvious that there are no limits to the possibility of creative imagination, but only the verification of technical experimentation and even more of the acceptance of the innovations by the market will state a line dividing viable new roads from fruitless imaginative efforts.

Chapter 2

Forces acting between road and wheel

2.1 Structure of pneumatic tires

The wheels of all modern motor vehicles are provided with pneumatic tires, which support the vehicle and transfer the driving power through the wheel-ground contact. They also provide the lateral forces which are needed in order to control the trajectory of the vehicle.

The rigid structure of the wheel, made by the disc and the rim, is surrounded by a compliant element, made by the tire and the tube. The latter can be absent in tubeless wheels, in which the tire fits airtight on the rim and contains the air. The tire is a complex structure, made by several layers of rubberized fabric, with a large number of cords running in the direction of the warp and only a few in that of the weft. The number of plies, their orientation, the formulation of the rubber and the material of the cords are widely variable: They are the parameters which give to each tire its peculiar characteristics.

A tire is designated by a group of 3 numbers and a letter (e.g. 6.00-16-4.50E) or by two numbers (e.g. 155 SR 15). In the first case the numbers indicate the section width, the diameter and the width of the rim in inches (Fig. 2.1) and the letter the type of the tire; in the second case the numbers state the section width in millimeters and the rim diameter in inches. A more complete designation includes three numbers and three letters, as in P 225/50 SR 15: The first letter stands for the destination (e.g., P indicates passenger cars), the first two numbers state the section width in millimeters and the aspect ratio H/W in percent, the following letter designate the speed rating (e.g. S stands for up to 180 km/h), the last letter the type of tire (R: radial, B: belted, D: bias) and the last number the rim diameter in inches.

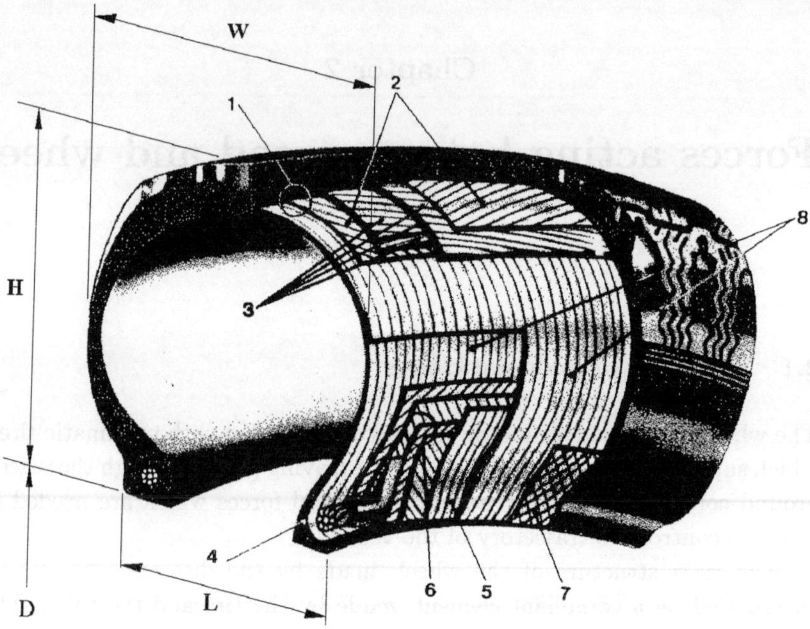

Fig. 2.1 Structure of a radial tire: 1, belt; 2, reinforcement plies at 45°; 3, main plies at 15°; 4, fold of the second ply; 5, fold of the first ply; 6, double fold at the toe; 7, bead; 8, carcass; D, rim diameter; H, section height; L, rim width; W, section width.

Independently from their use, structurally tires belong mainly to two types: or bias tires and radial tires (Fig. 2.1), although a sort of in-between type is constituted by belted tires. In the older bias tires the carcass is made by a number of plies whose reinforcement runs at an angle of 35°–40° with respect to the circumferential direction. In the case of belted tires, a number of plies run under the tread with a small angle, about 15° from the circumference. This belt gives the tire a higher circumferential stiffness. Radial tires are made by plies which are oriented in direction perpendicular to the circumference and by belt plies. This structure leads to more compliant sidewalls and to a tread band which is stiffer in circumferential direction.

Presently radial tires have almost completely substituted the other types, owing to their superior comfort and performance characteristics.

The main function of the tire is that of distributing the vertical load in a large enough area and to insure an adequate compliance, needed to absorb the irregularities of the road. It is essential that the compliance in different directions is suitably distributed: As already stated a radial tire for instance is very compliant in vertical direction, owing to the low stiffness of its side walls, while is very stiff in circumferential direction owing to the belt plies. The high riding comfort and good cornering force capabilities of radial tires is directly linked to this stiffness distribution.

2.2 Contact pressure and stiffness

In the study of the behaviour of the wheel it is useful to resort to the reference frame shown in Fig. 2.2. Its origin is in the centre of the area of contact and X'-axis is defined by the intersection of the ground plane with the mean plane of the wheel. Axis Z' is perpendicular to the ground and points upward[1] and consequently Y'-axis lies on the ground and points leftward.

The force the tire receives from the ground is assumed to be located at the centre of the contact area and can be decomposed along X', Y' and Z' axes, yielding the longitudinal force F_x, the lateral force F_y and the normal force F_z. Similarly, the moment the tire receives from the road in the contact area can be decomposed along the same directions, yielding the overturning moment M_x, the rolling resistance moment M_y and the aligning torque M_z. The moment applied to the tire from the vehicle about the spin axis is referred to as wheel torque T.

The angle between $X'Z'$ plane and the direction of the velocity of the centre of the tire contact is the sideslip angle α of the wheel and the angle between the plane $X'Z'$ and the mean plane of the wheel is the inclination angle γ. The positive directions of the angles are shown in the figure; in particular the inclination angle is positive when the top of the wheel leans towards the right of the driver (i.e., for a wheel of the left side if the top of the wheel leans inwards). While the inclination angle is referred to the direction normal to the road, the camber angle is defined with reference to the vertical direction and has opposite positive direction for right and left

[1]The terminology here followed is that recommended by SAE (Society of Automotive Engineers) in document J670e, Vehicle Dynamics Terminology, issued in July 1952 and last revised in July 1976. However in some cases the author chose to depart from some suggested definitions as in the case of the direction of Z' axis which in the mentioned document is directed downward.

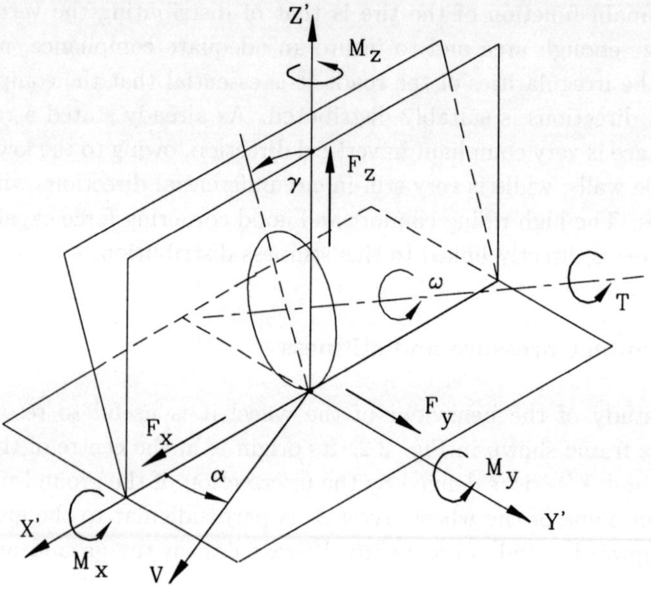

Fig. 2.2 Reference frame for the study of tire forces. Definition of forces, moments and of the slip and inclination angles.

wheels. In the present text the camber angle on horizontal road will be assumed to coincide with the inclination angle.[2]

Consider a small portion of the tire-road contact area. The force per unit area exerted by the tire on the road can be decomposed into a component perpendicular to the road and a tangential component. The first is the contact pressure σ_z while the other can be further decomposed in the directions of X' and Y' axes giving way to the components τ_x and τ_y. The resultant of the distributions of σ_z, τ_x and τ_y are the already defined normal, longitudinal and lateral forces F_z, F_x and F_y respectively.

These distributions are not constant and are strongly influenced by many factors as tire structure, load, inflation pressure etc. Some typical results obtained on a stationary wheel which is exerting no force in the $X'Y'$ plane are reported in Fig. 2.3. At the centre of the contact, the contact pressure σ_z is close to that of the air in the tire at the centre of the contact area, while at the sides it is higher. If the wheel is not rolling the distribution is symmetrical with respect to $Y'Z'$ plane and the resultant passes

[2]This is a further deviation from the mentioned SAE recommendation. The positive directions for camber angles are then considered the same for right and left wheels here.

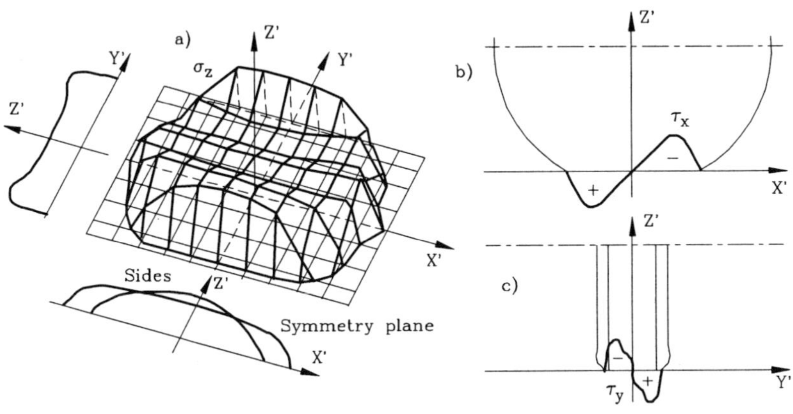

Fig. 2.3 Contact force distribution at the wheel-road contact. (a) Distribution of normal pressure σ_z. Tangential forces (b) τ_x on the plane of symmetry and (c) τ_y along X'-axis.

through the centre of the contact. Tangential forces do not vanish locally even when no force is exerted in $X'Y'$ plane, i.e. when their resultant is zero. In such case components τ_x are directed towards the centre of the contact area and the tire acts to "compact" the ground towards the centre of the contact. Component τ_y has the effect of "stretching" the ground outward.

The force-deflection characteristics of tires depend on many factors, like travelling speed, pressure, wear and many other. A strong difference can be found for this issue between bias-ply and radial tires, radial tires being less stiff at standstill in all directions.

The characteristics in direction perpendicular to the ground (force F_z versus deflection Z') for some tires are reported in Fig. 2.4. In Fig. 2.4a the curve obtained when removing the load has been reported together with that related to the application of the load. A hysteresis cycle can clearly be identified, denouncing the presence of damping in the motion along Z'-axis. This damping is usually at its maximum at standstill and decreases with rolling speed: The practice of neglecting it in the simulations of the motion of the vehicle is justified by this observation. In Fig. 2.4b the curves obtained for a radial and a bias-ply tire are compared.

A static tire rate can be defined as the tangent stiffness in any given equilibrium condition, i.e. at any given value of the load, inflation, pressure etc.

Similar plots can be obtained for forces in X' and Y' directions and

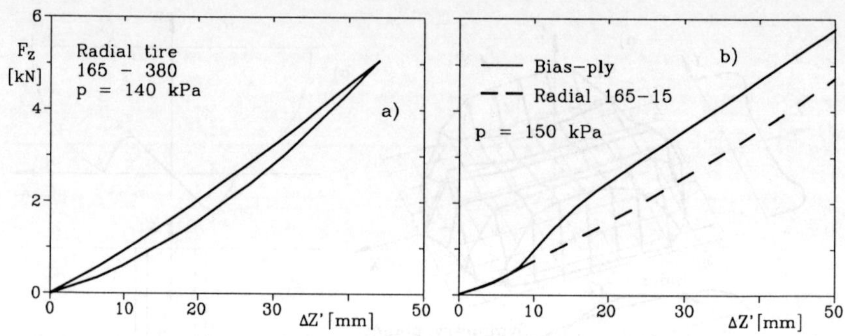

Fig. 2.4 Contact force F_z versus deflection $\Delta Z'$ for some tires at standstill (wheel not rolling) and in static conditions (load applied and removed slowly).

moments about Z' axis versus the corresponding displacements of the centre of the wheel (Fig. 2.5). In all cases the plots show a nonlinear behaviour and a hysteresis cycle; radial tires are generally less stiff than bias-ply tires of similar size. All characteristics are influenced by both the speed of rolling and the frequency of application of the force.

2.3 Rolling radius

Consider a wheel rolling on a level road with no braking or tractive moment applied to it and its mean plane perpendicular to the road. While the relationship between the angular velocity ω and the forward speed V of a rolling rigid wheel of radius R is simply $V = \omega R$, for a pneumatic tire an effective rolling radius R_e can be defined as the ratio between V and ω

$$R_e = V/\omega \; . \tag{2.1}$$

This amounts to define as effective rolling radius the radius of a rigid wheel which travels and rotates at the same speed of the pneumatic wheel.

The wheel-road contact is far from being a point-contact and the tread band is compliant also in circumferential direction; as a consequence radius R_e coincides neither with the loaded radius R_l nor with its unloaded radius R and the centre of instantaneous rotation is not coincident with the centre of contact A (Fig. 2.6).

Owing to the longitudinal deformations of the tread band, the peripheral velocity of any point of the tread varies periodically: When it gets near to

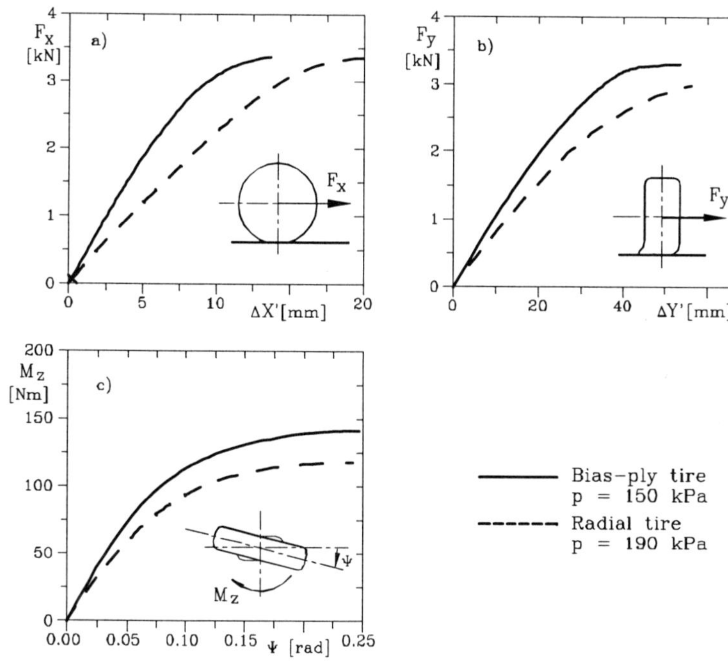

Fig. 2.5 Longitudinal force F_x (a), lateral force F_y (b) and aligning torque M_z (c) versus deflections $\Delta X'$ and $\Delta Z'$ and rotation about Z'-axis for a radial and a bias-ply tire at standstill (wheel non-rolling) and in static conditions (load applied and removed slowly).

the point in which it enters the contact zone it slows down and consequently a circumferential compression results. In the contact zone there is very limited sliding between tire and road: The peripheral velocity of the tread (relative to the wheel centre) in that zone coincides with the velocity of the centre of the wheel V. After leaving the contact zone, the tread regains its initial length and its peripheral velocity ωR is restored.

As a consequence of this mechanism, the spin speed of a wheel with pneumatic tire is smaller than that of a rigid wheel with the same loaded radius R_l and travelling at the same speed

$$R_l < R_e < R\,.$$

The centre of rotation of the wheel lies then under the surface of the road, at a short distance from it.

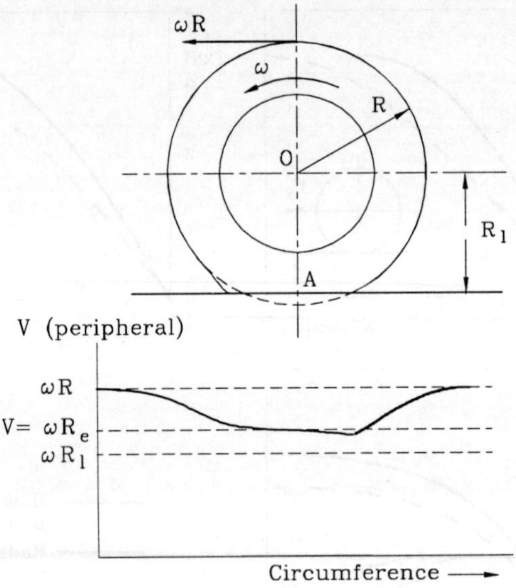

Fig. 2.6 Pneumatic wheel rolling on a flat road: Geometrical configuration and peripheral speed in the contact zone.

Owing to their lower vertical stiffness, radial tires have a lower loaded radius R_l than bias-ply tires with equal radius R but their effective rolling radius R_e is closer to the unloaded radius, as the tread is circumferentially stiffer. For instance in a bias-ply tire R_e can be about 96% of R while R_l is 94% of it; in a radial tire R_e and R_l can be respectively 98% and 92% of R.

The effective rolling radius depends on many factors, some of which are determined by the tire as the type of structure, the wear of the tread, and others by the working condition as inflation pressure, load, speed and others.

An increase of the vertical load F_z and a decrease of the inflation pressure p lead to similar results: A decrease of both R_l and R_e. With increasing speed, the tire expands under centrifugal forces, and consequently R, R_l and R_e increase. This effect is larger in bias-ply tires while, owing to the greater stiffness of the tread band, radial tires expand to a very limited, and usually negligible, extent. As it will be shown in the following sections, any tractive or braking torque applied to the wheel will cause strong variations of the effective rolling radius.

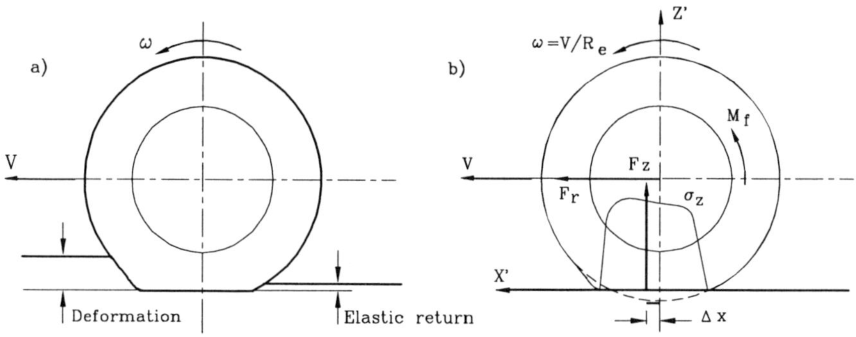

Fig. 2.7 (a) Wheel rolling on compliant soil: ground deformation and elastic return. (b) Forces and contact pressure σ_z in a rolling wheel.

2.4 Rolling resistance

Consider a wheel rolling freely on a flat surface. If both the wheel and the road were perfectly undeformable, there would be no resistance and consequently no need to exert a tractive force. In the real world however perfectly rigid bodies do not exist and both the road and the wheel are subject to deformation in the contact zone: During the motion new material enters continuously into this zone and is deformed, to spring back to its initial shape when leaving it. To produce this deformation it is necessary to spend some energy which is not completely recovered at the end of the contact zone due to the internal damping of the material.

This energy dissipation is what causes rolling resistance. It is then clear that it increases with increasing deformations and, mainly, with decreasing elastic return. A steel wheel on a steel rail has a lower rolling resistance than a pneumatic wheel and the motion on compliant soil causes greater resistance than that on a rigid surface. From this viewpoint, a wheel on compliant soil is always in the situation of a wheel that attempts to climb out of the pit it is digging by itself (Fig. 2.7).

In the case of pneumatic tires rolling on tarmac or concrete the deformations are almost only localized in the wheel and then the energy dissipated in the tires governs the phenomenon. Other mechanisms, like small sliding between road and wheel, aerodynamic drag on the disc and friction in the hub are responsible for a small contribution to the overall resistance, of the order of a few percent.

The distribution of the contact pressure, which at standstill was symmetrical with respect to the centre of the contact zone, becomes unsymmetrical when the wheel is rolling and the resultant F_z moves forward (Fig. 2.7b) producing a torque $M_y = -F_z \Delta x$ with respect to the rotation axis. Rolling resistance is due to this torque, together with the small contributions of the resistance in the hub and aerodynamic drag.

The two mentioned ways of seeing rolling drag are equivalent since resultant F_z is displaced forward to the centre of the contact owing to the energy dissipations occurring in the deformed parts of the wheel and possibly of the ground.

Rolling resistance is defined by the mentioned SAE document J670 as the force which must be applied to the wheel at the wheel centre with a line of action parallel to the X'-axis so that its moment with respect to a line through the centre of tire contact and parallel to the spin axis of the wheel will balance the moment of the tire contact forces about this line. Here this definition is used, with two small modifications. First thing the force is changed in sign, so that a true resisting force, i.e. a force acting in a direction opposite to the speed, is obtained. Second the aerodynamic drag moment on the wheel and the resisting torque applied to the hub are included, in order to include also these two effects in the overall rolling resistance.

To maintain a free wheel spinning a force at the wheel-ground contact is required and then some of the available traction is used: On the free wheel, to supply a torque which counteracts the total moment M_y, and on the driving wheels which must supply a tractive force against the rolling resistance of the former.

On driving wheels the driving torque is directly applied through the driving shafts to overcome rolling resistance moment. Rolling resistance of driving wheels thus does not involve forces acting at the road-wheel contact and does not use any of the available traction.

This is particularly important in the motion on compliant ground, which is usually characterized by high rolling drag and low available traction: If all wheels are driving wheels, rolling drag can be overcome directly by the driving torque; if some of the wheels are free, the traction the driving wheels can supply may not be sufficient to overcome the drag of the free wheels and motion may be impossible, even on level road.

Consider a free rolling wheel on level road with its mean plane coinciding with $X'Z'$-plane, i.e. with $\alpha = 0$, $\gamma = 0$ (Fig. 2.7). Assuming that no traction or braking moment other than moment M_f due to aerodynamic drag and bearing drag is applied to the wheel, the equilibrium equation in steady state rolling, solved in the rolling resistance F_r, is

$$F_r = \frac{-F_z \Delta x + M_f}{R_l} \; . \tag{2.2}$$

It must be noted that both rolling resistance F_r and drag moment M_f are negative.

In the case of driving wheels, moment M_f must be substituted by the difference $M_m - |M_f|$ between the driving and resistant moments. If such difference is positive and greater than moment $F_z \Delta x$, force F_r is positive, i.e. the wheel is exerting a driving action.

Equation (2.2) is of limited practical use, as Δx and M_f are not easily determined. For practical purposes, rolling resistance is usually expressed as

$$F_r = -f F_z \; , \tag{2.3}$$

where the rolling resistance coefficient f must be determined experimentally. The minus sign in Eq. (2.3) comes from the fact that traditionally the rolling resistance coefficient is expressed by a positive number.

Coefficient f depends on many parameters, as the travelling speed V, the inflation pressure p, the normal force F_z, the size of the tire and of the contact zone, the structure and the material of the tire, the working temperature, the road conditions and, last but not least, the forces F_x and F_y exerted by the wheel.

2.4.1 *Effects of the speed*

The rolling resistance coefficient f generally increases with the speed V of the vehicle, at the beginning very slowly and then at an increased rate (Fig. 2.8).

The law $f(V)$ can be approximated by a polynomial expression of the type

$$f = \sum_{i=0}^{n} f_i V^i \; . \tag{2.4}$$

Generally speaking, two terms of Eq. (2.4) are considered sufficient for approximating experimental data in a satisfactory way, at least up to the

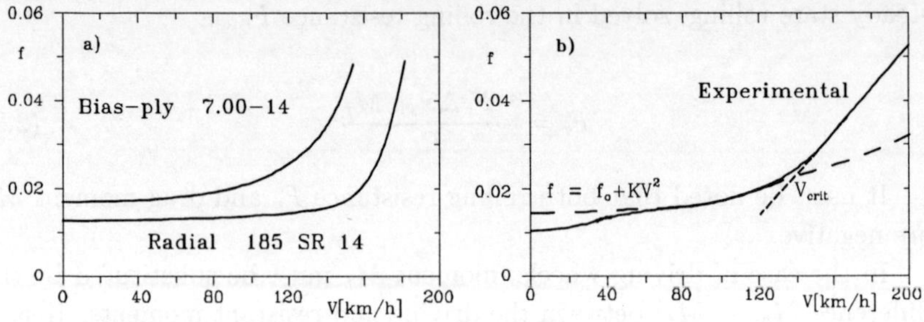

Fig. 2.8 Rolling coefficient versus speed. (a) Measured on a bias-ply and a radial tire; (b) experimental curve (radial tire 5.20-14 inflated at 190 kPa, with a load of 340 kN) compared with Eq. (2.6).

speed at which f starts to grow at a very high rate (Fig. 2.8).

The following expressions can be used

$$f = f_0 + KV \tag{2.5}$$

or

$$f = f_0 + KV^2 , \tag{2.6}$$

the second is generally preferred; it will be used throughout this book. The values of f_0 and K must be measured on any particular tire; as an example the tire of Fig. 2.8b is characterized, in the test conditions reported, by the values: $f_0 = 0.013$, $K = 6.5 \times 10^{-6}$ s^2/m^2.

The speed at which the curve $f(V)$ shows a sharp bend upward is generally said to be the "critical speed of the tire". Its presence can be easily explained by vibratory phenomena which take place in the tire at high speed, as the ones clearly visible in the pictures of a tire rolling at high speed against a drum reported in Fig. 2.9. The standing waves which propagate along the circumference of the tire from the contact zone are clearly visible. The tread band vibrates both in its plane and in the direction of the axis of the wheel.

The increase of rolling resistance which is linked with the occurrence of standing waves is easily explained by the fact that their wavelength is not much different from the length of the contact zone (Fig. 2.10); in the trailing part of the contact zone the tread then has a tendency to lift from the ground, or at least to decrease its pressure on it. The pressures concentrate

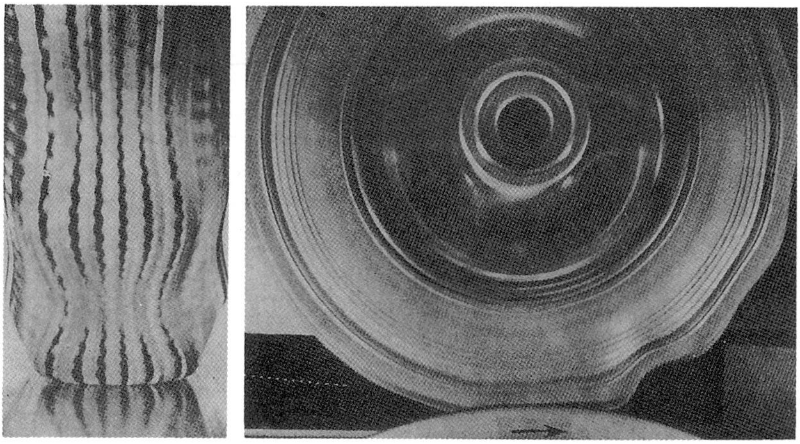

Fig. 2.9 Standing waves in a tire at high speed. (A. Morelli, Costruzioni automobilistiche in Enciclopedia dell'ingegneria, ISEDI, ilano, 1972).

consequently in the leading zone of the contact and their resultant moves forward, with an increase of the moment $F_z\Delta x$.

The critical speed of the tire, i.e. the speed at which such vibrations become important, must be considered as the speed at which the tire stops working in a regular way and consequently it should never be exceeded or even approached in the normal use of the vehicle. Above that speed strong overheating takes place; as most of the increase of the rolling power is converted into heat, the increase of temperature can quickly cause the destruction of the tire itself.

The critical speed is influenced by many parameters and it is one of the factors which must be taken into account in the choice of the tires for a particular vehicle.

2.4.2 *Effect of the structure of the tire and of the materials used*

The type of structure and the material used for the construction of the tire play an important role in determining the rolling resistance and the critical speed. Generally speaking, radial tires show a value of f about 20% lower than bias-ply types (Fig. 2.8) and a higher value of the critical speed. Even between tires of the same type, it is possible to observe significant differences in rolling resistance, as it is possible to optimize the structure (number of

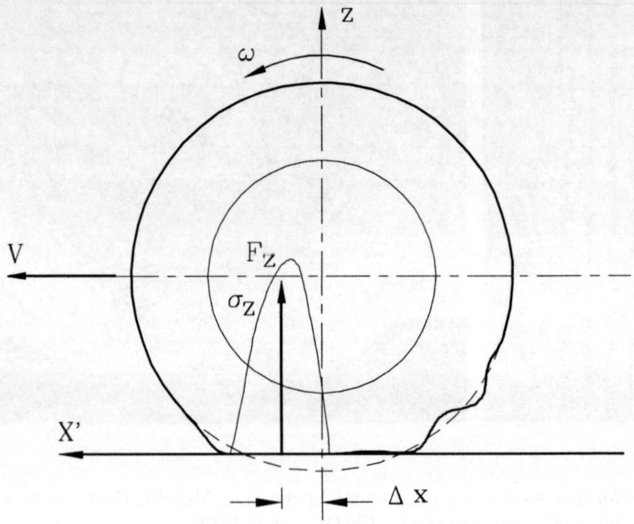

Fig. 2.10 Standing waves in a tire rolling above its critical speed. Distribution of the normal pressure.

plies, their orientation etc.) in order to obtain suitable characteristics for any particular application.

Main differences can be seen between tires for cars and for industrial vehicles. The latter are characterized by very low values of f_0, down to 0.005–0.008, and very limited or no increase of resistance with speed ($K \approx 0$), if not a slight decrease.

The nature of the material used has a great importance, as different rubber compositions are characterized by different values of the internal damping and by different dependences of the latter with the loading frequency. Natural rubbers have generally low damping, if compared with synthetic rubbers. This results in lower rolling resistance but also in lower critical speed.

Also the type and quantity of the other chemical components added to rubber have an important influence on damping and consequently on rolling resistance.

2.4.3 *Effect of wear*

In the case of bias-ply tires the rolling resistance decreases with wear, and its behaviour at high speed improves. This behaviour can be ascribed to the fact that deformations are localized in a small zone surrounding the contact zone and consequently hysteresis losses take place mainly in the tread band. Also vibratory phenomena interest mainly the zone immediately near the tread band. A decrease of the vibrating mass has the consequence of increasing the natural frequency and hence the critical speed.

In radial tires rolling resistance decreases with wear, but the behaviour at high speed can get worse. In radial tires deformations are more evenly distributed in the whole structure, as the stiffness of the sidewalls is low; with a decrease of the mass of the tread band, the centrifugal stiffening of the whole structure decreases, and vibratory phenomena become more important.

2.4.4 *Effect of the working temperature*

Internal damping of rubber decreases with increasing temperature and consequently rolling resistance, which is mainly due to hysteresis losses, decreases. Also that small part of rolling resistance which is due to localized sliding in the contact zone, decreases for the decrease of the friction coefficient.

The decrease of rolling resistance tends to stabilize the temperature of the tire as an increase of the temperature causes a decrease of power dissipation and consequently of the rate at which heat is generated within the tire.

Some curves $f(V)$ obtained at constant temperature are reported in Fig. 2.11a, while in Fig. 2.11b the rolling coefficient of the same tire maintained at each speed at the equilibrium temperature is shown. The curve is compared with that obtained by maintaining the tire at a temperature corresponding to low-speed running, as it occurs at start-up of the vehicle, before equilibrium temperature is reached. The decrease in time of the rolling resistance and the increase of the temperature in a tire rolling at 185 km/h is shown in Fig. 2.11c while in Fig. 2.11d the equilibrium temperature is plotted as a function of the speed. The last two plots have been obtained on a Nylon 7.25–13 radial tire, with $F_z = 4$ kN and $p = 150$ kPa, rolling against a roller of 2.5 m diameter. The temperature was measured using a thermocouple inset in the carcass and reflects the temperature of the material which is higher than that of the air. As the tests were performed on a roller, the increase of temperature and the resistance are somewhat higher than those which would be obtained on the road.

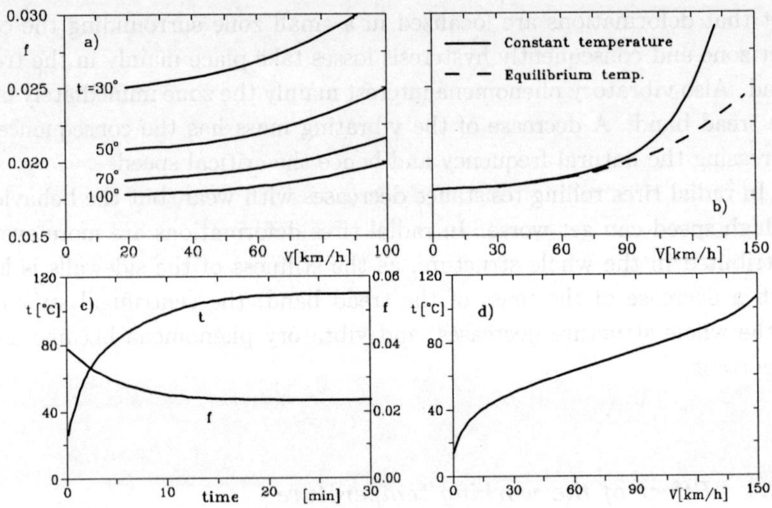

Fig. 2.11 (a) Rolling coefficient as a function of the temperature at constant speed. (b) Comparison between the law $f(V)$ at constant temperature and at the equilibrium temperature at each speed. (c) Decrease in time of the rolling resistance and increase of the temperature in a tire rolling at 185 km/h. (d) Equilibrium temperature as a function of the speed.

2.4.5 *Effect of inflation pressure and of the load*

Generally speaking an increase of inflation pressure or a reduction of the normal force F_z acting on the wheel cause the rolling resistance to decrease and the critical speed to increase. Also this effect tends to stabilize the temperature, as an increase of the latter causes an increase of the pressure which in turn decreases power dissipation and heat generation.

In order to take into account the influence of both load and pressure on the rolling resistance coefficient the following empirical formula suggested by the S.A.E. can be used

$$f = \frac{K'}{1000}\left(5.1 + \frac{5.5\times10^5 + 90F_z}{p} + \frac{1100 + 0.0388F_z}{p}V^2\right), \tag{2.7}$$

where coefficient K' takes the value 1 for conventional tires and 0.8 for radial tires. The normal force F_z, the pressure p and the speed V must be expressed respectively in N, N/m² (Pa) and m/s. The dependence of coefficient f on the speed V is the same as the one expressed by Eq. (2.6).

It must however be noted that for each tire the inflation pressure p

is determined as a function of the force F_z by design considerations, and that it is impossible to increase the pressure in order to reduce the rolling resistance.

2.4.6 *Effect of the size of tire*

The two geometrical parameters which have more influence on the rolling resistance are the radius of the tire and the aspect ratio H/W. An increase of the former and a decrease of the latter cause a decrease of rolling resistance and an increase of the critical speed. The use of small wheels, which are lighter, smaller and less expensive, is then justified for slow cars, and particularly for city cars.

The decrease of the aspect ratio is favorable as it causes an increase of the stiffness of the sidewalls and decreases the deformation under load, which in turn lower the hysteresis losses. Such ratio can be as low as 0.4 on radial tires used on fast modern cars, while values of 0.7–0.8 were the most common in the past.

2.4.7 *Effect of the nature and conditions of the road*

As a first approximation, the nature and conditions of the road are taken into account by choosing an appropriate value of f_0, i.e. by moving the curve $f(V)$ in the direction of the f-axis. Some values of f_0 for different road types are reported in Table 2.1.

2.4.8 *Effect of sideslip angle*

If the tire travels with a sideslip angle α, as it is the case any time it exerts a side force F_y or as a consequence of toe angle, a strong increase of rolling resistance can be expected. The force in the mean plane of the wheel increases but above all the transversal force F_y has a component which adds to the rolling resistance (Fig. 2.12). The rolling resistance is by definition the component of the force due to the road-tire contact directed as the velocity V; it can thus be expressed as

$$F_r = F_x \cos(\alpha) + F_y \sin(\alpha) \,. \tag{2.8}$$

If the component in the plane of symmetry of the wheel F_x were independent of the sideslip angle and the cornering force F_y were linear with it

Table 2.1 Values of f_0 for different types of road.

Road type and condition	f_0
Very good concrete	0.008 – 0.010
Very good tarmac	0.010 – 0.0125
Average concrete	0.010 – 0.015
Very good pavè	0.015
Very good Macadam	0.013 – 0.016
Average tarmac	0.018
Concrete in poor conditions	0.020
Good block paving	0.020
Average Macadam	0.018 – 0.023
Tarmac in poor conditions	0.023
Dusty Macadam	0.023 – 0.028
Good stone paving	0.033 – 0.055
Good natural paving	0.045
Stone pavè in poor conditions	0.085
Snow (50 mm layer)	0.025
Snow (100 mm layer)	0.037
Unmaintained natural road	0.080 – 0.160
Sand	0.150 – 0.300

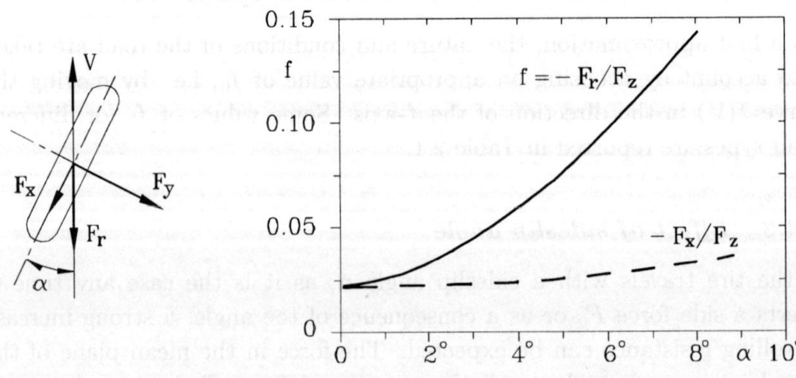

Fig. 2.12 Rolling resistance coefficient as a function of the slip angle α. Tire 7.50-14, $F_z = 4$ kN, $p = 170$ kPa.

($F_y = -C\alpha$, see Sec. 2.6), for small values of α the rolling resistance would follow a quadratic law

$$F_r = F_x - C\alpha^2 . \tag{2.9}$$

2.4.9 *Effect of the camber angle γ*

If the mean plane of the wheel is not perpendicular to the ground a component of the aligning torque M_z (see Sec. 2.6) contributes to rolling resistance. Equation (2.2) becomes

$$F_r = \frac{-F_z \Delta x \cos(\gamma) - M_z \sin(\gamma) + M_f}{R_l} \,. \tag{2.10}$$

This effect is usually very small, due to the fact that γ is usually small. It is however dependent on the sideslip angle α through the aligning torque M_z.

2.4.10 *Effect of tractive or braking forces*

Rolling resistance can also be defined when tractive or braking moments are applied to the wheel. In this case the power dissipated by rolling resistance $F_r V$ can be expressed as

$$|F_r|V = \begin{cases} |F_b|V - |M_b|\omega & \text{(braking)} \\ |M_t|\omega - |F_t|V & \text{(traction)} \end{cases} \tag{2.11}$$

where F_b, F_t, M_b and M_t are respectively the braking and tractive forces and moments. Equations (2.11) should be applied only in constant speed motion, since they do not include tractive (braking) moments needed to accelerate (decelerate) rotating parts.

The trend of the rolling resistance coefficient as a function of the longitudinal force F_x is shown in Fig. 2.13. The increase of rolling resistance is not negligible in case of strong longitudinal forces, particularly in the case of braking. This is due mainly to the fact that the generation of longitudinal forces is always accompanied by the presence of sliding in at least a part of the contact zone.

The minimum rolling resistance occurs when the wheels exert a low driving force[3] and can take a value as low as 75%-85% of that occurring in free rolling conditions. The fact that the rolling resistance decreases initially with the application of a driving force to increase steeply when the force increases would favour four wheel drive layouts in which all wheels

[3] D.J. Schuring, *Energy Loss of Pneumatic Tires Under Freely Rolling, Braking and Driving Conditions*, Tire Science and Technology, TSTCA, Vol. 4, No. 1, Feb. 1976, pp. 3-15.

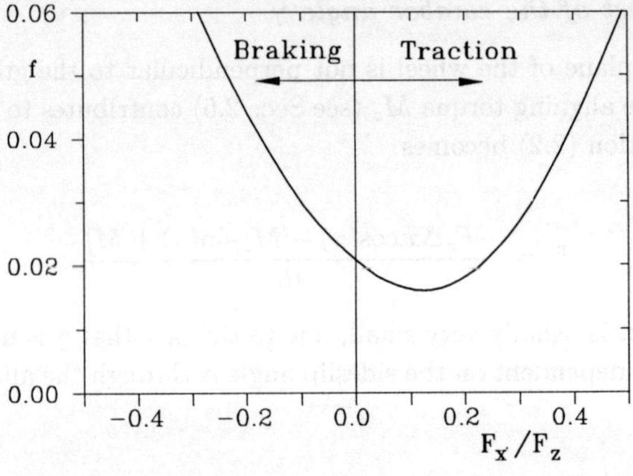

Fig. 2.13 Rolling resistance coefficient as a function of the tractive or braking force.

work under moderate driving forces instead of having some wheels in free rolling and some working with higher driving loads.

2.5 Tractive and braking forces

Consider a pneumatic wheel rolling on level road. If a braking moment M_b is applied to it, the distributions of normal pressure and longitudinal forces which result from that application are qualitatively sketched in Fig. 2.14a.

The tread band is circumferentially stretched in the zone which precedes the contact with the ground, while in free rolling the same part of the tire was compressed. The peripheral velocity of the tread band in the leading zone of the contact $\omega R'_e$ is consequently higher than that (ωR) of the undeformed wheel. The effective rolling radius R'_e, whose value R_e in free rolling was between R_l and R, grows towards R and, if M_b is large enough, becomes greater than R.

The instantaneous centre of rotation is consequently located under the road surface (Fig. 2.15). The angular velocity ω of the wheel is lower than that characterizing free rolling in the same conditions ($\omega_0 = V/R_e$).

In such conditions it is possible to define a "longitudinal slip" as

$$\sigma = \frac{\omega}{\omega_0} - 1 = \frac{v}{V}, \tag{2.12}$$

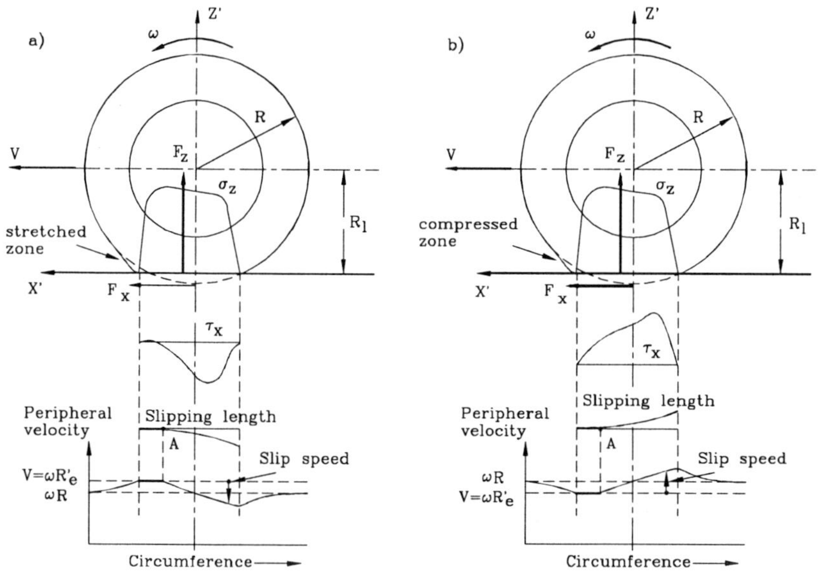

Fig. 2.14 Force distributions and peripheral velocity in a braking (a) and in a driving (b) wheel. Note that the equivalent rolling radius R'_e differs from R_e defined for free straight rolling conditions (force F_x lies on the ground).

where v is the linear speed at which the contact zone moves on the ground. The longitudinal slip is often expressed as a percentage; in the present book however the definition of Eq. (2.12) will be strictly adhered to.

If instead of braking, the wheel is driving, the leading part of the contact zone is compressed instead of being stretched (Fig. 2.14b). The value of the effective rolling radius R'_e is smaller than that characterizing free rolling and is usually smaller than R_l; the angular velocity of the wheel is greater than ω_0.

The slip defined by Eq. (2.12) is positive for driving conditions and negative for braking. The presence of the slip velocity[4] v does not mean however that there is an actual sliding of the contact zone as a whole. The peripheral velocity of the leading part of that zone is actually $V = \omega R'_e$, and consequently in that zone no sliding can occur. The speed of the tread band starts to decrease (in braking, increase in driving) and sliding begins

[4]The slip velocity is defined by SAE Document J670 as $\omega - \omega_0$, i.e. the difference between the actual angular velocity and the angular velocity of a free rolling tire. Here a definition based on a linear velocity rather than an angular velocity is preferred: $v = R_e(\omega - \omega_0)$.

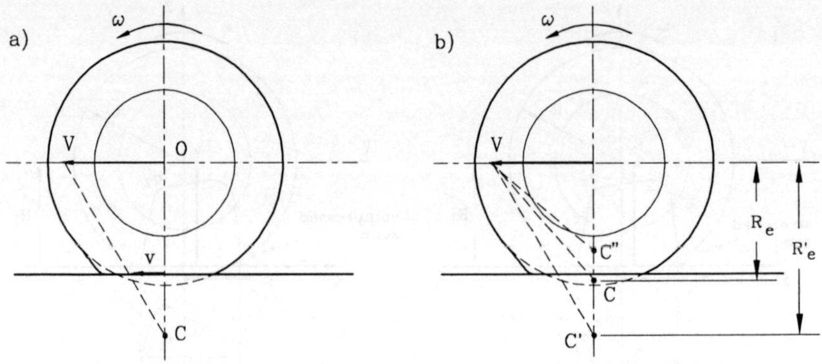

Fig. 2.15 (a) Braking wheel, centre of istantaneous rotation and slip speed. (b) Position of the centre of istantaneous rotation in free rolling C, braking C′ and driving C″.

only at the point indicated in Fig. 2.14 as point A. The slip zone, which interests only a very limited part of the contact zone for small values of σ, gets larger with increasing slip and, at a certain value of that parameter, reaches the leading part of the contact zone and global sliding of the tire occurs (Fig. 2.16a).

The longitudinal force F_x the wheel exchanges with the road is a function of σ. It vanishes when $\sigma = 0$ (free rolling conditions)[5] to change at high rate with almost linear law, for values of σ from −0.15 – −0.30 to 0.15 – 0.30. Outside this range, which depends on many factors, its absolute value decreases in braking up to the value $\sigma = -1$, which characterizes free sliding (locking of the wheel). Also in driving the force decreases above the stated range, but σ can have any positive value, up to infinity when the wheel spins while the vehicle is not moving.

As a first approximation, force F_x can be considered as roughly proportional to the load F_z, at equal value of σ. It is consequently useful to define a longitudinal force coefficient

$$\mu_x = \frac{F_x}{F_z} . \tag{2.13}$$

The qualitative trend of such coefficient is reported against σ in Fig. 2.16b.

Two important values of μ can be identified on the curve both in braking

[5]Actually free rolling is characterized by a very small negative slip, corresponding to the rolling resistance. This is however usually neglected when plotting curves $F_x(\sigma)$.

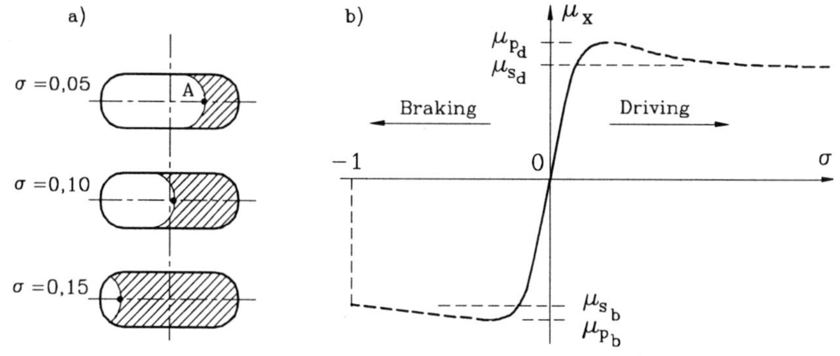

Fig. 2.16 (a) Slip zone at the tire-road contact with different values of the longitudinal slip σ. (b) Qualitative trend of the longitudinal force coefficient μ_x as a function of the longitudinal slip σ.

and in driving: The peak value μ_p and the value μ_s which characterizes pure sliding. The first is referred to as driving traction coefficient when the wheel is exerting a positive longitudinal force and as braking traction coefficient, which is usually reported in absolute value, in the opposite case. The second are the sliding driving traction coefficient and the sliding braking traction coefficient.

The part of the curve $\mu(\sigma)$ which lies beyond the range included by the two peak values, represented by a dashed line in Fig. 2.16b, is a zone of instability in the practical use of the vehicle. The equation of motion of a braking wheel is

$$J\frac{d\omega}{dt} = |F_x|R_l - |M_b| \; . \tag{2.14}$$

If a decrease of σ at constant speed V leads to a decrease of the absolute value of μ_x, it causes a decrease of the absolute value of F_x, i.e. of the force which maintains the rotation of the wheel. If this decrease of $|F_x|$ is not quickly followed by a decrease of the braking moment $|M_b|$ (and it is unrealistic to assume that the driver can react fast enough to release the brakes) a further slowing down of the wheel takes place, which in turn causes a further reduction of $|F_x|$.

Actually, when the optimum value of σ which is characterized by μ_{p_b}, is exceeded, the wheel locks in a very short time. In order to prevent the locking of wheels, devices generally defined as antilock or antiskid systems are now widely used.

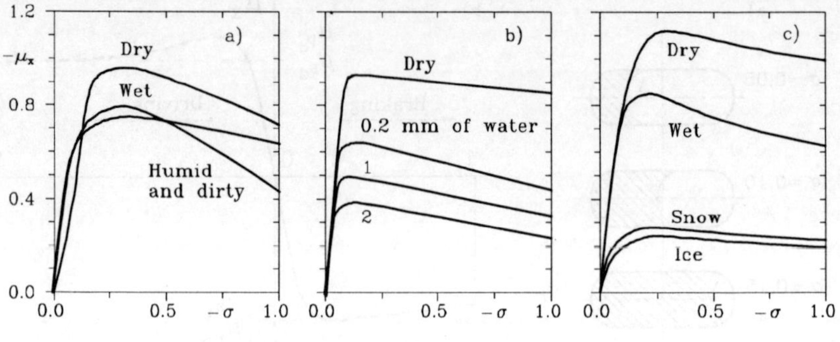

Fig. 2.17 Curves $\mu_x(\sigma)$ obtained in different conditions.

Such devices detect the deceleration of the wheel and, when it reaches a predetermined value, react decreasing the braking moment and avoiding the locking of the wheel. Antilock devices can operate on each wheel separately or, more often, on both wheels of an axle. Similarly, to avoid that a wheel slips under the effect of a driving torque applied to it, antispin devices limit the engine moment when the acceleration of the driving wheels exceeds a stated value.

The curves usually show a certain symmetry between the braking and driving conditions and often the maximum braking and driving forces are assumed to be equal. The values of function $\mu_x(\sigma)$ depend on a number of parameters, such as the type of tire, road conditions, speed, magnitude of the side force F_y exerted by the tire and many others. Moreover there is a significant difference between the curves obtained by different experimenters in conditions not exactly comparable. Some curves $\mu_x(\sigma)$ obtained in different conditions are reported in Fig. 2.17.

The maximum value of the longitudinal force decreases with increasing speed but this reduction is much influenced by operating conditions. Generally speaking, it is not very marked on dry road, while it is greater on wet surfaces. Also the difference between the maximum value (see Table 2.2) and the value related to sliding (Table 2.3) is more notable on wet roads (Fig. 2.18). Particularly dangerous conditions are encountered when the road is only partially wet and dirty: The behaviour can vary from spot to spot and the value at slip can be very much different from the maximum one.

The values shown in the tables must be regarded only as average indications as they are much sensitive. Note that in good conditions the longitudinal force can be almost equal to the load acting on the tire or even larger; the values reported however refer to standard tires used on passenger vehicles. High performance tires, particularly those used on racing cars, show peak values of μ_x which can be as high as 1.5 – 1.8 but even these tires do not reach very high values of longitudinal force coefficient in sliding condition; the difference between μ_p and μ_s is even larger. To reach very high performances particular formulations of the rubber must be used, which are characterized by a strong wear and consequently are restricted to competition tires. Note that radial tires almost always show better performances than bias-ply types.

Table 2.2 Maximum value of the longitudinal force coefficient (longitudinal traction coefficient) for different tires at a speed of 30 km/h. Tires 5.90-15; 6.00-15 and 165 R 15; F_z = 3kN, p = 160 kPa for bias-ply tires and 220 kPa for radial tires.

Tire	Road					
	Concrete		Tarmac			
	dry	wet	dry	wet	snow	ice
Radial	1.19	0.99	1.22	1.10	0.45	0.25
Bias-ply	1.13	0.84	1.02	1.07	0.27	0.24
Radial (snow)	1.04-1.12	0.62- 0.83	1.00- 1.09	1.00- 1.10	0.36- 0.47	0.24-0.44
Bias-ply (snow)	0.86-1.02	0.59- 0.70	0.81- 0.89	0.78- 1.02	0.41- 0.48	0.29- 0.37
Chains					0.60	0.40

Table 2.3 Values of the longitudinal force coefficient in sliding conditions (longitudinal sliding traction coefficient) for the same tires of the previous table.

Tire	Road					
	Concrete		Tarmac			
	dry	wet	dry	wet	snow	ice
Radial	0.95	0.73	1.03	0.90	0.43	0.16
Bias-ply	0.99	0.62	0.88	0.80	0.22	0.18
Radial (snow)	0.88- 1.00	0.50-0.61	0.87- 0.99	0.77- 0.93	0.35- 0.45	0.22- 0.41
Bias-ply (snow)	0.72- 0.90	0.47- 0.57	0.70- 0.78	0.67-0.84	0.39-0.47	0.29- 0.36

Tread wear has a great influence on the longitudinal forces, particularly at high speed. From Fig. 2.19 it can be assumed that the increase of μ_{p_b} due to wear can be, particularly at high speed, quite noticeable. It must however be noted that the figure refers to dry roads, as the presence of even a thin water layer on the road can change drastically these results.

When the road is wet, particularly if the water layer is thick, the tire can lift from the road surface as a result of hydrodynamic lift (hydroplaning). A thin layer of water can slip between the tire and the road thus reducing

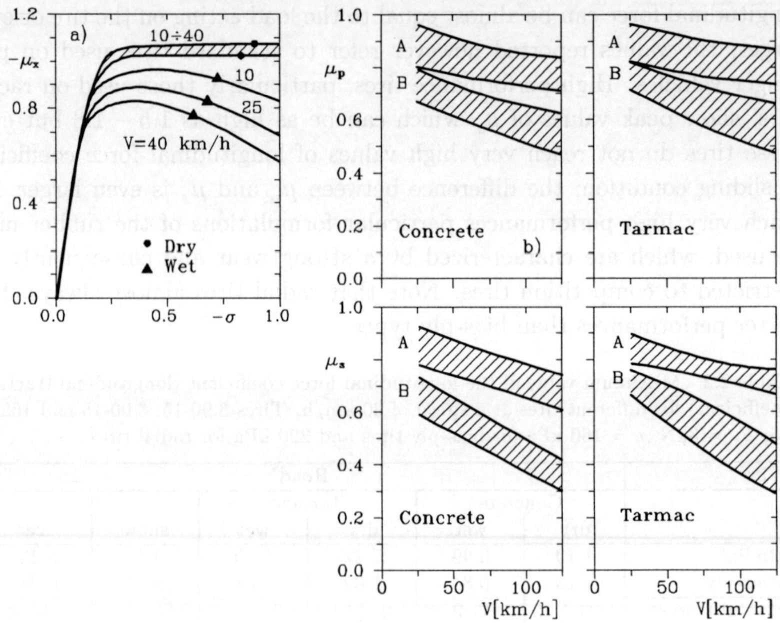

Fig. 2.18 Influence of speed on the values of μ_p and μ_s versus the speed on dry (A) and wet (B) road. Radial and bias-ply tires 6.40-13 and 6.00-15 with $F_z = 3.75$ kN.

the contact area (Fig. 2.20). With increasing speed the area of the contact zone further reduces, until a complete lifting of the tire takes place. True hydrodynamic lubrication conditions can be said to exist in this case and consequently the force coefficient or, better, the friction coefficient as in this condition sliding usually occurs, reduces to very low values, of the order of 0.05.

In order to avoid hydroplaning, or at least to postpone its occurrence, it is mandatory to evacuate water from the contact zone as quickly and effectively as possible. This can be done in two distinct ways: By making the road surface permeable or by using deep groves in the tread in both circumferential and transversal direction in order to allow a high flow rate.

The assumption, in a way implicit in the definition of the longitudinal force coefficient, that longitudinal forces are proportional to the normal force acting on the wheel is only a crude approximation. Actually the longitudinal force coefficient decreases with increasing load as shown in Fig. 2.21.

The curves $\mu_x(\sigma)$ can be approximated by analytical expressions. One

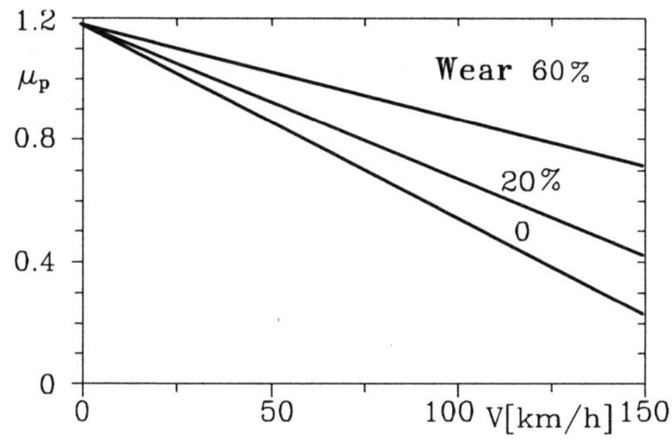

Fig. 2.19 Influence of tread wear on the maximum value of the longitudinal force coefficient with speed.

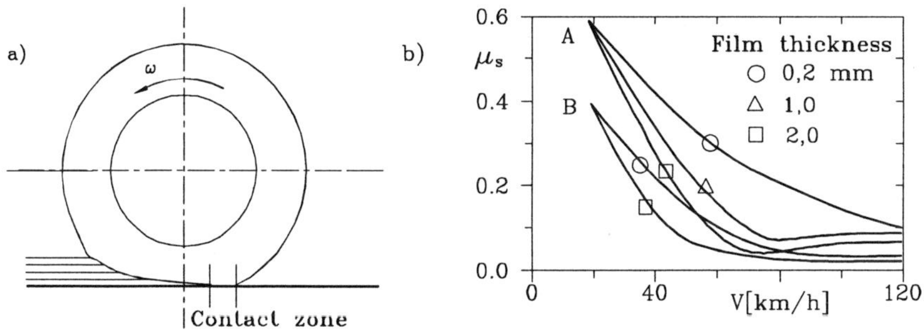

Fig. 2.20 Hydrodynamic lifting of the tire (hydroplaning). (a) Scheme. (b) μ_s as a function of speed on wet road. Tire 5.60-15 with tread (curve A) and with tread removed (curve B); $F_z = 3$ kN; $p = 150$ kPa.

of the formulas which can be used in the range $-1 < \sigma < 1$ is

$$\mu_x = A\left(1 - e^{-B\sigma}\right) + C\sigma^2 - D\sigma\ , \tag{2.15}$$

where

$$B = \left(\frac{K}{\alpha + d}\right)^{1/n}$$

is a factor which takes into account the interaction between the longitudinal slip σ and the sideslip α (see Sec. 2.7). The derivative in the origin $(\partial\mu_x/\partial\sigma)_{\sigma=0}$ is simply $AB - D$. Coefficients A, C, D, K, d and n must

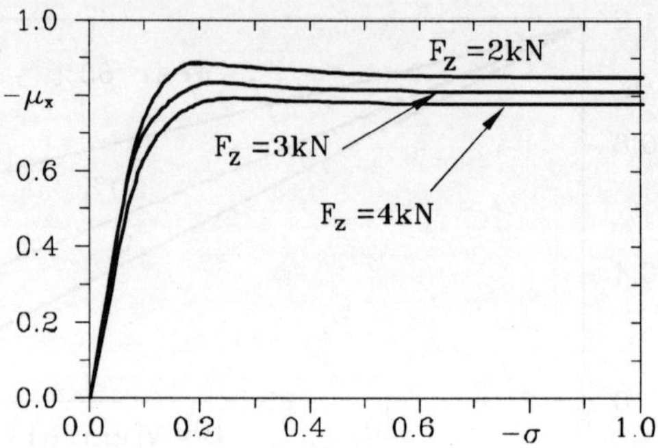

Fig. 2.21 Influence of normal force F_z on the curve $\mu_x(\sigma)$. Tire 6.00-15, $p = 170$ kPa, $V = 100$ km/h.

be obtained from the experimental curves and have no physical meaning. They depend not only on the road conditions but also on the load. A curve $\mu_x(\sigma)$ for a radial tire 145/80 R 13 4.5J obtained through Eq. (2.15) is reported in Fig. 2.22a, curve A.

A very good approximation of longitudinal force F_x as a function of the slip σ can be obtained through the empirical equation introduced by Pacejka[6] and known as "magic formula". Such mathematical expression allows one to express forces F_x and F_y and the aligning torque M_z as functions of the normal force F_z, of σ and of the sideslip (α) and camber (γ) angles.

The equation yielding the longitudinal force F_x, as a function of the slip σ, is

$$F_x = D \sin \left(C \arctan \left\{ B(1-E)(\sigma + S_h) + E \arctan \left[B(\sigma + S_h) \right] \right\} \right) + S_v \,, \tag{2.16}$$

where B, C, D, E, S_v and S_h are six coefficients which depend on the load F_z and on angle γ. They must be obtained from experimental testing and do not have any direct physical meaning. In particular, S_v and S_h have been introduced to allow nonvanishing values of F_x when $\sigma = 0$.

Coefficient D yields directly the maximum value of F_x, apart from the effect of S_v. The product BCD gives the slope of the curve for $\sigma + S_h =$

[6]E. Bakker, L. Lidner, H.B. Pacejka, *Tire Modelling for Use in Vehicle Dynamics Studies*, SAE Paper 870421; E. Bakker, H.B. Pacejka, L. Lidner, *A New Tire Model with an Application in Vehicle Dynamics Studies*, SAE Paper 890087.

0. The values of the coefficients are expressed as functions of a number of coefficients b_i which can be considered as characteristic of any specific tire, but depend also on road conditions and speed

$$C = b_0 \qquad D = \mu_p F_z \ ,$$

where for b_0 a value of 1.65 is suggested and

$$\mu_p = b_1 F_z + b_2 \ ,$$

$$BCD = \left(b_3 F_z^2 + b_4 F_z\right) e^{-b_5 F_z} \ ,$$

$$E = b_6 F_z^2 + b_7 F_z + b_8 \ ,$$

$$S_h = b_9 F_z + b_{10} \ , \qquad S_v = 0 \ .$$

If a symmetrical behaviour for positive and negative values of force X is accepted, this model can be used for both braking and driving. The curve is usually extended to braking beyond the point where $\sigma = -1$, to simulate a wheel rotating in backward direction while moving forward.

The coefficients introduced in Eq. (2.16) and the results obtained from it are usually expressed in non consistent units: Force F_z is in kN, longitudinal slip is expressed as a percentage and force F_x is in N.

A set of curves $F_x(\sigma)$ obtained for vertical loads F_z = 2, 4, 6 and 8 kN for a radial tire 205/60 VR 15 6J is shown in Fig. 2.22b. The values of the coefficients b_i are reported in Appendix 1.2.

A curve $\mu_x(\sigma)$ for a high performance radial tire 245/65 R 22.5 obtained from Eq. (2.16) is also reported in Fig. 2.22a, curve B. Note the very high peak value of the force coefficient.

The importance of the model expressed by Eq. (2.16) is mainly linked to the fact that tire manufacturers are increasingly giving the performances of their tires in terms of the coefficients to be introduced into it and in the similar expressions for the cornering force and the aligning torque. If this trend will consolidate, the "magic formula" will prove to be a simple and accurate model for tire behaviour and, which is even more important, one for which the data will be readily available.

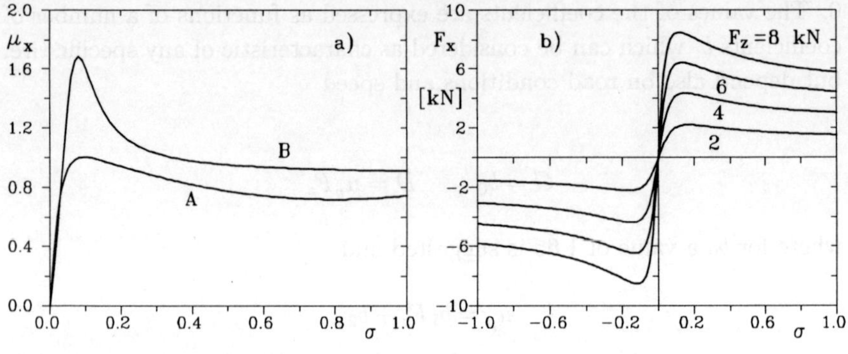

Fig. 2.22 (a) Curves $\mu_x(\sigma)$ for a tire 145/80 R 13 4.5J obtained through Eq. (2.15) (curve A) and a tire 245/65 R 22.5 obtained through Eq. (2.16) (curve B). (b) Curves $F_x(\sigma)$ for different values of the load obtained using Eq. (2.16) for a radial tire 205/60 VR 15 6J.

2.6 Cornering forces

In the previous section it was clear that a pneumatic tire can exert longitudinal forces only if deformations are present in the tread band and if the wheel has a nonvanishing longitudinal slip. In the same way the generation of cornering forces cannot be understood if no reference is made to the lateral deformations of the tire and to its sideslip angle: The generation of tangential forces in the road-wheel contact is directly linked with the compliance of the tire.

The fact that the wheel has a sideslip angle, i.e. is not in pure rolling, does not mean that in the contact zone the tire slips on the road: Also in this case, as seen for longitudinal forces, the compliance of the tire allows the tread to move, relatively to the centre of the wheel, with the same velocity as the ground. However some localized sliding between the wheel and the road can be present and, with increasing sideslip angle, they become more and more important, until the whole wheel is in actual, macroscopic sliding.

If the velocity of the centre of the wheel does not lie in its mean plane, i.e. if the wheel travels with a sideslip angle, the shape of the contact zone is quite distorted (Fig. 2.23). Consider a point belonging to the mean plane on the tread band. Upon approaching the contact zone it tends to move in a direction parallel to the velocity V, relatively to the centre of the wheel, and consequently goes out of the mean plane.

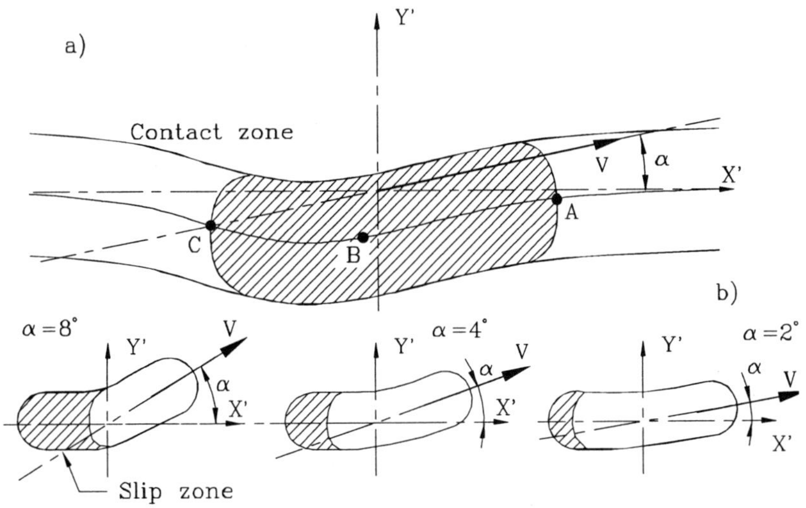

Fig. 2.23 Wheel-road contact when a sideslip angle is present. (a) Contact zone and path of a point belonging to the mean plane of the tread band. (b) Contact and slip zones at various sideslip angles α.

After touching the ground at point A, it continues following the direction of the velocity V (for an observer fixed to the ground, it remains still) until it reaches point B. At that point, the elastic forces pulling it towards the mean plane are strong enough to overcome those due to the friction on the road, forcing it to slide on the road and to deviate from its path. This sliding continues for the remaining part of the contact zone, until point C is reached. The contact zone can thus be divided into two parts: A leading zone in which no sliding occurs and a trailing one in which the tread slips towards the mean plane. This second zone grows with the sideslip angle (Fig. 2.23b), until it invades the whole contact zone and the wheel actually slips on the ground.

The lateral deformations of the tire are plotted in a qualitative way in Fig. 2.24, together with the distribution of σ_z, τ_y, and of the lateral velocity. The resultant F_y of the distribution of side forces is not applied at the centre of the contact zone but at a point which is located behind it at a distance t. Such distance is defined as "pneumatic trail".

The moment $M_z = F_y t$ is the aligning moment as it tends to force the mean plane of the wheel towards the direction of the velocity V. The absolute value of the side force F_y grows almost linearly at first as α increases,

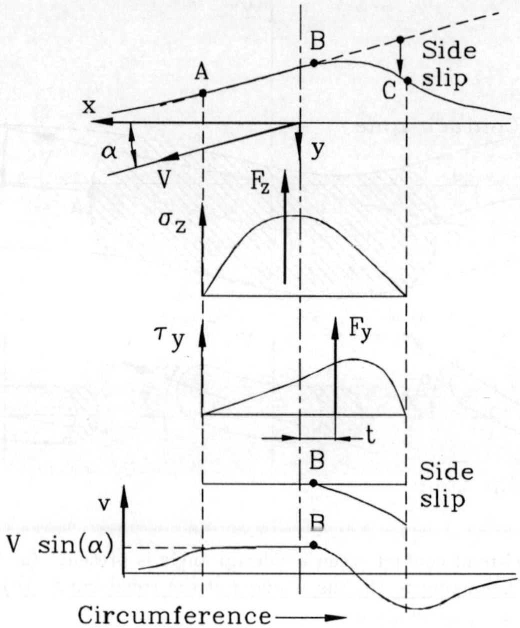

Fig. 2.24 Lateral deformations, distribution of σ_z and τ_y, sliding and lateral velocities in a tire rolling with a slip angle α.

then, when the limit conditions of sliding are approached, in a slower way. Eventually it remains constant, or decreases slightly, when sliding conditions are reached.

The side force F_y is plotted as a function of α for the cases of a radial and a bias-ply tire in Fig. 2.25a. Radial tires show a "stiffer" behaviour than bias-ply ones for what side forces are concerned, as they require smaller sideslip angles to produce the same side force.

With increasing sideslip angle, τ_y is more evenly distributed and the pneumatic trail decreases. The aligning moment is consequently the product of a force which increases with α and a distance which decreases; its trend is consequently of the type shown in Fig. 2.25b. At high values of α, M_z can change direction, as is shown in the figure.

The side force coefficient $\mu_y = F_y/F_z$ is often used for the side force. Its maximum value, usually defined as lateral traction coefficient, is written as μ_{y_p} and the value taken in sliding conditions as μ_{y_s}.

Both force F_y and moment M_z depend on many factors, besides the

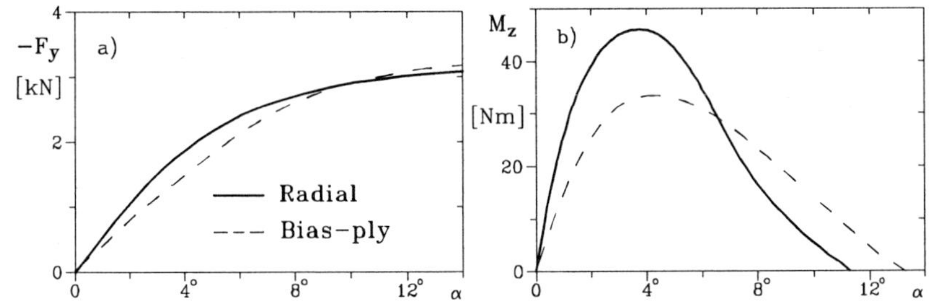

Fig. 2.25 Side force F_y and aligning moment M_z for tires of the same size but different type. Tire 5.60-13; $F_z = 3$kN, $p = 170$ kPa; $V = 40$ km/h.

angle α, as normal force F_z, speed, pressure p, road conditions etc. The lateral behaviour of the tire can be summarized in a single diagram, the so-called Gough diagram, in which the side force F_y is plotted against the self aligning torque M_z with F_z, α and t as parameters. The Gough diagrams for two different tires are shown in Fig. 2.26.

Alternatively, the "carpet plot" can be drawn; the force is plotted against the slip angle and the various curves at different F_z are just superimposed, translated by a quantity proportional to the normal force with respect to each other. Also the aligning torque can be plotted in this way. Carpet plots are shown in Fig. 2.34.

At increasing speed, the curve $F_y(\alpha)$ lowers, mainly in the part corresponding to the higher values of the sideslip angle. The linear part remains almost unchanged (Fig. 2.27). Also the pneumatic trail t decreases with increasing speed and consequently the aligning torque shows a decrease which is more marked than that of the side force.

The decrease of F_y, t and M_z is more pronounced in the case of bad road conditions. As far as hydrodynamic lifting (hydroplaning) is concerned, the same considerations seen for the longitudinal force F_x can be repeated for the side force F_y. The decrease of the aligning torque which goes together with the decrease of the side force should warn the driver of the approaching of the loss of traction. A poorly designed tire however could maintain strong aligning moments even in conditions of reduced transversal traction, leading the driver into error.

The presence of a camber angle produces, even if no slip angle is present, a lateral force. It is usually said camber thrust or camber force, as distinct from cornering force, due to sideslip angle alone. The camber force added to the cornering force give the total side or lateral force. The camber force

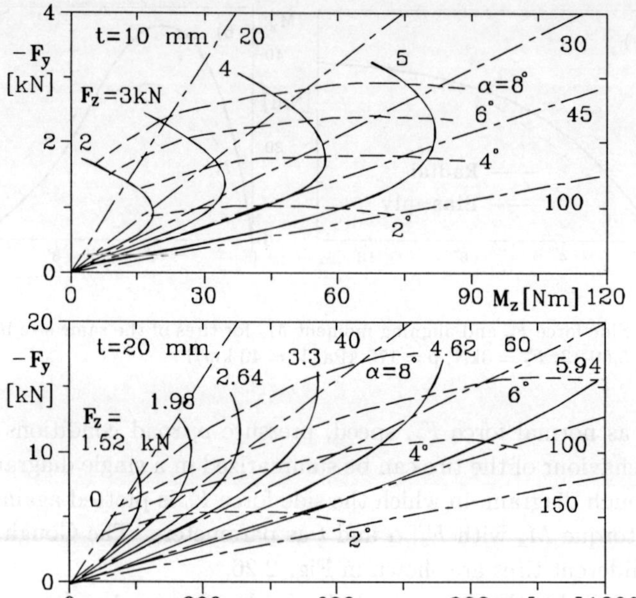

Fig. 2.26 Gough diagrams. (a) Radial tire for passenger vehicle 5.60-16; $p = 180$ kPa, $V = 50$ km/h; (b) Bias-ply tire for heavy truck 12.0-20, $p = 675$ kPa.

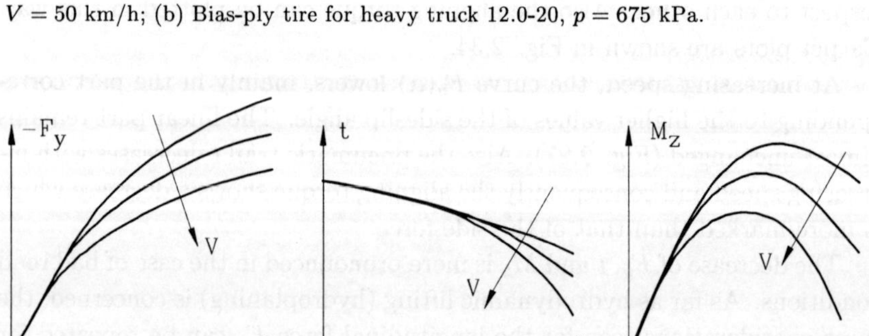

Fig. 2.27 Qualitative curves $F_y(\alpha)$, $M_z(\alpha)$ and $t(\alpha)$ at different speeds.

is usually far smaller than the cornering force, at least at equal values of angles α and γ. It depends on the load F_z, is practically linear with it (Fig. 2.28), and is strongly dependent on the type of tire considered.

The camber thrust is usually applied in a point which leads the centre of the contact zone, producing a small moment M_{z_γ}. It is usually neglected,

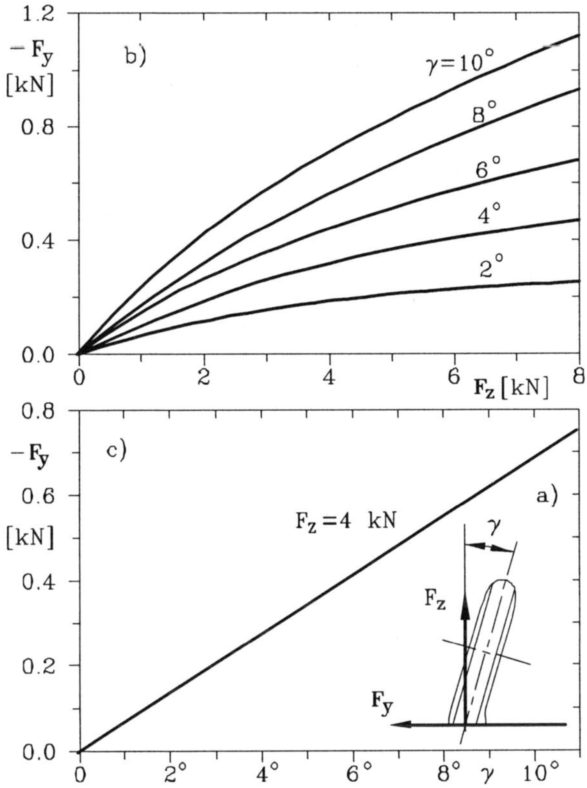

Fig. 2.28 Camber thrust. (a) sketch; note that the force F_y is negative and hence is directed in a direction opposite to that shown. Force as a function of the normal load (b) and of the camber angle (c). Tire 6.40-13, $p = 200$ kPa.

owing to its very small value. Bias-ply tires usually produce greater camber thrusts and moments than radial ones.

Usually both sideslip and camber are simultaneously present. As an example, the effect of camber on the force due to sideslip for two different values of the load is shown in Fig. 2.29. The camber thrust is more noticeable at low sideslip than when the sideslip is large, particularly when the wheel is less loaded. The aligning torque under the combined effect of sideslip and camber for a radial tire is shown in Fig. 2.30.

Ideally, when both sideslip and camber angle are equal to zero the lateral force and the aligning torque should be vanishingly small. This is in practice not true for the following reasons. Firstly, the lateral behaviour of

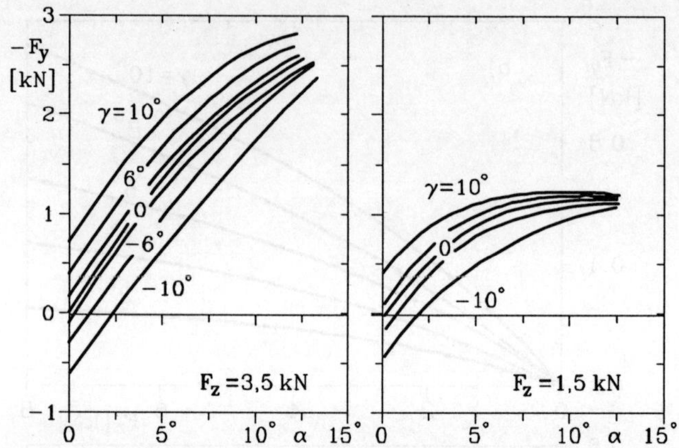

Fig. 2.29 Side force as a function of the sideslip angle at various camber angles for two different values of the normal force. Tire 5.20-13, $p = 160$ kPa, $V = 40$ km/h.

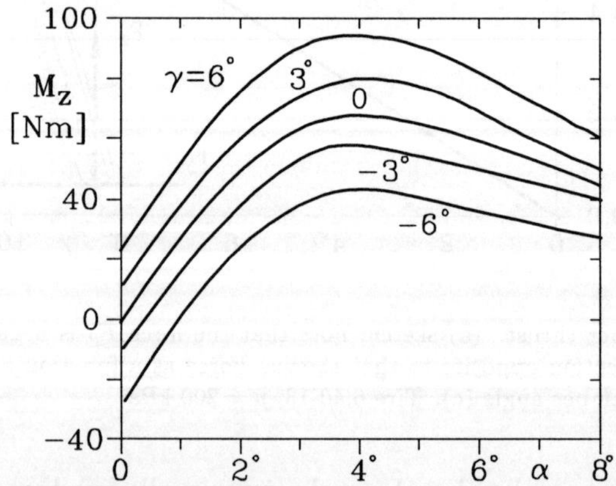

Fig. 2.30 Aligning torque as a function of the sideslip angle at various camber angles. Tire 185 HR 14, $p = 230$ kPa, $F_z = 3.5$ kN.

tires exhibits a hysteresis, in such a way that when the zero sideslip angle condition is reached from a condition in which a force was exerted on the right, a small residual force in the same direction remains and the same holds when the wheel is centreed from the opposite direction. This can give a feeling of lack of precision of the steering system and compel the driver to continuous corrections.

Moreover, the centre of the hysteresis cycle is not at the point in which both angle and force are equal to zero: Owing to lack of geometrical symmetry, a tire working in symmetrical conditions produces a side force. A first effect is due to a possible conicity of the outer surface of the tire: A conical drum would roll on a circular path whose centre coincides with the apex of the cone. Conicity is due to lack of precision during the manufacturing process and hence is linked with manufacturing quality control; its direction is random and its amount changes from tire to tire of the same model. If a tire is turned on the rim, the direction of the conicity is reversed, as is the force it causes when the tire is rolled along a straight path.

Another unavoidable lack of symmetry is linked with the angles of the various plies and their stacking order; the effect it causes is called ply steer. If the wheel is rolled free ply steer causes it to roll along a straight line angled with respect to the plane of symmetry; if the wheel rolls with no sideslip angle the generation of a side force results. If a tire is turned on the rim the direction of the force due to ply steer is not reversed. As it is caused by a factor included in the tire design, unlikely the effect of conicity, that of ply steer is very consistent between tires of the same model.

While conicity can be included into the models of the tire only in a statistical way, ply steer is one of the peculiarities of each tire and can be accounted for with precision. Note that while these effects are usually considered as a nuisance, opposite ply steer of the wheels of a given axle can even be used as a substitute of toe-in; while the latter increases the rolling resistance the first one has no effect on it.

Generally speaking, the lateral force offset is subdivided into two parts: The part which does not change sign when the direction of rotation is reversed is said to be ply-steer force, while the part that changes sign is said to be conicity force.

As already stated, for low values of the sideslip angle the cornering force increases linearly with α. The slope $\partial F_y/\partial\alpha$ of the curve in the origin is usually defined as "cornering stiffness" or "cornering power" and written as C. Since the cornering stiffness is expressed as a positive number while, at least in the initial part of the curve $F_y(\alpha)$ the derivative $\partial F_y/\partial\alpha$ is always negative, the cornering force can be expressed, for low values of α, as

$$F_y = -C\alpha \,. \tag{2.17}$$

Expression (2.17) is quite useful to study the dynamic behaviour of vehicles under the assumption of small sideslip angles, as it actually occurs

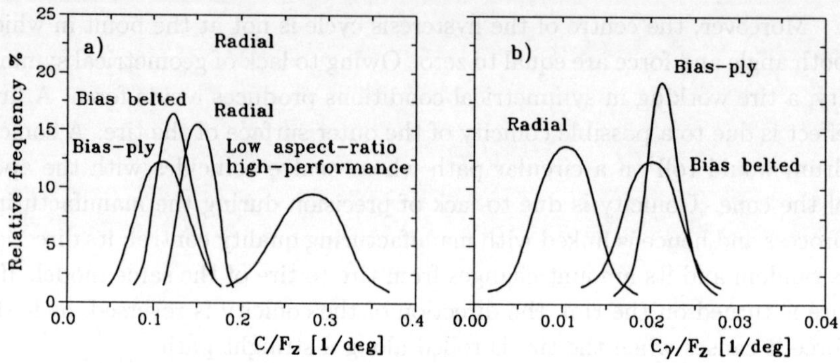

Fig. 2.31 Frequency distribution of (a) cornering stiffness coefficient and (b) camber stiffness coefficient for passenger car tires.

in normal driving conditions. In particular, it is essential in the study of the stability of linearized models.

The ratio between the cornering stiffness and the normal force is usually referred to as cornering stiffness coefficient (the term cornering coefficient is also used but SAE recommendation J670 suggests to avoid it for clarity). For bias-ply tires it is of the order of 0.12 $\deg^{-1} = 6.9 \text{ rad}^{-1}$ and for radial tires of the order of 0.15 $\deg^{-1} = 8.6 \text{ rad}^{-1}$. As the cornering coefficient changes much from tire to tire, it is interesting to note the distribution of its value for tires of different types (Fig. 2.31a).

In the same way the camber stiffness can be defined as the slope of the curve $F_y(\gamma)$ for $\gamma = 0$: $C_\gamma = \partial F_y/\partial\gamma$. Note that the camber thrust produced by a positive camber angle is negative and hence the camber stiffness is negative. The ratio between the camber stiffness and the normal force is usually referred to as camber stiffness coefficient. This coefficient is higher for bias-ply tires than for radial tires: In the first case an average value is of the order of 0.021 $\deg^{-1} = 1.2 \text{ rad}^{-1}$ and in the second is of the order of 0.01 $\deg^{-1} = 0.6 \text{ rad}^{-1}$. Also the distribution of the camber coefficient for tires of different types is of some interest (Fig. 2.31b).

The camber stiffness coefficient is interesting in the case a wheel rolls on a road with a transversal slope with its midplane remaining vertical: In this case there is a component of the weight which is directed downhill and the camber thrust which is directed uphill. The net effect can be in one direction or the other depending on the magnitude of the camber stiffness coefficient: The downhill component of the weight is $W\sin(\alpha_t) \approx W\alpha_t$, where α_t is the transversal inclination of the road while the camber thrust is equal to the weight multiplied by the camber stiffness coefficient and the angle. It is clear that if the value of the camber stiffness coefficient is larger than one (measured in rad^{-1}), as it occurs for bias ply tires, the net force is directed uphill; the opposite occurs for radial tires. This situation occurs when a rut is present in the road: A radial tire tends to track in the bottom while a bias-ply tire tends to climb out of the rut.

To include camber thrust into the linearized model, Eq. (2.17) can be modified as

$$F_y = -C\alpha + C_\gamma \gamma \,. \tag{2.18}$$

It can be used with confidence for values of α up to about 4° and of γ up to 10°.

Also the self aligning moment can be expressed by a linear law

$$M_z = (M_z)_{,\alpha}\alpha \,, \tag{2.19}$$

where $(M_z)_\alpha$ is the derivative $\partial M_z/\partial\alpha$ computed for vanishingly small α and γ and is defined as aligning stiffness coefficient or simply aligning coefficient.

Here again the effect of the camber angle can be included by modifying Eq. (2.19) as

$$M_z = (M_z)_{,\alpha}\alpha + (M_z)_{,\gamma}\gamma \,, \tag{2.20}$$

where $(M_z)_{,\gamma}$ is the derivative $\partial M_z/\partial\gamma$ computed for vanishingly small α and γ, but the second effect is so small that it is usually neglected.

Equation (2.20) supplies a good approximation of the aligning torque for a range of α far more limited than that in which Eq. (2.17) holds. It must however be noted that the importance of the aligning torque in the study of the behaviour of the vehicle is limited and consequently a precision lower than that required for side forces can be accepted. Practically, a good approximation of the aligning torque is important only when studying the steering mechanism.

The aligning stiffness coefficient due to sideslip angle is of about 0.01 m/deg (Nm/N deg) for bias ply tires and of 0.013 m/deg for radial tires while that due to camber (aligning camber stiffness coefficient) is approximately of 0.001 m/deg for the first ones and of 0.0003 m/deg for the latter.

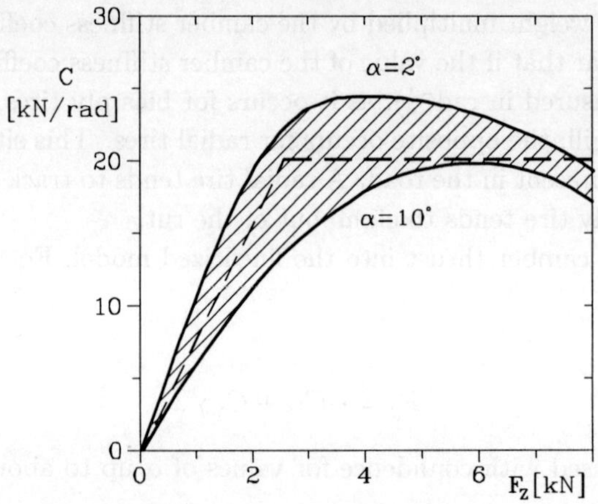

Fig. 2.32 Cornering stiffness as a function of the load F_z (the curve labelled $\alpha = 10°$ is related to a sort of "secant" stiffness).

A small aligning moment is due to the curvature of the path even if the sideslip angle is equal to zero; however this effect is not negligible only if the radius of the trajectory is very small, of the order of a few meters, and consequently it is present only in low speed manoeuvres. It can be important for the dimensioning the steering system for the mentioned conditions.

The definition of the cornering coefficient implies that the cornering stiffness is linear with the normal load F_z; actually the cornering stiffness behaves in this way only for low values of force F_z and then increases to a lesser extent (Fig. 2.32). When the limit value has been reached it remains constant or slightly decreases. It is often expedient to approximate the cornering stiffness as a function of the load with two straight lines, the second of which is horizontal. Note that in the figure the line corresponding to a sideslip angle of 2° refers to the true cornering stiffness while the other curve ($\alpha = 10°$) is related to a sort of "secant" stiffness.

When the need for a more detailed numerical description of the lateral behaviour of a tire arises, there is no difficulty, at least in theory, to approximate the experimental law $F_y(\alpha, \gamma, F_z, p, V, ...)$ and the similar relationship for the aligning torque, using the algorithms which are common in numerical analysis. This approach can be used with success in the numerical simulations of the behaviour of the vehicle, even if it is often quite expensive in terms of time needed for data preparation and computation. A problem which is common to many numerical approaches like this is that of requiring a great amount of experimental data, which are often difficult, or costly, to obtain.

Polynomial approximations, with terms including the third power of the slip angle α can be used.

As already stated, Eq. (2.16) can also be used to express the cornering force and the aligning moment as function of the various parameters.

In the case of the side force, the "magic formula" is

$$F_y = D\sin\left(C\arctan\left\{B(1-E)(\alpha+S_h)+E\arctan\left[B(\alpha+S_h)\right]\right\}\right)+S_v\ , \tag{2.21}$$

where the product of coefficients B, C and D yields directly the cornering stiffness. The values of the other coefficients are

$$C = a_0 \qquad D = \mu_{y_p} F_z\ ,$$

where a value of 1.30 is suggested for a_0 and $\mu_{y_p} = a_1 F_z + a_2$,

$$E = a_6 F_z + a_7\ ,$$

$$BCD = a_3 \sin\left[2\arctan\left(\frac{F_z}{a_4}\right)\right](1 - a_5|\gamma|)\ ,$$

$$S_h = a_8\gamma + a_9 F_z + a_{10}\ ,$$

$$S_v = a_{11}\gamma F_z + a_{12} F_z + a_{13}\ .$$

To obtain a better description of the camber thrust, the constant a_{11} is often substituted by the linear law

$$a_{11} = a_{111} F_z + a_{112}\ .$$

Coefficients S_h and S_v account for ply steer and conicity forces.

Similarly, in the case of the aligning torque the formula is

$$M_z = D\sin\left(C\arctan\left\{B(1-E)(\alpha+S_h)+E\arctan\left[B(\alpha+S_h)\right]\right\}\right)+S_v\ , \tag{2.22}$$

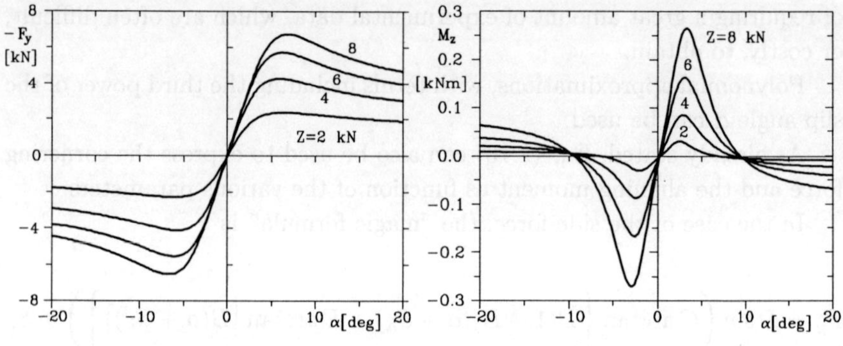

Fig. 2.33 Curves $F_y(\alpha)$ and $M_z(\alpha)$ obtained by using the "magic formula" (2.21) and (2.22). Radial tire 205/60 VR 15 6J.

$$C = c_0 , \qquad D = c_1 F_z^2 + c_2 F_z ,$$

where a value of 2.40 is suggested for c_0,

$$E = (c_7 F_z^2 + c_8 F_z + c_9)(1 - c_{10}|\gamma|) ,$$

$$BCD = (c_3 F_z^2 + c_4 F_z)(1 - c_6|\gamma|)e^{-c_5 F_z} ,$$

$$S_h = c_{11}\gamma + c_{12} F_z + c_{13} ,$$

$$S_v = (c_{14} F_z^2 + c_{15} F_z)\gamma + c_{16} F_z + c_{17} .$$

Also in this case the units introduced into the "magic formula" (2.21) and (2.22) are usually not consistent: The load F_z is expressed in kN, angles α and γ are in degrees, F_y and M_z are obtained in N and Nm respectively.

The curves $Y(\alpha)$ and $N_a(\alpha)$ for values of the vertical load F_z equal to 2, 4, 6 and 8 kN for a radial tire 205/60 VR 15 6J are shown in Fig. 2.33. The values of coefficients a_i and c_i are reported in Appendix A.2.

The magic formula can be used to model the behaviour of the tire in a more complete way than to express the direct relationship of the side force and aligning moment with the sideslip angle. As an example, the carpet plots of the side force, the side force coefficient, the aligning moment, the camber thrust and the camber moment are reported in Fig. 2.34, together with the Gough diagram, for the same tire of Fig. 2.33.

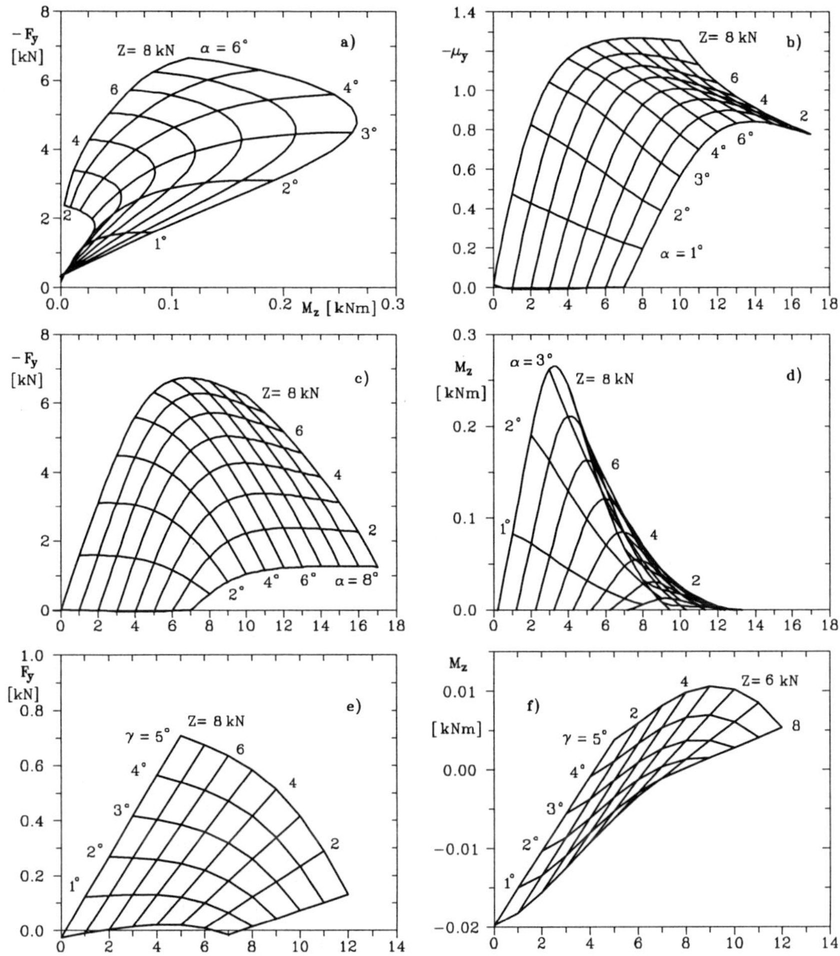

Fig. 2.34 (a) Gough diagram and carpet plots of the (b) side force coefficient, (c) side force, (d) aligning moment, (e) camber thrust and (f) camber moment for a radial tire 205/60 VR 15 6J obtained through the "magic formula".

It is also possible to build structural models of the tire to express the forces it exerts by taking into account the deformations and stresses their structure is subjected to. Apart from very complex numerical models, mainly based on the finite element method, which allow one to compute the required characteristics but are so complex that they are of little use in vehicle dynamics computations, it is possible to resort to simplified models, dealing with the tread band as a beam or as a string on elastic foundations[7]. These models allow one to obtain interesting results, particularly from a qualitative viewpoint, as they link the performance of the tire with its structural parameters, but their quantitative precision is usually smaller than that of empirical models, in particular of those based on the "magic formula" which is increasingly becoming a standard in tire modelling.

2.7 Interaction between longitudinal and side forces

The considerations seen in the preceding sections apply only in the case in which longitudinal and side forces are generated separately. If the tire produces simultaneously forces in X' and Y' directions the situation can be different as the traction used in one direction limits that available in the other.

By applying a driving or braking force to a tire which has a certain sideslip angle, the cornering force reduces and the same applies to the longitudinal force a tire can exert if it is called to exert also a lateral force.

An example is shown in Fig. 2.35: By setting the tire at a given sideslip angle and applying a braking torque, the shape of the curve $\mu_x(\sigma)$ is deeply changed. While the value in slipping conditions μ_s is almost unchanged, the peak value μ_p decreases to a greater extent.

It is then possible to obtain a polar diagram of the type shown in Fig. 2.36 in which the force in Y' direction is plotted versus the force in X' direction for any given value of the sideslip angle α. Each point of the curves is characterized by a different value of the longitudinal slip σ. In a similar way it is possible to plot a curve $F_y(F_x)$ at constant σ.

Strictly speaking, the curves are not exactly symmetrical with respect to F_y-axis: Usually tires develop the maximum value of the force F_y when they exert a very light longitudinal braking force and a slight braking force can actually increase the lateral force exerted at a moderate sideslip angle.

Together with the curves $F_y(F_x)$ two other curves $M_z(F_x)$ are also reported in Fig. 2.36. An application of driving forces results in an increase of the aligning torque, while a braking force causes the aligning moment to decrease to the point that it changes its sign when the braking limit is approached. This effect is destabilizing as it tends to increase the sideslip angle.

[7] See, for instance, J. R. Ellis, *Vehicle Dynamics*, Business books Ltd., London, 1969; G. Genta, *Meccanica dell'autoveicolo*, Levrotto & Bella, Torino, 1993.

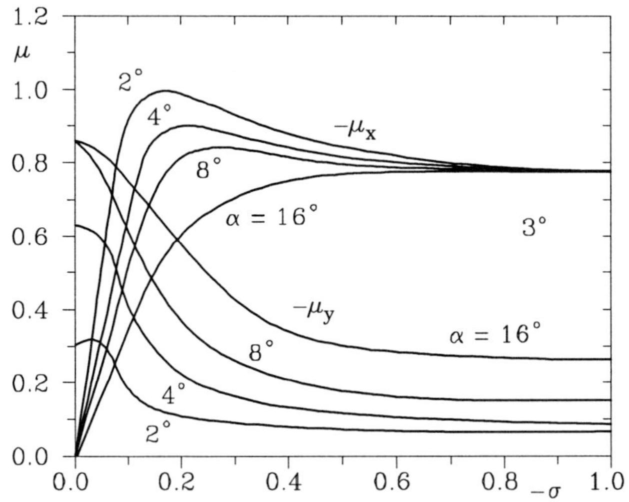

Fig. 2.35 Longitudinal and lateral force coefficients as functions of the longitudinal slip for given values of the sideslip angle.

A set of experimental curves $F_y(F_x)$ at constant α is shown in Fig. 2.37a.

If F is the total force exerted on the wheel by the road while F_x and F_y are its components, the resultant force coefficient can be expressed as

$$\mu = \frac{F}{F_z} = \sqrt{\mu_x^2 + \mu_y^2} \,. \tag{2.23}$$

The various curves plotted for different values of α are enveloped by the polar diagram of the maximum force the tire can exert. If it were a circle, the so-called friction circle, as in simple models it can be assumed to be, the maximum force coefficient would be independent of the direction.

Actually, not only the value of μ_x is greater than that of μ_y but, as already stated, there is some difference in longitudinal direction between driving and braking conditions. The envelope, as well as the whole diagram, is a function of many parameters. Apart from the already mentioned dependence on the type of tire and road conditions, there is a strong reduction of the maximum value of force F with the speed, which is particularly strong on wet road (Fig. 2.38).

A model allowing to approximate the curves $F_y(F_x)$ at constant α with simple functions can be quite useful. This can be obtained by using the

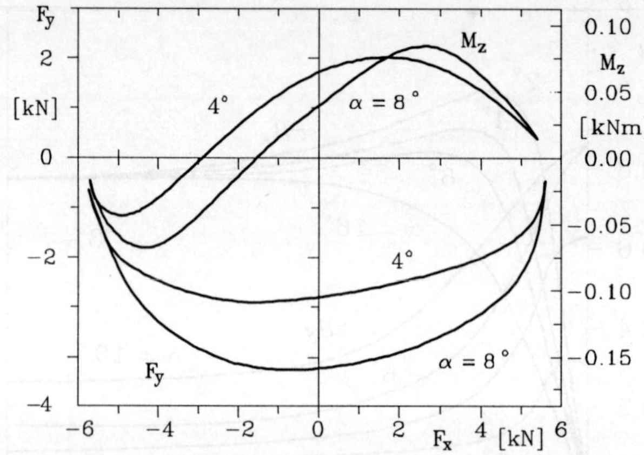

Fig. 2.36 Polar diagrams of the force exerted on the wheel with constant sideslip angle and aligning moment as a function of the longitudinal force.

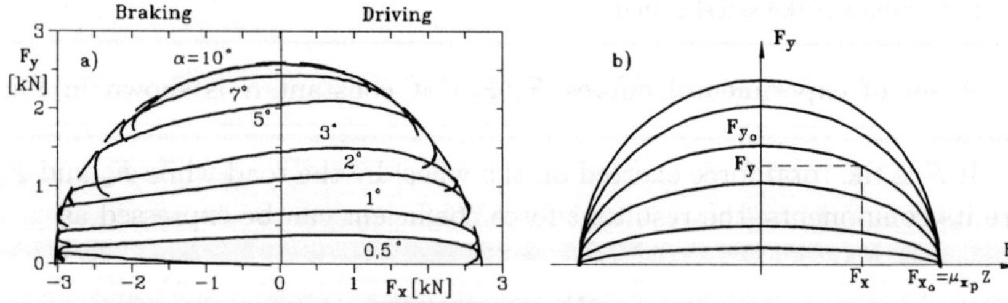

Fig. 2.37 Polar diagrams of the force exerted on the wheel with constant sideslip angle. (a) Experimental plots; (b) elliptic approximation.

elliptical approximation (Fig. 2.37b)

$$\left(\frac{F_y}{F_{y0}}\right)^2 + \left(\frac{F_x}{F_{x0}}\right)^2 = 1\,, \tag{2.24}$$

where forces F_{y0} and F_{x0} are respectively the force F_y exerted, at the given sideslip angle, when no force F_x is exerted and the maximum longitudinal force exerted at zero sideslip angle. The envelope curve is then elliptical, the friction ellipse.

If Eq. (2.24) is used in order to express function $F_y(F_x)$, the cornering stiffness of a tire which is exerting a longitudinal force F_x can be expressed

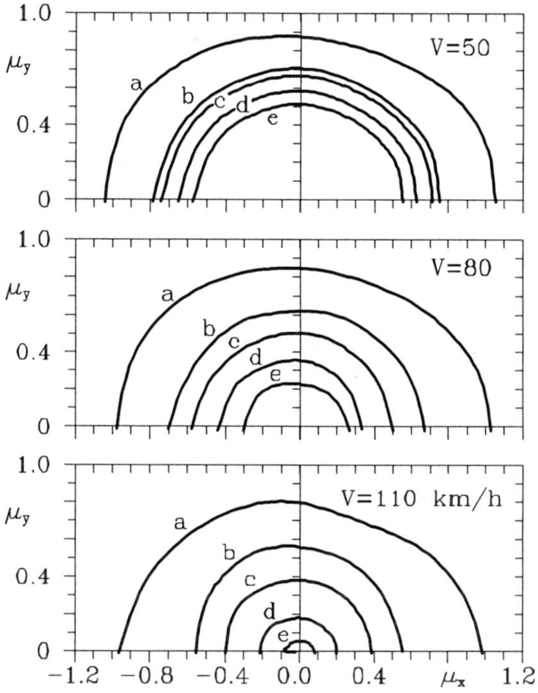

Fig. 2.38 Envelope curve of the curves $F_y(F_x)$ at different speeds and on different road conditions. (a) Dry road and (b) with 0.2 mm; (c) 0.5 mm; (d) 1 mm and (e) 2 mm of water.

as a function of the cornering stiffness C_0 (i.e. the cornering stiffness when no longitudinal force is produced) by the expression

$$C = C_0\sqrt{1 - (F_x/\mu_p F_z)^2}\,, \tag{2.25}$$

where force F_{x_0} has been substituted by $\mu_p F_z$.

Although a rough approximation, particularly for the case in which the longitudinal force approaches its maximum value (the differences between the curves of Fig. 2.37a and those of Fig. 2.37b are evident), the elliptical approximation is often used for all the cases in which the very concept of cornering stiffness is useful.

The empirical model expressed by equations (2.16) and (2.21) can be modified to allow for the interaction between longitudinal and lateral forces in a better way than that of computing separately the two forces and then

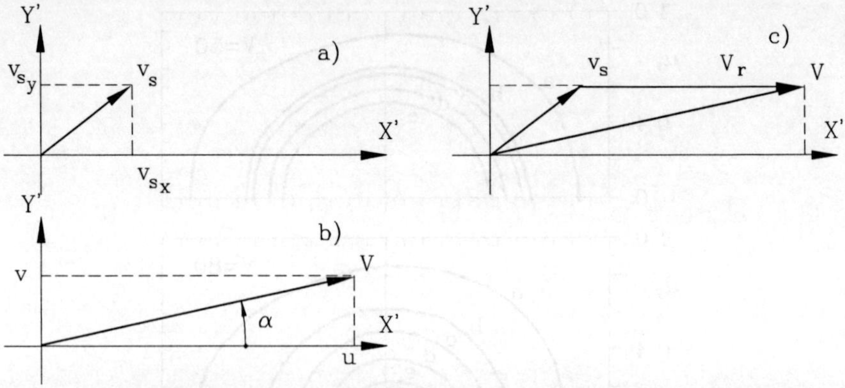

Fig. 2.39 (a) Slip velocity v_s, decomposed along axes X' and Y'; (b) velocity of the centre of the wheel V and its components u and v along the same axes; (c) definition of the relative velocity V_r. Note that V_r has a direction parallel to X'-axis.

using the elliptic approximation.

With reference to Fig. 2.39 it is possible to define a longitudinal slip σ_x and a lateral slip σ_y

$$\begin{cases} \sigma_x = v_{s_x}/V_r \\ \sigma_y = v_{s_y}/V_r \; . \end{cases} \tag{2.26}$$

In the case the sideslip angle α vanishes, the longitudinal slip σ is defined as $\sigma = -v_{s_x}/u$. It then follows that

$$\begin{cases} \sigma_x = \dfrac{v_{s_x}}{u - v_{s_x}} = \dfrac{-\sigma}{1+|\sigma|} \\ \sigma_y = \dfrac{v_{s_y}}{u - v_{s_y}} = \dfrac{\tan(\alpha)}{1+|\sigma|} \; . \end{cases} \tag{2.27}$$

Note that the longitudinal component of slip σ_x so defined does not coincide with the longitudinal slip σ even when the sideslip angle α vanishes.

Force F_y does not vanish when the sideslip angle α goes to zero owing to the presence of coefficients S_v and S_h. This occurs even in case of vanishing camber angle γ and is due to ply steer and conicity. It is then possible to define a sideslip angle $\delta\alpha$ as the angle at which force F_y is equal to zero. As this angle is small, it can be approximated as

$$\delta\alpha = -S_h - \frac{S_v}{BCD} \; . \tag{2.28}$$

The sideslip velocity corresponding to the sideslip angle $\delta\alpha$ is

$$\delta v_{s_y} = V_r \tan(\delta\alpha) \approx V_r \delta\alpha \; . \tag{2.29}$$

The total lateral slip speed is obviously $\delta v_{s_{ytot}} = v_{s_y} + \delta v_{s_y}$.

In a similar way, the longitudinal slip which causes force F_x to vanish is $\delta\sigma$. As in this case $S_v = 0$, it follows that $\delta\sigma = -S_h$. The longitudinal slip velocity corresponding to the slip $\delta\sigma$ is

$$\delta v_{s_x} = V_r \frac{-\delta\sigma}{1+\delta\sigma} \approx -V_r \delta\sigma \; . \tag{2.30}$$

The total longitudinal slip speed is then $\delta v_{s_{xtot}} = v_{s_x} + \delta v_{s_x}$. The longitudinal and transversal components of the total slip are so defined

$$\begin{cases} \sigma_{x_{tot}} = \dfrac{-\sigma}{1+|\sigma|} - \delta\sigma \\ \sigma_{y_{tot}} = \dfrac{\tan(\alpha)}{1+|\sigma|} + \delta\alpha \; . \end{cases} \tag{2.31}$$

The curves $F_x(\sigma_{x_{tot}})$ and $F_y(\sigma_{y_{tot}})$, obtained respectively when there is only longitudinal or side slip, can be defined as "original basic curves". Similarly an original basic curve $M_z(\sigma_{y_{tot}})$ can be defined for the aligning moment.

If both longitudinal and side slip are present at the same time the total combined slip $\sigma_{tot} = \sqrt{\sigma_{xtot}{}^2 + \sigma_{ytot}{}^2}$ is defined. Two reference values of the forces F_x and F_y are then obtained from the original basic curves in correspondence to the combined total slip σ_{tot} and the actual values of the forces are obtained by multiplying them by the ratios of the components of the total slip in X' and Y' directions and σ_{tot}

$$\begin{cases} F_x = \dfrac{\sigma_{x_{tot}}}{\sigma_{tot}} F_x(\sigma_{tot}) \\ F_y = \dfrac{\sigma_{y_{tot}}}{\sigma_{tot}} K_y(\sigma_{tot}) \; . \end{cases} \tag{2.32}$$

This way of computing $F_x - F_y$ interaction has however one drawback: The values of the longitudinal slip and the side slip which cause the tire to start sliding are quite different and there can be cases in which the wheel slips in longitudinal direction but not in the transversal one. This appears to be physically inconsistent, since when a wheel starts to slip it is global in this situation. To allow for this consideration it is possible to define nondimensional slip components $\sigma_x^* = \sigma_{x_{tot}}/\sigma_{x_m}$ and $\sigma_y^* = \sigma_{y_{tot}}/\sigma_{y_m}$ by dividing the slip components by the values of σ_{x_m} and σ_{y_m} for which forces F_x and F_y reach their maximum values.

If F_{x0} and F_{y0} are the values of the forces computed from the original basic curves in correspondence to the nondimensional values σ_x^* and σ_y^*, the nondimensional values of the forces are

$$\begin{cases} F_{x0}^* = F_{x0} - \varepsilon(F_{x0} - F_{y0})\left(\dfrac{\sigma_y^*}{\sigma^*}\right)^2 \\ F_{y0}^* = F_{y0} - \varepsilon(F_{y0} - F_{x0})\left(\dfrac{\sigma_x^*}{\sigma^*}\right)^2 , \end{cases} \tag{2.33}$$

where $\varepsilon = \min(\sigma^*, 1)$.

The actual values of the forces are then obtained from Eq. (2.32)

$$\begin{cases} F_x = \dfrac{\sigma_x^*}{\sigma^*} F_{x0}^* \\ F_y = \dfrac{\sigma_y^*}{\sigma^*} F_{y0}^* . \end{cases} \tag{2.34}$$

In a similar way the interaction between the aligning moment M_z and force F_x can also be modelled, considering also the effect of the moment of force F_x and that due to the camber angle γ. Some other coefficients, which must be evaluated experimentally, enter this part of the model.

This way of accounting for the interaction between F_x and F_y forces is quite complicated and not well theoretically justified, which is actually not required for empirical models. It yields reasonable results, far better than those obtained through the elliptic approximation. However there are cases in which the curves $F_y(F_x)$ have unrealistic shapes and the above mentioned formulae must be used with care. The research work on this aspect of the model is still in progress, aiming to a simpler and more reliable empirical formulation.

A plot of the type of that in Fig. 2.37a is shown in Fig. 2.40.

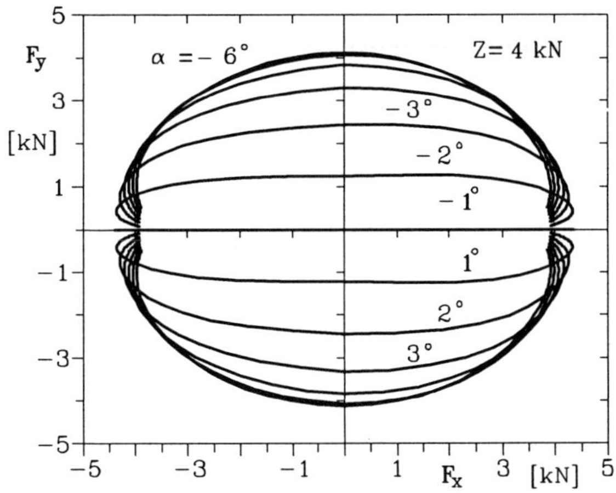

Fig. 2.40 Interaction between F_x and F_y forces, empirical model.

2.8 Dynamic behaviour of tires

The dynamic behaviour of the tire is quite important in determining both comfort and stability of the vehicle. Although the strong interactions between the vibrational behaviour of the tires and that of the suspensions and of the suspended mass of the vehicle suggest that the dynamic study should be performed on the complete vehicle, it is at any rate interesting to study the behaviour of the tire alone, at least to obtain the data which will later be introduced in more complex models.

As the stiffness of the tire is greater than that of the suspension, the behaviour of the former is generally not important in low frequency motions (1–3 Hz), for which the tire can be modelled as a rigid body.

In an intermediate field of frequencies (10–20 Hz) the tire can be considered as a massless deformable element introduced into the system and it is possible to define a dynamic stiffness, in both vertical and transversal direction. It is possible to show that the tire should be characterized by a low vertical stiffness, in order to minimize vertical displacements of the suspended mass on irregular road surface, and high transversal stiffness, to react with small displacements to side forces applied to the vehicle. As a general rule, the stiffness of radial tires is lower than that of bias-ply tires; the lower vertical stiffness is an advantage of radial tires while the disadvantage of a lower transversal stiffness is usually reduced by the higher

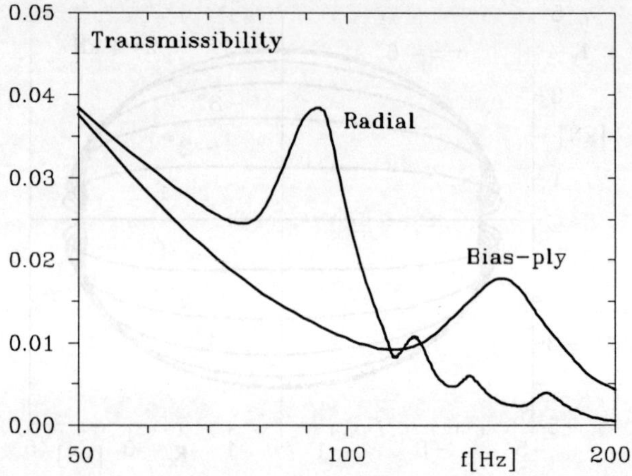

Fig. 2.41 Transmissibility, defined as ratio between the amplitude of the reponse and that of the excitation, for a radial and a bias ply tire 155-15 as function of the frequency. A frequency range between 50 and 200 Hz is reported; the resonance of the unsprung mass on the compliance of the tire occurs at a far lower frequency.

cornering stiffness.

At higher frequencies (over 50 Hz) the tire vibrates in its own modes. Radial tires have a marked resonance peak in the range 60 – 90 Hz depending on the tire dimensions and other less important resonance peaks at higher frequencies. Bias-ply tires have a resonance at higher frequency, with a lower magnification ratio.

The transmissibility, defined as the ratio between the amplitude of the response and that of the excitation, for a radial and a bias ply tire are reported in Fig. 2.41. The figure refers the vertical response to a vertical motion of the contact point. Similarly, the horizontal transmissibility can also be measured.

The resonance of the unsprung mass on the compliance of the tire occurs at a far lower frequency; the first resonance peak of the radial tire shown in the figure is related to a mode in which the tread band moves more or less as a rigid body owing to the elasticity of the sidewalls. The subsequent modes, in which the tread band vibrates following a shape with two, three and more lobes are far less excited and more heavily damped.

If the geometrical parameters (slip and camber angles) or the forces in X' and Z' directions are variable during motion, the values of the side force and of the aligning moment are at any instant different from the ones

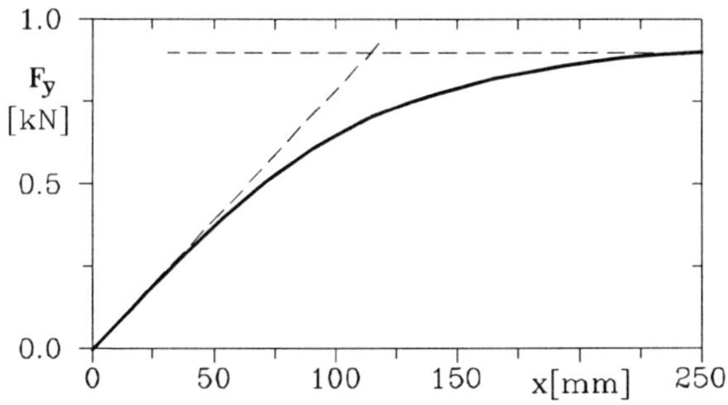

Fig. 2.42 Lateral force exerted by a tire set at a given sideslip angle at standstill and then rolled on, as a function of the distance. Tire 5.50×12, $F_z = 3.5$ kN, $p = 138$ kPa.

which would characterize a stationary condition with the same values of all parameters. As an example, if a tire is tilted about the vertical axis at standstill and then it is allowed to roll, the side force reaches the steady-state value only after a certain time, after rolling for a certain distance (Fig. 2.42), usually referred to as relaxation length. This effect is usually not noticeable in normal driving as the time delay is very small, but the fact that there is a delay between the setting of the sideslip angle and the force generation is very important in dynamic conditions.

If the sideslip angle is changed with harmonic law in time, the side force and the aligning torque follow the sideslip angle with a certain delay, function of the frequency, and their value is lower than that obtained in quasi-static conditions, i.e. with very low frequency.

If the frequency is not very high, at the speeds encountered in normal driving the average values are not much lower than those characterizing static conditions, but a certain phase lag between the sideslip angle and the F_y force remains (Fig. 2.43). The plots reported in the figure have been obtained for a given tire and are not easily generalized, particularly at high frequency. If a resonance of the tire occurs (as the phase of the aligning torque seems to suggest in the plot) the response is no longer dependent on the ratio V/λ but on the frequency and, to a lesser extent, on the speed separately.

More important for what practical applications are concerned is the case in which the load F_z applied by the wheel on the ground is variable, as is the case of rolling on uneven road (Fig. 2.44). The frequency can be very high, if the speed is high enough, and the decrease of lateral force due to dynamic effects can be large. In the figure the law $z(t)$ of the vertical

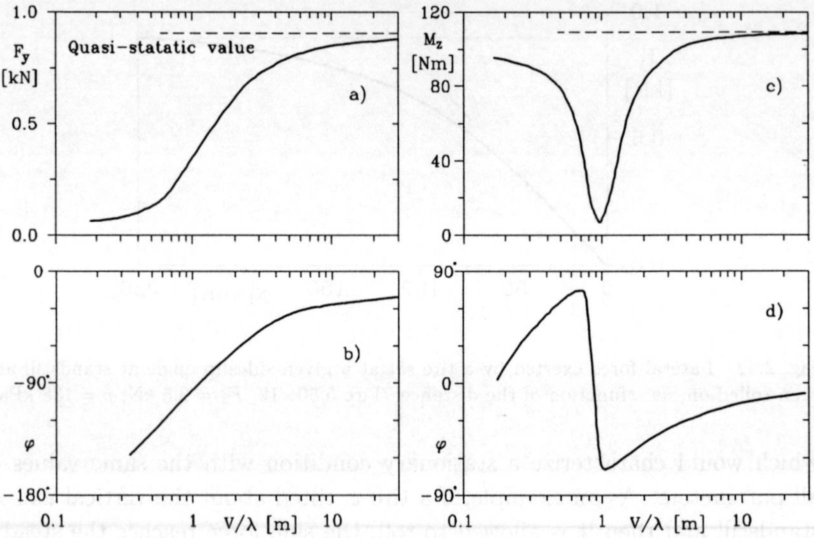

Fig. 2.43 Lateral force and aligning torque generated by a tire working with a sideslip angle changing harmonically in time between $-4°$ and $4°$, as a function of the ratio V/λ between the forward velocity and the frequency of the harmonic law $\alpha(t)$, i.e. of the wavelength. The peak values (a), (b) and the phase lag (c), (d) of the force and the moment are reported. Tire 7.50×14, $F_z = 4.8$ kN, $p = 165$ kPa.

displacement of the hub of the wheel is harmonic with a frequency of about 7 Hz while the response $F_y(t)$ is more complicated, with even an inversion of sign occurring at each cycle. The decrease of the average value of the lateral force at increasing frequency is shown in Fig. 2.44b.

Strictly linked with the dynamic behaviour of the tire are the self-excited vibrations of the tire and the whole steering mechanism known as "wheel shimmy". Such vibrations are today more of historical interest, as modern vehicles are free from this problem which was very important about half a century ago when it represented an actual danger in the automotive and aeronautical fields.

2.9 Experimental study of the characteristics of tires

The characteristics of tires can be measured using road or laboratory tests. The most common testing machines for tires are based on a drum which simulates the road; the tire can roll on the outer surface or on the inner

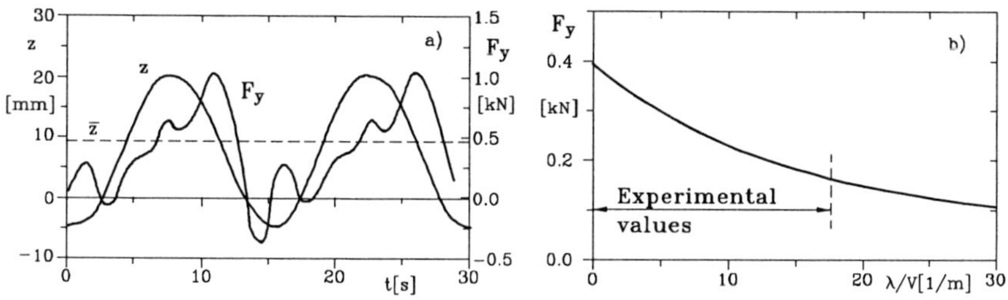

Fig. 2.44 (a) Lateral force generated by a tire working with constant slip angle but with the hub moving vertically with harmonic law $z(t)$. (b) Average value of the lateral force F_y as a function of the ratio between the circular frequency λ of the law $\alpha(t)$ and the speed V.

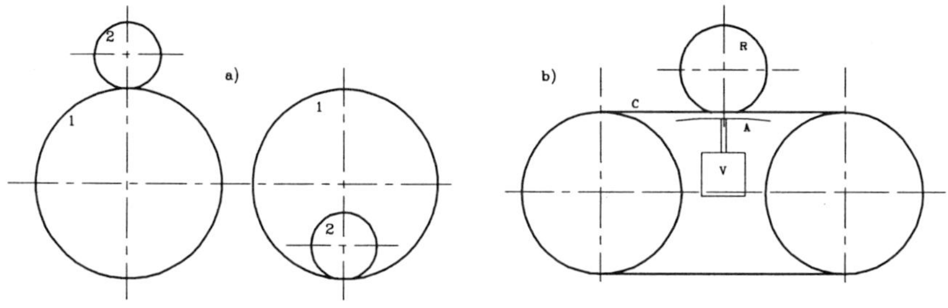

Fig. 2.45 Tire testing machines. (a) Drum machines, with positive or negative curvature. 1: drum; 2: tire under test. (b) Belt machine. A: belt support; C: belt; R: wheel under test; V: shaker.

surface of the drum (Fig. 2.45a). The actual conditions are in a way intermediate between those encountered on the two types of test machines.

To avoid the differences between the contact condition occurring in drum machines and the actual ones, more modern machines use a steel belt (Fig. 2.45b). The belt is kept flat by a hydrostatic bearing, which can be connected to a shaker to simulate the motion on uneven road. The simulation of the road surface is easier on drum than on belt machines, on which the surface is usually a plain metal surface.

Other machines use a flat disc: In this case the contact surface is flat but the tire works with a ground moving along a circular path, generating a side force even with no sideslip. For low speed and low duration tests the wheel can roll also against a moving platform. With platforms it is possible to simulate easily different road surfaces.

In all testing machines the tire is fitted on a hub which is provided by a device able to measure the three components of the force and the three components of the moment. The various driving and braking conditions can be simulated by using two different motors for the wheel and the "road" and the wheel can be maintained at any sideslip and camber angle.

Results of laboratory tests must be validated by road tests, which are performed using particular test vehicles. Road testing of tires require instrumented vehicles, usually provided with the same instrumentation common on testing machines. The wheel under testing is usually lowered from the vehicle, which rolls on its own wheels; when the first exerts lateral forces also the wheels of the vehicle work with a sideslip angle, which must be accounted for in measuring the sideslip angle of the wheel under test. The power needed for the test vehicle is very high if high speed testing in strong braking or driving conditions is required.

If only the rolling resistance has to be measured, it is possible to tow a whole vehicle or a trailer with a single wheel on level road with a bar provided of a load cell. If the vehicle or the trailer are kept within a box to avoid aerodynamic drag, the rolling resistance, which includes the bearing drag, can be measured with precision, even if the aerodynamic drag on the disc of the wheel is accounted for in an approximated way. Note that a slope of the road of only 0.1% is sufficient to introduce an error of 10% on the rolling resistance coefficient.

Rolling resistance can also be measured using a spin down test: The tire and the testing machine are accelerated at a given speed and the motor disconnected. As the inertia of the system and the overall drag of the machine are known, (the second at least approximately), from the measurement of the speed as a function of time it is possible to compute the rolling drag. The problem here is that the drag is measured in conditions which are far from steady-state working. In particular, the temperature is continuously changing, leading to a drag which is higher than that characterizing steady-state conditions at the beginning of the test (high speed) and lower at the end (low speed).

Chapter 3

Road vehicle aerodynamics

3.1 General considerations

The study of the aerodynamic forces and moments is performed with reference to the following frames (Fig. 3.1):

- *Earth-fixed axis system XYZ*. It is a right-hand reference frame fixed on the road. In the following sections it will always be regarded as an inertial frame, even if strictly speaking it is not such as it moves together with the Earth: The inertial effects due to its motion are so small that they can be neglected in all phenomena studied in motor vehicle dynamics. Axes X and Y lie in a horizontal plane while axis Z is vertical pointing upwards[1].
- *Vehicle axis system xyz*. It is a right-hand reference frame fixed to the vehicle and moving with it. x-axis lies in the symmetry plane of the vehicle in an almost horizontal direction[2], z-axis lies in the symmetry plane, is perpendicular to x-axis and points upwards and y-axis is perpendicular to the other two[3].
- *Wind axis system $x''y''z''$*. It is a right-hand reference frame fixed to the vehicle and moving with it used in the study of aerodynamic forces. It is similar to the vehicle fixed frame, with the difference that axis x'' has the direction of the resultant air velocity vector V_r of the vehicle with respect to the ambient air (defined below)

[1]Recommendation SAE J670 and ISO/TC 22/SC9 standard state that Z-axis is vertical and points downwards. Note that in the present text the direction of Y and Z axes is opposite to that suggested in the mentioned standard.

[2]The mentioned standard states that x-axis is contained in the plane of symmetry of the vehicle, is "substantially" horizontal and points forward.

[3]Here again there is a deviation from the mentioned standard.

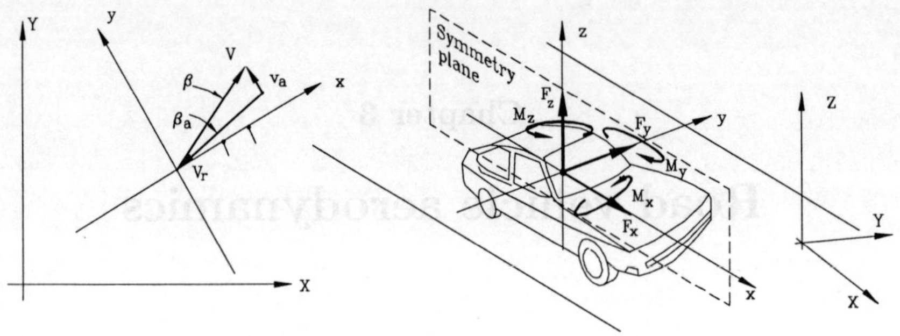

Fig. 3.1 Reference frames for the vehicle. (a) Projection on the ground (plane XY) of the absolute, relative and wind velocities; (b) forces and moments acting on the vehicle.

changed of sign. Obviously $-V_r$ coincides with the absolute velocity V only if the air is still, i.e. if the ambient wind velocity v_a is zero. z''-axis is contained in the symmetry plane, perpendicular to x''-axis and y''-axis is perpendicular to the other two.

The angle between the projection on XY plane of x''-axis and that of the velocity vector V is the "sideslip angle" β of the whole vehicle; sometimes referred to as the attitude angle. As it was the case for the sideslip angle of the tire α, it is positive when vector V points to the left of the driver (in forward motion).

The ambient wind velocity v_a is defined as the horizontal component of the air velocity relative to the earth-fixed axis system in the vicinity of the vehicle. The ambient wind angle is the angle between X-axis and thc ambient wind velocity.

The resultant air velocity V_r is the difference between the ambient wind velocity and the projection of the absolute velocity V of the vehicle. Note that it points towards the negative x''-axis: As it is common in the study of aerodynamic forces, the situation to which reference is made is that of the wind tunnel, with the object at rest and the air moving.

The angle between the projection of x-axis on the ground and x''-axis is the aerodynamic angle of sideslip β_a. Note that angles β and β_a are usually referred to the centre of mass, but can be referred to any specific point of the vehicle.

The air surrounding a road vehicle exerts on any point P of its surface a force per unit area

$$\vec{t} = \lim_{\Delta S \to 0} \frac{\Delta \vec{F}}{\Delta S} , \tag{3.1}$$

where ΔS and $\Delta \vec{F}$ are respectively the area of a small surface surrounding point P and the force acting on it.

The force per unit area $\vec{t}$ can be decomposed in a pressure force $\vec{t_n} = p\vec{n}$, acting in direction perpendicular to the surface, and a tangential force $\vec{t_t}$ laying on the plane tangent to the surface. The latter is due to the fluid viscosity.

These force distributions, once integrated on the whole surface, result in an aerodynamic force, which is usually applied in the centre of mass of the vehicle

$$\begin{cases} F_{x_a} = \int_S \vec{t_t} \times \vec{i} dS + \int_S \vec{t_n} \times \vec{i} dS \\ F_{y_a} = \int_S \vec{t_t} \times \vec{j} dS + \int_S \vec{t_n} \times \vec{j} dS \\ F_{z_a} = \int_S \vec{t_t} \times \vec{k} dS + \int_S \vec{t_n} \times \vec{k} dS \end{cases} \tag{3.2}$$

and in a resultant moment

$$\begin{cases} M_{x_a} = -\int_S z\vec{t_t} \times \vec{j} dS + \int_S y\vec{t_t} \times \vec{k} dS - \int_S z\vec{t_n} \times \vec{j} dS + \int_S y\vec{t_n} \times \vec{k} dS \\ M_{y_a} = -\int_S x\vec{t_t} \times \vec{k} dS + \int_S z\vec{t_t} \times \vec{i} dS - \int_S x\vec{t_n} \times \vec{k} dS + \int_S z\vec{t_n} \times \vec{i} dS \\ M_{z_a} = -\int_S y\vec{t_t} \times \vec{i} dS + \int_S x\vec{t_t} \times \vec{j} dS - \int_S y\vec{t_n} \times \vec{i} dS + \int_S x\vec{t_n} \times \vec{j} dS . \end{cases} \tag{3.3}$$

Both the aerodynamic force and moment are usually decomposed with reference to frame xyz: The components of the aerodynamic force are referred to as longitudinal F_{x_a}, lateral F_{y_a} and normal F_{z_a} forces while those of the moment are the rolling M_{x_a}, pitching M_{y_a} and yawing M_{z_a} moments.

If the aerodynamic force is decomposed along the axes of frame $x''y''z''$, as it is more common in aeronautics, the components would have been referred to as drag D, side force S and lift L.

In the present text aerodynamic forces will always be referred to frame xyz which is centreed in the centre of mass of the vehicle. However in wind tunnel testing the exact position of the centre of mass is usually unknown

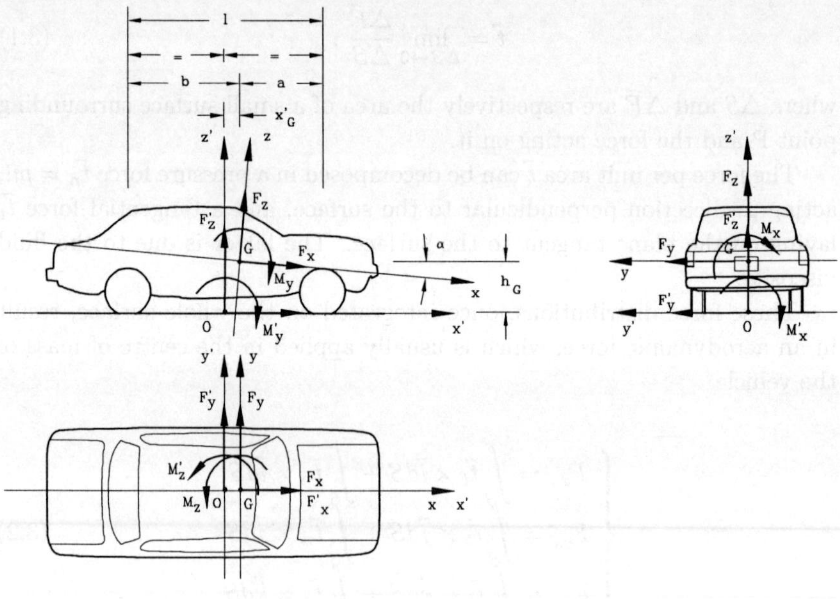

Fig. 3.2 Reference frame often used to express aerodynamic forces in wind tunnel tests.

and the forces are referred to a frame which is immediately identified. Moreover, the position of the centre of mass of the vehicle depends also on the loading, while aerodynamic forces are often assumed to be independent of it, although a change of the load of the vehicle can affect its attitude on the road and hence the value of aerodynamic forces and moments.

The frame often used to express forces and moments for wind tunnel tests is a frame centreed in a point on the symmetry plane located at mid-wheelbase, with x'-axis laying on the ground in the plane of symmetry of the vehicle and y'-axis laying also on the ground (Fig. 3.2). As the resultant air velocity V_r lies in a horizontal plane, angle α is the aerodynamic angle of attack. From the definition of x axis it is a small angle and is often assumed to be equal to zero. Note that from the definitions used here for the reference frames it follows that α is positive when x-axis points downwards.

The forces and moments expressed in xyz frame are related to those expressed in $x'y'z'$ frame (the latter are indicated with a prime) by the relationships

$$\begin{cases} F_x = F'_x \cos(\alpha) - F'_z \sin(\alpha) \\ F_y = F'_y \\ F_z = F'_x \sin(\alpha) + F'_z \cos(\alpha) \end{cases} \tag{3.4}$$

$$\begin{cases} M_x = M'_x + F_y h_G \\ M_y = M'_y - F_x h_G + F_z x'_G \\ M_z = M'_z - F_y x'_G \,. \end{cases} \tag{3.5}$$

Distance x'_G is the coordinate of the centre of mass with reference to $x'y'z'$ frame and is positive if the centre of mass is forward than mid-wheelbase ($a < b$).

At standstill the only force exerted by air is the aerostatic force, which acts in vertical direction and equals the weight of the displaced fluid. It reaches non-negligible values only for very light and large bodies and it is completely neglected in aerodynamics.

If air were an inviscid fluid, i.e. if its viscosity were nil, no tangential forces could act on the surface of the body and it can be demonstrated that no force could be exchanged between the body and the fluid, apart from aerostatic forces, at any relative speed as the resultant of the pressure distribution always vanishes. This result, due to D'Alembert, was formulated in 1744[4] and then again in 1768[5] . It is since known as the D'Alembert Paradox.

In the case of a fluid with no viscosity, the pressure p and the velocity V can be linked by Bernoulli equation

$$p + \frac{1}{2}\rho V^2 = \text{constant} = p_0 + \frac{1}{2}\rho V_0^2 \,, \tag{3.6}$$

where p_0 and V_0 are the values of the ambient pressure and of the velocity far enough upstream from the body.

The values of the ambient pressure, together with those of the density, temperature, and kinematic viscosity at altitudes of interest in road vehicle technology are reported in Table 3.1 from ICAO standard atmosphere.

The density at temperatures and pressures different from p_a and T_a in standard conditions can be computed as

[4]D'Alembert, *Traité de l'équilibre et du moment des fluides pour servir de suite un traité de dynamique*, 1774.

[5]D'Alembert, *Paradoxe proposé aux geometres sur la résistance des fluides*, 1768.

Table 3.1 Pressure, temperature, density and kinematic viscosity of air at various altitudes, from ICAO standard atmosphere. Only the part of the table related to altitudes of interest for road vehicles is reported.

z [m]	p [kPa]	T [K]	ρ [kg/m^3]	ν [m^2/s]
-500	107.486	291.25	1.2857	13.97×10^{-6}
0	101.325	288.16	1.2257	14.53×10^{-6}
500	95.458	284.75	1.1680	15.10×10^{-6}
1000	89.875	281.50	1.1123	15.71×10^{-6}
1500	84.546	278.25	1.0586	16.36×10^{-6}
2000	79.489	275.00	1.0070	17.05×10^{-6}
2500	74.656	271.75	0.9573	17.77×10^{-6}
3000	70.097	268.50	0.9095	18.53×10^{-6}

$$\rho = \rho_a \frac{p}{p_a} \frac{T_a}{T} , \tag{3.7}$$

where temperatures are absolute.

Note that Bernoulli equation, which holds along any streamline, has been written without the gravitational term, the one linked with aerostatic forces. It states simply that the total energy is conserved along any streamline.

An example of the D'Alembert Paradox is shown in Fig. 3.3, where the cross section of a cylinder of infinite length, whose axis is perpendicular to the direction of the velocity V_r of the fluid, is represented. The streamlines open around the body and thc local velocity of the fluid increases on its sides, leading to a decrease of pressure as described by Bernoulli Equation. On the front of the body there is a point (actually in the case of the cylinder it is a line) which divides the part of the flow which goes "above" the body from that going "below" it. At this point, known as stagnation point, the velocity of the fluid reduces to zero and the pressure reaches its maximum, the value of the total pressure $p_{tot} = p_0 + 1/2\rho V^2$.

As there is no viscosity, no energy is dissipated, and when the fluid slows down again, after reaching the maximum velocity in the point in which the width of the body is maximum, the pressure is fully recovered: The pressure distribution is symmetrical and no net force is exchanged between the fluid and the body. This holds for any possible shape, provided that the viscosity is exactly nil.

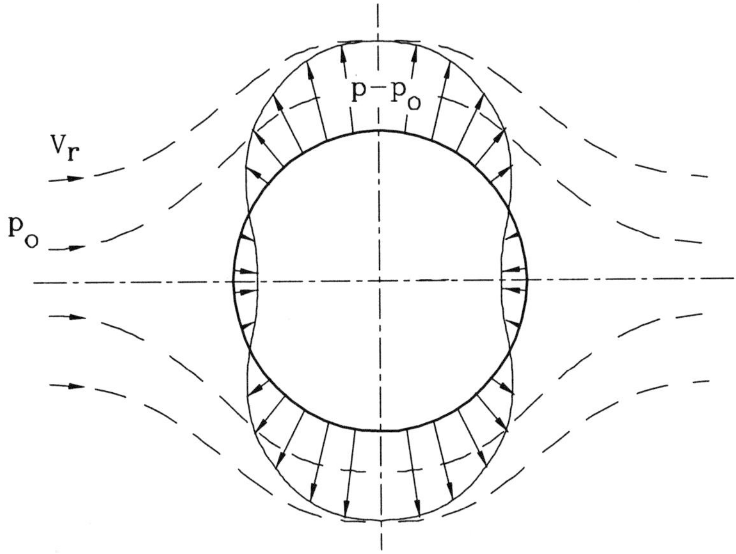

Fig. 3.3 Streamlines and pressure distribution on a circular cylinder whose axis is perpendicular to the fluid. Case of a fluid with no viscosity.

No fluid has actually zero viscosity and the Paradox is not applicable to any real fluid. Viscosity has a twofold effect: It causes the tangential forces which give way to the so-called friction drag and modifies the pressure distribution, whose resultant is no longer equal to zero. The latter effect, which for fluids with low viscosity is generally more important than the former, generates the lift, the side force and the pressure drag. The direct effects of viscosity (i.e. the tangential forces) can usually be neglected while its modifications on the aerodynamic field must be at any rate accounted for.

Owing to viscosity, the layer of fluid in immediate contact with the surface tends to adhere to it, i.e. its relative velocity vanishes, and the body is surrounded by a zone in which there are strong velocity gradients. This zone is usually referred to as the "boundary layer" (Fig. 3.4) and all viscous effects are concentrated in it. The viscosity of the fluid outside the boundary layer is usually neglected and Bernoulli equation can be used in this region.

Note that the thickness of the boundary layer increases as the fluid in it loses energy owing to viscosity and slows down. If the fluid outside the boundary layer increases its velocity a negative pressure gradient along the

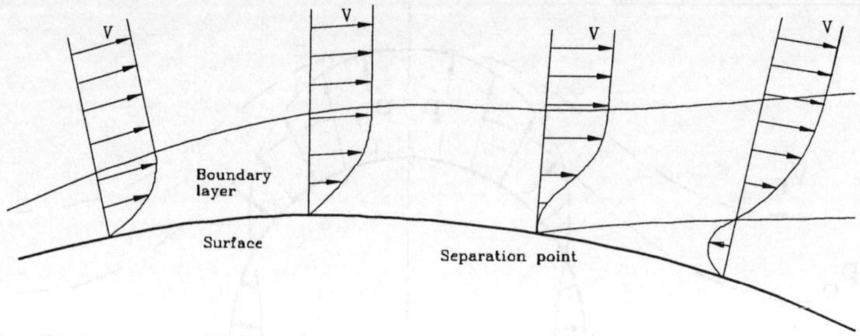

Fig. 3.4 Boundary layer: Velocity distribution in direction perpendicular to the surface. The separation point is also represented.

separation line between the external flow and the boundary layer is created, and this decrease of pressure in a way helps the flow within the boundary layer contrasting its slowing down. On the contrary, if the outer flow slows down, the pressure gradient is positive and the airflow in the boundary layer is hampered.

At any rate, at a certain point the flow in the boundary layer can stop and a zone of stagnant air can form in the vicinity of the body: The flow separates from the surface possibly starting the formation of a wake.

If the velocity distribution outside the boundary layer were known, it were possible to compute the pressure distribution at the interface between the boundary layer and the external fluid. Provided that the boundary layer is very thin, and this is the case except where the flow is detached from the surface, the pressure on the surface of the body can be assumed to be equal to that occurring at the outer surface of the boundary layer and then the aerodynamic forces and moments could be computed by integrating the pressure distribution. While this can be applied for computing the lift of streamlined objects, for blunt bodies, as the ones studied by road vehicle aerodynamics, and for the drag very little can be obtained along these lines.

To generalize the results obtained by experimental testing, mainly performed in wind tunnels, the aerodynamic force F and moment M are expressed as

$$F = \frac{1}{2}\rho V_r{}^2 S C_f \ , \qquad M = \frac{1}{2}\rho V_r{}^2 S l C_m \ , \tag{3.8}$$

where forces and moments are assumed to be proportional to the dynamic pressure of the free current $\rho V_r{}^2/2$, to a reference surface S (in the expression of the moment a reference length l is also present) and to nondimensional coefficients C_f and C_m to be experimentally determined.

Such coefficients depend on the geometry and position of the body and on two nondimensional parameters, the Reynolds and the Mach numbers. The first, defined as

$$\mathcal{R}_e = \frac{Vl}{\nu} ,$$

where ν is the kinematic viscosity of the fluid (see Table 3.1), is a parameter which indicates the relative importance of the viscous and inertial effects in determining the aerodynamic forces. If its value is low the former are of great importance while if it is high aerodynamic forces are mainly due to the inertia of the fluid. In this latter case (for vehicles, if $\mathcal{R}_e > 3,000,000$) the dependence of the aerodynamic coefficients on the Reynolds number is very low and can be neglected. This is usually the case for road vehicles, at least for speeds in excess of 30–40 km/h.

The Mach number is the ratio between the airspeed and the speed of sound. It has very little importance in road vehicle aerodynamics as at the speeds attained by all vehicles, except record crafts, its value is low enough to have no effects on aerodynamic coefficients.

The reference surface S and length l are arbitrary, to the point that in some cases a surface not existing physically, as the power 2/3 of the displacement for airships, was used. They express simply the dependence of aerodynamic forces on the square of the dimensions of the body and that of the moments on their cube. It is however clear that the numerical values of the coefficients depend on the choice of S and l, which must be clearly stated. In the case of road vehicles the surface is that of the cross section, with some uncertainty on whether the frontal projected area or that of the maximum cross section has been used (Fig. 3.5). The mentioned SAE recommendation states that the first one, which should include the tires and the underbody parts, must be used. A simple but sometimes rough way of obtaining its value is

$$S = \psi t h_1 , \tag{3.9}$$

where coefficient ψ takes values between 0.85 and 0.95.

The reference length l is usually the wheelbase, but in the expressions of moment M_x the track t is often used.

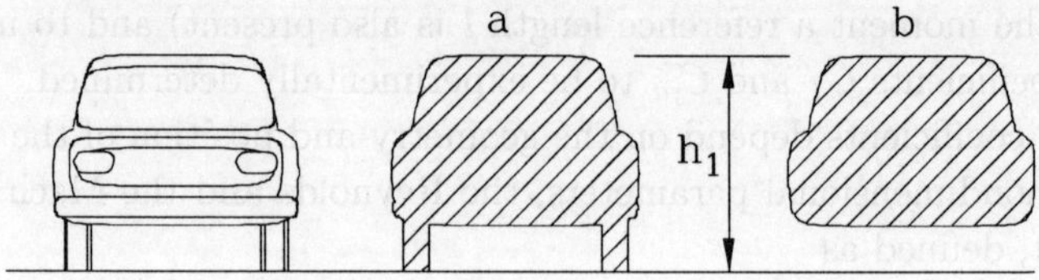

Fig. 3.5 Defnitions of surface S and of the overall height h_1.

The aerodynamic coefficients used in motor vehicle aerodynamics are those of the forces and moments decomposed along the vehicle axis system xyz: The longitudinal force coefficient C_x, the side force coefficient C_y, the normal force coefficient C_z, the rolling moment coefficient C_{M_x}, the pitching moment coefficient C_{M_y}, the yawing moment coefficient C_{M_z}.

3.2 Aerodynamic field

Consider a saloon car as the one sketched in Fig. 3.6. As usual in aerodynamics, refer to a "wind tunnel" situation, i.e. consider the vehicle as stationary while the air flows around it.

The stream has a stagnation point at A, where the flow divides below and above the vehicle; in the vicinity of A the pressure takes the value p_{tot}. In the vicinity of B the pressure takes values lower than the total pressure and even lower than the ambient pressure p_0, as the velocity increases, as shown in Fig. 3.7b in which the pressure distribution is reported in terms of pressure coefficient

$$c_p = \frac{p - p_0}{\frac{1}{2}\rho V_r^2} = 1 - \frac{V^2}{V_r^2} . \tag{3.10}$$

Note that the pressure coefficient is negative if the pressure is lower than the ambient pressure.

After point C, located between B and the lower edge of the windscreen, the flow detaches from the surface, to attach again at point D located on the windscreen. Between points C and D a separation bubble is formed. The pressure in such turbulent zone is fairy high and it is reasonable to locate there the intakes for cabin ventilation (Fig. 3.7). The separation bubble can be reduced by reducing the inclination of the windscreen, which can be done only up to a limit as it can reduce the visibility, or by increasing

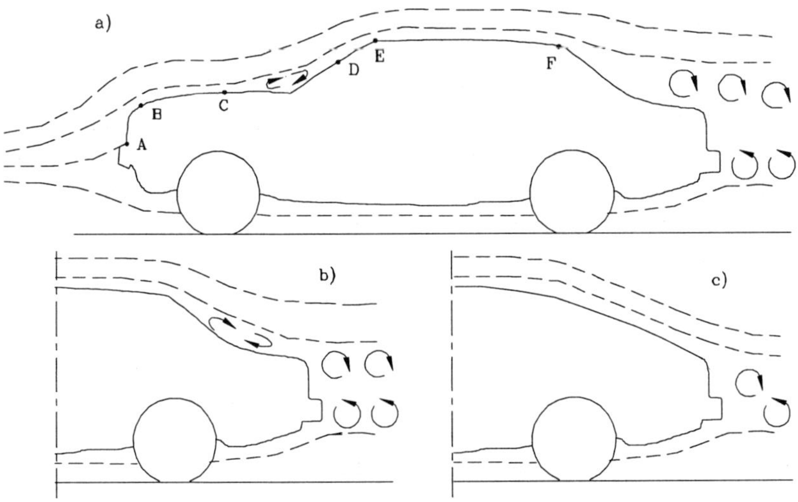

Fig. 3.6 Streamlines about a passenger vehicle in the symmetry plane.

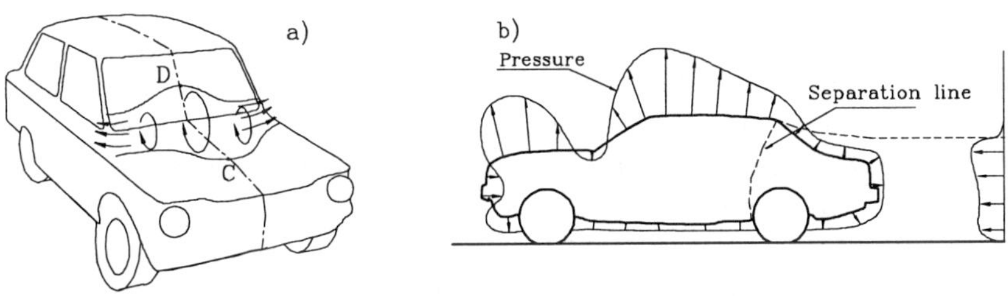

Fig. 3.7 (a) Separation bubble on the windscreen of a car. (b) Pressure distribution on the symmetry plane of a saloon car and in the wake.

the transversal curvature of the windscreen and of the hood. A curved windscreen is effective in reducing the drag but has a cost and also a weight which are greater than those of a simple, flat one.

On the roof the pressure is again low, with a distribution which depends on its shape and curvature. At the end of the roof the flow must slow down and the pressure should raise. In these conditions the flow easily detaches and any surface irregularity can trigger the formation of the wake. In Fig. 3.6a the separation point has been located at the rear edge of the roof. There are cases in which the flow attaches again to the back of the boot, giving way to a second separation bubble (Fig. 3.6b).

In the case of fastback cars with a sloping enough back part, the flow can remain attached up to the end of the body, giving way to a very small wake (Fig. 3.6c). The two situations are shown in the pictures of Fig. 3.8, obtained by visualizing the streamlines using smoke in a wind tunnel test.

The streamlines shown in Fig. 3.6 describe the situation occurring in the plane of symmetry. Outside this plane, the flow is no more two-dimensional and tends to surround the vehicle also at the sides. This effect is generally beneficial as it tends to reduce all aerodynamic forces and must be encouraged giving a suitable curvature in transversal direction to all surfaces. As already stated, point C can be moved further backwards by allowing the air to flow to the sides of the hood by lowering the fenders and giving them a curved shape; point D can be lowered by using a curved windscreen. This results in a reduced separation bubble (Fig. 3.7).

The tri-dimensional flow on the back of the vehicle can cause the development of strong vortices, as shown by tests on slanted blocks (Fig. 3.9). If angle α in the figure is lower than a critical value (about 62°) the flow separates abruptly while for higher values the flow becomes strongly tri-dimensional and the streamlines which flow along the sides wind up in two large vortices while those flowing of the roof are deflected downwards and follow the tail of the vehicle. The flow, which is of the type shown in Fig. 3.17, in the symmetry plane is similar to that shown in Fig. 3.6c.

The wake is smaller but this does not mean that the drag is lower: The pressure in the vortices is low and is the pressure of the very fast flow on the centre of the tail of the vehicle is also low; the overall pressure behind the vehicle can be even lower than that characterizing a large wake due to a small angle α.

The flow under the vehicle can be quite complicated and depends on many factors as the distance between vehicle and ground and the presence of a fairing of the bottom of the body. Wind tunnel simulations can be misleading as in the actual situation the ground is stationary with respect to the air, at least if there is no wind, and not with respect to the vehicle as occurs in wind tunnels.

In the actual situation, starting from the stagnation point A the boundary layer gradually thickens (Fig. 3.10). Outside the boundary layer the velocity of the flow is different from that characterizing the free stream, i.e., the flow is no more at rest with respect to the ground, and from point G a second boundary layer appears also on the ground. Depending on the distance between the vehicle and the ground, the two boundary layers can

a)

b)

Fig. 3.8 Streamlines in the symmetry plane about two fastback cars. In (a) the flow detaches at the end of the roof while in (b) it remains attached up to the end of the trunk.

meet in H or can remain separated. In the first case the flow is blocked and the air under the vehicle tends to move with it, giving way to another boundary layer starting from L. Between H and L a vortex can result. In the second case the flow between the vehicle and the ground lowers aerodynamic lift, or if it is fast enough causes a negative lift, and lowers the drag, also because the pressure in the wake is increased.

All improvements which facilitate the flow under the vehicle have these effects: Either the distance between vehicle and ground is increased or the bottom is given a curved shape, in longitudinal or transversal direction, or the bottom is supplied with a smooth fairing which covers the mechanical

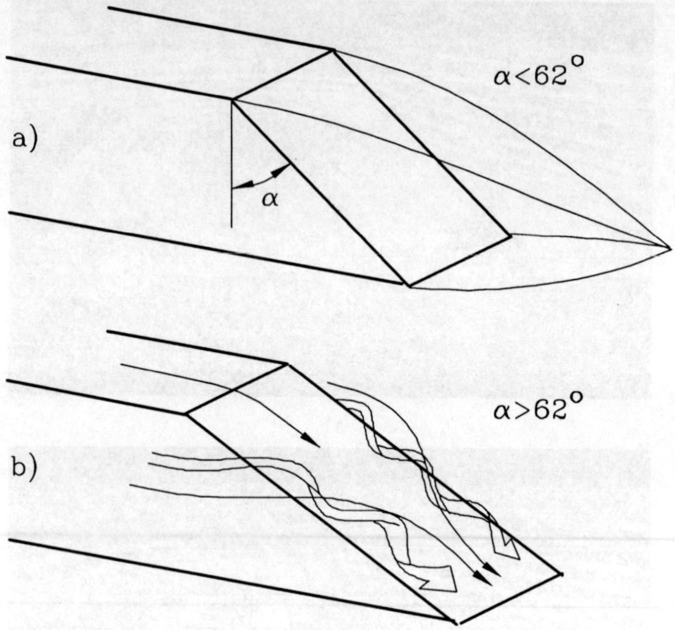

Fig. 3.9 Flow on the back of slanted blocks.

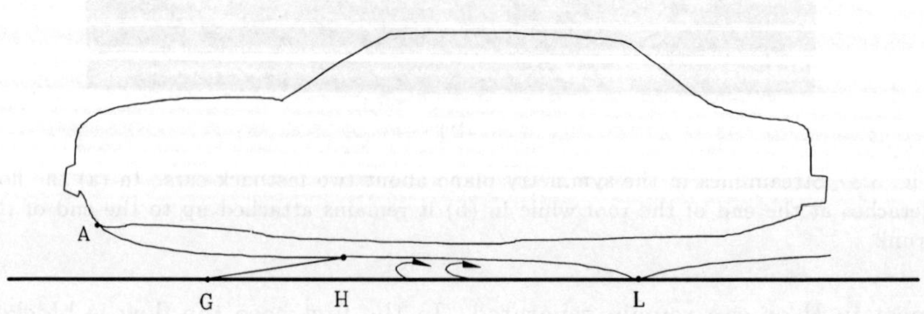

Fig. 3.10 Flow below the vehicle. Boundary layer formation.

elements which are usually in the airflow. The last device can succeed in reducing the drag up to about 10–15%, but is seldom used in passenger cars as it is more difficult to reach the mechanical elements making maintenance more costly.

Two effects can modify the airflow around the vehicle and make it more complicated: Wheel rotation and the presence of internal flows. As the

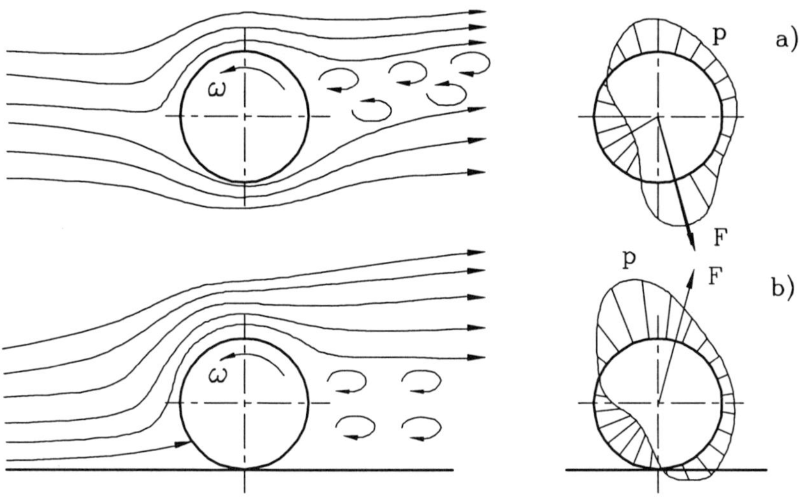

Fig. 3.11 Streamlines, pressures and aerodynamic force acting on a wheel, modelled as a rotating cylinder, far from the ground (a) and in contact with it (b).

drag coefficient of a rolling wheel exposed to the airflow is about 0.45, it would seem that only if the drag coefficient of the vehicle is lower than that value is there an advantage in inserting the wheels within the body. This is not true, as the presence of uncovered rotating wheels changes the whole aerodynamic field and causes a very strong drag: Uncovered wheels are present only in racing cars when rules explicitly dictate so. In Formula 1 racers up to 45% of aerodynamic drag can be ascribed to the wheels.

Consider a cylinder rotating and moving in directions consistent with those of a rolling wheel (Fig. 3.11a). It generates a drag and a lift (Magnus effect) which is directed downwards. If the wheel is in contact with the ground, however, the streamlines are completely changed by the presence of the latter and the lift becomes positive. The wake is larger and the drag coefficient increases; both the size of the first and the value of the second depend on the speed (Fig. 3.11b).

The situation shown in Fig. 3.11b is typical of racing cars with uncovered wheels: The position of the aerodynamic devices aimed to generate negative lift must be studied with care keeping into account the effect of the rotating wheels on the aerodynamic field. In all other cases the wheels are covered and this reduces their aerodynamic force. A sketch of the streamlines around a partially covered wheel is shown in Fig. 3.12 together with

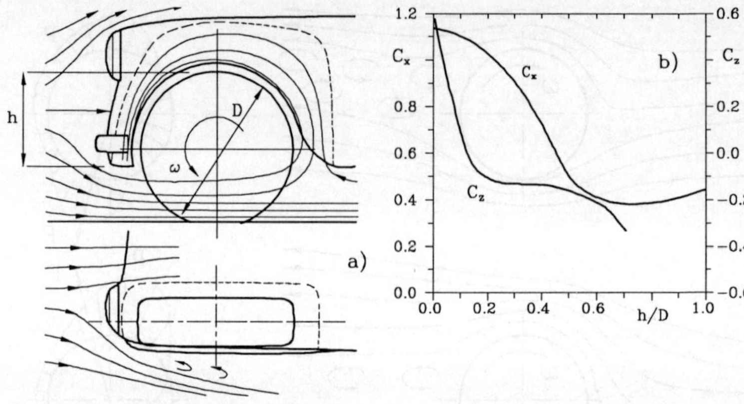

Fig. 3.12 (a) Flow in the cavity around a covered wheel. (b) Aerodynamic coefficients of the wheel as functions of ratio h/D.

a plot of coefficients C_x and C_z versus ratio h/D between the amount of covered wheel and its diameter. The curves are experimental and, particularly that related to C_z not very reliable owing to the method used to simulate the presence of the ground, but the results are at least qualitatively significant. The advantage of covering the wheel, without exceeding a value $h/D = 0,5 - 0,7$, is clear. The values of C_x are generally very high, particularly if compared with those of an insulated wheel and the increase of the drag when the wheel is much covered can be explained with viscous effects within the fender.

Another cause of the deviation of the aerodynamic field from that shown in Fig. 3.6 is the presence of internal flows. There are usually two separtate flows inside the vehicle: A cooling flow in the engine compartment and a flow in the passenger compartment. The second one is of lesser importance: If the intake is set at the base of the windscreen and the outlet is in a zone in the wake the result can be that of reducing slightly the drag, as it reduces the pressure in the separation bubble and increases that in the wake.

A larger amount of air is needed for engine cooling. A good solution would be to use a radiator of the type common in water-cooled aircraft piston engines, in which a diffusor slows down the flow which is driven through the heat exchanger and is then accelerated again in a converging duct (Fig. 3.13a). The diffusor should be long enough to allow the flow to be slowed down without separation (a slope of about 7° has been found to be a practical maximum) and the fan should operate only at speeds lower

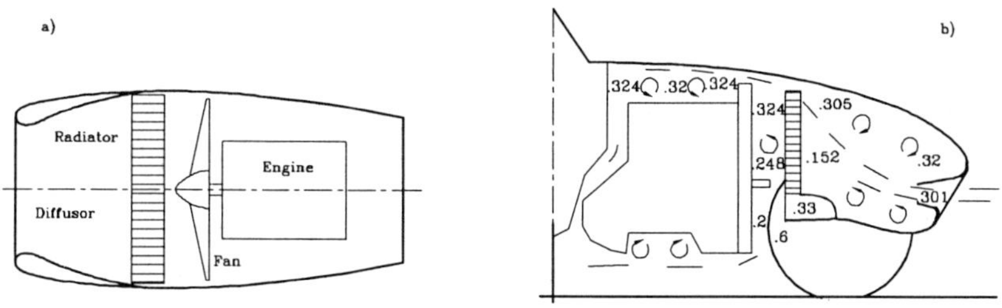

Fig. 3.13 (a) Ideal radiator. (b) Flow in the engine compartment of a sport car and values of the pressure coefficient.

than those for which the system has been designed.

In practice this solution cannot be used, at least on normal vehicles: A system of that type would be too long to be accommodated in the hood and in practice there is a short diffusor, whose opening is too large to allow a good attached flow, followed by a radiator and then the flow goes directly in the engine compartment without being further guided. The internal flow then mixes with the flow passing under the vehicle in a very disordered way. This situation is sketched in Fig. 3.13b, in which the values of the pressure coefficient measured on the symmetry plane of a model in scale 1:8 of a sports car are reported. The complexities needed for obtaining a well guided flow, separated from the external flow, are considered not worth the added cost and weight and the difficulties it would add to maintenance operations in the engine compartment.

As the presence of the internal flow in the engine compartment can account for about 13% to 20% of the total drag, aerodynamic testing should always be performed on models which reproduce also the inside of the engine compartment or, better, on the actual vehicle, with open air intakes. As the engine temperature affects the internal flow, aerodynamic testing should be done with the engine at running temperature.

3.3 Aerodynamic drag

As already stated, aerodynamic drag is the component of the aerodynamic force acting in the direction of the relative velocity while force F_x acts in the direction of x-axis. While in the case of bodies with high aerodynamic efficiency (defined as the ratio between lift and drag) the two can be very different[6], in the case of road vehicles the difference between the two is small, to the point that sometimes they are confused and force F_x is referred to as drag. Note that in many cases drag is considered positive when directed backwards, which is inconsistent with the general conventions on forces.

Aerodynamic drag can be considered as the sum of three terms: Friction drag, shape drag and induced drag. Also coefficient C_x can be similarily considered as the sum of the three corresponding terms.

3.3.1 *Friction drag*

Friction drag is the resultant of the tangential forces acting on the surface

$$\int_S \vec{t}_t \times \vec{i} dS .$$

As it is practically impossible to measure the friction drag on a body with complex geometry, reference is usually made to flat plates, where the only drag present is friction drag. Friction drag coefficient C_f referred to the "wet" surface, i.e. to the surface exposed to the fluid, is plotted versus Reynolds number, computed with reference to the length of the plate, in Fig. 3.14.

The two straight lines are referred to a laminar and a turbulent flow in the boundary layer. They are approximated by the empirical relationships

$$C_f = \frac{1.328}{\sqrt{\mathcal{R}_e}} , \quad C_f = \frac{0.074}{\sqrt[5]{\mathcal{R}_e}} , \tag{3.11}$$

respectively for the two cases.

The flow is laminar if it is free from vorticity and there is no mixing between adjacent streamlines. The vortices which are present in a turbulent boundary layer are very small but they cause a mixing and a strong energy transfer within the layer. If the fluid is free from vorticity when it enters into contact with the plate, a laminar flow is maintained up to values of about 500,000, provided that surface irregularities do not trigger turbulence. If the Reynolds number is higher, at least a part of the plate has a turbulent flow on it; the transition is shown in Fig. 3.14 to occur where the local Reynolds

[6]In the case of a wing at high angle of attack, force F_x can even be positive, i.e. pointing forward, while drag is obviously always directed backwards.

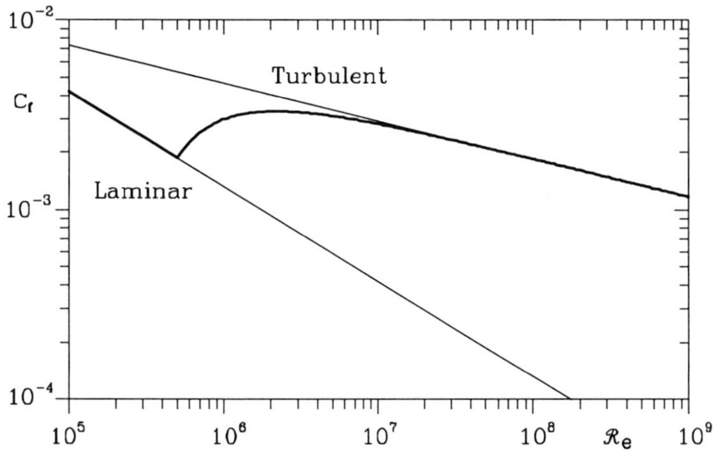

Fig. 3.14 Friction coefficient referred to the wet surface versus Reynolds number.

number, computed with the distance from thc lcading edge, reaches a value of 500,000.

In the case of streamlined bodies it is expedient to maintain a laminar boundary layer as long as possible, to reduce friction drag. However, in the case of blunt bodies, it often occurs that a laminar boundary layer results in a higher drag than a turbulent one. This is due to the fact that in a laminar layer the fluid which is in immediate contact with the surface receives less energy from adjacent layers and tends to slow down more quickly. Particularly in cases where the flow outside the boundary layer slows down and then the pressure increases, this causes a thickening of the boundary layer which eventually results in the detachment of the flow and the formation of a wake. This eventally occurs also in the case of turbulent layer, but the energy exchanges due to fluid mixing within the boundary layer help to maintain the flow attached to the surface for a longer distance.

The drag coefficient of a sphere is plotted as a function of the Reynolds number in Fig. 3.15 together with a sketch of the streamlines for the cases of laminar and turbulent flow.

The flow around motor vehicles is always turbulent, also owing to the presence of vortices in the air near the ground. The percentage of the drag due to friction is usually low, of the order of 10% of the total aerodynamic drag.

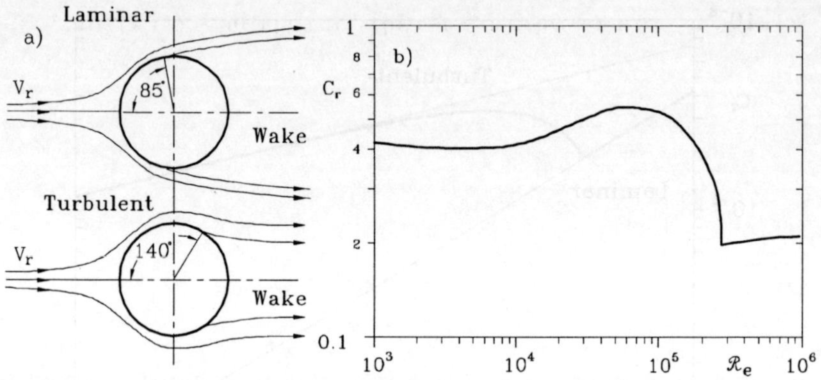

Fig. 3.15 (a) Qualitative sketch of the streamlines around a sphere. (b) Drag coefficient of a sphere as a function of the Reynolds number.

3.3.2 *Induced drag*

Induced drag is that portion of aerodynamic drag which is linked with the generation of lift. In aeronautics it plays the same role which in motor vehicle dynamics is played by rolling resistance: Is responsible for the energy which is dissipated to support the vehicle during motion.

In the case of road vehicles aerodynamic lift is not needed, actually it is a nuisance, and the induced drag should be reduced to a minimum by having a vanishing lift. An exception is the negative lift produced by aerodynamic devices aimed to increase the normal force of the vehicle on the ground: In this case induced aerodynamic drag adds to the increased rolling resistance.

To understand the origin of induced drag reference can be made to the theory of high aspect ratio (the ratio between the span and the chord) wings due to Prandtl. This theory can be applied in many cases to the wings which produce negative lift in racing cars. The lift of a wing is directly linked with a difference of fluid velocity between the upper and the lower surface of the wing, which causes a difference of pressure and ultimately a lift force. The difference of velocity can be thought as a vortex which is superimposed to the uniform airflow (Figure 3.16a).

If the wing had an infinite span, all sections would be in the same situation and a two-dimensional flow results: No induced drag is produced. In the case of an actual finite-span wing the vortex cannot vanish at the tips of the wing and its core is simply deflected backwards, originating a vortex

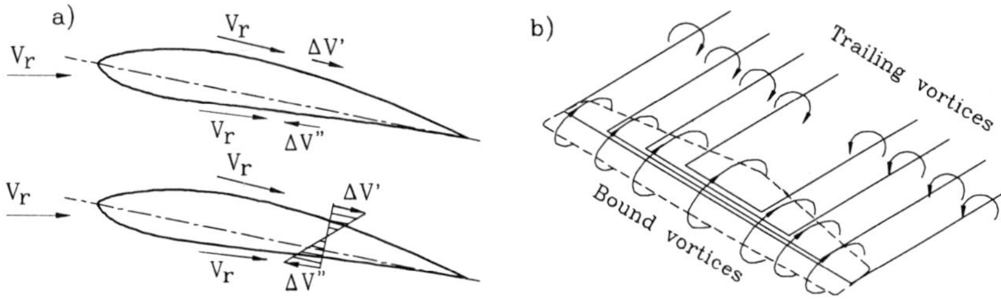

Fig. 3.16 Vorticity in a lifting wing. (a) attached vortex; (b) trailing vortices.

of the type shown in Figure 3.16b: Or better, as the vorticity is not constant along the wing, a set of such vortices. The energy dissipation needed for the creation of the trailing vortices explains the presence of the induced drag. Any device which reduces trailing vortices, as tip plates or particularly designed wing tips, is effective in reducing induced drag. Trailing vortices are sometimes well visible at the tips of the wings of racing cars.

From the theory of high aspect ratio wings it can be obtained that induced drag is proportional to the square of the lift, or which is equivalent, that the induced drag coefficient is proportional to the square of the lift coefficient. However, in the case of low aspect ratio wings and, above all, blunt bodies, this proportionality does not hold any more. Also the presence of the ground can modify the pattern of vortices. It has been suggested that in the case of road vehicles it is not possible to define an induced drag and that the term vortex drag is preferred[7]. In any case the vortices which are created behind a vehicle (Fig. 3.17) are linked with the generation of lift and a reduction of lift always causes a reduction of the overall aerodynamic drag.

If the shape of the body is well streamlined and the zones in which the flow is detached are small, it is possible to resort again to the methods of theoretical aerodynamics, like the lifting plate theory[8], to compute both the lift and the aerodynamic drag.

[7] R.T. Jones, Discussion on T. Morel, *The effect of base slant on the flow pattern and drag of three-dimensional bodies with blunt ends*, in Aerodynamic drag mechanism of bluff bodies and vehicles, Plenum Press, New York, 1978.

[8] A. Morelli, *Metodo teorico per la determinazione della distribuzione di portanza su un veicolo*, ATA, Sept. 1974.

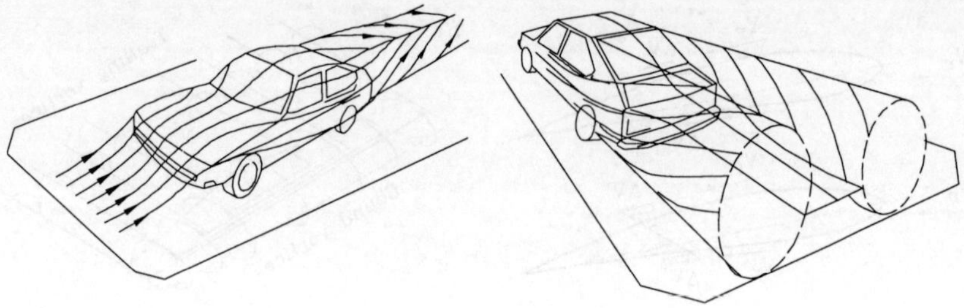

Fig. 3.17 Vortex pattern behind a road vehicle.

3.3.3 *Shape drag*

Shape drag is what remains of the drag if the contributions due to friction and induced drag are removed and, in the case of road vehicles, it is mainly due to the wake. The pressure in the wake is low and fairly constant and hence shape drag can be approximately evaluated as the product of the wake pressure by the projection on yz plane of the area exposed to it: The shape of the part of the vehicle which is in the wake has little importance. This statement must not be misunderstood: The shape of the tail of the vehicle is important to assess where the wake starts, but once this issue is solved, only the extension of the wake matters. Any geometrical irregularity can precipitate the detachment of the flow and the wake formation, particularly if it is located in a zone in which the flow slows down.

While in aeronautics it is practically important to separate the three components of the drag, as they can be readily identified and evaluated separately and then recombined to obtain a polar lift-drag diagram, in the case of road vehicles the three contributions are much less distinct and one of them, namely shape drag, is predominant. The distinction is therefore more useful to understand drag generation than to obtain quantitative predictions.

3.3.4 *Aerodynamic drag reduction*

Since the beginning of motor vehicle technology several attempts aimed to reduce aerodynamic drag were performed. The shapes which were developed for aircrafts and mainly for airships were adopted, often in a naive way as the attempt to streamline the body was often accompanied by mechanical parts completely exposed to the wind. An example of this approach

Fig. 3.18 Electric vehicle *Jamais Contente* built to break the barrier of 100 km/h. In 1898 it actually set a world speed record at 105.85 km/h.

is shown in Fig. 3.18: The electric vehicle *Jamais Contente* built to break the barrier of 100 km/h. In 1898 it actually set a world speed record at 105.85 km/h.

The shape which proved to produce the lowest ratio between the aerodynamic drag and the volume was the straight circular slender body with a ratio diameter over length equal to about 0.3. For a Reynolds number of about 10^7 its drag coefficient is of 0.05. However it is difficult to obtain a suitable car body from it and, moreover, it is optimal only if the motion takes place far from the ground. The presence of the latter causes the flow to change substantially (Fig. 3.19a) and the value of C_x is far higher up to values about 0.15 at distances from the ground typical of motor vehicles.

If the vehicle would move in contact with the ground the best shape would be that of half of a slender body, as shown in Fig. 3.20, consideration which seems to have inspired several designs of the past. However the distance of the vehicle from the ground cannot vanish and this solution leads to quite high values of the drag.

If the axis of the slender body is curved a lower resistance in the motion in proximity of the ground results (Fig. 3.19b) with an optimum value of the camber ratio a/l existing for each value of the distance from the ground. For a value of the distance $h/D = 0.1$ the optimum value of the camber ratio is about 10%. However the difficulties of adapting a slender body to a vehicle and to house in it the mechanical components and the wheels

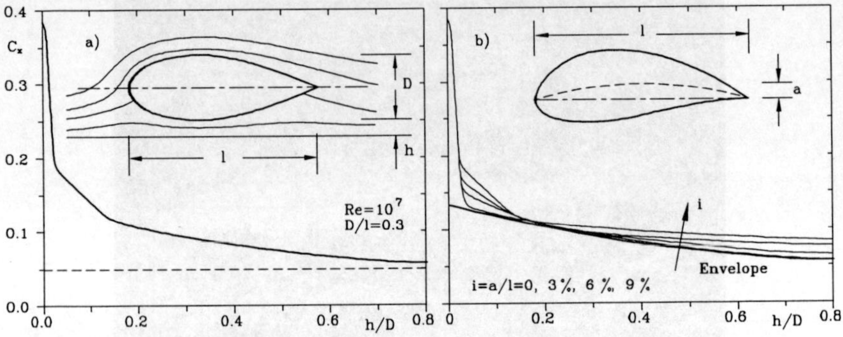

Fig. 3.19 (a) Streamlines around a straight slender body and values of C_x versus the distance from the ground. (b) Values of C_x versus the distance from the ground for cambered slender bodies with different values of the camber.

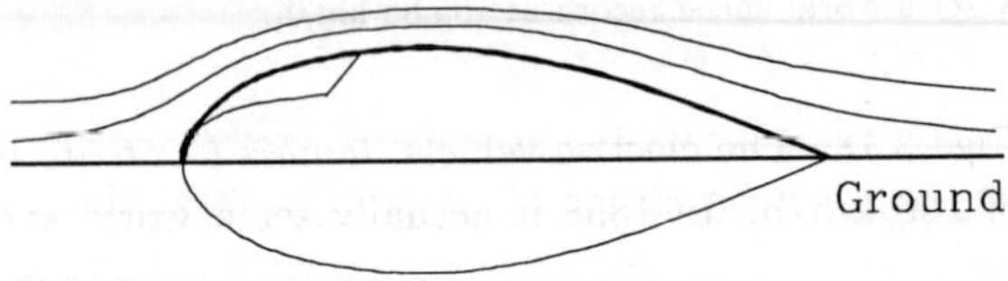

Fig. 3.20 Streamlines around half a slender body at zero distance from the road. Sketch of a possible vehicle obtained in this way.

remain.

The results obtained by Lay in 1933 through a series of wind tunnel tests performed on modular models are still interesting. His basic shapes were a flat plate perpendicular to the current (he found a value of C_x larger than 1.2), a slender body (whose C_x was measured at 0.08) a rectangular box ($C_x = 0.86$) and a vehicle of his times ($C_x = 0.6$).

He then discovered that a slight rounding of the corners of the box resulted in a decrease of C_x to 0.46. By fitting different front and rear parts to the vehicle model he saw that only by shaping suitably both parts aerodynamic drag could be reduced but that the shaping of the rear part is more important than that of the front part. Also, using the shapes tested by Lay, to obtain very low drag it is necessary to accept a very long vehicle.

Ten years earlier, in 1923, Jaray obtained a patent in Germany for a car body made by a rectangular cambered stub wing with a slender body superimposed to it (Fig. 3.21a). This approach, named J-shape, allowed to

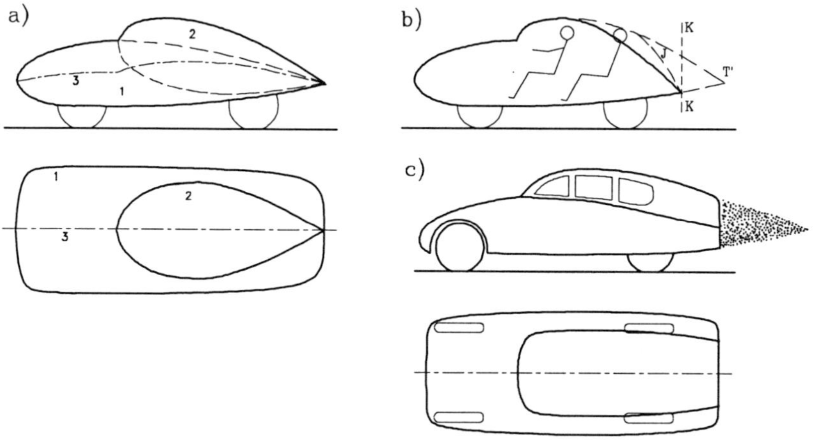

Fig. 3.21 Streamlined car bodies following J and K shapes. (a) J-shape; (b) problems linked to the length of J-shapes and cutting of the tail; (c) K-shape: One of the drawings from the original patent by Koënig.

house easily the wheels and other mechanical components, but the problems related to the length of the vehicle, if a sufficient height for the passengers in the rear seats was required, were not solved. The centreline of the body was also quite curved, which results in non-negligible lift and induced drag. However the J-shape can be easily identified in many vehicles starting from the fifties, like the Lancia Stratos, Citroen DS and many coupé built by Porsche.

In 1937 a new approach was patented almost simultaneously by Kamm and by Koënig and Fachsenfeld Reinhard. From the observations that to obtain a good height in the end part of the J-shape without a very long vehicle a shape which is prone to produce a large wake is obtained and that the shape of the part of the vehicle which is in the wake has little influence, they suggested that the long streamlined tail of the J-shape could be cut altogether, following line KK of Fig. 3.21b. The result is shown in Fig. 3.21c, from the original patent[9].

The cut does not affect shape drag, if the part cut off was already in the wake, and likely reduces lift and induced drag. This statement is a rough approximation, as any change in a part of the shape changes all the aerodynamic field, but the use of the K-shape allowed to reduce the drag of many passenger vehicles. Many cars of the seventies had essentially a K-

[9] Brevetto industriale n. 352583 - Carrozzeria per automobili.

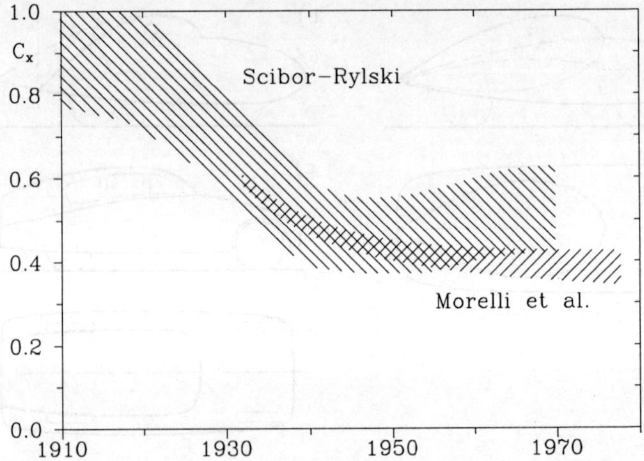

Fig. 3.22 Experimental values of C_x for passenger vehicles versus the year of construction.

shape, like the Citroen GS and CX, Lancia Beta and Gamma, Alfa Romeo Alfasud, Rover 3 liters and many others.

The aerodynamic evolution of passenger vehicles is summarized in Fig. 3.22: The two hatched zones refer respectively to cars with a generally good aerodynamic shape, many of which built following the J- or K-shape[2] and to a larger sample containing cars of all types[3]. There is no contradiction between them: While the better vehicles showed a constant progress towards low drag, the availability of more powerful engines and the low cost of energy caused a decrease of the drive towards better aerodynamics in the fifties, with an average increase of the drag of cars.

Some values of both coefficient C_x and of the product SC_x, which allows a more direct comparison, of more modern cars are reported in Table 3.2[4]. By comparing the values in the table with those reported in Fig. 3.22 the progress which occurred in the eighties, mainly linked with the increase of the cost of energy which took place a decade earlier, is clear. At any rate it must be noted that, with the exception of few cases, the values of C_x are very rarely lower than 0.35, with many cars having a value between 0.4 and 0.5.

[2] A. Morelli, L. Fioravanti, A. Cogotti, *Sulla forma della carrozzeria di minima resistenza aerodinamica*, ATA, Nov. 1976.

[3] A.J. Scibor Ryilski, *Road Vehicle Aerodynamics*, Pentech Press, London, 1975.

[4] H.P. Leicht, *Kanal voll* , Auto-motor sport, 18/1986.

Table 3.2 Values of coefficient C_x, surface S and product SC_x for some European cars.

	$C_X S$	C_X	S [m2]		$C_X S$	C_X	S [m2]
Lancia Y10	0.57	0.33	1.76	Opel Corsa SR	0.61	0.35	1.73
Fiat Uno	0.62	0.34	1.83	VW Polo	0.65	0.38	1.70
Renault 5	0.67	0.37	1.80	Austin Metro	0.67	0.39	1.73
Peugeot 205	0.68	0.39	1.74	Fiat Panda	0.70	0.41	1.70
Citroen Visa	0.70	0.40	1.75	Ford Fiesta	0.73	0.41	1.76
Renault 4	0.90	0.49	1.83				
Opel Kadett GSi	0.60	0.32	1.88	Peugeot 309	0.64	0.34	1.86
VW Golf GL	0.65	0.34	1.89	Mercedes 190 E	0.65	0.34	1.89
Renault 21	0.66	0.34	1.94	Ford Sierra XR 4i	0.67	0.34	1.98
VW Golf GTI 16V	0.67	0.35	1.91	Citroen BX	0.68	0.36	1.91
VW Jetta CL	0.68	0.36	1.89	VW Passat GL	0.70	0.37	1.90
Fiat Ritmo	0.70	0.37	1.88				
Opel Omega	0.58	0.28	2.06	Mercedes 200	0.60	0.29	2.07
Audi 100	0.62	0.30	2.05	Renault 25	0.62	0.31	2.03
Ford Scorpio	0.70	0.35	2.02	Fiat Croma	0.70	0.34	2.04
Lancia Thema	0.73	0.36	2.06	Honda Prelude 16V	0.76	0.41	1.84
Alfa 90	0.77	0.40	1.92	Citroen CX	0.78	0.40	1.96
Mitsubishi Galant	0.79	0.40	1.98				
Ferrari Testarossa	0.61	0.33	1.85	Mercedes 190 E2.3	0.64	0.33	1.94
Porsche 944 turbo	0.65	0.35	1.89	VW Scirocco 16V	0.68	0.38	1.78
Porsche 911 Carrera	0.68	0.38	1.77	Mitsubishi Starion T	0.69	0.37	1.84
Alfa Romeo GTV	0.71	0.40	1.77	Jaguar XJ-S	0.73	0.40	1.83
Porsche 928 S	0.77	0.39	1.96	Audi Quattro	0.80	0.43	1.86
BMW M 635 CSi	0.80	0.40	2.00				

A method, often referred to as shape optimization, is now widely used. It is based on subsequent detail modifications to achieve drag reductions which can be quite substantial. The drag coefficient of the base shape is measured and a number of specific details of the car body are chosen. One of them is modified and the wind tunnel test is repeated, continuing to modify until a minimum of the drag is obtained. The work is then repeated for all the chosen details; an example is shown in Fig. 3.23 where the base shape and five details are shown. The thin line gives the initial configuration while the thick one describes the optimized shape. By operating in this way the drag was reduced of about 21%, while a more substantial reduction of 33% could be achieved only by introducing modifications which changed to a larger extent the overall appearance of the car.

The results obtained by changing the inclination of the rear part of a hatchback car is shown in Fig. 3.23: If the angle between the rear window and the horizontal is larger than 35° the value of C_x is 0.4. For a low value of the angle it is possible to reduce the drag, with a coefficient of about 0.34, but there is a region, around an angle of 28°, in which large

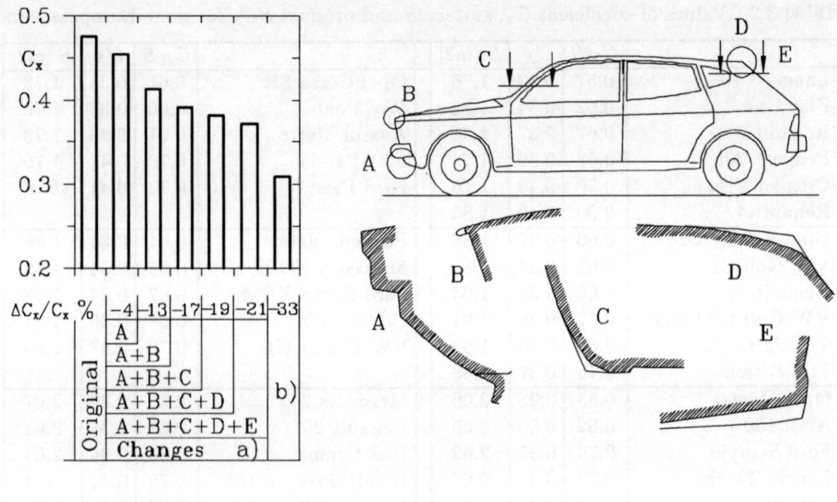

Fig. 3.23 Shape optimization method. Definition of the five details used to optimize the shape and to reduce C_x coefficient. (a) Optimized shape, (b) modified shape.

vortices produce a substantial increase of drag, up to $C_x = 0.44$. This result confirms the already mentioned results obtained from slanted rectangular blocks.

Alternatively, it is possible to develop from theoretical considerations ideal shapes aimed to reduce drag to a minimum and then to modify these ideal shapes to adapt them to motor vehicle use. An example of this procedure is shown in Fig. 3.25 in which both the ideal shape and the car derived from it are represented. The ideal shape has been obtained by stating that the lift and pitching moment must be zero, the positive pressure gradients must be as low as possible, the cross section of the body must vary slowly in shape and area and its contour must be as much rounded as possible.

The value of C_x of the ideal body in the vicinity of the ground proved to be as low as 0.049, the same as that of a slender body located at an infinite distance from the ground. The vehicle obtained from it had a value of C_x of only 0.23, while maintaining a good internal space for the occupants, the luggage and the mechanical components, not unlike a regular saloon car.

As a result of the research work performed in the eighties in many countries it is possible to state that it is possible to achieve values of coefficient C_x as low as 0.25 without sacrificing much in terms of internal space and general characteristics of the vehicle. It is also clear however that the lower the drag coefficient is, the lower are the advantages of further reductions in

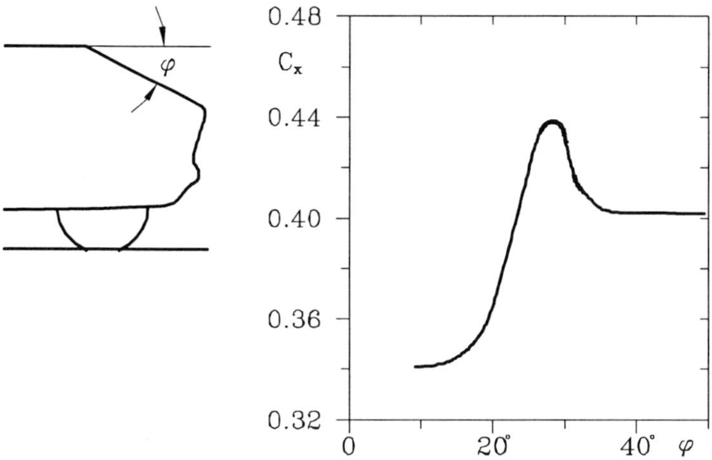

Fig. 3.24 Coefficient C_x as a function of the angle between the rear window of a given car and the horizontal.

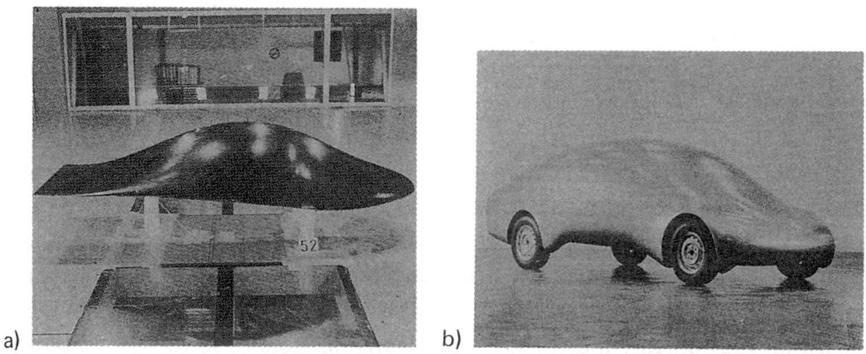

Fig. 3.25 Ideal shape and actual shape for a very low-drag research vehicle shown in the wind tunnel.

terms of fuel consumption, particularly as the average use of road vehicles occurs at speeds which are lower than those at which aerodynamic drag is the most important form of resistance to motion. As the actual average speed of driving depends mostly on issues unrelated to the design of vehicles themselves (road conditions, laws and their enforcement etc.), no further considerations on the drag of passenger vehicle will be made.

Industrial vehicles were, even in the recent past, mostly designed with

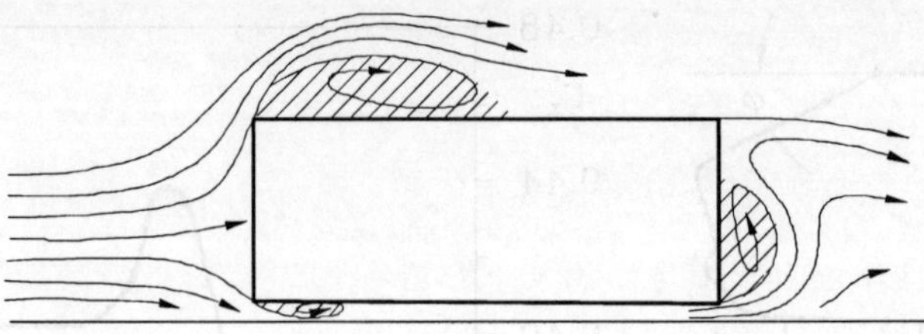

Fig. 3.26 Streamlines around a square box with blunt edges, at a distance from the ground equal to 0.06 D, where D is the diameter of a circle having the same cross-sectional area of the body.

little concern about their aerodynamic characteristics. The low speed and the high value of the ratio between the mass and the frontal area makes the power needed to overcome aerodynamic drag a small fraction of the total power needed for motion. However the higher speeds reached on highways also by industrial vehicles and the higher concern about energy saving lead to many studies aimed to reduce the drag also in this field.

In the case of single-body vehicles, as buses, vans and nonarticulated trucks without trailer the basic shape is essentially a square box. If the edges are blunt the value of C_x is in the range 0.82–0.86, mostly owing to the fact that the flow is much separated and that the wake is very large (Fig. 3.26).

From this figure it is clear that any change of the shape which allows to reduce flow separation is very beneficial: Simply by slightly rounding the edges, the drag coefficient reduces almost by half, reducing to about 0.45 and, with a better profile of the front part, values in the range 0.4–0.43 can be obtained for buses and vans. It is difficult to further lower this value. With fairings on the underside it is perhaps possible to lower it by about 0.05 and many other devices like the ones shown in Fig. 3.27, aimed to increase the pressure in the wake, have been tried without success.

In the case of articulated vehicles two stagnation points of the flow can be present, one on the cab and one on the trailer and a flow occurs in the space between the two bodies (Fig. 3.28a). The drag depends largely on their distance, with values of C_x spanning from about 0.72 when they are in contact to about 0.93 when the distance is about 0.6 times the diameter of a circle with the same area of the cross section. To reduce drag the flow between cab and trailer must be blocked, to have a single stagnation point; the simplest way is to put a vertical flat plate on the roof of the former. The flat plate must cause the flow to attach to the trailer in correspondence to

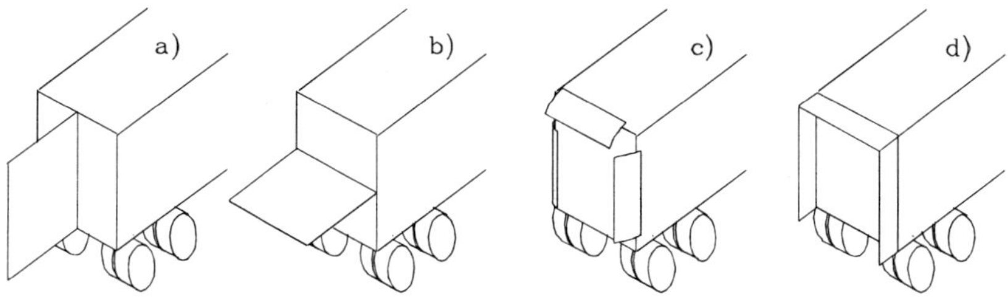

Fig. 3.27 Devices aimed to increase the pressure in the wake of industrial vehicles. Only device (d) caused slight drag reductions.

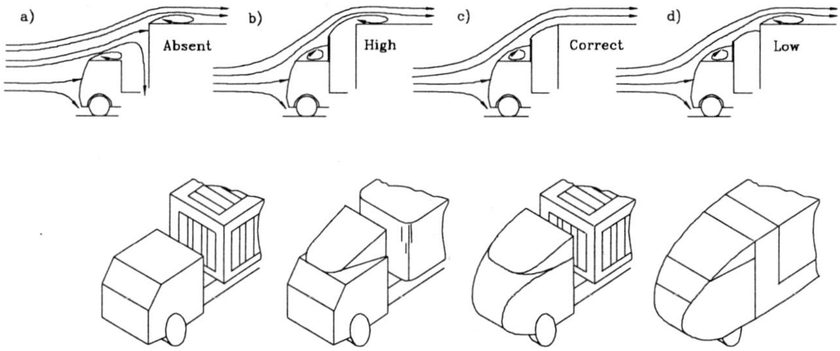

Cx	0,863	0,568	0,560	0,480
ΔCx /Cx	—	34%	35%	44%

Fig. 3.28 (a) Flow on the front part of an articulated truck without and a flat plate on the cab. (b) Reduction of the drag of articulated vehicles; values obtained on half-size models.

the front upper edge: If it is too low or too high the effect is an increase of drag due to separation of the flow and the formation of a separation bubble (Fig. 3.28a). This is minimized by rounding the edge or the trailer and by using a shaped deflector instead of a flat plate.

Some results obtained on half scale models are shown in Fig. 3.28b; they show that with just few modifications, include a deflector, a large reduction is readily obtained, while further improvements are difficult to achieve and require a global streamlining of the vehicle. A value of C_x of about 0.5 can thus be obtained; lower values, down to 0.24 as obtained on a 1/10 scale model, could be achieved by using a complete fairing of the underside.

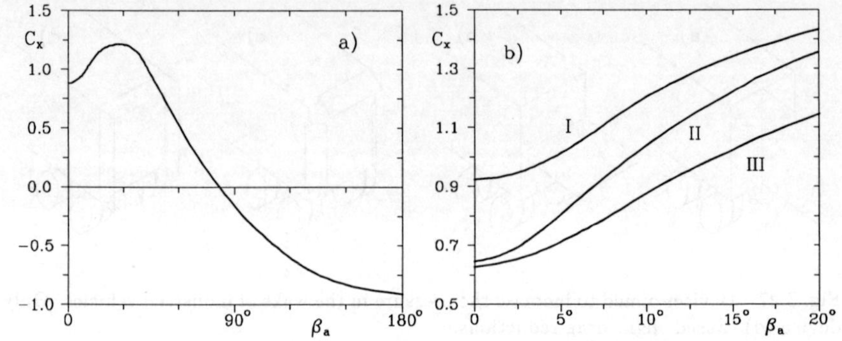

Fig. 3.29 (a) Coefficient C_x versus sideslip angle β_a for an articulated truck. (b) Effects of drag reducing devices on the curve $C_x(\beta_a)$: I no deflector; II flat plate deflector; III deflector and round front edges of the trailer.

In the case of articulated vehicles the aerodynamic sideslip angle is important as the flow between the cab and the trailer, which can be eliminated as described above, can also occur in transversal direction. For this reason the drag of articulated vehicles increases with aerodynamic sideslip angle β_a more than that of other vehicles and the advantage due to the use of deflector plates tends to vanish when it reaches values of about 20° (Fig. 3.29). Only using shaped deflectors in connection with the rounding of the front edges of the trailer allows one to obtain a reduction of the drag when a sidewind is present.

3.4 Aerodynamic lift and pitching moment

Apart from the increase of drag due to induced drag, aerodynamic lift must be avoided as it reduces the load on the tires and consequently the forces which the vehicle can exert on the ground; moreover this reduction is strongly dependent on the speed. In the case of vehicles with high power/weight ratio it is possible to use negative aerodynamic lift to allow to transfer all the power through the road-wheel contact. The same holds for increasing the cornering forces.

Also aerodynamic pitching moment M_y must be as small as possible, since it causes strong variations of the forces exerted by the wheels on the road, which depend on speed. With reference to Fig. 3.30, the pitching moment is positive when it acts to increase the load on the front wheels. As the aerodynamic drag is applied in the centre of mass which is at a

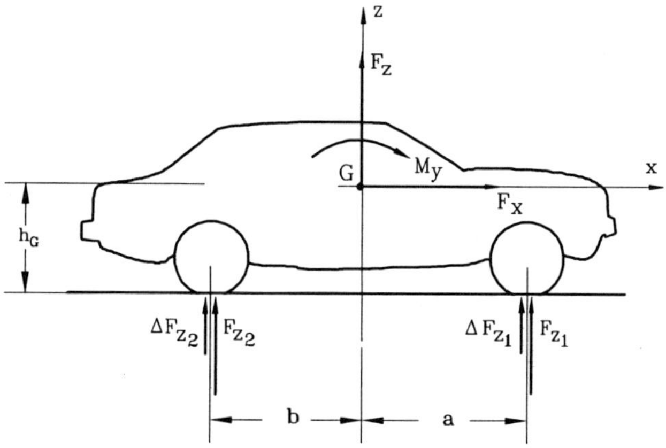

Fig. 3.30 Longitudinal load transfer due to aerodynamic pitching moment and lift. Forces F_z and ΔF_{z_i} are the forces the vehicle receives from the ground; a positive ΔF_z indicate an increase of load.

distance h_G from the ground, the longitudinal load transfer on a vehicle with two axles is

$$\begin{cases} \Delta Z_1 = \frac{1}{2}\rho V^2 S\left(C_{M_y} - \frac{h_G}{l}|C_x| - \frac{b}{l}C_z \right) \\ \Delta Z_2 = \frac{1}{2}\rho V^2 S\left(- C_{M_y} + \frac{h_G}{l}|C_x| - \frac{a}{l}C_z \right) . \end{cases} \tag{3.12}$$

While it is usually impossible to compute the drag using the methods of theoretical aerodynamics, there are methods for the computation of lift and pitching moments which are of widespread use in aeronautics. However their application in motor vehicle technology is questionable, owing to the presence of a large wake and of large parts of the body on which the flow is detached. A particular version of the method of the lifting surface, derived from the one developed for low aspect ratio wings, has been devised for motor vehicle use[5]; it can be used for an evaluation of the lift and pitching moment of the vehicle provided that the wake is small enough. The vehicle is substituted by a surface, whose projected shape on the ground is that of the vehicle and its profile is that of the line connecting the centres of the cross sections of the vehicle (Fig. 3.31).

[5]A. Morelli, *Metodo teorico per la determinazione della distribuzione di portanza su un veicolo*, ATA, Sept. 1974.

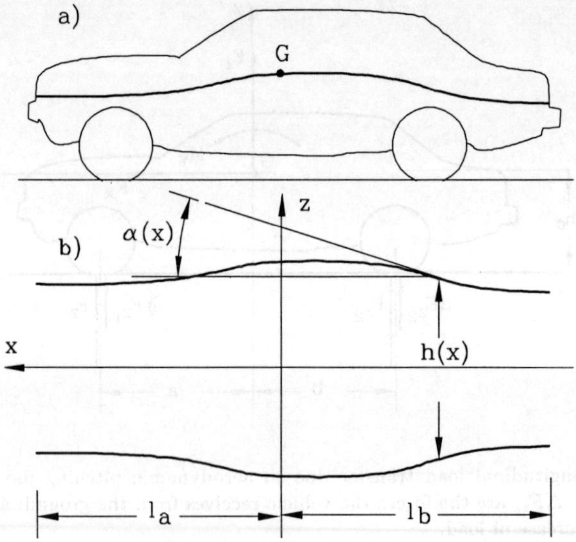

Fig. 3.31 Lifting surface method for the computation of the lift and the pitching moment: (a) vehicle, (b) pair of lifting surfaces which substitute the vehicle.

To simulate the presence of the ground a second surface is located below it, in a specular position. Let $b(x)$, $h(x)/2$ and $\alpha(x)$ represent respectively the width, the distance from the ground and the slope of the surface ($2\tan(\alpha) = dh/dx$), the difference Δp between the pressure operating on the upperside and the underside of a point of coordinate x of the surface located on the symmetry plane is

$$\Delta p = \frac{1}{4}\rho V_r^2 \frac{d}{dx}\left(\frac{\alpha b^2}{h}\right) . \tag{3.13}$$

Δp is assumed to have a parabolic dependence on the y coordinate, leading to the general expression

$$\Delta p = -\rho V_r^2 \frac{\partial}{\partial x}\left[\frac{\alpha(4y^2 - b^2)}{4h}\right] . \tag{3.14}$$

The lift and the pitching moment can then be obtained by integrating the difference of pressure over the whole surface

$$\begin{cases} F_z = \displaystyle\int_{-l_b}^{l_a} \int_{-b/2}^{b/2} \Delta p \, dx \, dy \\ M_y = \displaystyle\int_{-l_b}^{l_a} \int_{-b/2}^{b/2} \Delta p \, x \, dx \, dy \,. \end{cases} \tag{3.15}$$

The lift and pitching moment coefficients are then

$$\begin{cases} C_z = \dfrac{1}{S} \displaystyle\int_{-l_b}^{l_a} b^2 \left[\dfrac{b}{3} \dfrac{d}{dx} \left(\dfrac{\alpha}{h} \right) + \dfrac{\alpha}{h} \dfrac{db}{dx} \right] dx \\ C_{M_y} = \dfrac{1}{Sl} \displaystyle\int_{-l_b}^{l_a} b^2 \left[\dfrac{b}{3} \dfrac{d}{dx} \left(\dfrac{\alpha}{h} \right) + \dfrac{\alpha}{h} \dfrac{db}{dx} \right] x dx \,. \end{cases} \tag{3.16}$$

These expressions are clearly approximated: Apart from the usual approximations linked with the lifting surface method, the distance h between the surfaces has been assumed to be small if compared with both its length and its width, angle α is small enough to allow the linearization of its trigonometrical functions and the induced airspeed in any transversal plane is influenced only by the vortex distribution in the same plane.

If the body is streamlined enough and the wake is small, this approach allows to compute the lift and the pitching moment and, even more important, gives important information on the pressure distribution, allowing to modify the shape to avoid strong positive pressure gradients. It has been used with success to identify the shape shown in Fig. 3.25a: In the case of the ideal body the good streamlining and the virtual absence of detached flow allow the methods of analytical aerodynamics to be used with confidence.

Aerodynamic lift and pitching moment depend on the position of the vehicle on the ground, mainly on the aerodynamic angle of attack α, which in the following pages is assumed to coincide with the pitch angle θ. Lift can be considered as varying linearly for small changes of α (or θ) and both C_z and $\partial C_z/\partial\theta$ must be measured in the wind tunnel. The same holds for the pitching moment. A plot of the moment coefficient C_{M_y} versus angle θ for 3 different vehicles is reported in Fig. 3.32; the values of the derivative $\partial C_{M_y}/\partial\theta$ (indicated as $(C_{M_y})_{,\theta}$) for small movements about the reference position are also reported. Note that the moment and its derivative are mostly negative; this is a general rule.

To reduce the lift and in some cases to make it negative many passenger vehicles nowadays are provided with spoilers on the rear part of the body

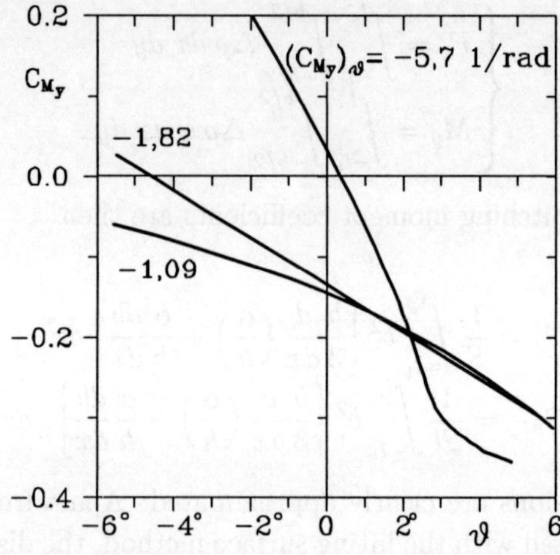

Fig. 3.32 Pitching moment coefficient C_{M_y} as a function of angle θ for three different vehicles.

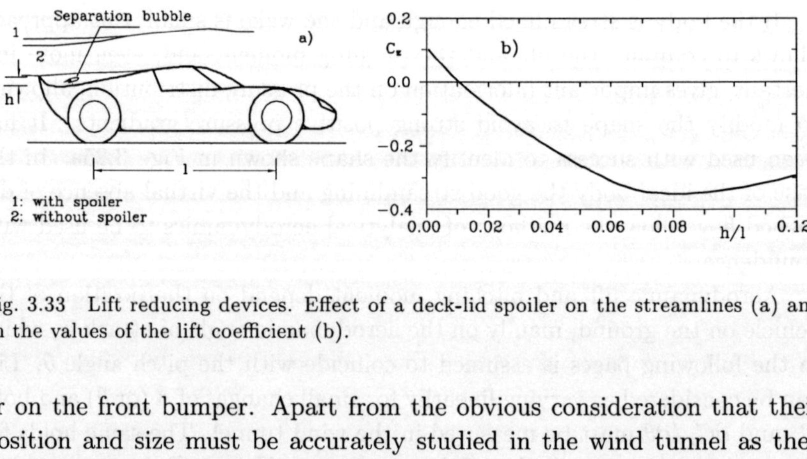

Fig. 3.33 Lift reducing devices. Effect of a deck-lid spoiler on the streamlines (a) and on the values of the lift coefficient (b).

or on the front bumper. Apart from the obvious consideration that their position and size must be accurately studied in the wind tunnel as they are useless if located in the wake or in other zones in which the flow is detached, they must be located in such a way to avoid to give way to pitching moments.

A spoiler usually create some shape drag, as it increases the size of the wake, but can be effective in reducing the total aerodynamic force F_x owing to the reduction of the induced drag (Fig. 3.33).

Since spoilers cause an increase of the pressure on the tail of the vehicle, they usually create a positive pitching moment. This moment must

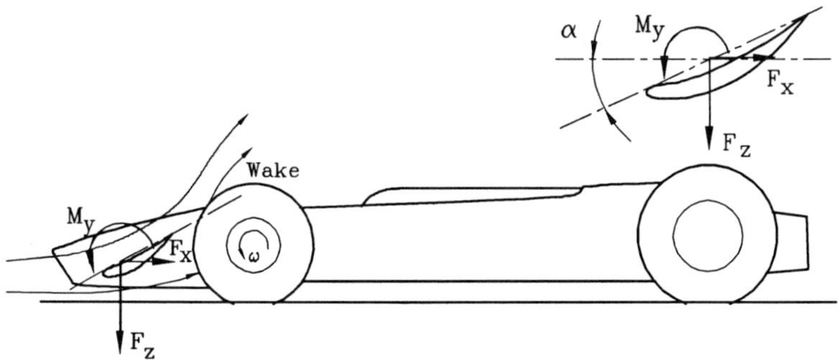

Fig. 3.34 Forces and moments acting on the wings of a racing car. Interference between wings and rotating wheels.

be compensated by another surface positioned near the stagnation point, usually referred to as bumper spoiler. Its presence is usually beneficial also on the drag.

In racing cars strong negative lift forces are obtained by using wings and by a suitable aerodynamic design of the whole body. As aerodynamic devices have a very large impact on safety, they are regulated in a very detailed way: Owing to the frequent change of such rules very little can be said in general.

The airflow in the zone in which these wings are located is much influenced by the rotation of the wheels and their actual angle of attack, i.e. the angle between the surface and the direction of the impinging current, can be quite different from the geometrical angle of attack (Fig. 3.34). Each one of the wings produces a lift, which is negative, a drag, usually quite strong, and a pitching moment. They must combine in such a way that the total pitching moment acting on the car is not too large.

The whole body of a racing car can be designed to produce negative lift; in particular if the pressure under the vehicle is lower than atmospheric pressure, strong downwards forces can be collected on the underside.

3.5 Side force, rolling and yawing moments

If the vehicle has a symmetry plane and is in a symmetrical position with respect to the airflow, i.e. if the roll and the aerodynamic sideslip angles are equal to zero, the side force F_y, the rolling moment M_x and the yawing moment M_z vanish. In general, what matters is their rate of change with

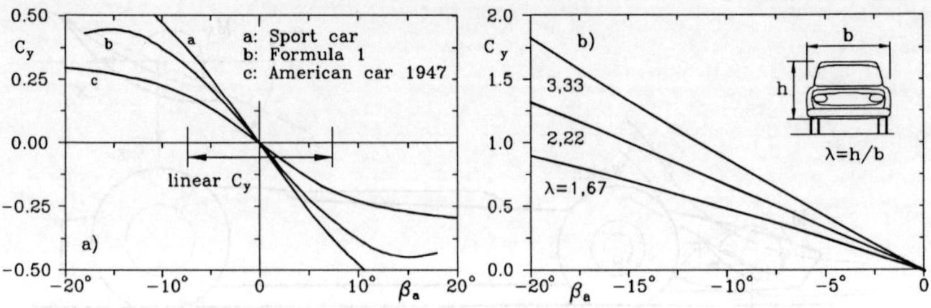

Fig. 3.35 Coefficient C_y versus angle β_a. (a) Typical values for vehicles of different types; (b) dependence from the ratio width/height of the vehicle.

angle β_a and, sometimes, roll angle ϕ. In the case of racing cars with uncovered wheels, they can also be produced by offset steering wheels and it is important to study their variation with the steering angle δ.

For small variations of the mentioned parameters about zero, coefficients C_y, C_{M_x} and C_{M_z} can be approximated with linear functions and their derivatives $(C_y)_{,\beta_a}$, $(C_{M_x})_{,\beta_a}$, etc. can be considered constant.

Some typical curves $C_y(\beta_a)$ are reported in Fig. 3.35a. The slope of the curve in the origin is –2.2 rad^{-1} for a typical American saloon car and –2.85 rad^{-1} for a sport car. For a first approximation evaluation of the slope $(C_y)_{,\beta_a}$ (in rad^{-1}), the following formula has been suggested

$$(C_y)_{,\beta_a} = \frac{\text{lateral area}}{\text{frontal area}}\left(0.005 + 0.0019 n_f\right), \tag{3.17}$$

where n_f is a numerical factor which must be obtained from experimental results on vehicles similar to the one under study.

As already stated, $(C_y)_{,\delta_a}$ is usually small, except for racing cars. On some Formula 1 racers a value of 1.37 rad^{-1}[6] has been recorded.

Even more important than the side force, the aerodynamic yawing moment M_z plays a key role in the dynamics of high speed driving. Some typical laws $C_{M_z}(\beta_a)$ are reported in Fig. 3.36a: Usually it is negative for positive sideslip angle, which amounts to state that the side force is applied in a forward position with respect to the centre of mass. This arrangement, which occurs very often, is unstabilizing at high speed; the value of C_{M_z} can be large enough to cause force F_y to be applied in front of the vehicle.

A nonlinear expression for coefficient C_{M_z} is

[6] A.J. Scibor Rylski, *Road Vehicle Aerodynamics*, Pentech Press, London, 1975.

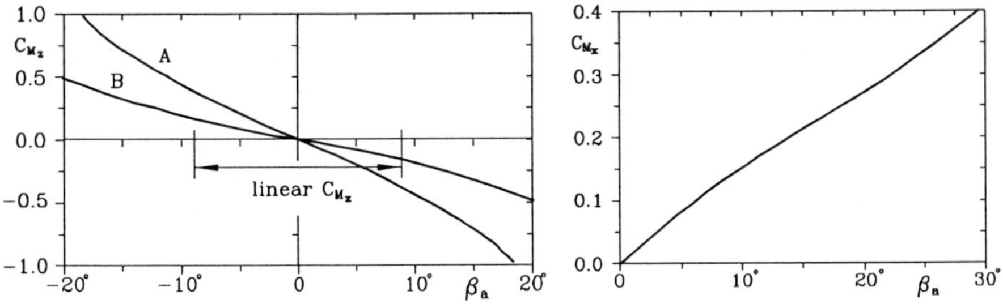

Fig. 3.36 (a) Typical laws $C_{M_z}(\beta_a)$ for a well streamlined car (A) and a car with a less careful aerodynamic design (B). (b): Rolling moment coefficient C_{M_x} versus β_a for a typical saloon car.

$$C_{M_z} = -\left[K - \sin^2(\beta_a)\right]\tan(\beta_a) , \tag{3.18}$$

in which constant K, whose obvious meaning is the value of $(C_{M_z})_{,\beta_a}$ in the linear part of the plot, must be obtained from vehicles similar to the one under study. MIRA[7] suggests for $(C_{M_z})_{,\beta_a}$ the expression

$$(C_{M_z})_{,\beta_a} = -\frac{1}{100}\left[\frac{\text{lateral area}}{\text{frontal area}}(0.208 + 0.0655 n_f - 0.00508 n_f^2)\right] , \tag{3.19}$$

where n_f must again be obtained from experimental data.

For $(C_{M_z})_{,\delta}$ the same considerations seen for $(C_y)_{,\delta}$ hold; an order of magnitude for Formula 1 racers is a value of 0.46 rad^{-1}.

A plot of the aerodynamic rolling moment coefficient versus the sideslip angle for a typical saloon car is reported in Fig. 3.36b. From the graph a value of the derivative $(C_{M_x})_{,\beta_a}$ for the linear part of 1.05 rad^{-1} can be obtained.

3.6 Study of aerodynamic forces and moments

Traditionally, the tool for the study of aerodynamic forces and moments is the wind tunnel. Modern wind tunnels used for road vehicle aerodynamics

[7] R.G.S. White, *A rating method for assessing vehicle aerodynamics side force and yawing moment coefficients*, MIRA Rep. n. 1, 1970.

are specialized devices, quite different from those used in aeronautics. Wind tunnel tests of cars are almost always performed on real-size vehicles, as it is very difficult to simulate with the required precision internal flows on reduced scale models.

Wind tunnels for vehicular use are divided in aerodynamic and climatic tunnels: The first ones are aimed to simulate the aerodynamic field for the measurement of aerodynamic forces and moments, of pressure distribution, etc. while the latter are used to simulate the motion in various climatic conditions, usually with lower precision in the simulation of the aerodynamic field. Sometimes the two can be combined in a single setup, but at the cost of added complexity.

If the aerodynamic circuit of the tunnel is closed the motor must supply to the flow only an energy equal to the losses while if it is open new air must be continuously accelerated and the power needed is larger than the actual power of the airstream

$$\frac{1}{2}\rho V^3 A \,,$$

where A is the cross section of the test chamber. An efficiency can be defined as

$$\eta = \frac{\frac{1}{2}\rho V^3 A}{P} \,, \tag{3.20}$$

where P is the power of the motors. Clearly this efficiency is not a true efficiency in the usual sense as it can be greater than unity in the case of closed aerodynamic circuit.

Open tunnels are simpler, smaller and cheaper but are energetically less efficient. However it must be considered that in closed-circuit devices the circulating air must be cooled and, if low temperature tests are required, the power of the refrigerators can be higher than that of the motors themselves. While in the aeronautical field most wind tunnels are of the closed type, modern plants for vehicular use are mostly open, perhaps with air recirculation within the building to avoid the need of filtrating large quantities of air and to reduce noise pollution to the environment. Recirculation can also raise the energetic efficiency. The cross section of a wind tunnel for vehicular aerodynamics is shown in Fig. 3.37 while the plans of some modern wind tunnels are reported in Fig. 3.38.

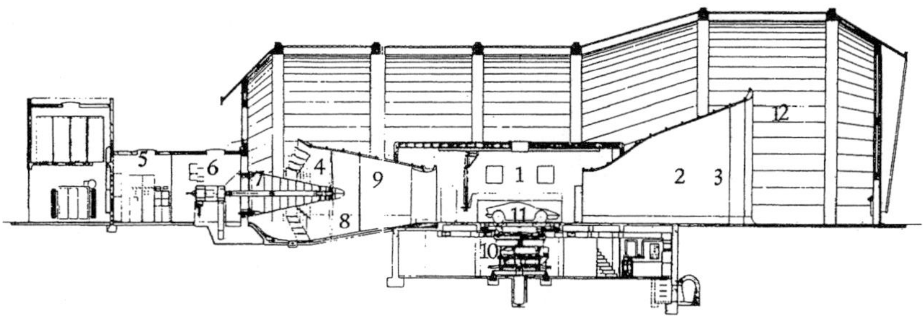

Fig. 3.37 Skematic cross-section of the Pininfarina wind tunnel. 1) test chamber; 2) intake cone; 3) turbulence generators; 4) outlet cone; 5) power conditioning units; 6) motor; 7) propeller shaft; 8) propeller; 9) nets; 10) six-component force transducer; 11) vehicle under test; 12) building.

The test chamber can be open or closed. While in the case of aeronautical wind tunnels the test chamber is mostly closed, those for motor vehicle study are often of the open type. Open test chambers usually allow one to obtain the same precision in the simulation of the aerodynamic field with a smaller cross-section of the airflow, or, better, with a higher ratio between the cross section of the object under test and that of the flow. However they have a lower energetic efficiency, so a tradeoff between the decreased size of the plant, which is strictly linked to the area of the airstream, and the decreased efficiency must be done. Also the shape of the test chamber must be designed with care, as it is important to achieve the best results with a size as small as possible.

Peculiar difficulties encountered in wind tunnel test of vehicles are linked to the presence of the ground, which should move with respect to the vehicle, and with the rotation of the wheels. In spite of many attempts to solve this problem and of the use of sophisticated devices, most testing is performed with a standard arrangement in which the vehicle is simply supported on the floor of the tunnel, or, better, of the balance fixture, with stationary wheels. In the case of standard cars this does not introduce large errors, except for what the study of the flow under the vehicle is concerned.

For the measurement of the drag alone it is possible to perform road tests in which the vehicle is allowed to coast down from various speeds on level road: From the deceleration law and the knowledge of the rolling resistance and of all other forms of drag it is possible to obtain good estimates of the aerodynamic drag. Road tests can be useful to calibrate results obtained

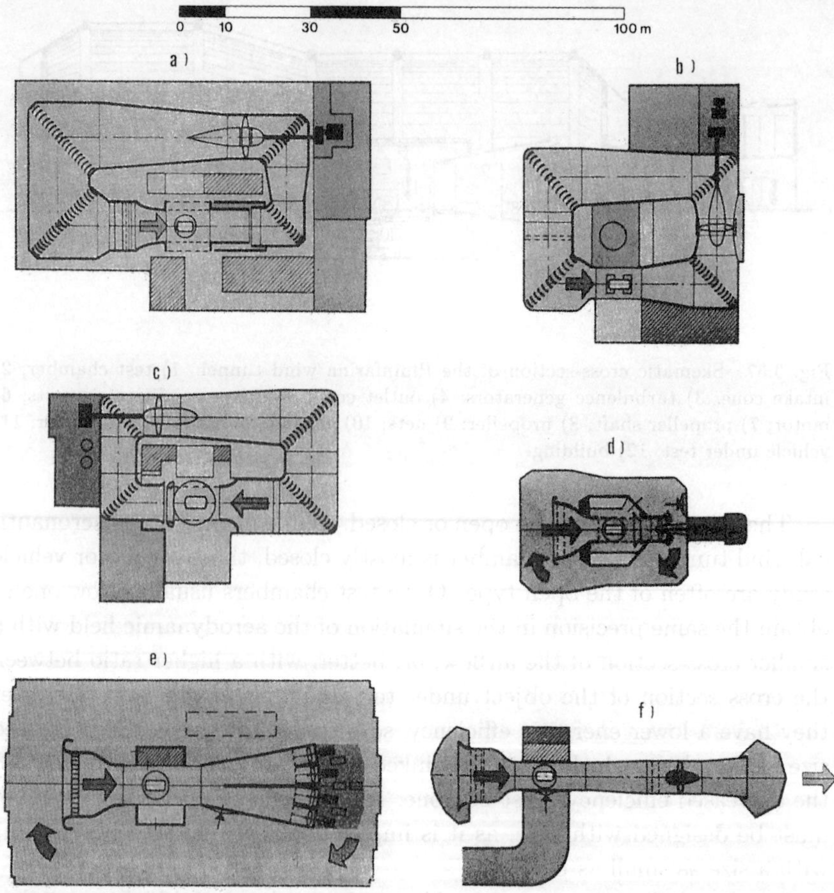

Fig. 3.38 Plans of six wind tunnels for full-scale testing of road vehicles. (a) Volkwagen, Wolfsburg, Germany; (b) Ford, Detroit, USA; (c) F.K.F.S., Stuttgard, Germany; (d) Pininfarina, Torino, Italy; (e) M.I.R.A., Lindley Nuneaton, G.B., (f) Nissan, Oppama Yokosuka, Japan.

in the wind tunnel, to assess the importance of the relative velocity of air and ground and of the rotation of wheels.

In the last twenty years a rapid progress of computational aerodynamics allowed to simulate numerically aerodynamic fields of increasing complexity. Much work has been performed to adapt the methods of numerical aerodynamics from aeronautics to vehicle technology and good success has been obtained in spite of the great complexities due mainly to the detached flow and the large wake.

Numerical aerodynamics is mainly based on the discretization of the aerodynamic field, through the finite difference, finite elements or boundary elements approach. Also combined approaches, allowing to exploit the advantages of the different methods, are possible.

The surface of the body is subdivided in a number of elements (panels); if the first two of the mentioned methods are used the whole aerodynamic field, up to a certain distance from the vehicle, must be discretized as well. The equations are then written in discrete form, this step usually resulting in a very large number of equations and unknown. The main difficulty is to compute the points in which the flow detaches from the body; this aspect leads to a strongly nonlinear problem which requires an iterative solution.

The large computational work makes it necessary to use very powerful computers and a great deal of computer time on the supercomputers used in the motor industry is devoted to aerodynamic simulation. The increasing power of computers and the progress in the field of computational aerodynamics are making these methods a very powerful and widespread tool, both for the direct evaluation of pressure distributions and forces and moments and as a support to the interpretation of the results obtained in wind tunnel testing.

Chapter 4

Longitudinal dynamics

4.1 Load distribution on the ground

4.1.1 *Vehicles with two axles*

Consider the vehicle as a rigid body or, in case of articulated vehicles or vehicles with a trailer, as made by a number of rigid bodies. A vehicle with more than three wheels is statically indeterminate unless the presence of the suspensions or the compliance of the body is considered. However, if the system is symmetrical with respect to xz plane[1], a two-axle vehicle can be considered as a beam on two supports and the normal forces F_{z_1} and F_{z_2} on the axles can be easily determined.

With the vehicle at standstill on level road the normal forces are

$$\begin{cases} F_{z_1} = mg\epsilon_{0_1} \\ F_{z_2} = mg\epsilon_{0_2} \end{cases} \quad \text{where} \quad \begin{cases} \epsilon_{0_1} = b/l \\ \epsilon_{0_2} = a/l \, . \end{cases} \tag{4.1}$$

Equation (4.1) can also be used to find the position of the centre of gravity of the vehicle by simply measuring the load on the ground on the two axles: $a = lF_{z_2}/mg$.

The forces acting on a two-axle vehicle moving on straight road with longitudinal grade angle α (positive when moving uphill) are sketched in Fig. 4.1. Note that x-axis is assumed to be parallel to the road surface.

Taking into account also the inertia force $-m\dot{V}$ acting in x direction on the centre of mass, the dynamic equilibrium equations for translations in x

[1]In the present section, where the longitudinal dynamics is studied, a complete symmetry with respect to xz plane is assumed: The loads on each wheel are respectively $F_{z_1}/2$ and $F_{z_2}/2$ for the front and the rear wheels. To simplify the equations x-axis is assumed to be parallel to the road surface.

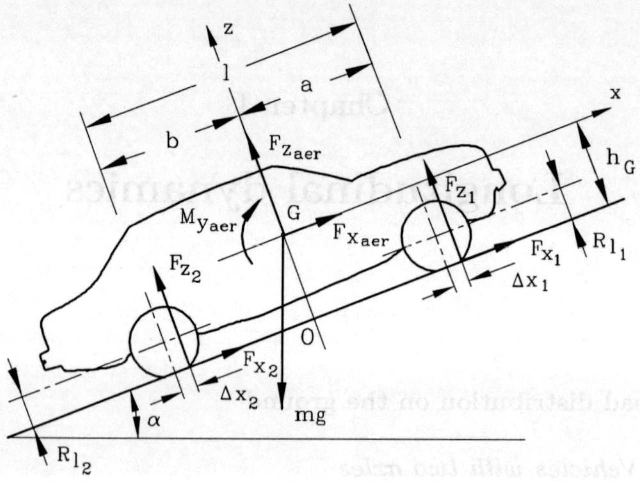

Fig. 4.1 Forces acting on a vehicle moving on an inclined road.

and z direction and rotations about point O are

$$\begin{cases} F_{x_1} + F_{x_2} + F_{x_{aer}} - mg\sin(\alpha) = m\dot{V} \\ F_{z_1} + F_{z_2} + F_{z_{aer}} - mg\cos(\alpha) = 0 \\ F_{z_1}(a + \Delta x_1) - F_{z_2}(b - \Delta x_2) + mgh_G\sin(\alpha) - M_{aer} + |F_{x_{aer}}|h_G = -mh_G\dot{V} \, . \end{cases} \tag{4.2}$$

If all the rolling resistance is ascribed completely to the forward displacement of the resultant F_{z_i} of contact pressures σ_z, distances Δx_i can be easily computed as

$$\Delta x_i = R_{l_i} f = R_{l_i}(f_0 + KV^2) \, . \tag{4.3}$$

Except the case of vehicles with different wheels on the various axles, as in the case of F-1 racers, the values of Δx_i are all equal.

The second and third Eq. (4.2) can be solved in the normal forces acting on the axles, yielding

$$\begin{cases} F_{z_1} = mg\dfrac{(b-\Delta x_2)\cos(\alpha) - h_G\sin(\alpha) - K_1V^2 - \dfrac{h_G}{g}\dot{V}}{l+\Delta x_1 - \Delta x_2} \\ F_{z_2} = mg\dfrac{(a+\Delta x_1)\cos(\alpha) + h_G\sin(\alpha) - K_2V^2 + \dfrac{h_G}{g}\dot{V}}{l+\Delta x_1 - \Delta x_2} \end{cases} \tag{4.4}$$

where

$$\begin{cases} K_1 = \dfrac{\rho S}{2mg}\left[C_x h_G - lC_{M_y} + (b-\Delta x_2)C_z\right] \\ K_2 = \dfrac{\rho S}{2mg}\left[-C_x h_G + lC_{M_y} + (a+\Delta x_1)C_z\right]. \end{cases}$$

The values of Δx_i are usually quite small (in particular their difference is usually equal to zero) and can be neglected. If considered, they introduce a further weak dependence of the vertical loads on the square of the speed, owing to the term KV^2 in the rolling resistance.

Example 4.1

Compute the force distribution on the ground of the small car of Appendix 1.1 at sea level, with standard pressure and temperature, in the following conditions

a) at standstill on level road;

b) driving at 100 km/h on level road;

c) driving at 70 km/h on a 10% grade;

d) braking with a deceleration of 0.4 g on level road at a speed of 100 km/h.

The air density in the mentioned conditions is 1.2258 kg/m^3.

a) Using Eq. (4.1), the static load distribution between the axles is

$$\epsilon_{0_1} = 0.597, \qquad \epsilon_{0_2} = 0.403.$$

The forces acting on the axles are then

$$F_{z_1} = 4863 \text{ N}, \qquad F_{z_2} = 3280 \text{ N}.$$

b) From Eq. (4.3), at 100 km/h = 27.78 m/s the value of Δx is 4.6 mm for all tires. This value is so small that it could be neglected; it will however be considered in the following computations.

Constants K_1 and K_2 are easily computed

$$K_1 = 8.505 \times 10^{-6} \text{ s}^2/\text{m}, \qquad K_2 = -5.869 \times 10^{-5} \text{ s}^2/\text{m}.$$

The forces acting on the axles are then

$$F_{z_1} = 4820 \text{ N}, \qquad F_{z_2} = 3491 \text{ N}.$$

c) A 10% grade corresponds to a grade angle $\alpha = 5.7°$. Operating in the same way, at 70 km/h = 19.44 m/s the value of Δx is 4.0 mm for all tires. The other results are

$$K_1 = 8.490 \times 10^{-6} \text{ s}^2/\text{m}, \qquad K_2 = -5.867 \times 10^{-5} \text{ s}^2/\text{m}.$$

$$F_{z_1} = 4643 \text{ N}, \qquad F_{z_2} = 3542 \text{ N}.$$

d) The acceleration is $\dot{V} = -3.924 \text{ m/s}^2$. As the speed is the same as in case b), the same values for Δx, K_1 and K_2 hold. The forces are

$$F_{z_1} = 5498 \text{ N}, \qquad F_{z_2} = 2813 \text{ N}.$$

4.1.2 *Vehicles with more than two axles*

If more than two axles are present, even in symmetrical conditions the system remains statically indeterminate and it is necessary to take into account the compliance of the suspensions (Fig. 4.2a). The equilibrium equations (4.2) still hold, provided that the terms $F_{x_1} + F_{x_2}$, $F_{z_1} + F_{z_2}$ and $F_{z_1}(a+\Delta x_1) - F_{z_2}(b-\Delta x_2)$ are substituted by $\sum_{\forall i} F_{x_i}$, $\sum_{\forall i} F_{z_i}$ and $\sum_{\forall i} F_{z_1}(x_i+\Delta x_i)$, where distances x_i are positive for axles located forward than the centre of mass and negative otherwise.

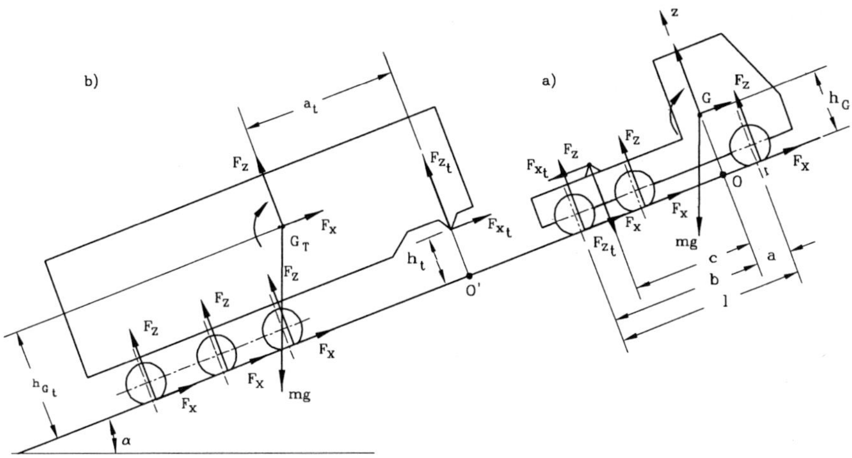

Fig. 4.2 Forces acting on an articulated vehicle moving on an inclined road. (a) Tractor or Vehicle with more than two axles; (b) trailer.

For the computation of the normal loads on the ground a number $(n-2)$ of equations, where n is the total number of axles, must be added. They simply express the condition that the displacements of the various suspensions are compatible

$$\frac{b+x_i}{K_1}F_{z_1}+\frac{a-x_i}{K_n}F_{z_n}-\frac{l}{K_i}F_{z_i}-\frac{b+x_i}{K_1}(F_{z_1})_0+ \\ -\frac{a-x_i}{K_n}(F_{z_n})_0+\frac{l}{K_i}(F_{z_i})_0=0\ , \tag{4.5}$$

$$\text{for } i=2,\dots,n-1\ ,$$

where K_i is the linearized stiffness of the i-th suspension, possibly including the compliance of the tires, $(F_{z_i})_0$ is the normal force of the same axle in any reference condition, $a=x_1$, $b=-x_n$ and $l=a+b$. The mentioned reference condition can be referred to any value of the load or any position of the centre of gravity, provided that the values of the linearized stiffnesses are the same as those in the actual condition. If the springs are linear and the suspensions are such that if the body is lifted there is a position in which all wheels just touch the ground (neglecting the weight of the axles) forces $(F_{z_i})_0$ can all be set to zero.

Equations (4.5), together with the second and third Eq. (4.2) form a set of n equations which can be solved to yield the n normal forces acting on the axles. Care must be given to the fact that in practice forces F_{z_i} can never become negative: If a negative value is obtained it means that the relevant axle leaves contact with the ground and the computation must be repeated after setting to zero the force due to the relevant axle. The procedure is repeated until no negative force is present.

4.1.3 *Articulated vehicles*

In the case of articulated vehicles with a tractor with two axles and trailers with no more than a single axle each, the computation is straightforward.

In the general case of multi-axle vehicles (Fig. 4.2), the equilibrium equations of the tractor are

$$\left\{\begin{array}{l}\displaystyle\sum_{i=1}^{n} F_{x_i} - F_{x_t} + F_{x_{aer}} - mg\sin(\alpha) = m\dot{V}\\ \displaystyle\sum_{i=1}^{n} F_{z_i} - F_{z_t} + F_{z_{aer}} - mg\cos(\alpha) = 0\\ \displaystyle\sum_{i=1}^{n} F_{z_i}(x_i + \Delta x_i) + F_{z_t}c + F_{x_t}h_t + mgh_G\sin(\alpha) - M_{aer} +\\ \qquad +|F_{x_{aer}}|h_G = -mh_G\dot{V}\ ,\end{array}\right. \tag{4.6}$$

where forces F_{x_t} and F_{z_t} are those the tractor exerts on the trailer, as in the figure, the number of axles of the tractor is assumed to be n, the moments are computed with reference to the point O and the aerodynamic forces and moments are those exerted on the tractor only.

Similarly, the equilibrium equation of the trailer are

$$\left\{\begin{array}{l}\displaystyle\sum_{i=1}^{m} F_{x_i} + F_{x_t} + F_{x_{T_{aer}}} - m_T g\sin(\alpha) = m_T\dot{V}\\ \displaystyle\sum_{i=1}^{m} F_{z_i} + F_{z_t} + F_{z_{T_{aer}}} - m_T g\cos(\alpha) = 0\\ \displaystyle\sum_{i=1}^{m} F_{z_i}(x_i + \Delta x_i) - F_{x_t}h_t + m_T g h_{G_T}\sin(\alpha) + m_T g a_T\cos(\alpha) - M_{T_{aer}} +\\ \qquad +|F_{x_{T_{aer}}}|h_{G_T} = -m_T h_{G_T}\dot{V}\ ,\end{array}\right. \tag{4.7}$$

where the number of axles of the trailer is assumed to be m, the moments are computed with reference to the point O$'$, the aerodynamic forces and moments are those exerted on the trailer only and x_i are the coordinates of the axle in the reference frame centreed in O′. Note that all x_i are usually negative.

The last two equations (4.6), together with the last two equations (4.7) are sufficient only on level road at standstill, when force F_{x_t} vanishes. If it is different from zero also the first equation (4.6) must be used. However the forces F_{x_i} it contains are not known since they depend on the normal forces F_{z_i}. A simple iterative scheme can be used, computing the normal forces with $F_{x_t} = 0$ and then repeating the computation until a stable solution is found. If the wheels of the trailer exert driving or braking forces, these forces must also be introduced in the computation.

If the tractor has more than two axles or the trailer has more than one, additional equations must be introduced. For the tractor $(n-2)$ equations, n being the number of axles, are equations (4.5) while the $(m-1)$ equations for the trailer, where m is the number of its axles are

$$\begin{gathered}\frac{(a+c)(x_m-x_i)}{lK_1}F_{z_1}+\frac{(b-c)(x_m-x_i)}{lK_n}F_{z_n}+\frac{x_i}{K_{t_m}}F_{z_{tm}}-\frac{x_m}{K_{t_i}}F_{z_{t_i}}+\\-\frac{(a+c)(x_m-x_i)}{lK_1}(F_{z_1})_0-\frac{(b-c)(x_m-x_i)}{lK_n}(F_{z_n})_0+\\-\frac{x_i}{K_{t_m}}(F_{z_{tm}})_0+\frac{x_m}{K_{t_i}}(F_{z_{t_i}})_0=0\,,\end{gathered}\tag{4.8}$$

$$\text{for } i=1,\ldots,m-1\,,$$

where K_{t_i} and $F_{z_{t_i}}$ are the linearized stiffness of the i-th suspension of the trailer and the force acting on it and $(F_{z_{t_i}})_0$ is the normal force in the same axle in any reference condition.

The number of unknowns and equations is then equal to the total number of axles plus one, as also the normal force the two parts of the vehicle exchange is unknown. Force F_{x_t}, when does not vanish, must be computed iteratively as seen above.

Example 4.2

Compute the force distribution on the ground of the five-axle articulated truck of Appendix 1.4 at sea level, with standard pressure and temperature, in the following conditions

a) at standstill on level road;

b) at standstill on a 10% grade;

c) driving at 70 km/h on a 10% grade;

The air density in the mentioned conditions is 1.2258 kg/m^3.

a) The static load distribution on level road can be computed directly, as the horizontal force exchanged between the two parts of the vehicle vanishes. The unknowns are six, the loads of the five axles and the vertical force exchanged between tractor and trailer. They can be computed from the following set of linear equations

$$\begin{bmatrix} 1 & 1 & 0 & 0 & 0 & -1 \\ 1.175 & -2.31 & 0 & 0 & 0 & 1.86 \\ 0 & 0 & 1 & 1 & 1 & 1 \\ 0 & 0 & -6.135 & -7.395 & -8.715 & 0 \\ -1.070\times 10^{-3} & -1.109\times 10^{-4} & 4.054\times 10^{-3} & 0 & -4.446\times 10^{-3} & 0 \\ -5.474\times 10^{-4} & -6.087\times 10^{-5} & 0 & 4.054\times 10^{-3} & -5.359\times 10^{-3} & 0 \end{bmatrix}$$

$$\times \begin{Bmatrix} F_{z_1} \\ F_{z_2} \\ F_{z_{t_1}} \\ F_{z_{t_2}} \\ F_{z_{t_3}} \\ F_{z_t} \end{Bmatrix} = \begin{Bmatrix} 70100 \\ 0 \\ 313900 \\ -1597900 \\ 0 \\ 0 \end{Bmatrix} .$$

The forces acting on the axles are then of 58.660 kN, 105.700 kN, 80.060 kN, 83.600 kN and 56.050 kN. The force at the tractor-trailer connection is 94.210 kN.

b) A 10% grade corresponds to a grade angle $\alpha = 5.7^\circ$. Also in this case the load distribution can be computed directly as the horizontal force exchanged between the two parts of the vehicle do not depend on the normal forces. Operating in the same way as in the previous case, the forces acting on the axles are 45.050 kN, 115.720 kN, 78.900 kN, 84.490 kN and 58.000 kN. The forces at the tractor-trailer connection are 90.980 kN in vertical direction and 31.236 kN in horizontal direction.

c) At 70 km/h = 19.44 m/s the value of Δx is 3.7 mm for all tires. In this case, owing to rolling resistance, an iterative solution must be obtained. However the convergence is very fast, as only 5 iterations are needed to reach a difference between the results at the i-th and at the $(i-1)$-th iteration smaller than 10^{-6} in relative terms. The other results are not much different from those obtained in the previous case: The forces on the axles are 43.980 kN, 116.970 kN, 78.710 kN, 84.440 kN and 58.060 kN; those at the tractor-trailer connection are 91.150 kN in vertical direction and 33.440 kN in horizontal direction.

Note that the matrix of the coefficients of the relevant set of equations is the same in all cases.

4.2 Steady state motion on straight road

4.2.1 *Total resistance to motion*

Consider a vehicle moving at constant speed on a straight and level road. The forces which must be overcome to maintain its speed constant are aerodynamic drag and rolling resistance. With increasing speed the importance of the former grows; at a given value of the speed aerodynamic drag becomes more important than rolling resistance. This speed is lower for small cars while for larger vehicles, particularly for trucks at full load, rolling resistance is the main form of drag. This is also due to the fact that usually the mass of the vehicle grows with its size more rapidly than the area of its cross section.

If the road is not level, the component of weight acting in a direction parallel to the velocity V, i.e. the grade force, must be added to the resistance to motion: It becomes far more important than all other forms of drag even for moderate values of the grade (Fig. 4.1). Taking into account also aerodynamic lift, the total resistance to motion, or road load, as it is commonly referred to, can be written in the form

$$R = \left[mg\cos(\alpha) - \frac{1}{2}\rho V^2 S C_z\right](f_0 + KV^2) + \frac{1}{2}\rho V^2 S C_x + mg\sin(\alpha)\ , \tag{4.9}$$

i.e.

$$R = A + BV^2 + CV^4\ , \tag{4.10}$$

where

$$A = mg\left[f_0\cos(\alpha) + \sin(\alpha)\right]\ ,$$

$$B = mgK\cos(\alpha) + \frac{1}{2}\rho S[C_x - C_z f_0]\ ,$$

$$C = -\frac{1}{2}\rho SKC_z \,.$$

The last term in Eq. (4.10) becomes important only at very high speed in case of vehicles with strong negative lift: It is usually neglected except for racing cars.

Since the grade angle of roads open to vehicular traffic is usually not very large, it is possible to assume that $\cos(\alpha) \approx 1$ and $\sin(\alpha) \approx \tan(\alpha) \approx i$, where i is the grade of the road. In this case coefficient B is independent of the grade of the road and $A \approx mg(f_0 + i)$ depends linearly on it.

The power needed to move at constant speed V is simply obtained by multiplying the road load given by Eq. (4.10) by the value of the velocity

$$P_r = VR = AV + BV^3 + CV^5 \,. \tag{4.11}$$

Motion at constant speed is possible only if the power available at the wheels at least equals the required power given by Eq. (4.11). This means that the engine must supply a sufficient power, taking into account also the losses in the transmission, and that the road-wheel contact must be able to transmit it.

Example 4.3

Plot the curves of the road load of the car of Appendix 1.1 on level road and on a 10% grade. Plot the curve of the power needed for constant speed driving on level road.

The results obtained through Eq. (4.9) are reported in Figs. 4.3 and 4.4.

Example 4.4

Plot the curves of the road load of the articulated truck of Appendix 1.4 on level road and on a 10% grade.

The results obtained through Eq. (4.9) are reported in Fig. 4.5. Note that in this case aerodynamic drag amounts to a relatively small portion of the road load and that on a 10% grade the grade force is very high.

4.2.2 *Characteristics of internal combustion engines*

As most road vehicles are powered by reciprocating internal combustion engines, their characteristics will be summarized in the present section.

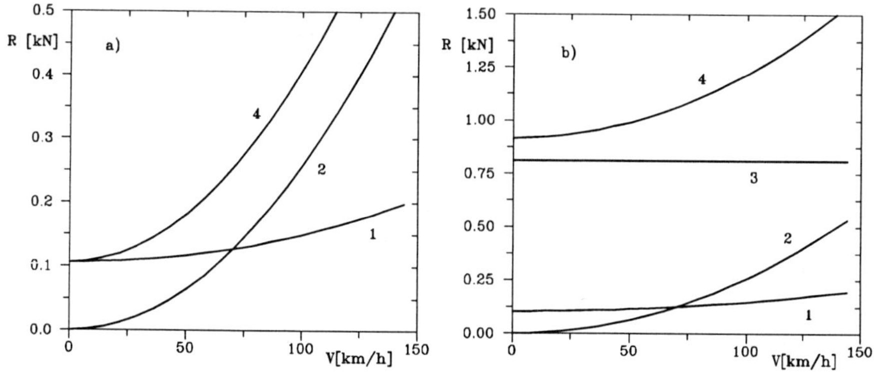

Fig. 4.3 Rolling resistance (curve 1), aerodynamic drag (2), grade force (3) and total road load (4) for the small car of Appendix 1.1 on level road (a) and on a 10% grade (b).

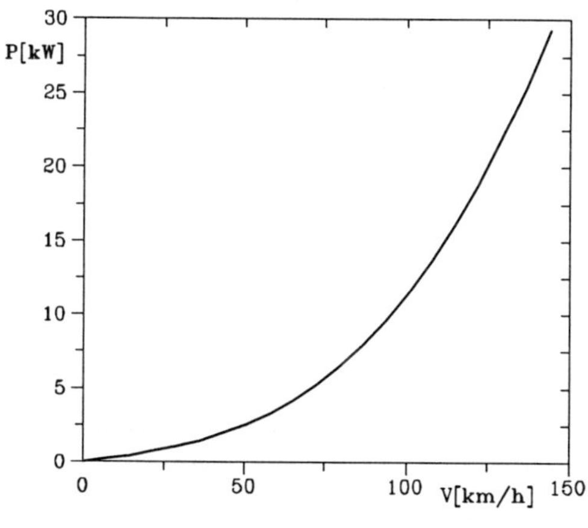

Fig. 4.4 Power needed for constant speed driving on level road for the small car of Appendix 1.1.

Apart from the action of the throttle control, the power supplied by the engine depends mainly on the rotational speed. The performance of an internal combustion engine is usually summarized in a single map plotted in a plane whose axes are the rotational speed ω_e and the power P_e (Fig. 4.6). Often the former is reported in rpm and the latter in HP; in the present text however S.I. units, i.e. rad/s and W (or kW) will be used.

The plot is limited by the curve $P_e(\omega_e)$ which expresses the maximum

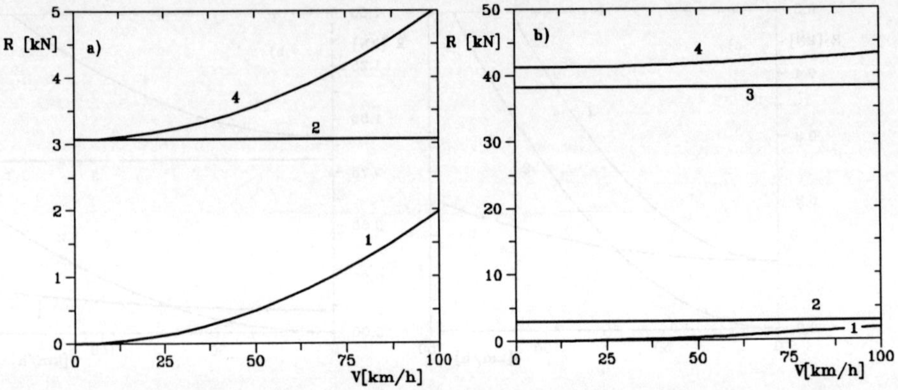

Fig. 4.5 Aerodynamic drag (curve 1), rolling resistance (2), grade force (3) and total road load (4) for the articulated truck of Appendix 1.4 on level road (a) and on a 10% grade (b).

power that the engine can supply as a function of the speed. Such curve is typical of any particular engine and must be obtained experimentally. However, for a first modelization of the vehicle, it is possible to approximate it with a polynomial, usually with terms up to the third power

$$P_e = \sum_{i=0}^{3} P_i \omega_e^i \,. \tag{4.12}$$

The values of coefficients P_i can easily be obtained from the experimental curve of an engine similar to the one under study, possibly by scaling it with the values of the maximum power and maximum speed expected. In the literature it is possible to find some values of the coefficients which can be used as a first rough approximation. M.D. Artamonov *et al.*[2] suggest the following values

$$P_0 = 0 \,, \qquad P_3 = -P_{max}/\omega_{max}^3$$

for all types of internal combustion engines and

$$P_1 = P_{max}/\omega_{max} \,, \qquad P_2 = P_{max}/\omega_{max}^2 \,,$$

[2]M.D. Artamonov et al. *Motor vehicles, fundamentals and design*, Mir, Moscow, 1976.

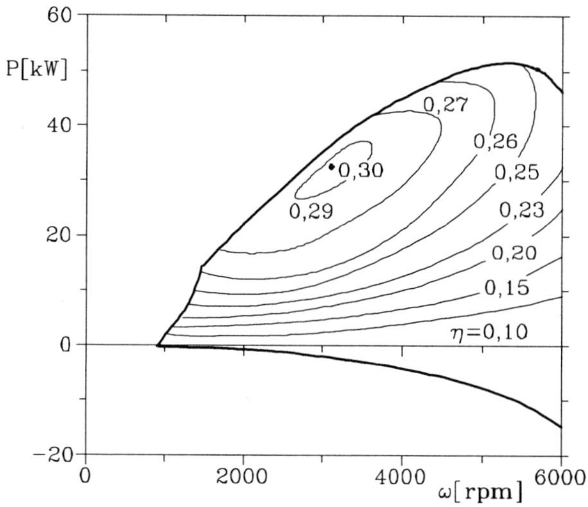

Fig. 4.6 Map of a spark ignition internal combustion engine, with constant efficiency curves.

for spark ignition engines,

$$P_1 = 0.6P_{max}/\omega_{max} \ , \qquad\qquad P_2 = 1.4 \ , P_{max}/\omega_{max}^2$$

for indirect injection diesel engines and

$$P_1 = 0.87P_{max}/\omega_{max} \ , \qquad\qquad P_2 = 1.13P_{max}/\omega_{max}^2$$

for direct injection diesel engines. In these formulae ω_{max} is the speed at which the power reaches its maximum value.

The driving torque of the engine is simply

$$T_e = \frac{P_e}{\omega_e} \ , \tag{4.13}$$

or, if coefficient P_0 vanishes,

$$T_e = \sum_{i=1}^{3} P_i \omega_e^{i-1} \ . \tag{4.14}$$

All points which lie below the maximum power curve are possible working condition of the engine, when operating with the throttle control par-

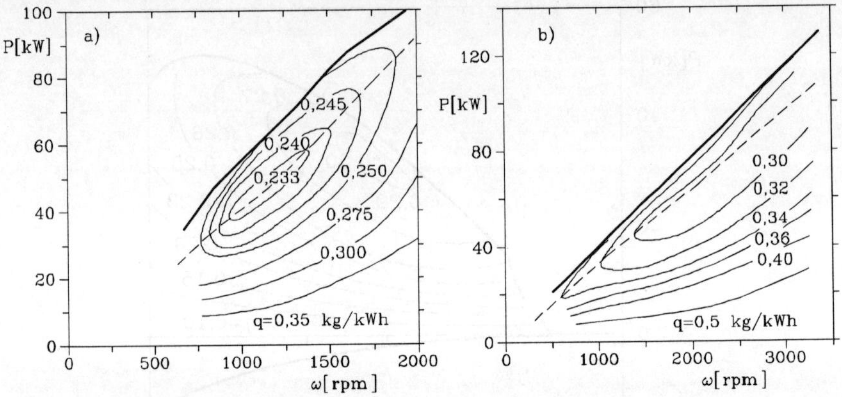

Fig. 4.7 Map of a diesel internal combustion engine (a) and of a spark ignition engine (b), with constant specific fuel consumption curves. A line connecting the points characterized by minimum consumption has also been plotted.

tially closed. The plot stops on the left at a velocity below which smooth working is not possible.

All possible working conditions are characterized by a value of the efficiency of the conversion of the chemical energy of the fuel into mechanical energy at the output shaft. On the map a set of curves connecting the points characterized by some selected values of the efficiency are usually plotted (Fig. 4.6). As clearly seen from the figure, the efficiency of a spark ignition engine reaches its maximum in conditions very close to operation with wide open throttle at a speed close to that characterized by maximum torque and reduces quickly with decreasing power at any value of the speed. In case of Diesel engines this reduction is less marked.

Instead of mapping the efficiency, often curves for constant specific fuel consumption $q = 1/H\eta_e$, were H is the thermal value of the fuel, are reported (Fig. 4.7).

The correct S.I. units for the specific fuel consumption is kg/J, i.e. s^2/m^2, while the common practical units are still g/HPh or g/KWh. If the thermal value of the fuel is equal to 4.4×10^7 J/kg, it follows that

$$q = 2.272 \times 10^{-8}/\eta_e \text{ [kg/J]} = 60.16/\eta_e \text{ [g/HPh]} = 81.79/\eta_c \text{ [g/kWh]}.$$

Example 4.5

Compute the coefficients of a third degree polynomial which approximates the engine power curve of the vehicle of Appendix

1.1. Compare the curve so obtained with the experimental curve and that obtained through the coefficients supplied by Artamonov. Plot on the same graph the engine torque curve and that of the specific fuel consumption.

By reading from the graph points equispaced of 250 rpm and using a standard least-squares routine, the following equation is obtained

$$P = -10.628 + 0.1506\omega - 9.5436 \times 10^{-5}\omega^2 - 5.0521 \times 10^{-8}\omega^3 ,$$

where ω is expressed in rad/s and P is expressed in kW.

Using the coefficients supplied by Artamonov for spark ignition engines the equation is

$$P = 0.7024\omega + 1.290 \times 10^{-4}\omega^2 - 2.369 \times 10^{-7}\omega^3 .$$

The relevant curves are plotted in Fig. 4.8. Note that both expressions approximate quite well the experimental curve, even if the coefficients are different.

The internal combustion engine is connected to the wheels through a transmission system which includes a clutch (or a torque converter) a number of gear wheels and some joints. The efficiency η_t of the transmission must be accounted for and the power available at the wheels is

$$P_a = P_e\eta_t. \tag{4.15}$$

The efficiency of the transmission can be estimated easily, although only roughly, in the case of a mechanical transmission while in the case a torque converter is present it depends on the working conditions of the latter.

In the first case a fixed efficiency can be given to each pair of gear wheels and to each joint, with values of 0.97 for cylindrical gears, 0.95 for conical or hypoid gears and 0.99 for universal or homokinetic (constant velocity) joints being often used. These values are only rough averages, as the efficiency of gear wheels depends on factors like lubrication condition, temperature, wear and precision of setup, and above all the power which is transmitted. However, as the efficiency is quite high, some changes in the power losses in the transmission do not affect much the available power. Consider for example the transmission layouts shown in Fig. 4.9. In the

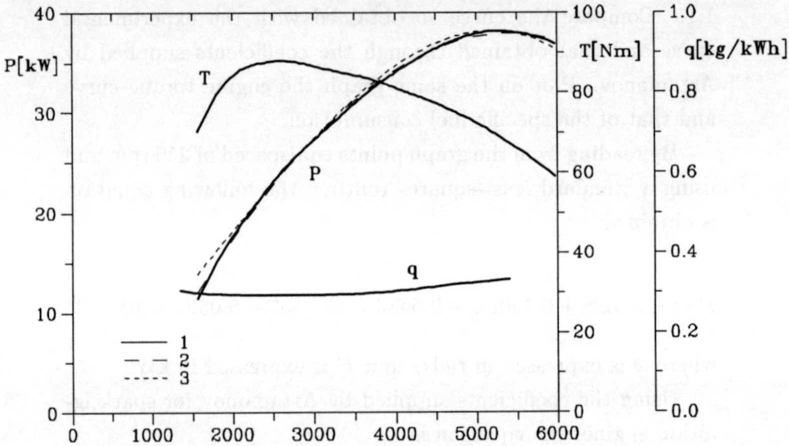

Fig. 4.8 Engine power curve for the car of Appendix 1.1. (1) Experimental curve, (2) third-power least square fit, (3) third power polynomial with coefficients computed as suggested by Artamonov *et al.* Also the torque and the specific fuel consumption are reported as functions of the speed.

case of the layout (a) for a passenger vehicle with front engine and rear drive the value of the efficiency is 0.87 in all lower gears, when the power flows through two pairs of gear wheels in the gearbox and 0.93 in top gear when the input and output shafts of the gearbox are directly connected. Intermediate values of about 0.91 or 0.93 can be found for layouts with engine on the driving wheels respectively for transversal and longitudinal (Fig. 4.9b) engine position. Note that the internal gears of the differential are not accounted for since in straight running they have no relative velocity.

An expression of the efficiency which can be used to account for its dependence on the speed but neglects its dependence with the transmitted torque is

$$\eta_t = \left(0.96 - 7.07 \times 10^{-4} V - 2.9 \times 10^{-5} V^2\right) \left\{0.998[1 - 0.007(n_g - i)] - 1.965 \times 10^{-4}(2.08^{n_g - i})V\right\}, \tag{4.16}$$

where the speed is expressed in m/s, n_g is the number of gears and i is the particular gear used. The first term describes the efficiency of a typical drive axle while the second that of a typical gearbox.

The relationship between the angular velocity of the engine and the

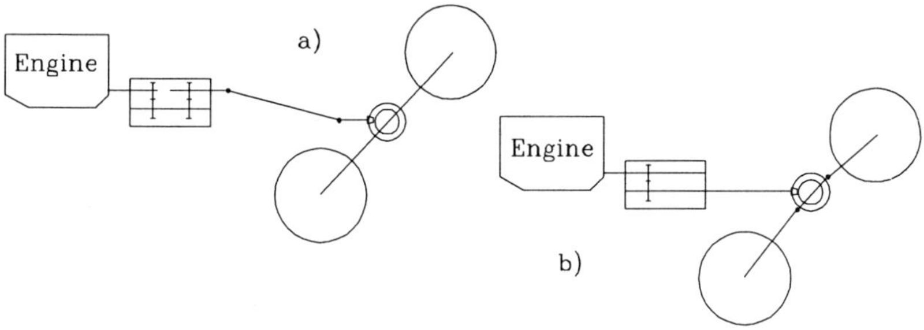

Fig. 4.9 Transmission layouts for vehicles with front engine and rear drive (a) and with the engine on the driving wheels in longitudinal position (b).

velocity of the vehicle is simply given by

$$V = \omega_e R_e \tau_g \tau_f \ , \tag{4.17}$$

where τ_g and τ_f are the overall transmission ratios (defined as the ratio between the velocities of the output and the input shafts) of the gear box and the axle respectively. They are usually smaller than 1.

Once that the transmission ratios have been stated, the curve of the power available at the wheels can be expressed as a function of the speed of the vehicle and plotted together with that of the power needed for motion at constant speed.

The situation is more complicated if a torque converter or a continuously variable transmission (CVT) is used. Two parameters, namely the torque ratio (ν) and the speed ratio (τ_c) between the output and the input shafts, can be defined. The latter is obviously equivalent to the transmission ratio and has a value smaller than 1. A plot of the efficiency (η_c) and the torque ratio of a torque converter versus the speed ratio is reported in Fig. 4.10.

Accounting also for the presence of a torque converter, the power available at the wheels and the relationship between the speed of the vehicle and that of the engine are

$$P_a = \eta_t \eta_c P_e \ , \qquad V = \omega_e R_e \tau_g \tau_f \tau_c \ . \tag{4.18}$$

Note that if the curves of the required and the available power are plotted on a logarithmic diagram, changing the overall transmission ratio results in a translation along the V-axis of the engine curve and changing

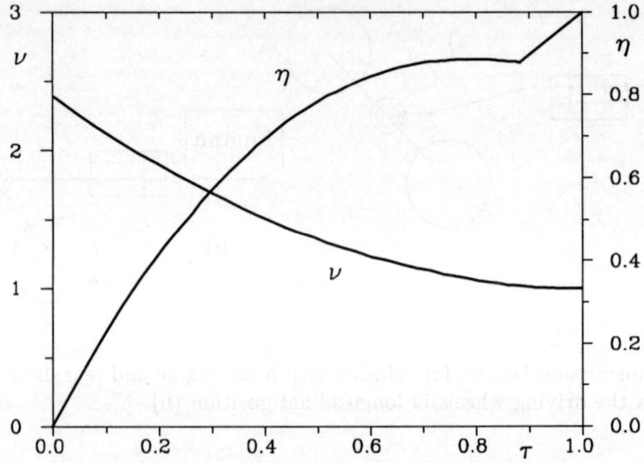

Fig. 4.10 Efficiency η_c and torque ratio ν versus the speed ratio τ_c for a torque converter.

the efficiency of the transmission results in a translation along the P-axis (Fig. 4.11). If a torque converter or a CVT is present, the position of the curve is a function of ratio τ_c and then it is possible to identify a zone in the VP plane in which all possible working conditions are included.

4.2.3 *Maximum power which can be transferred to the road*

The power needed to overcome the road load must be transferred through the road-wheel contact. As it increases with increasing speed and grade of the road, there is a limit to the maximum speed which can be reached and the maximum grade which can be managed due to the maximum driving force the vehicle can exert, even if no limit would exist on the power supplied by the engine.

The maximum power which can be transferred by the vehicle is

$$P_{max} = V \sum_{\forall i} F_{z_i} \mu_{i_p} , \tag{4.19}$$

where the sum is extended to all the driving wheels.

If the maximum longitudinal force coefficient μ_{i_p} and the load acting on the driving wheels were independent of the speed, the maximum power would increase linearly with V. The optimum engine characteristic $P_e(\omega_e)$ for a vehicle with a transmission with fixed ratio would be a linear character-

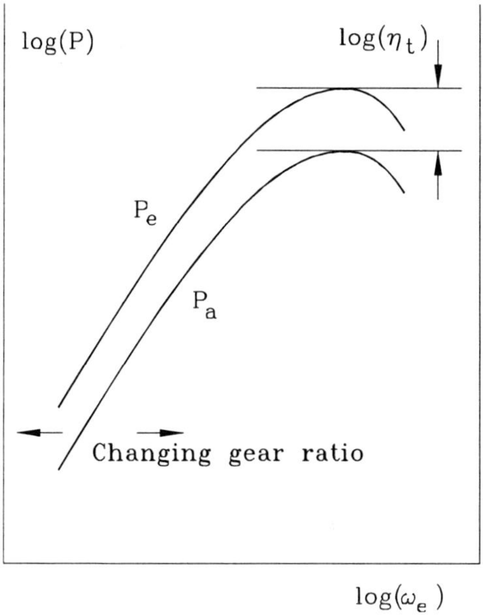

Fig. 4.11 Curves of maximum engine power and power available at the wheels plotted with logarithmic scales. Changing the efficiency of the transmission and the overall transmission ratio.

istic. This is however not the case as the situation is far more complicated.

Consider firstly the case of a vehicle with two axles, all of which are driving and assume that all wheels work with the same longitudinal slip, i.e. that the values of μ_i are all equal. This situation will be referred here to as "ideal driving force".

The maximum power which can be transferred to the road is then

$$P_{max} = V\mu_{i_p}\left(mg\cos(\alpha) - \frac{1}{2}\rho V^2 SC_z\right) . \tag{4.20}$$

To model in a very simple way the decrease of driving force occurring with increasing speed, it is possible to use a linear law

$$\mu_{i_p} = c_1 - c_2 V . \tag{4.21}$$

The maximum grade which can be managed at low speed is immediately computed. Assuming that rolling resistance of all wheels, which are driving wheels, is due to the forward displacement of the normal force F_z at the road-wheel contact and hence is overcome directly by the driving torque exerted by the engine, the only road load which must be overcome at the wheel-road contact at low speed is the grade load. By equating the power required for motion (Eq. (4.11) to the maximum power which can be transferred (Eq. 4.20), the maximum grade is readily obtained

$$\tan(\alpha_{max}) = i_{max} = \mu_{x_p} . \tag{4.22}$$

In a similar way the maximum speed can be obtained. By equating the power required for motion at constant speed (except that related with rolling resistance) to Eq. (4.20) and using Eq. (4.21) for expressing the decrease of the available driving force with the speed, the maximum speed can be obtained from the cubic equation

$$C_z c_2 V^3 + \left(C_z c_1 + C_x\right) V^2 - \frac{2mg}{\rho S}\left[(c_1 - c_2 V)\cos(\alpha) - \sin(\alpha)\right] = 0 . \tag{4.23}$$

The values of the slope and speed so obtained are only ideal reference values, as were obtained assuming that the longitudinal slip of all wheels were equal.

The forces acting on the driving axles can be computed using the procedure shown in Section 4.1. Generally speaking they depend on the static load distribution but also on the speed and the acceleration.

In the case of a vehicle with two axles, both driving, the ratio $K_T = F_{x_1}/F_{x_2}$ between the driving force at the front wheels and that at rear wheels is usually a constant of the vehicle. If the wheels have all the same diameter it coincides with the ratio between the driving torques supplied to the two axles. Assume that the two axles have tires of the same type and work on patches of road with the same characteristics. If $K_T F_{z_2}/F_{z_1}$ is greater than unity the limit conditions occur at the front wheels. At the onset of slipping the power which can be transferred to the road is then

$$P_{max} = V \mu_{x_p} F_{z_1} \frac{1 + K_T}{K_T} . \tag{4.24}$$

By plotting the maximum power obtained by Eq. (4.24) versus the speed together with the power required given by Eq. (4.10) multiplied by V, the maximum speed at which the vehicle is able to transfer enough power to maintain its speed is readily obtained. It must be remembered that in this computation the rolling resistance must be neglected in the computation of the required power.

If on the contrary $K_T F_{z_2}/F_{z_1}$ is smaller than unity the limit conditions occur at the rear wheels and the maximum power which can be transferred to the road is

$$P_{max} = V\mu_{x_p} F_{z_2}(1 + K_T) . \tag{4.25}$$

If not all axles are driving the power which can be transferred to the ground is smaller. Aerodynamic drag increases the load on rear wheels, as does a positive grade of the road: The power which can be transferred by a vehicle with rear wheel drive thus increase with speed due to drag and with the slope. Aerodynamic moment and lift have different effects depending on the sign of the moments and the position of the centre of mass. The maximum power is then

$$P_{max} = V\mu_{x_p} F_{z_1} , \qquad P_{max} = V\mu_{x_p} F_{z_2} , \tag{4.26}$$

respectively for the cases of front and rear wheel drive. In the computation of the power needed for constant speed driving only the rolling resistance of the free wheels must be accounted for; this is easily done by introducing F_{z_2} or F_{z_1}, respectively for front and rear wheel drive, in the expression of the resistance to motion instead of the total load on the ground.

Also the maximum grade which can be managed is easily obtained. As in this case the speed can be set to zero, it follows, for an all-wheel drive vehicle

$$\begin{aligned} \tan(\alpha_{max}) = i_{max} &= \frac{b\mu_{x_p}\left(1 + \frac{1}{K_T}\right)}{l + h_G\mu_{x_p}\left(1 + \frac{1}{K_T}\right)} , \\ \tan(\alpha_{max}) = i_{max} &= \frac{a\mu_{x_p}(1 + K_T)}{l + h_G\mu_{x_p}(1 + K_T)} , \end{aligned} \tag{4.27}$$

respectively if the rear wheels slip first, i.e. if

$$K_T < \frac{b - h_G \tan(\alpha_{max})}{a + h_G \tan(\alpha_{max})},$$

or if this condition does not hold. The value of μ_{x_p} is that for vanishingly small speed. Note that it is unlikely, as it requires an abnormally low value of K_T, that the rear wheels are in a critical condition on a very steep grade.

For a vehicle with only one driving axle Eqs. (4.27) could still be used, the first one for front wheel drive and the second for rear wheel drive, but they do not include the rolling resistance of the free wheels. If this effect is accounted for, they are modified as

$$\begin{aligned} \tan(\alpha_{max}) = i_{max} &= \frac{b\mu_{x_p} - af_0}{l + h_G(\mu_{x_p} + f_0)}, \\ \tan(\alpha_{max}) = i_{max} &= \frac{a\mu_{x_p} - bf_0}{l - h_G(\mu_{x_p} + f_0)}. \end{aligned} \tag{4.28}$$

Example 4.6

Plot the curves of maximum transmissible and required power for the vehicle of Appendix 1.1 on dry and wet road. Compute the maximum speed and the maximum grade which can be managed. Repeat the computations assuming that the same vehicle has rear wheel drive, without changing the static load distribution on the ground.

Assume that $c_1 = 1.1$ and $c_2 = 6 \times 10^{-3}$ s/m on dry road and $c_1 = 0.8$ and $c_2 = 8 \times 10^{-3}$ s/m on wet road.

The curves of the maximum transmissible power are reported in Fig. 4.12, together with those of the required power. The vehicle of the example can thus reach a maximum speed of 225 km/h (wet road) or 308 km/h (dry road) for reasons only linked with the wheel driving force. The computations were repeated assuming that the driving wheels are the rear ones. In this case the maximum power which can be transferred to the ground at low speed is lower than in the previous case as the static load distribution was stated in order to obtain good performances with front wheel drive. By increasing the speed the load on the rear wheels increase and eventually get larger than that on the front wheel. On dry road the performance in

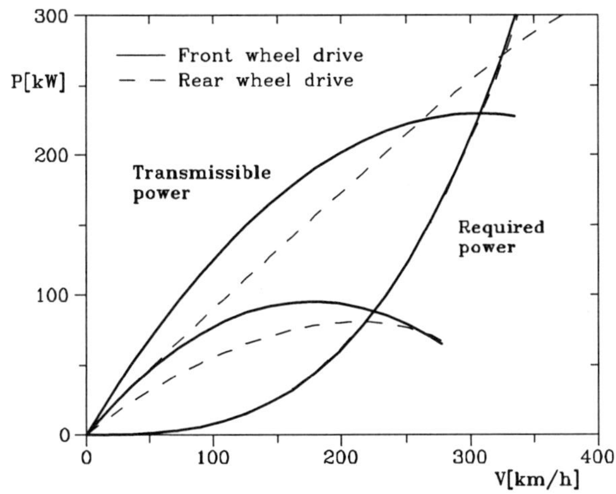

Fig. 4.12 Maximum transmissible power and required power on level road in the case of Example 4.6.

terms of maximum speed is then better for the vehicle with rear wheel drive, despite the fact that at standstill the front wheels are loaded by about 60% of the weight.

The values of the maximum speed for rear wheel drive are of 218 km/h on wet road or 328 km/h on dry road.

Note that the required power curve includes only the rolling resistance of the non-driving wheels and then is slightly different in the two cases. Also note that the curves do not take into consideration the load transfer due to acceleration, so for speeds lower than the maximum speed, where the vehicle would accelerate if the maximum power is applied, they are not realistic.

The maximum grade angle which can be managed for reasons only linked with the wheel driving force can be computed using Eq. (4.28), obtaining 28.0° on dry road and 22.1° on wet road, corresponding to grades of 53.1% and 40.6% respectively. If the driving wheels were the rear ones the values would have been 29.4° (56.3%) and 20.6° (37.6%).

In the case of rigid axles in which the final gear is directly mounted on the axle and the propeller shaft is in longitudinal direction, the drive torque T_d applied to the axle causes a transversal load shift between the driving

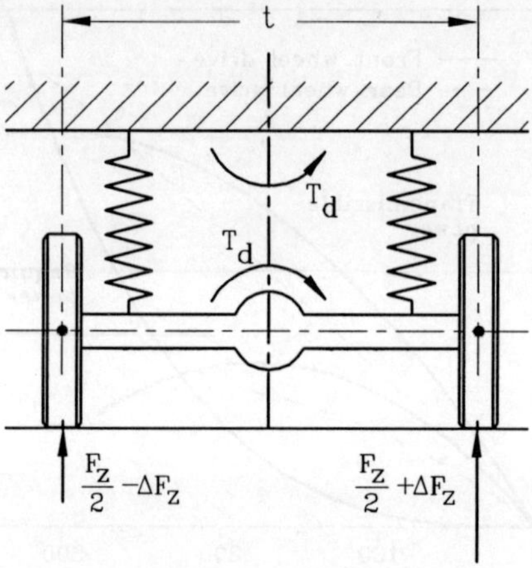

Fig. 4.13 Transversal load shift due to the driving torque T_d.

wheels of the same axle.

With reference to Fig. 4.13 the load shift ΔF_z could be determined easily as

$$\Delta F_z = \frac{T_d}{t_i} \qquad \text{where} \qquad T_d = F_x R_l \tau_f \,, \tag{4.29}$$

F_x is the longitudinal force exerted by the axle on the ground and τ_f is the gear ratio of the final drive, defined as the ratio between the speed of the wheels and that of the propeller shaft (it is usually smaller than unity).

Equation (4.29) is not however usually correct as under the action of the driving torque the vehicle body is subject to a roll rotation, which in turn produces an added torque on the axle through the suspension system. If the roll stiffness of the i-th suspension is K_{t_i}, the roll angle is $\phi = -T_d / \sum_{\forall i} K_{t_i}$. The torque exerted on the axle is then equal to $\phi K_t = -T_d K_t / \sum_{\forall i} K_{t_i}$, where K_t is the roll stiffness of the relevant suspension.

The load shift is thus

$$\Delta F_z = \frac{F_x R_l \tau_f}{t_i} \left(1 - \frac{K_t}{\sum_{\forall i} K_{t_i}} \right) . \tag{4.30}$$

If the vehicle has a standard differential gear, the maximum driving force which can be exerted by the driving axle is equal to twice that which can be exerted by the less loaded wheel, i.e.

$$F_{x_{max}} = \mu_p(F_z - 2\Delta F_z) \ . \tag{4.31}$$

If on the contrary a locking differential is used, within the limits of the assumption that the force coefficient μ_p is independent of the load, the transversal load shift does not affect the maximum driving force.

Example 4.7

Consider the articulated truck of Appendix 1.4. Compute

a) the maximum driving force at a constant speed of 70 km/h on level road;

b) the same as in a), but on a 10% grade;

c) the maximum grade which can be managed at 10 km/h.

All the above computations must be performed taking into account the transversal load shift and repeated for the case of a locking differential. Assume that the maximum longitudinal force coefficient is $\mu_p = 1$.

a) At 70 km/h = 19.44 m/s the load on the driving axle is 106.940 kN while the required driving force is 3.187 kN. Keeping into account the gear ratio of the final drive, the driving torque on the axle is of 344 Nm, yielding a roll angle of 2.67°. The transversal load shift is $\Delta F_z = 96.4N$ and the maximum longitudinal force is of 106.75 kN. This value compares with that of 106.94 which could be exerted if a locking differential were used, showing that the latter would improve only marginally the ability of exerting longitudinal forces in this case.

b) At 70 km/h on a 10% grade the load on the driving axle is 116.97 kN and the required driving force is 91.15 kN, corresponding to a driving torque on the axle of 4453 Nm. A very large roll angle, namely 34.6° results from the values of the stiffness of the axles, but this is an unrealistic result as for large torques the nonlinear nature of the suspensions would limit rotations. Assuming that the stiffness distribution between the suspensions in the nonlinear range is the same as in the linear range, the transversal load shift is $\Delta F_z = 1249N$, yielding a maximum longitudinal force of 114.47 kN; if a locking differential were used a force of 116.97 kN would have been exerted.

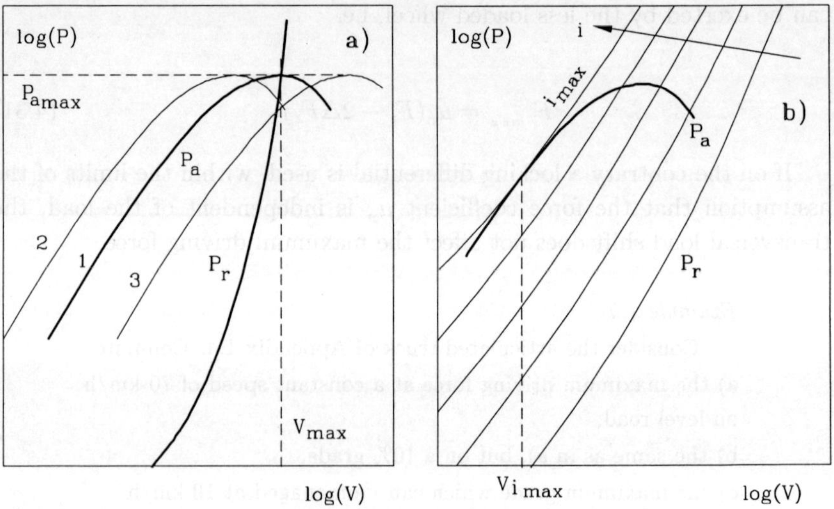

Fig. 4.14 Maximum speed (a) and maximum slope (b) for a vehicle with internal combustion engine.

c) By computing the force required for motion and the maximum force which can be exerted by the driving wheels at 10 km/h = 2.78 m/s for different values of the grade it is possible to find the value of the latter at which the two are equal. This procedure allows one to find the maximum value of the grade as 34.9%, i.e. a grade angle of 19.2°.

Note that the driving torque is very large on that grade and the suspensions operate clearly outside their linear range: The load shift can thus be far smaller than that computed. If no load shift was accounted for a value of the grade of 37.8%, i.e. a grade angle of 20.7° would have been found.

4.2.4 *Maximum performances allowed by the engine*

The maximum speed which can be reached on level road with a given transmission ratio can be found by intersecting the curves of the available power at the wheels and of the required power on level road. The transmission ratio which causes this intersection to occur at the maximum available power allows to reach the highest speed that can be attained by a given vehicle-engine combination (Fig. 4.14a).

The computation of the maximum speed and of the gear ratio τ_g at the gearbox allowing to reach it is straightforward. By intersecting the required power curve with the horizontal straight line $P = P_{e_{max}}\eta_t$, a fifth degree equation is obtained

$$AV + BV^3 + CV^5 = P_{e_{max}}\eta_t \ , \tag{4.32}$$

whose solution yields directly the maximum value of the speed.

If aerodynamic lift is neglected (actually it is sufficient to neglect the contribution to rolling resistance proportional to the square of the speed due to lift), the equation is cubic and its solution can be obtained in closed form

$$V_{max} = A^* \left(\sqrt[3]{B^* + 1} - \sqrt[3]{B^* - 1} \right) \ , \tag{4.33}$$

where

$$A^* = \sqrt[3]{\frac{P_{e_{max}}\eta_t}{2mgK + \rho S C_X}} \ ,$$

$$B^* = \sqrt{1 + \frac{8m^3g^3f_0^3}{27P_{e_{max}}^2\eta_t^2(2mgK + \rho S C_X)}} \ .$$

Once the maximum speed has been obtained, the gear ratio allowing to reach it is

$$\tau_g\tau_f = \frac{V_{max}}{R_e(\omega_e)_{P_{max}}} \ , \tag{4.34}$$

where $(\omega_e)_{P_{max}}$ is the engine speed at which the peak power is obtained.

The transmission ratio of the gearbox, which in top gear is usually close to 1, can be fixed and consequently the gear ratio τ_f at the final drive can be computed.

Note that this procedure is based on the assumption that the intersection in Fig. 4.14a occurs at the peak of the engine power curve. This can however occur only in one given condition as the load, but also the rolling resistance coefficient and even the air density affect the road load curve. Air density also affects the engine power curve. If the intersection occurs in the descending branch of the curve (situation 2 in Fig. 4.14a) the vehicle is said to be "undergeared", i.e. the overall transmission ratio is "too short". Conversely, if the intersection occurs in the ascending branch of the curve (situation 3 in Fig. 4.14a) the vehicle is "overgeared" and the overall transmission ratio is "too long".

The first situation can be purposely obtained to improve the acceleration and grade performances of the vehicle, while the second allows a reduction of the fuel consumption. The degree of undergearing λ_u can be defined as

$$\lambda_u = \frac{(\omega)_{V_{max}}}{(\omega)_{P_{max}}} . \tag{4.35}$$

It is greater than unity if the vehicle is undergeared and lower in case of overgearing.

There are thus two ways of choosing the top gear ratio: One has already been stated, namely a "fast" gear ratio, with a degree of undergearing equal to about unity, i.e. chosen in order to reach the maximum speed. A different approach is that of using a longer overgeared ratio, with the goal of reducing fuel consumption (see below). Practically this trade-off is typical of five speed transmissions: Either the maximum speed is reached in fifth gear or in fourth gear, the fifth being an overdrive gear. Note that only in the case of vehicles with high power/weight ratio this strategy works: In low powered vehicles this "economy" gear would be very difficult to use as any increase of the required power, e.g. due to a slight slope, headwind, etc. would compel to shift to a shorter gear. In this case undergearing can be a necessity.

The maximum slope which can be managed with a given gear ratio can be obtained by plotting the curves of the required power at various values of the slope and looking for the curve which is tangent to the curve of the available power (Fig. 4.14b). The slope so obtained is however only a theoretical result, as it can be managed only at a single value of the speed: If the vehicle travels at a higher speed, it slows down as the power is not sufficient, but also if its speed is smaller the power is insufficient and the vehicle slows down further: The condition is therefore unstable and the vehicle stops.

To be able to manage with safety a certain slope, the curve of the available power must be above that of the required power in a whole range of speeds, starting from a value low enough to assure that also starting on that slope is possible. To choose a value of the gear ratio of the bottom gear allowing to start on a given grade it is possible to state a reference speed and to compute the gear ratio in such a way that at that speed the P_a and P_r curves intersect.

As the vehicle is moving at low speed, only the first term of the required power curve needs to be accounted for. As the power developed by the engine can be written in the form

$$P_e = T_e\omega_e = T_e\frac{V}{R_e\tau_g\tau_f}\,, \tag{4.36}$$

where T_e is the engine torque, the equilibrium condition allows to compute the overall gear ratio

$$\tau_g\tau_f = \frac{T_e\eta_t}{R_e mg[f_0\cos(\alpha) + \sin(\alpha)]}\,. \tag{4.37}$$

The value of the engine torque to be introduced into Eq. (4.37) can be the maximum torque available at the minimum engine speed, possibly multiplied by a number smaller than one for safety. The mass of the vehicle must be that at full load, including the maximum trailer which the vehicle is allowed to tow. For the grade, values of 33% or even 25% can be considered, but it must be kept in mind that in some cases, as some ferry ramps or private garage ramps, very steep gradients can be encountered.

Another consideration in the choice of the gear ratio for the bottom gear is that of assuring a regular working of the engine at a speed chosen to avoid a prolonged use of the clutch in very low speed driving.

Once the ratios of the bottom and top gears have been chosen, the intermediate ones can be stated using different criteria. The simplest one is that of setting them in geometric sequence, i.e., stating that the ratios between two subsequent gear ratios are all equal. Operating in this way the available power curves on the $P(V)$ logarithmic plot are all equispaced.

There may be some advantages to have the curves a bit more close to each other in the high speed range, in such a way that the third gear (in a four speed gearbox) is closer to the fourth. If this is required, it is possible to set in a geometric sequence not the gear ratios but their ratios. This can give a feeling of sport driving.

The choice of the transmission ratios is much influenced by considerations which are beyond the scope of the present section, being mostly linked with the acceleration performance of the vehicle. This aspect will be dealt with in Section 4.4. As the values of the gear ratios have a large influence on the various performances of the vehicle and above all on the feeling that the driver has of them, the trade-off which dominates their choice is also a matter of subjective impressions and of traditions of various manufacturers. The market sector a manufacturer aims to can have more influence in deciding the whole matter than technical considerations alone.

Example 4.8

Chose the overall top gear ratio for the car of Appendix 1.1 to reach the maximum speed in the load condition indicated. Chose the bottom gear ratio to start on a 33% grade with a safety margin of 1.1 with respect to the maximum engine torque. Compare the ratio obtained with those indicated in the Appendix.

Equation (4.32), solved numerically, yields a maximum speed of 42.6 m/s = 153.4 km/h. The overall transmission ratio $\tau_g\tau_f$ which allows the intersection between the two curves on the $P(V)$ plane to occur at the peak power is 0.3044. If the value of 22/21 = 1.048 is accepted for the top gear ratio, the transmission ratio of the final drive is 0.2906, which can be approximated as 18/62 with an error of about 0.08%.

The actual ratio of the final drive is 0.284. By computing the maximum speed with this value of the transmission ratio, a value of 41.2 m/s = 148.36 km/h is found. The top speed is reached at 5147 rpm, yielding a degree of undergearing λ_u = 0.99.

The overall transmission ratio of the bottom gear can be found using Eq. (4.37). By dividing the maximum engine torque by a factor 1.1, a value of 0.1056 is obtained, corresponding to a value of the gearbox ratio of 0.3639. This value is far longer than the actual one (0.2154), as the computation has been performed with the vehicle quite unloaded.

4.2.5 *Fuel consumption at constant speed*

The energy needed to travel at constant speed can be immediately computed by multiplying the power required for constant speed driving by the time

$$E = P_r t = \frac{P_r d}{V} , \tag{4.38}$$

where d is the distance travelled. Note that Eq. (4.38) gives the energy required at the wheels: To obtain the energy actually required it must be divided by the various efficiencies (transmission, engine, etc.).

If the efficiency of the engine η_e and the thermal value H of the fuel are known, the fuel consumption can be computed. Introducing the expression (4.9) for the total road load into the expression for the power, the fuel consumption per unit distance Q is

$$Q = \frac{A + BV^2 + CV^4}{\eta_t \eta_e H \rho_f} , \tag{4.39}$$

where ρ_f is the density of the fuel, introduced to obtain the consumption in the terms of volume of fuel per unit of distance. In S.I. units it is measured in m^3/m, while liters per 100 km is a more practical, although not consistent, unit. Often the reciprocal of Q, expressed in km per liter or miles per gallon, is used.

From Eq. (4.39), if the aerodynamic lift is neglected, the fuel consumption would be a quadratic function of the speed if the efficiency of the engine could be considered as a constant. This is however not the case as the efficiency of the engine is strongly influenced by its rotational speed and above all by the power the engine is required to supply.

To compute the consumption Q the simplest procedure is to obtain the power required at the wheels as a function of the speed and hence to compute the power the engine must supply to travel at constant speed ($P_e = P_r/\eta_t$). Once the transmission ratio has been stated the rotational speed of the engine is known and hence the working point on the map of the engine is located. From it the efficiency η_e or, which is the same, the specific consumption $q = H/\eta_e$ is obtained and the fuel consumption can be computed as

$$Q = \frac{q P_r}{\eta_t V \rho_f} . \tag{4.40}$$

The curves $Q(V)$ are of the type shown in Fig. 4.15. They usually have a minimum at low speed, which is reached in conditions in which the engine works at low power with low efficiency. As the condition in which the engine

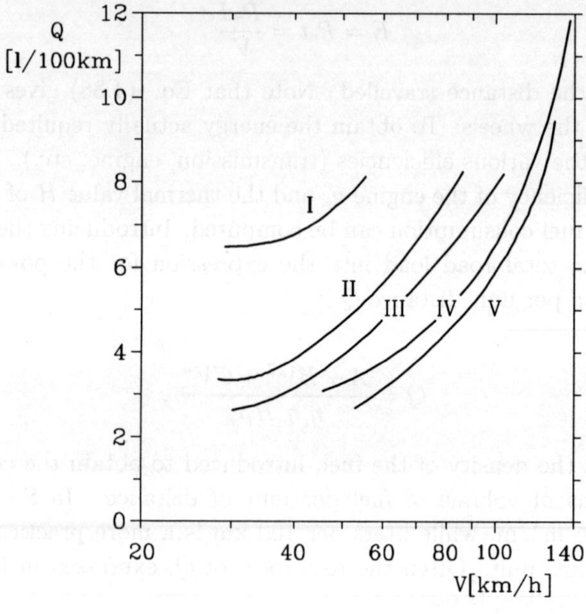

Fig. 4.15 Fuel consumption with different gear ratios at constant speed on level road. Passenger vehicle with five-speeds gear box.

works depends on the overall transmission ratio, the fuel consumption is also influenced largely by the value of the gear ratio. Usually the longer the ratio, the lower the consumption, as a "long" ratio allows to use the engine at low speed in conditions which are near to the maximum power, where the specific fuel consumption is low.

As already stated, a transmission ratio which is longer than that allowing to reach the maximum speed can be used. It is possible to choose it in such a way that at a certain cruise speed, e.g. equal to 3/4 of the top speed, the curve of the required power crosses that of the maximum efficiency: The fuel consumption at that speed is consequently the minimum possible value with the added advantages of a reduction of the noise and of the engine wear due to the reduced engine speed. Obviously the performances in terms of maximum speed, acceleration and gradeability are reduced with respect to those available with a shorter gear ratio.

If a CVT is used, it is possible to control it in such a way that at any speed the engine works in the conditions of maximum efficiency, i.e. at any speed the working point on the map lies on the curve of the maximum efficiency. However this is really expedient only if the increase of the efficiency so obtained is greater than the loss of efficiency, with respect to that of a simpler transmission, due to the use of the CVT. Moreover, the control law for the transmission ratio of the CVT is a trade-off among different requirements, which take into account also acceleration and gradeability.

Example 4.9

Plot the fuel consumption curve in top gear for the car of Appendix 1.1.

The map of the engine is reported in Fig. 4.16a. In the same plot the curves of the power required at the engine, i.e. that of the power required at the wheels divided by the transmission efficiency, are plotted for the different gear ratio. The curves identify the working conditions of the engine.

The points at which the curve of the power required in top gear intersects the curves at constant specific fuel consumption are reported in the first two columns of the following table

ω [rpm]	P [kW]	q [g/HPh]	V [km/h]	Q [l/100km]	$1/Q$ [km/l]
2083	3.819	400	60.05	4.65	21.52
3157	9.711	300	91.02	5.85	17.10
4135	19.610	250	119.20	7.51	13.31
4664	27.152	240	134.46	8.85	11.30
5320	37.876	250	153.36	11.28	8.87

The other columns list the specific fuel consumption, the speed and the fuel consumption (in l/100 km) and its reciprocal (in km/l). A value of 730 kg/m^3 has been used for the density of the fuel. The fuel consumption is also reported in Fig. 4.16b.

4.3 Vehicle take-off from rest

Since internal combustion engines cannot operate below a minimum speed ω_{min}, the vehicle cannot slow down below the speed $V_{min} = \omega_{min} R_e \tau_f \tau_g$ with the engine connected to the driving wheels. Either a torque converter or a friction clutch must be used, both for starting and stopping the vehicle and to facilitate the shifting of gears.

The starting manoeuvre can be easily simulated in an approximated way by accepting the following assumptions:

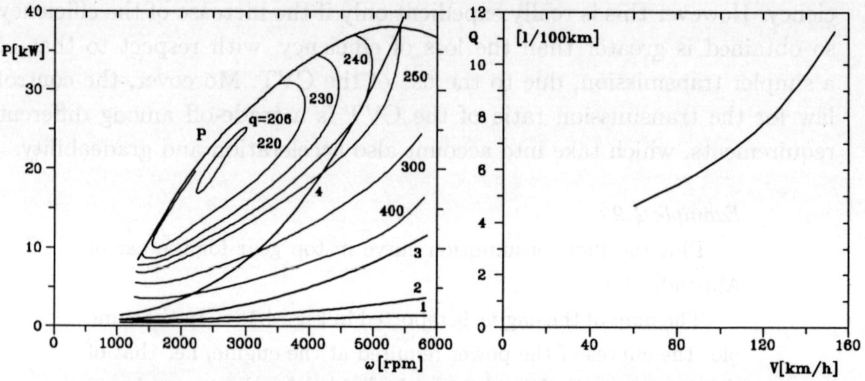

Fig. 4.16 Fuel consumption in top gear for the car of Example 1. (a) Map of the engine with superimposed the curves of the power required at the engine in various gears (1: bottom gear; 2,3: intermediate gears; 4: top gear). The specific fuel consumption is reported in g/CVh. (b) Fuel consumption in l/100 km as a function of the speed.

a) The manoeuvre is started with the engine running at a speed ω_{e_0} and the clutch control is released gradually from time $t = 0$ to time $t = t_i$ in such a way that the torque it transmits T_c increases linearly in time from 0 to the maximum value it can handle in slipping conditions T_c^* and then remains constant until time t_s when no more slipping occurs;
b) the engine torque is maintained constant at the value T_e;
c) if the vehicle starts on a sloping road, it is kept stationary by some external means until the clutch torque is sufficient to start the motion;
d) the longitudinal slip of the wheels is small;
e) the terms in V^2 and V^4 of the road load are neglected owing to the low speed at which the manoeuvre is performed.

The vehicle can be modelled as two moments of inertia, one to model the engine J_e and one to model the vehicle J_v (Fig. 4.17a). The first one includes the moment of inertia of the engine, up to the flywheel, while the moments of inertia of the clutch disks, of the shaft entering the gearbox, of all the rotating parts, reduced to the engine shaft, and the mass of the vehicle as "seen" from the engine are included in the second. For the computational details, see Section 4.4.

Torque T_e, which has been assumed to be constant, acts on the moment of inertia J_e. On J_v a drag torque T_d is acting, whose value is simply

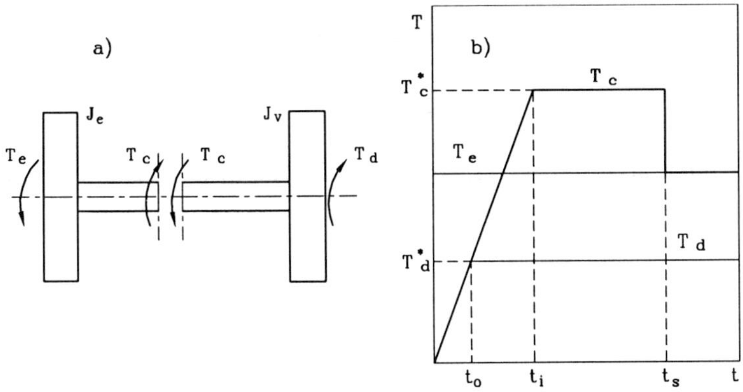

Fig. 4.17 (a) Model of the vehicle for the starting manoeuvre. (b) Time history of the torques acting on the vehicle.

$$T_d^* = mg\left[f_0 \cos(\alpha) + \sin(\alpha)\right] R_e \frac{\tau_f \tau_g}{\eta_t} , \tag{4.41}$$

when the vehicle is moving. When the vehicle is stationary, at the beginning of the starting manoeuvre, the drag torque is simply equal to the torque the clutch is supplying

$$T_d = \min(T_d^*, T_c) . \tag{4.42}$$

The maximum torque which the clutch can transfer to the vehicle T_d^* is usually slightly larger, by 10% to 20%, than the maximum engine torque.

The torques acting on the system are plotted versus time in Fig. 4.17b. The manoeuvre can thus be subdivided into three phases: From $t = 0$ to $t = t_0 = t_i T_d^*/T_c^*$, in which the vehicle is at standstill, as the torque transferred by the clutch is not yet sufficient to overcome the drag. The engine speeds up.

From $t = t_0$ to $t = t_i$, the clutch slips, the vehicle accelerates and the engine initially continues to speed up but when T_c^* becomes greater than T_e, it starts to slow down.

From $t = t_i$ to $t = t_s$, the clutch continues to slip until time t_s, when the transmission starts to behave as a rigid system and the acceleration continues as will be seen in Section 4.4.

The equation of motion of the system is simply

$$\begin{cases} \dot{\omega}_e = \dfrac{T_e - T_c}{J_e} \\ \dot{\omega}_v = \dfrac{T_c - T_d}{J_v} \end{cases} . \tag{4.43}$$

By integrating Eq. (4.43) separately for the three phases, the following time histories for the engine and for the vehicle are obtained

$$\begin{aligned} &\omega_e = \omega_{e_0} + \frac{1}{J_e}\left(T_e t - \frac{T_c^*}{2t_i}t^2\right) && \text{for } 0 < t < t_i \\ &\omega_e = \omega_{e_0} + \frac{1}{J_e}\left[t(T_e - T_c^*) + \frac{t_i T_c^*}{2}\right] && \text{for } t_i < t < t_s \end{aligned} \tag{4.44}$$

$$\begin{aligned} &V = 0 && \text{for } 0 < t < t_0 \\ &V = R_e\tau_f\tau_g\omega_v = \frac{1}{J_v}\left(\frac{T_c^* t^2}{2t_i} - T_d^* t + \frac{T_d^{*^2} t_i}{2T_c^*}\right) && \text{for } t_0 < t < t_i \\ &V = R_e\tau_f\tau_g\omega_v = \frac{1}{J_v}\left[t(T_c^* - T_d^*) + \frac{t_i}{2T_c^*}\left(T_d^{*^2} - T_c^{*^2}\right)\right] && \text{for } t_i < t < t_s . \end{aligned} \tag{4.45}$$

The starting time t_s can be defined as the time at which the clutch stops slipping: $\omega_v = \omega_e$. By equating the two angular velocities it follows that

$$t_s = \frac{2J_e J_v T_c^* \omega_{e_0} + T_c^{*^2} t_i (J_v - J_e) - t_i T_r^{*^2} J_e}{2T_c^* \left[J_e(T_c^* - T_r^*) + J_v(T_c^* - T_e)\right]} . \tag{4.46}$$

To make the subsequent acceleration of the vehicle possible, the angular velocity of the engine at time t_s must be in excess of the minimum one at which it can work regularly, otherwise it stops. This can occur if the values of ω_{e_0} or of T_e are too low or if the clutch engages too fast (t_i too low). If $t_s < t_i$ the vehicle completes the starting manoeuvre before the clutch is fully engaged: There is no problem in that, but Eq. (4.46) fails to yield a correct value of t_s.

During the manoeuvre the engine delivers an energy equal to the difference between its kinetic energy at times 0 and t_s added to the energy it produces in that time interval

$$E_e = \int_0^{t_s} T_e \omega_e dt + \frac{1}{2} J_e \left(\omega_{e_0}^2 - \omega_{e_s}^2 \right) . \tag{4.47}$$

Similarly, the vehicle receives the energy

$$E_v = \int_0^{t_s} T_d \omega_v dt + \frac{1}{2} J_v \omega_{v_s}^2 . \tag{4.48}$$

The difference $E_c = E_e - E_v$ yields the energy which is dissipated by the clutch during the starting manoeuvre. It is strictly linked to the quantity of friction material which is removed from the disc of the clutch, i.e. with the wear of that element.

The overall efficiency of the clutch is

$$\eta_c = \frac{E_v}{E_e} . \tag{4.49}$$

The space travelled during the take-off manoeuvre can be computed by integrating the speed in time. As the vehicle speed follows a pattern which is roughly quadratic, it can be approximated as $V_s t_s / 3$.

Example 4.10

Simulate a starting manoeuvre for the car of Appendix 1.1. Assume that the manoeuvre is started at 2000 rpm with the engine supplying a torque equal to 60% of the maximum torque while the clutch can transfer a torque equal to 120% of the maximum torque. Assume that $t_i = 0.5$ s, but repeat the computations for $t_i = 0.2$ s and $t_i = 0.8$ s.

With simple computation it follows that the moment of inertia simulating the vehicle is $J_v = 0.2113$ kg m^2 and that $T_d^* = 1.829$ Nm, $T_e = 52.2$ Nm, $T_i^* = 104.4$ Nm and $\omega_{e_0} = 209.4$ rad/s. The results are reported in Fig. 4.18.

The angular velocity of the flywheel simulating the vehicle at the end of the manoeuvre is 160.4 rad/s, corresponding to a vehicle speed $V = 2.561$ m/s $= 9.22$ km/h. The engine speed, 1532 rpm, is low but sufficient for allowing it to accelerate the vehicle. The time t_s is 0.59 s. The other results are $E_e = 8.52$ kJ, $E_v = 2.78$ kJ, $E_c = 5.74$ kJ, and $\eta = 33\%$. The space travelled up to time t_s is $s_s = 0.494$ m.

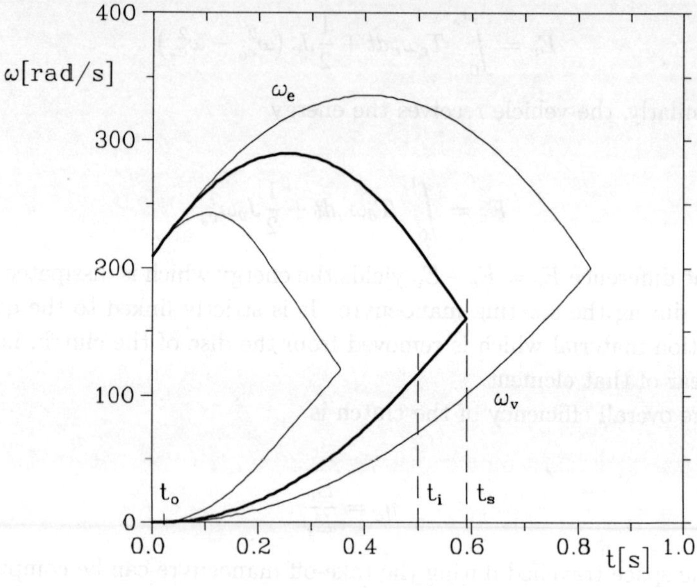

Fig. 4.18 Angular velocities of the engine and of the flywheel which simulates the vehicle during a starting manoeuvre. Results for t_i = 0.5 s, 0.2 s and 0.8 s.

For the manoeuvre with t_i = 0.8 s the results are respectively ω_{v_s} = 200.0 rad/s, V_s = 3.193 m/s = 11.5 km/h, ω_{e_s} = 1910 rpm, t_s = 0.83 s, E_e = 12.8 kJ, E_v = 4.45 kJ, E_c = 8.37 kJ, η = 35% and s_s = 0.863 m.

Finally, assuming t_i = 0.2 s the results are respectively ω_{v_s} = 120.1 rad/s, V_s = 1.918 m/s = 6.90 km/h, ω_{e_s} = 1147 rpm, t_s = 0.36 s, E_e = 5.08 kJ, E_v = 1.63 kJ, E_c = 3.45 kJ, η = 36% and s_s = 0.257 m.

Note that in all cases the efficiency of the clutch is lower than the value 0.5 which is often assumed. Actually it would be 0.5 if the engine rotates at constant speed with no drag moment acting on the inertia which has to be accelerated.

The assumptions made are quite rough, particularly those on the laws $T_e(t)$ and $T_c(t)$. However the results allow one at least to obtain reference values which are independent of the actual behaviour of the driver.

In the case of an automatic transmission the computation can be performed in the same way: Equations (4.43) still hold, with the only difference that now the torque T_{c_e} applied to the engine shaft is not equal to the torque T_{c_v} applied to the inertia simulating the vehicle. Their ratio is a function of the speed ratio ω_v/ω_e between the speeds of the output and input shafts of the torque converter. The torque applied to the engine shaft now does not depend on an imposed law simulating the behaviour of the driver but must be obtained from the characteristics of the torque converter, usually in the form of a plot of the so-called K-factor defined as

$$K = \frac{\omega_e}{\sqrt{T_{c_e}}}$$

plotted against the speed ratio ω_v/ω_e.

No closed form integration is usually possible, but there is no difficulty to integrate numerically the equations of motion: At each step the angular velocity of both input and output shafts of the torque converter are known and, as a consequence, both the input torque and the torque ratio of the converter are known. The accelerations of both engine and vehicle can be computed and the integration can proceed in a straightforward way.

4.4 Acceleration

If the curve of the required power lies, at a certain speed, below that of the power available at the wheels, the difference $P_a - P_r$ between the two is the power which is available to accelerate the vehicle.

Consider a vehicle with a mechanical transmission with a number of different gear ratios. During an acceleration a number of rotating elements (wheels, transmission, the engine itself) must increase their angular velocity and it is expedient to write an equation linking the engine power with the kinetic energy $\mathcal{T}$ of the vehicle

$$\eta_t P_e - P_r = \frac{d\mathcal{T}}{dt} \,. \tag{4.50}$$

Note that the engine power P_e should be that provided in non steady-state running; owing to the different time scale of the acceleration of the crankshaft and the thermodynamic cycle, the error introduced by using the values obtained from the steady-state map is negligible. Also, the efficiency of the transmission should not be considered when dealing with the power needed to accelerate the inertia of the engine, which is accelerated directly by the engine torque. Also the error introduced in this way is negligible.

Once that the transmission ratio has been chosen, Eq. (4.17) gives the relationship between the speed of the vehicle and the rotational speed of the engine. Similar relationships can be used for the other rotating elements which must be accelerated when the vehicle speeds up.

The kinetic energy of the vehicle can then be expressed as

$$\mathcal{T} = \frac{1}{2}mV^2 + \frac{1}{2}\sum_{\forall i} J_i\omega_i^2 = \frac{1}{2}m_e V^2 , \tag{4.51}$$

where the sum extends to all rotating elements which must be accelerated when the vehicle speeds up. The term m_e is the equivalent or apparent massof the vehicle, i.e. the mass of an object that when moving at the same speed of the vehicle has the same total kinetic energy. Usually it is written in the form

$$m_e = m + \frac{J_w}{R_e^2} + \frac{J_t}{R_e^2\tau_f^2} + \frac{J_e}{R_e^2\tau_f^2\tau_g^2} , \tag{4.52}$$

where J_w is the total moment of inertia of the wheels, which are assumed to have the same radius and hence to rotate at the same speed, and of all elements rotating at their speed, J_t is the moment of inertia of the propeller shaft and of all elements of the transmission and J_e is the moment of inertia of the engine, the clutch and all the elements rotating at speed ω_e. To account for the fact that the engine is accelerated directly, at least in an approximated way, the last term is sometimes multiplied by η_t. The modifications to Eq. (4.52) to take into account the presence of different wheels on different axles are obvious.

Of the three last terms the first is usually small, the second is negligible while the third can become very important, particularly in low gear. As only the last term depends on the transmission ratio at the gearbox, the equivalent mass can be written in the form

$$m_e = F + \frac{G}{\tau_c^2} , \tag{4.53}$$

where $F = m + J_w/R_e^2 + J_t/R_e^2\tau_f^2$ and $G = J_e/R_e^2\tau_f^2$.

As the equivalent mass is a constant, once that the gear ratio has been chosen, Eq. (4.50) yields

$$\eta_t P_e - P_r = m_e \frac{dV}{dt} \, . \tag{4.54}$$

Equation (4.54) holds only in the case of constant equivalent mass. If a CVT or a torque converter is used, the overall transmission ratio, and hence the equivalent mass, changes in time and the equation should be modified as

$$\eta_t P_e - P_r = m_e V \frac{dV}{dt} + \frac{1}{2} V^2 \frac{dm_e}{dt} \, . \tag{4.55}$$

The correction present in Eq. (4.55) is however usually very small, as the equivalent mass does not change very quickly.

From Eq. (4.54) the maximum acceleration the vehicle is capable at the various speeds is immediately obtained

$$\left(\frac{dV}{dt}\right)_{max} = \frac{\eta_t P_e - P_r}{m_e V} \, , \tag{4.56}$$

where the engine power P_e is the maximum power the engine can deliver at the speed ω_e related to speed V.

The plot of the maximum acceleration versus the speed for a passenger vehicle with a four speed gearbox is reported in Fig. 4.19.

The minimum time needed to accelerate from speed V_1 to speed V_2 can be computed by separating the variables in Eq. (4.56) and integrating

$$T_{V_1 \to V_2} = \int_{V_1}^{V_2} \frac{m_e}{\eta_t P_e - P_r} V dV \, . \tag{4.57}$$

The integral must be performed separately for each velocity range in which the equivalent mass is constant, i.e., the gearbox works with a fixed transmission ratio. Although it is possible to integrate analytically Eq. (4.57) if the maximum power curve is a polynomial, numerical integration is usually performed.

A graphical interpretation of the integration is shown in Fig. 4.20: The area under the curve $V m_e/(\eta_e P_e - P_r) = 1/a$ versus V is the time required for the acceleration.

The speeds at which gear shifting must occur to minimize acceleration time are readily identified on the plot $1/a(V)$. As the area under the curve

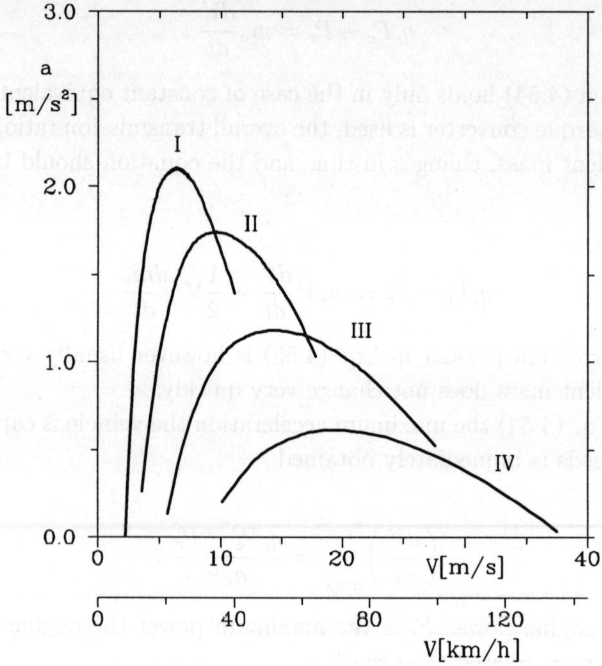

Fig. 4.19 Maximum acceleration as a function of the speed. Vehicle with a 4-speed gearbox.

is the acceleration time or the time to speed, the area must be minimized and gears must be shifted at the intersection of the various curves. If they do not intersect, the shorter gear must be used up to the maximum engine speed.

A criterion for choosing the gear ratios can also be evolved. The lower envelope of the curves (dashed line in the figure) does not depend on the transmission ratios and can be thought as the curve which can be followed using a CVT having the same efficiency of the gearbox and optimized to obtain the maximum acceleration. The area under the dashed curve is the minimum time to speed under ideal conditions.

The areas between the dashed and the continuous lines account for the time which must be added due to the presence of a finite number of speeds: The transmission ratios can be chosen in such a way to minimize this area. By increasing the number of speeds the acceleration time is reduced, as the actual curve gets nearer to the ideal dashed line. However, at each gear

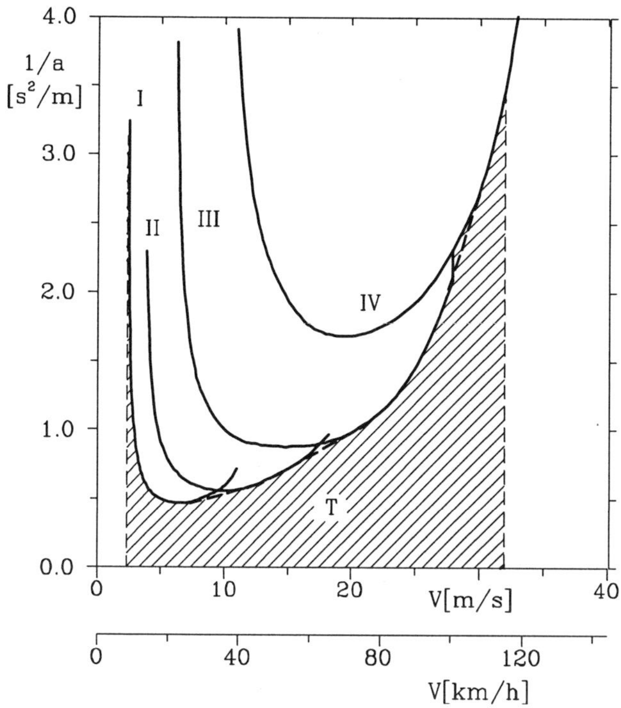

Fig. 4.20 Function $1/a(V)$ and search for the optimum speeds for gear shifting. The hatched area is the time to speed.

shifting there is a time in which the clutch is disengaged and consequently the vehicle does not accelerate: Increasing the number of speeds leads to an increase of the number of gear shifting and thus of the time wasted without acceleration. This restricts the use of a high number of gear ratios.

The speed-time curve at maximum power, which can be easily obtained by integrating Eq. (4.57) is reported in Fig. (4.21). The actual curve, obtained by adding the time needed for gear shifting, is also reported. The speed is assumed to be constant during gear shift.

By further integration it is possible to obtain the distance needed to accelerate to any value of the speed

$$s_{V_1 \to V_2} = \int_{t_1}^{t_2} V dt \,. \tag{4.58}$$

It is however possible to obtain directly the acceleration space, by writ-

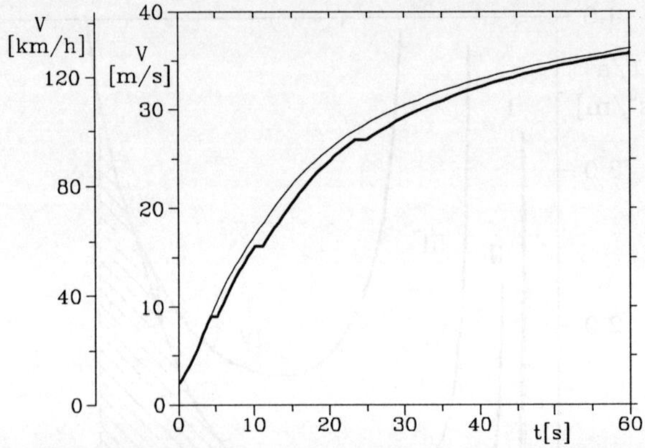

Fig. 4.21 Speed-time curve for the vehicle studied in the previous figures.

ing the acceleration as

$$a = \frac{dV}{dt} = \frac{dV}{dx}\frac{dx}{dt} = V\frac{dV}{dx} \ . \tag{4.59}$$

By separating the variables and integrating it follows

$$s_{V_1 \to V_2} = \int_{V_1}^{V_2} \frac{V}{a} dV = \int_{V_1}^{V_2} \frac{m_e}{\eta_t P_e - P_r} V^2 dV \ . \tag{4.60}$$

Note that sometimes instead of modelling the vehicle as an equivalent mass which is accelerated along the road, it is modelled as an equivalent moment of inertia attached to the flywheel of the engine, as seen in the previous section. Its value is

$$J_e = F' \tau_c^2 + G' \ , \tag{4.61}$$

where $F' = F R_e^2 \tau_f^2$ and $G' = J_e$.

The acceleration curves can thus be obtained in terms of acceleration of the engine instead of acceleration of the vehicle.

It is possible to choose the gear ratio of the bottom gear in a way which optimizes the acceleration at low speed. When the transmission ratio is shortened the torque available at the wheels increases, the equivalent mass also increases and then it is not convenient, from the viewpoint of the acceleration, to use transmission ratios which are too short.

Assuming that the engine torque T_e is constant and discarding the terms in V^3 and V^5 in the required power as at low speed their contribution is negligible, Eq. (4.56), written for the case of level road, yields

$$\left(\frac{dV}{dt}\right)_{max} = \frac{\eta_t T_e \omega_e - AV}{m_e V} = \frac{\eta_t T_e - A R_e \tau_f \tau_g}{R_e \tau_f \tau_g \left[F + \frac{G}{\tau_g^2}\right]} . \tag{4.62}$$

By differentiating Eq. (4.62) with respect to τ_g and equating the derivative to zero, a quadratic equation in τ_g yielding the value of the gear ratio which maximizes the acceleration is obtained. If the resistance to motion is neglected, which is reasonable on level road when dealing with strong accelerations, the value of the optimum gear ratio is

$$(\tau_g)_{opt} = \sqrt{\frac{G}{F}} \approx \sqrt{\frac{J_e}{m R_e^2}} . \tag{4.63}$$

The last value has been obtained by neglecting the terms containing the inertia of the wheels and of the transmission in the expression of the equivalent mass. Note that the value so obtained is that leading to equal contributions of the mass of the vehicle and of the inertia of the engine in the equivalent mass.

The value of the transmission ratio obtained with this criterion is however too short: Usually it yields driving torques which exceed the maximum torque to be transmitted by the driving wheels without slipping.

Example 4.11

Plot the acceleration curve for the vehicle in Appendix 1.1 and compute the time needed to reach 100 km/h. Compute also the time needed to travel for 1 km from standstill. Assume that the time needed for gear shifting is 0.5 s and that the takeoff manoeuvre follows the results obtained in Example 4.10.

Constants F and G are F = 855.2 kg and G = 15.96 kg, leading to the following values of the equivalent mass m_e = 1199 kg = 1.45 m in I gear, m_e = 975 kg = 1.18 m in II gear, m_e = 897 kg = 1.08 m in III gear, m_e = 870 kg = 1.05 m in IV gear.

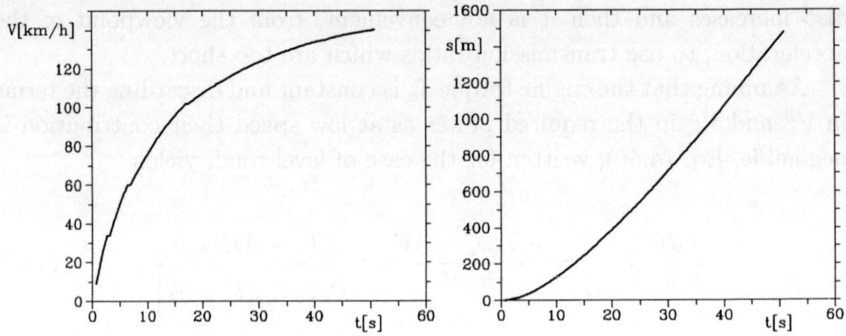

Fig. 4.22 Speed and distance travelled as functions of space during a full power acceleration. The initial take-off manoeuvre has also been considered.

If the equivalent moment of inertia approach were followed, its values would have been: $J_e = 0.296$ kg m^2 in I gear, $J_e = 0.692$ kg m^2 in II gear, $J_e = 1.823$ kg m^2 in III gear, $J_e = 5.085$ kg m^2 in IV gear.

The results of the numerical integration yielding the speed and the distance travelled as functions of time during an acceleration are reported in Fig. 4.22. They were computed starting from the results obtained in Example 4.10 with a time $t_i = 0.5$ s, namely a time of 0.59 s, a speed of 9.22 km/h and a distance of 0.494 m.

The engine power has been introduced in the computation through the best-fit third degree polynomial found in Example 4.5 and the speeds at which gear shifting occurs have been determined as the minimum value between that corresponding to the maximum speed of the engine (6000 rpm) and the speed at which the acceleration obtainable in the following gear equals that obtainable with the gear under consideration. They occur at 5784 rpm (V=34.3 km/h) for the bottom gear, at 6000 rpm (V=60.6 km/h) for the second gear and at 6000 rpm (V=102.3 km/h) for the third gear.

The time to reach a speed of 100 km/h is 16.3 s and that needed to reach the 1 km mark is 38.1 s.

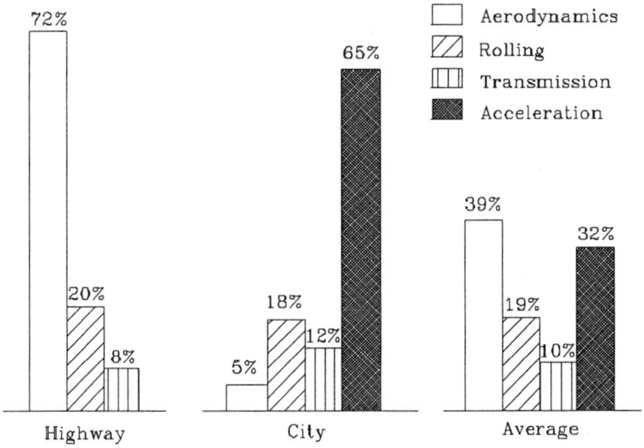

Fig. 4.23 Energy required for motion in two different driving conditions.

4.5 Fuel consumption in actual driving conditions

Fuel consumption has been previously studied at constant speed. In actual driving conditions, particularly in city traffic, frequent accelerations and slowing down occur, with occasional stops. The energy needed for accelerating the vehicle can thus become an important fraction of the total energy requirements.

In these conditions it is impossible to obtain simple results: The only way is to assume a given driving cycle, often deduced from experimental results, and to perform a numerical simulation in which the time histories of the various parameters of the motion are computed. An example of the results so obtained is presented in Fig. 4.23, where the relative importance of the various forms of drag is shown in two different driving cycles. Although they have been obtained for a given vehicle (a medium-size saloon car) the qualitative trend is general. While in highway driving aerodynamic drag is very important, in city driving most of the energy is spent to accelerate the vehicle.

The average has been computed using statistical data about the relative incidence of highway and city driving with reference to typical European conditions. From the average it is clear that the reduction of the mass of the vehicle, which affects both rolling resistance and acceleration power, is more important than aerodynamic streamlining and that the possibility of

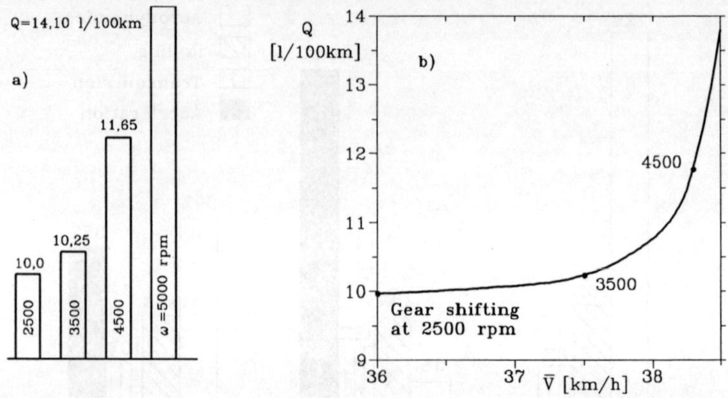

Fig. 4.24 and of the average speed (b) on the fuel consumption in city driving.

recovering energy during braking, which allows a partial recovering of the energy spent for the acceleration of the vehicle, would lead to important energy savings.

Numerical simulations also allow one to study the effects of the driving style on the fuel consumption. Also in city driving it is expedient to use the engine at the lowest speed which is consistent with its regular working and particularly to maintain it in the vicinity of the maximum torque and maximum efficiency speed. Prolonged use of low gears increases consumption without increasing much the average speed (Fig. 4.24).

In all developed countries the total amount of energy used by motor vehicles is a substantial fraction of the total energy consumption. Past values of this percentage for Italy and United States are of about 17% (12% for passenger cars) and 25% respectively. However, taking into account that the total dependence on liquid hydrocarbon fuels is far higher in Italy than in the United States, motor vehicles are responsible for about 23% and 53% respectively of the total oil consumption. Note that most of the energy consumption due to road vehicles occurs in city traffic (Table 4.1).

A parameter which has a certain importance in evaluating the energetic efficiency of vehicles is the specific tractive force, i.e. the nondimensional parameter

$$\frac{P_{max}}{mgV_{max}} = \frac{P_{max}t}{mgd}$$

computed by dividing the maximum power of the propulsion system (or,

Table 4.1 Energy used in Italy by road vehicles (reference year 1975).

Type of transportation	Energy [kcal]	Percentage
Passengers (individual)	120×10^{12}	12.20 %
		City: 7.23 %
		Roads 3.69 %
		Highways: 1.28 %
Passengers (collective)	5×10^{12}	0.50 %
Goods light	13×10^{12}	1.30 %
Goods medium	17×10^{12}	1.70 %
Goods heavy	14×10^{12}	1.40 %
TOTAL	169×10^{12}	17.10 %

better, the total installed power) by the maximum weight of the vehicle and its maximum speed. Assuming that the vehicle uses its maximum power when travelling at the maximum speed, it can be seen as the ratio between the driving force at top speed, which in steady-state operation equals the total drag, and the weight, being so a sort of friction coefficient. It can also be interpreted as the mechanical energy supplied by the engine to carry the unit weight at a unit distance.

The values of the specific tractive force for various types of vehicles are reported as functions of the speed in Fig. 4.25. Each one of the curves was obtained by considering many vehicles of the same type and plotting a point for each one of them on a plane $P_{max}/mgV_{max}(V_{max})$. The lower envelope of such clouds of points are assumed as the characteristic line of that type of vehicles and reported on the plot.

This procedure is not ideal as the maximum power is considered together with the maximum speed, while usually the most efficient working conditions from an energetic viewpoint can be quite different. Also, the total mass of the vehicle is considered while it would be more correct to use the payload.

All curves lie above a line which is straight in a logarithmic plot. Such line seems to indicate the optimum conditions for any vehicle, independently from the type or the mechanism used to supply the supporting forces. The plot of Fig. 4.25b, which is an enlargement of the zone for ground vehicles, shows that vehicles with trailer or made by several units are more energetically efficient and can even go beyond the limit line.

In spite of all its limitations, parameter P_{max}/mgV_{max}, seen as specific tractive force, gives an immediate evaluation of the relative energetic efficiency of the various types of vehicles and the distance from the limit line

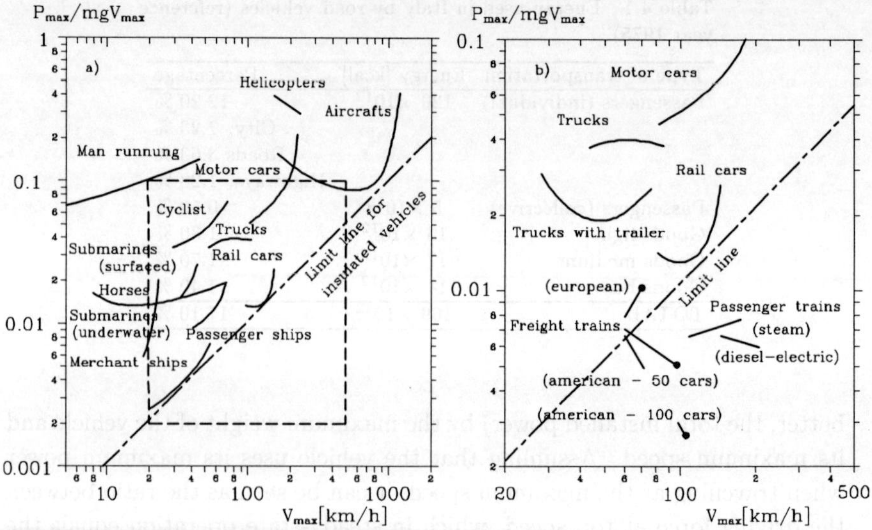

Fig. 4.25 Specific tractive force P_{max}/mgV_{max} as a function of the maximum speed for various types of vehicles. (a) General plot; (b) enlargement of the zone of interest for ground vehicles.

suggests the possible margins for improvements.

4.6 Electric and hybrid vehicles

The only case of road vehicles receiving the energy needed for motion from outside while travelling is that of trolleybuses. In all other instances road vehicles must carry onboard an energy supply. In the majority of cases the energy is accumulated in the form of chemical energy of a fuel, but it is possible to store the energy required for motion as electrochemical energy (electrical batteries) or, even if only few cases have been attempted and even fewer have entered commercial service, kinetic energy (flywheels) or elastic energy (springs). These forms of energy storage are compared in Table 4.2.

Vehicles in which two or more different types of energy storage devices are present are defined as hybrid vehicles. They must not be confused with bimodal vehicles which can work either with energy supplied from the outside or with energy stored onboard. A trolleybus with batteries is a bimodal vehicle, while a bus with internal combustion engine and batteries is a hybrid vehicle.

Table 4.2 Onboard energy storage. Energy density E/m, power density P/m and general characteristics (data for electrochemical energy refer to lead-acid batteries).

Energy stored	Chemical	Electrochemical	Elastic	Kinetic
E/m [Wh/kg]	10.000 – 12.000	10 – 40	2 – 10	6 – 20
P/m [W/kg]	Engine dependent	10 – 100	High	Very high
Efficiency	0.2 – 0.3	0.6 – 0.85	0.7 – 0.9	0.7 – 0.95
Reversibility	None	Possible		
Pollution	In the site of utilization	In the site of generation		
Dependence from liquid hydrocarbons	Almost complete	The primary source can be different		

The advantages of using the chemical energy of a fuel are so large that since the beginning of the century this form of energy storage has largely dominated the field. The advantages of simplicity of refuelling and above all the very high energy density, even taking into account the low efficiency of the energy conversion which involves the passage through thermal energy (except if fuel cells are used) are impressive. The energy density is in the range of 30 – 50 times that of the other storage devices.

The power density depends only on the power/mass ratio of the engine, which for modern reciprocating internal combustion engines is very high. The relevant technology has been developed for more than a century without interruptions and is still evolving. Other types of thermal engines, as rotary engines, gas turbines and steam engines have been tried several times, sometimes with technical success, but have never challenged the success of the former.

In the last decades however some drawbacks have started to be felt as important, particularly for urban driving. The first one is undoubtedly the unavoidable pollution of the engine, which can be reduced only up to a certain point, and at the cost of increasing operational costs and fuel consumption and lowering performances. The second is the impossibility of recovering the braking energy, which in city driving is an important percentage of the total. A last consideration is that the engine is used in urban driving in conditions which are very far from optimal, making worse all pollution and consumption problems.

Many governmental or local administrations recently passed acts which tend to limit the access to city centres of vehicles with internal combustion engine and this trend is going to develop in the future. The California Air

Resources Board stated, for example, that the percentage of *zero emission vehicles* must pass from 2% of the total of vehicles sold in the state in 1998 to 10% in 2003. The market for vehicles powered by alternative energy storage systems is then bound to increase in the near future. Apart from pollution, this will also produce benefits from the energetic point of view.

The most common alternative to the use of internal combustion engines is to use electrochemical accumulators in connection with electric motors. Their main drawback is the impossibility to exploit the accumulators with high energy and power density simultaneously. This is particularly true for lead-acid batteries, whose energy density decreases rapidly with increasing power density, i.e., with increasing discharge current. Also their efficiency and life decreases in the same way. Research in the field of batteries for automotive traction is very active, as it is generally understood that only progress in this field will actually make possible to solve some of the problems caused by the use of motor vehicles in urban areas, which are now felt as social problems.

Some of the alternatives considered today are reported in Table 4.3. Future progress seems at present more linked with the possibility of mass producing batteries with adequate performances at costs compatible with automotive use than with the development of batteries with the required characteristics.

Table 4.3 Main characteristics of some battery types for automotive use (M.J. Riezenman, *The great battery barrier*, IEEE Spectrum, Nov. 1992). a): Constant current 3 hours discharge. b): Cycles with 80% discharge depth. c): 100% discarge depth in urban cycle. d): 80% discharge.

Type	E/m [a] [Wh/kg]	P/m [b] [W/kg]	Efficiency	Life[c] [cycles]
Sodium-sulphur	81	152	91 %	592
Sodium-sulphur	79	90	88 %	795
Litium-sulphides	66	64	81 %	163[d]
Zinc-bromine	79	40	75 %	334
Nichel-zinc	67	105	77 %	114
Nichel-metal hydrides	54	186	80 %	333
Nichel-metal hydrides	57	209	74 %	108
Nichel-metal hydrides	55	152	80 %	380
Nichel-iron	51	99	58 %	918

The recent advances in power electronics have made it possible to use electric motors of simpler and more efficient type, e.g. induction or even synchronous A.C. machines instead of classical D.C. motors, and to control them in a very efficient way.

The advantages are mainly linked to the possibility of moving the pollution from the place of utilization of the vehicle to that of the power generation, with the better pollution control of power stations versus small engines, and to make regenerative braking possible. The performance of the latter is however decreased by the losses in both the engine and the batteries and above all by the difficulties for the batteries to accept the high power bursts occurring in braking. The disadvantages are also well known: The reduced range and duration of batteries and their high mass. However, even today, the performance of electric vehicles are sufficient for urban use.

From an energy viewpoint their advantages are still in doubt: When the primary source is a fossil fuel, in spite of the greater efficiency of the primary conversion and regenerative braking, the overall consumption is comparable with that of internal combustion engines. The very fact that the thermo-mechanical conversion occurs far from the vehicle makes it impossible to use waste heat for heating, and this makes the energy balance worse.

Elastic energy can be stored in a solid or in a gas. In the first case the energy density of the device is

$$\frac{E}{m} = \alpha_1 K \frac{\sigma^2}{\rho E} , \tag{4.64}$$

where α_1 and K are coefficients linked to the ratio of the mass of the energy storage elements and that of the whole device and to the shape of the storage element and the stress distribution, σ is the maximum stress in the energy storing element and E is the Young's modulus of the material. Material with very high strength (spring steel) or low stiffness (elastomers) must be used. The latter are particularly well suited, as some of them can be stretched up to 500% with a good fatigue life and limited energy losses.

The use of a compressed gas, while considered for fixed installations, has several disadvantages for vehicular uses, due the lower efficiency, high mass of the container of the pressurized fluid and burst danger. Hydraulic accumulators, in which the energy is stored in the walls of an elastomeric vessel full of fluid, have been suggested and tested in connection with hydraulic motors and pumps. The efficiency can be very high and safety can be easily insured but the energy density is still too low for primary onboard energy storage. They are on the contrary good candidates for hybrid vehicles.

The energy density of a kinetic energy accumulator can be expressed as

$$\frac{E}{m} = \alpha_1 \alpha_2 K \frac{\sigma}{\rho} , \tag{4.65}$$

where α_1, α_2 and K are coefficients linked with the ratio of the mass of the flywheel and that of the whole system, to the depth of discharge actually performed and to the shape and the stress distribution, σ is the maximum stress in the energy storing element and ρ is the density of the material.

Apart from some applications, as the city buses built by Oerlikon in the fifties and actually used in public service, flywheels are now considered for use in hybrid systems. Their potentially high power density makes them very suitable for supplying short bursts of power for acceleration or for storing braking energy.

Some possible schemes for hybrid vehicles are the following:
a) internal combustion engine - electric accumulator,
b) internal combustion engine - elastic accumulator,
c) internal combustion engine - flywheel,
d) electric accumulator - flywheel,
e) internal combustion engine - electric accumulator - flywheel.

The first three systems are similar, at least in principle. The thermal engine supplies the average power, working in conditions which can be optimized from the viewpoint of the efficiency or the pollution. A trade-off between these requirements can be made. When the duty cycle includes frequent accelerations and braking, the advantages of disconnecting the instantaneous power requirements from the working conditions of the thermal engine and to make possible regenerative braking are large. The possibility of using a far smaller engine allows one to keep the mass and the cost of the system within the limits of conventional ones or even to obtain mass and cost savings.

Solution (a) appears to be the worst, as the electric accumulators work exactly in the way which should be avoided, as they are called to supply high power for short times; nevertheless they are often considered, particularly if pollution reduction is considered more important than lowering the fuel consumption.

Solution (b) can be used in hydrostatic transmission; owing to the cost of the latter, is mainly considered for large city buses.

Solution (c) allows the use of mechanical transmission, although the requirement of an efficient CVT with large range of transmission ratios is not easy to meet. The very high efficiency and power density of flywheels

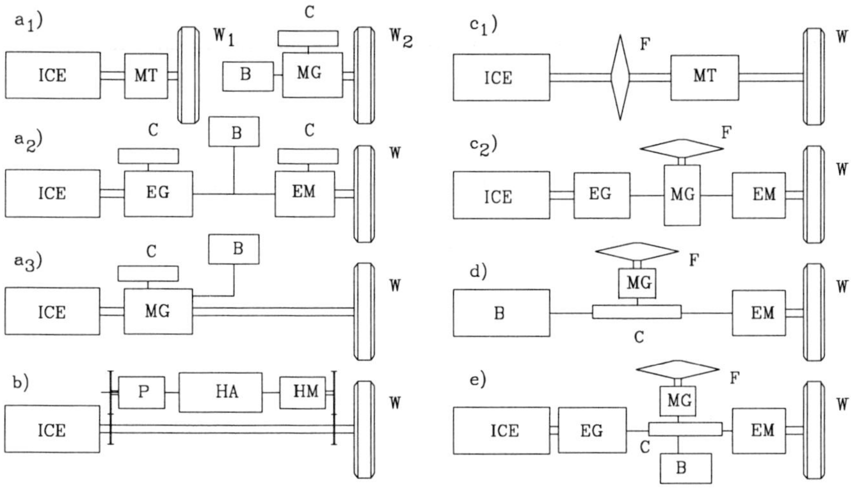

Fig. 4.26 Some possible schemes of hybrid vehicles B, batteries; C, control unit; EG, electric generator; F, flywheel; HA, hydraulic accumulator; HM, hydraulic motor; ICE, internal combustion engine; MG electric motor/genertor; MT, mechanical transmission; P, pump; W wheels.

can be exploited.

Solution (d) is very interesting, as the flywheel manages the power peaks occurring during acceleration and regenerative braking, allowing the use of batteries with low power density, thus increasing the efficiency, and hence the range of the vehicle, and the life cycle of the batteries.

Solution (e) combines the advantages of (a) and (d): The batteries work in optimal conditions, and hence a smaller mass of batteries than in (a) is required. The presence of the batteries allows a far larger engine-off range than in (c), to cope with conditions in which the use of an internal combustion engine is not allowed (it behaves as a *zero-emission vehicle*) while the latter allows a practically unlimited range outside these conditions.

Some possible schemes of hybrid vehicles are reported in Fig. 4.26.

4.7 Braking on straight road

4.7.1 *Braking in ideal conditions*

Apart from cases in which the vehicle slows down under the braking effect of the engine, which can dissipate a non-negligible power (lower part of the graph of Fig. 4.6) and from regenerative braking in electric and hybrid vehicle, braking is performed in all modern vehicles on all wheels.

Ideal braking can be defined as the condition in which all wheels brake with the same longitudinal force coefficient μ_x.

The study of the braking forces which the vehicle can exert follows closely that in Section 4.2.3, with the only obvious difference that braking force, as well as the corresponding force coefficient and the longitudinal slip, are negative. The vertical forces between the vehicle and the ground can be computed using the equations in Section 4.1 remembering that also the acceleration is now negative.

The total braking force F_x is thus

$$F_x = \sum_{\forall i} \mu_{x_i} F_{z_i} , \tag{4.66}$$

where the sum is extended to all the wheels. The longitudinal equation of motion of the vehicle is then

$$\frac{dV}{dt} = \frac{\sum_{\forall i} \mu_{x_i} F_{z_i} - \frac{1}{2}\rho V^2 S C_X - f \sum_{\forall i} F_{z_i} - mg\sin(\alpha)}{m} , \tag{4.67}$$

where m is the actual mass of the vehicle and not the equivalent mass, and α is positive for uphill grades. The rotating parts of the vehicle are directly slowed down by the brakes and hence do not enter the evaluation of the forces exchanged between vehicle and road. They must be accounted for when assessing the required braking power of the brakes and the energy which must be dissipated.

In a simplified study of braking aerodynamic drag and rolling resistance can be neglected, as they are usually far smaller than braking forces. Also, rolling resistance can be considered as causing a braking moment on the wheel more than a braking force directly on the ground. As in ideal braking all force coefficients μ_{x_i} are assumed to be equal, the acceleration is

$$\frac{dV}{dt} = \mu_x \left[g\cos(\alpha) - \frac{1}{2m}\rho V^2 S C_Z \right] - g\sin(\alpha) . \tag{4.68}$$

In case of level road, for a vehicle with no aerodynamic lift, Eq. (4.68) reduces to

$$\frac{dV}{dt} = \mu_x g \, . \tag{4.69}$$

The maximum deceleration in ideal conditions can be obtained by introducing the maximum negative value of μ_x into Eq. (4.68) or (4.69).

The assumption of ideal braking implies that the braking torques applied on the various wheels are proportional to the forces F_z, if the radii of the wheels are all equal. As will be seen later, this can occur in only one condition, unless some sophisticated control device is implemented to allow braking in ideal conditions.

If μ_x can be assumed to remain constant during braking, the motion of the vehicle occurs with constant acceleration, and the usual formulae hold

$$t_{V_1 \to V_2} = \frac{V_2 - V_1}{|\mu_x| g} \, , \qquad s_{V_1 \to V_2} = \frac{V_2^2 - V_1^2}{2|\mu_x| g} \, . \tag{4.70}$$

The time and the space to stop the vehicle from speed V are then

$$t_{stop} = \frac{V}{|\mu_x| g} \, , \qquad s_{stop} = \frac{V^2}{2|\mu_x| g} \, . \tag{4.71}$$

The time needed to stop the vehicle increases linearly with the speed while the space increases quadratically.

To compute the forces F_x the wheels must exert to perform an ideal braking manoeuvre, forces F_z on the wheels must be computed first. This can be done using the formulae seen in Section 4.1. However, for vehicles with low aerodynamic vertical loading, as all commercial and passenger vehicles, with the exception of racers and some sports cars, aerodynamic loads can be neglected. Also drag forces can be neglected and, in case of a two-axle vehicle, the equations reduce to

$$F_{z_1} = \frac{m}{l} \left[gb\cos(\alpha) - gh_G \sin(\alpha) - h_G \frac{dV}{dt} \right] \, , \tag{4.72}$$

$$F_{z_2} = \frac{m}{l} \left[ga\cos(\alpha) + gh_G \sin(\alpha) + h_G \frac{dV}{dt} \right] \, . \tag{4.73}$$

From Eq. (4.67)

$$\frac{dV}{dt} = \frac{\mu_{x_1} F_{z_1} + \mu_{x_2} F_{z_2}}{m} - g \sin(\alpha) , \tag{4.74}$$

since the values of μ_x are all equal in ideal braking, the values of longitudinal forces F_x are

$$F_{x_1} = \mu_x F_{z_1} = \mu_x \frac{mg}{l} \left[b \cos(\alpha) - h_G \mu_x \right] , \tag{4.75}$$

$$F_{x_2} = \mu_x F_{z_2} = \mu_x \frac{mg}{l} \left[a \cos(\alpha) + h_G \mu_x \right] . \tag{4.76}$$

By eliminating μ_x using Eqs. (4.75) and (4.76), the following relationship between F_{x_1} and F_{x_2} is readily obtained

$$(F_{x_1} + F_{x_2})^2 + mg \cos^2(\alpha) \left(F_{x_1} \frac{a}{h_G} - F_{x_2} \frac{b}{h_G} \right) = 0 . \tag{4.77}$$

The plot of Eq. (4.77) in F_{x_1},F_{x_2} plane is a parabola whose axis is parallel to the bisector of the second and fourth quadrants if $a = b$ (Fig. 4.27). The parabola is thus the locus of all pairs of values of F_{x_1} and F_{x_2} leading to ideal braking.

Actually, only a part of the plot is interesting: That with negative values of the forces (braking in forward motion) and with braking forces actually achievable, i.e. with reasonable values of μ_x (Fig. 4.28).

On the same plot it is possible to draw the lines with constant μ_{x_1}, μ_{x_2} and acceleration. On level road, the first two are straight line passing respectively through points B and A, while the lines with constant acceleration are straight lines parallel to the bisector of the second quadrant.

Note that the forces are related to each axle and not to each wheel: In the case of axles with two wheels their values are then twice the values referred to the wheel.

The moment to be applied to each wheel is approximately equal to the braking force multiplied by the loaded radius of the wheel: If the wheels have equal radii the same plot holds also for the braking torques. If this condition does not apply the scales are just multiplied by two different factors and the plot is distorted, but remains essentially unchanged.

To perform a more precise computation, the rolling resistance should be accounted for, which is a small correction, and the torque needed for

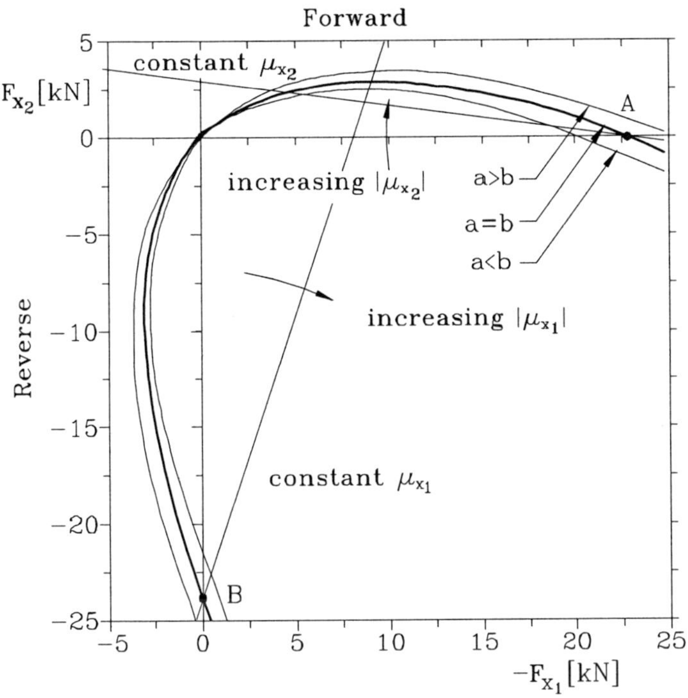

Fig. 4.27 Braking in ideal conditions. Relationship between F_{x_1} and F_{x_2} for vehicles with the centre of mass at mid-wheelbase, forward and backward of that point. Plots obtained with $m = 1000$ kg; $l = 2.4$ m, $h_G = 0.5$ m, level road.

decelerating the rotating inertias should be added. This correction is important only in the case of driving wheels and braking in low gear, but in this case the braking effect of the engine, which is even more important and has opposite sign, should be considered.

As already stated, the law linking F_{x_1} to F_{x_2}, i.e. M_{b_1} to M_{b_2} to allow braking in ideal conditions depends on the mass and the position of the centre of mass. For passenger vehicles it is possible to plot the lines for the minimum and maximum load and to assume that all conditions are included between them; for industrial vehicles the position of the centre of mass can vary to a larger extent, and a larger set of load conditions should be considered. The curves for three different types of passenger vehicles are reported in Fig. 4.29 as an example. On the same plot the curve $M_{b_2}(M_{b_1})$ defined by CEE standards and the lines at constant acceleration are also reported.

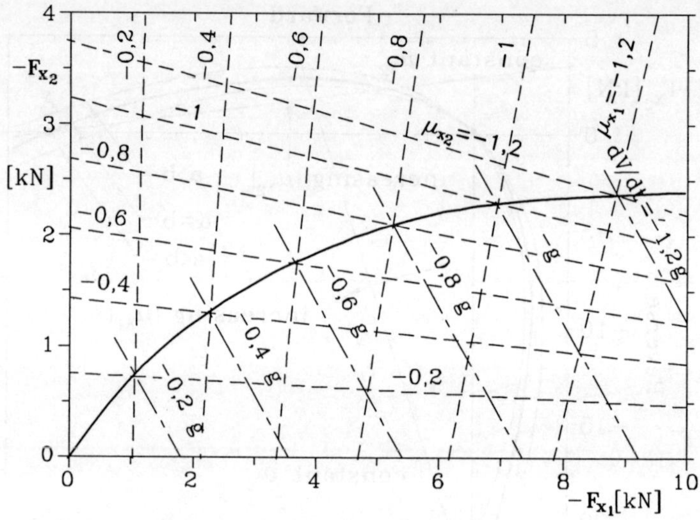

Fig. 4.28 Enlargement of the useful zone of the plot of Fig. 4.27. Also the lines with constant μ_{x_1}, μ_{x_2} and acceleration are reported.

4.7.2 *Braking in actual conditions*

The relationship between the braking moments at the rear and front wheels is in practice different from that stated in order to comply the conditions to obtain ideal braking and is imposed by the parameters of the actual braking system of the vehicle.

A ratio $K_B = M_{b_1}/M_{b_1}$ between the braking moments at the front and rear wheels can be defined. If all wheels have the same radius, its value coincides with the ratio between the braking forces[3]. For each value of the deceleration a value of K_B which allows the braking to take place in ideal conditions can be easily found from the plot of Fig. 4.28.

K_B depends on the actual layout of the braking system, and in some simple cases is almost constant. In case of hydraulic braking systems, the braking torque is linked to the pressure in the hydraulic system by a relationship of the type

[3]Note that this statement neglects the braking moment needed to decelerate rotating parts. This can be overcome by considering M_b as the part of the braking moment which causes braking forces on the ground; to it the fraction of the braking moment needed to decelerate the wheels and the transmission must be added.

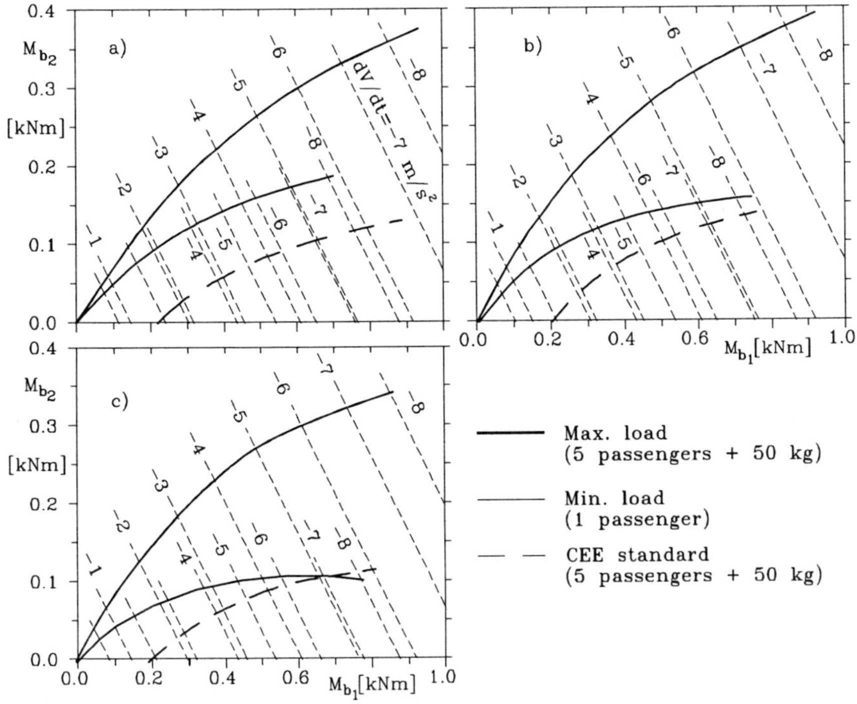

Fig. 4.29 Plots $M_{b_2}(M_{b_1})$ for ideal braking. (a) typical plot for a rear drive car with low ratio h_G/l; (b) typical plot for a front drive saloon car with higher ratio h_G/l; (c) plot for a small front drive car, sensitive to the load conditions and with high value of ratio h_G/l.

$$M_b = \epsilon_b(Ap - Q_m) \, , \tag{4.78}$$

where ϵ_b, sometimes referred to as efficiency of the brake, is the ratio between the braking torque and the force exerted on the braking elements and hence has the dimensions of a length, A is the area of the pistons, p is the pressure and Q_m is the restoring force due to the springs, when they are present.

The value of K_B is thus

$$K_B = \frac{\epsilon_{b_1}(A_1p_1 - Q_{m_1})}{\epsilon_{b_2}(A_2p_2 - Q_{m_2})} \, , \tag{4.79}$$

or, if no spring is present as in the case of disc brakes,

$$K_B = \frac{\epsilon_{b_1} A_1 p_1}{\epsilon_{b_2} A_2 p_2} \,. \tag{4.80}$$

In the case of disc brakes ϵ_b is almost constant and is, as a first approximation, the product of the average radius of the brake, the friction coefficient and the number of braking elements acting on the axle, as braking torques are again referred to the whole axle. In this case, if the pressure acting on the front and rear wheels is the same, the value of K_B is constant and depends only on geometrical parameters.

The behaviour of drum brakes is more complicated, as restoring springs are present and the dependence of ϵ_b on the friction coefficient is more complex. Shoes can be of the *leading* or of the *trailing* type. In the first case the braking torque increases more than linearly with the friction coefficient and there is even a value of the friction coefficient for which the brake sticks and the wheel blocks altogether.

For trailing shoes the opposite occurs and ϵ_b increases less than linearly with the friction coefficient.

The efficiency of the brakes is a complex function of both temperature and velocity and, during braking, it can change for a combined effect of these factors. When the brake heats up there is usually a decrease of the braking torque, at least initially. Later an increase due to the reduction of the speed can restore the initial values. This "sagging" in the intermediate part of the deceleration is more pronounced in drum than in disc brakes. With repeated braking the overall increase of temperature can lead to a general "fading" of the braking effect.

If K_B is constant, the characteristic line on the plane M_{b_1}, M_{b_2} is a straight line through the origin (Fig. 4.30).

The intersection of the characteristics of the braking system with the curve yielding ideal braking defines the conditions in which the system succeeds to perform in ideal conditions. On the left of point A, i.e. for low values of the deceleration, the rear wheels brake less than the required quantity and the value of μ_{x_2} is smaller than that of μ_{x_1}. If the limit conditions occur in this zone, i.e. for roads with poor traction, the front wheels lock first.

On the contrary all working conditions beyond point A are characterized by $\mu_{x_2} > \mu_{x_1}$ and the rear wheels brake more than required, i.e., the braking capacity of the front wheels is underexploited. In this case when the limit

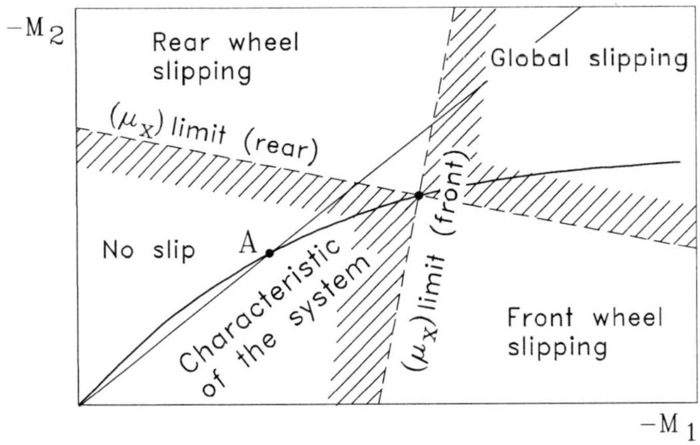

Fig. 4.30 Conditions for ideal braking, characteristic line for a system with constant K_B and zones in which the front or the rear wheels lock. In the case shown the value of $(\mu_x)_{limit}$ is high enough to cause sliding beyond point A.

conditions are reached, the rear wheels lock first, as in the case of Fig. 4.30.

From the viewpoint of the handling of the car it is advisable that $\mu_{x_2} < \mu_{x_1}$, as this increases the stability of the vehicle; the characteristics of the braking system should lie completely below the line for ideal braking. Locking of the rear wheels is a condition which must be avoided as it triggers directional instability.

In A the ideal conditions are obtained: If the limit value of the longitudinal force coefficient occurs at that point, simultaneous locking of all wheels occurs. The values of ratio K_B for which the ideal conditions occur at a given value of the longitudinal force coefficient μ_x^* is immediately computed

$$K_B^* = \frac{b + h_G|\mu_x^*|}{a - h_G|\mu_x^*|} \,. \tag{4.81}$$

It is possible to define an efficiency of braking as the ratio between the acceleration obtained in actual conditions and that occurring in ideal conditions, obviously at equal value of the coefficient μ_x of the wheels whose longitudinal force coefficient is higher

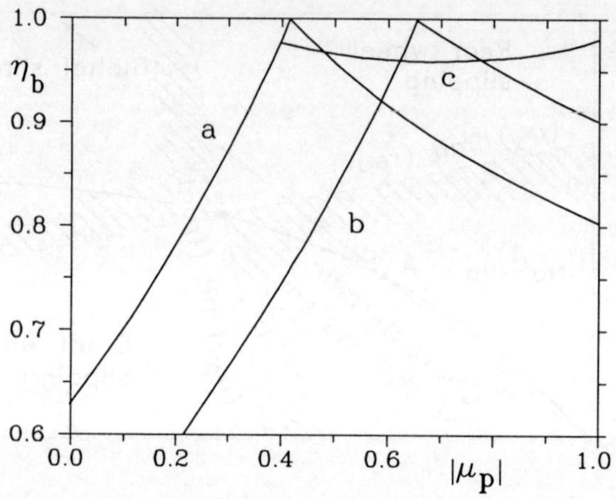

Fig. 4.31 Braking efficiency η_b as a function of the limit value of μ_x for a vehicle without (a) and (b) and with (c) pressure proportioning valve.

$$\eta_b = \frac{(dV/dt)_{actual}}{(dV/dt)_{ideal}} = \frac{(dV/dt)_{actual}}{\mu_x g} , \tag{4.82}$$

where the last expression holds only on level road for a vehicle with negligible aerodynamic loading. With simple computations it is possible to demonstrate that in the latter case the braking efficiency is

$$\eta_b = \min\left\{\frac{a(K_B+1)}{l-\mu_p h_G(K_B+1)}, \frac{b(K_B+1)}{lK_B+\mu_p h_G(K_B+1)}\right\} . \tag{4.83}$$

The first value holds when the rear wheels lock first (above point A in Fig. 4.30), the second one when the limit conditions are reached at the front wheels first.

A typical plot of the braking efficiency versus the peak braking force coefficient is plotted in Fig. 4.31.

The value of the maximum longitudinal force coefficient μ_p at which the condition $\eta_b = 1$ must hold can be stated and the value of ratio K_B can be easily computed. For values of $|\mu_p|$ lower than the chosen one the rear wheels lock first while for higher values locking occur at front wheels. Once K_B is known, the braking system can easily be designed.

The curve $\eta_b(\mu_x)$ can be plotted by assigning increasing values to the pressure in the hydraulic system, computing K_B and then the values of μ_x

and η_b referred to the front and rear wheels. The result is of the type shown in Fig. 4.31, curve (a) or (b).

Operating in this way the rear wheels lock when the road is in good conditions. To postpone the locking of the rear wheels curves of the type of line (b) can be used, but this reduces the efficiency when the road is in poor conditions.

To avoid locking of the rear wheels without lowering the efficiency at low values of μ_x, a pressure proportioning valve, i.e. a device which reduces the pressure in the rear brake cylinders when the overall pressure in the system increases above a given value, can be used. A possible law is a linear reduction of the pressure on the rear brakes with increasing pressure in the front ones above a certain pressure p_i

$$\begin{array}{ll} p_2 = p_1 & \text{if } p_1 \leq p_i \\ p_2 = p_i + \rho_c(p_1 - p_i) & \text{if } p_1 > p_i \end{array} , \tag{4.84}$$

where ρ_c is a characteristic constant of the valve. Pressure p_i and constant ρ_c must be chosen in such a way that the device starts acting when the efficiency η_b gets near unity and that the reduction of the rear pressure is just right not to cause locking of the rear wheels but also not high enough to lower substantially the efficiency (see Fig. 4.31, curve (c)).

To comply with these conditions in all loading conditions of the vehicle, p_i and, possibly, ρ_c must vary following the load. A possible way to do this is to monitor the load on the rear axle, e.g. by monitoring the vertical displacement of the rear suspension.

The characteristic line in the F_{x_1}, F_{x_2} plane of a device which operates following this line is reported in Fig. 4.32.

Antilock systems (ABS) act directly to reduce the pressure in the hydraulic cylinders of the relevant brakes when the need of reducing the braking force arises. Modern devices are based on wheel speed sensors which allow to compare the instantaneous speed of the wheels and the speed which corresponds to the velocity of the vehicle. If a slip which exceeds the allowable limits is detected, the device acts to reduce the braking torque, restoring appropriate working conditions. However, once the wheel has resumed low slip conditions, the device allows the braking torque to increase again to the previous value and incipient locking can occur.

The brakes operate then in a cyclic way, with subsequent interventions of the ABS system, thus maintaining the longitudinal force coefficient near its maximum value (Fig. 4.33). Different devices however can operate in

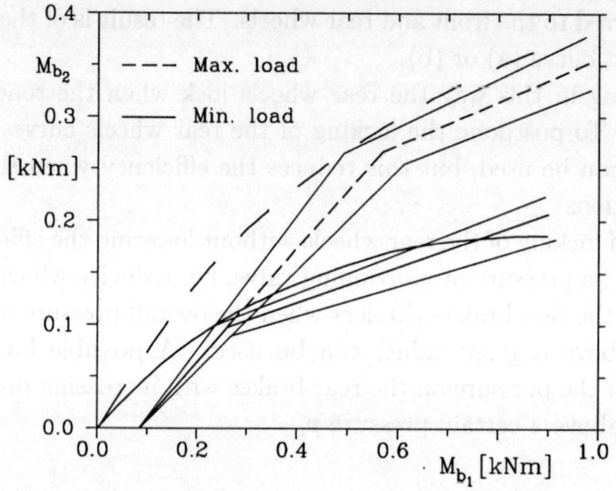

Fig. 4.32 Characteristic of a braking system in which a pressure proportioning valve operating following Eq. (4.84) is present. To take into account the variability of the parameters of the system, mainly the friction coefficient, a band of characteristics has been considered instead of a single line. The ideal braking lines at the two different load conditions have also been plotted.

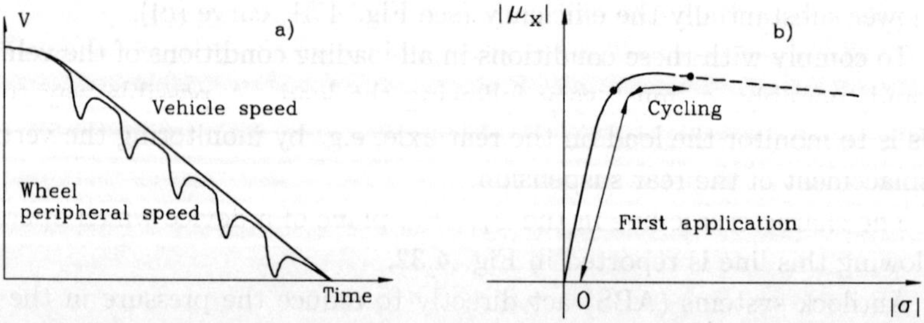

Fig. 4.33 Operation of antilock systems. (a) Time history of the vehicle speed and the peripheral velocity of the wheels during a braking manoeuvre with ABS intervention. (b) Zone of the curve $\mu_x(\sigma)$ in which the ABS device maintains the longitudinal force coefficient.

different ways, both for what the hardware characteristics and the control algorithms are concerned.

The above braking efficiency holds only in case of rigid vehicles. If the presence of suspensions is accounted for, the load transfer from the rear to the front wheels does not occur instantaneously and at the beginning of braking the vertical loads on the wheels are the same as those at constant speed. The body of the vehicle then starts to dive[4] and the load on front wheels increases, until steady state conditions are reached and the loads take the values of Eqs. (4.72) and (4.73). At the beginning of the manoeuvre the load on the rear wheels is higher and the locking of the rear wheels is more difficult: This consideration explains the practice of giving short brake pulses, effective when modern braking systems designed to avoid rear wheels locking were not used.

4.7.3 *Braking power*

The instantaneous power the brakes must dissipate is

$$|P| = |F_x|V = V\left|\frac{dV}{dt}m_e + mg\sin(\alpha)\right| , \tag{4.85}$$

where all forms of drag have been neglected.

The brakes cannot dissipate it directly; they usually work as a heat sink, storing some of the energy in the form of thermal energy and dissipating it in due time. Obviously care must be exerted to design the brakes in such a way that they can store the required energy without reaching too high temperatures and to ensure adequate ventilation for cooling. The average value of the braking power must at any rate be lower than the thermal power the brakes can dissipate.

Nevertheless the road conditions and, above all, the driving style can cause a very intense heating of the brakes. The metal part can be subject to strong thermal stressing and also the braking fluid can overheat. To give an order of magnitude of the temperatures involved, some temperature measurements taken on the front discs, the rear drums and the braking fluid of a car on a downhill mountain road and a hilly road are reported in Fig. 4.34.

Two reference conditions are usually considered: Driving on continuous acceleration-braking cycles and downhill running in which the speed is kept constant with the use of brakes. In the first case the average braking power firstly increases with the average speed, as the total energy to be dissipated increases but then, at high speed it decreases again as the acceleration time increases far more than the braking energy when the vehicle is approaching its maximum speed. The acceleration-deceleration cycling is usually the

[4]Anti-dive arrangements will be described in section 6.1.3.

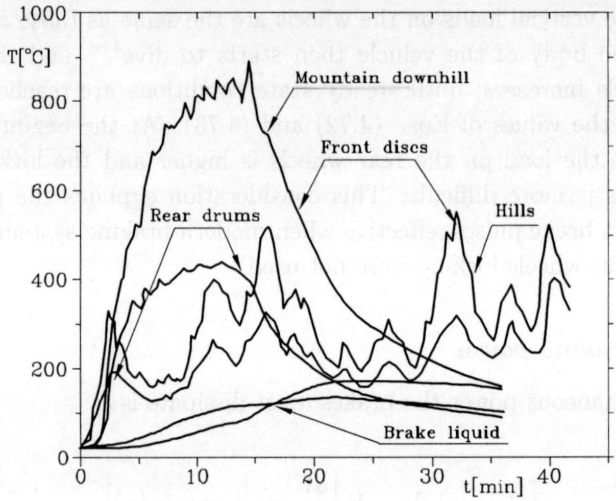

Fig. 4.34 Time history of the temperature of the brakes and of the braking fluid during testing of a car on different roads.

critical condition for passenger vehicles and, above all, for sports cars. Note that this way of driving is typical of racing conditions, where the engine is almost always used at full power unless the brakes are applied, with deceleration at the limit of slipping.

The second condition leads to an average power, which now coincides with the instantaneous power as no heat accumulation in the brakes can be allowed in long downhill driving, increasing linearly with speed at constant slope, until aerodynamic drag becomes large enough to help keeping the speed constant. This condition is critical in industrial vehicles, which must handle with care downhill grades, keeping the speed low by using engine drag and auxiliary systems, sometimes electromagnetic brakes.

In the case the vehicle has a device which allows one to perform regenerative braking, the latter must be able to deal with the instantaneous braking power. As this condition would impose very heavy conditions on the regenerative braking device, usually braking energy is recovered only during gradual slowing down and emergency braking is performed through conventional brakes.

4.8 An outline on standards for braking systems

As the characteristics of the braking system have a great impact on the safety of road vehicles, there are many standards which deal with them. Only a few points of CEE standards (International CEE laws Dir. 79/489, 18 April 1979 and International ECE laws E/ECE/324 and E/ECE/TRANS/505, 24 August 1982) will be dealt with here. The standards usually distinguish between devices for service, emergency and parking braking. The first ones must act on all wheels, with a suitable distribution of the braking force between the axles. The force must be applied symmetrically with reference to the symmetry plane of the vehicle (xz plane)[5].

A minimum efficiency is stated for each vehicle type. The efficiency is measured in terms of the distance travelled between the instant the driver operates the brakes and the one at which the vehicle stops. The braking distance s (measured in m) must be

$$s \leq aV + \frac{V^2}{b} , \tag{4.86}$$

where V is the vehicle speed, in km/h and constants a and b take the following values:
$a = 0.1$ for motor cars (category M1)
$a = 0.15$ for all other motor vehicles
$b = 150$ for cars (category M1)
$b = 130$ for all other passenger vehicles (categories M2 and M3)
$b = 115$ for freight vehicles (categories N1, N2 and N3).

The values of the deceleration obtained with the mentioned values of b are respectively of 5.8 m/s^2, 5 m/s^2 and 4.4 m/s^2. For each category the speed at which the test must be performed and the maximum force which must be exerted on the pedal are prescribed.

As the test must be performed on a road with good conditions, to state a value of the braking distance amounts to stating a minimum value of the braking efficiency η_b. If, for example, $|\mu_p|$=0.8, the mentioned values of the deceleration correspond to values of the efficiency equal to 0.74, 0.64 and 0.56 respectively. Note that such values are quite low, and can be obtained even without devices aimed to control the pressure with particular laws.

[5]A possible exception are Vehicle Dynamics Control (VDC) systems in which differential braking of the wheels is used to produce yaw moments (see Chapter 5).

The requirements must however be met in any load condition.

The braking forces must be subdivided between the axles in such a way to guarantee a deceleration

$$\left|\frac{dV}{dt}\right| \geq g\left[0.1+0.85(|\mu_x|-0.2)\right] , \tag{4.87}$$

or, in terms of braking efficiency

$$\eta_b \geq 0.85 - \frac{0.07}{|\mu_x|} . \tag{4.88}$$

Such efficiency must be granted in such a way that the rear wheels use less of their longitudinal force capacity than front wheels, i.e. $\mu_{x_1} > \mu_{x_2}$. This criterion, which holds for cars, must be applied strictly for $0.15 < |\mu_x|\eta_b < 0.8$, can be overlooked when $0.3 < |\mu_x|\eta_b < 0.4$ provided that η_{b_2}, i.e. the efficiency computed with reference to the rear wheels, is below 1.005.

For vehicles of other categories the force coefficient at the rear wheels is required to be lower than that of the front wheels only when $0.15 < |\mu_x|\eta_b < 0.30$, while when the road conditions are better ($|\mu_x|\eta_b > 0.30$) braking with larger μ_x at the rear wheels is allowed.

The above mentioned conditions allow one to plot a curve on M_{b_1}, M_{b_2} plane. The condition that the front wheels must slip first, i.e. that $\mu_{x_1} > \mu_{x_2}$, allows one to write

$$F_{x_1} = \mu_x F_{z_1} = \mu_x m \left[g\frac{b}{l} + \frac{h_G}{l}\left|\frac{dV}{dt}\right|\right] . \tag{4.89}$$

The corresponding curve on the F_{x_1}, F_{x_2} can be easily obtained by stating a set of values of μ_x and for each one of them computing the deceleration through Eq. (4.87), the force at the front axle using Eq. (4.89) and those at the rear axle in such a way that the total braking force corresponds to that which insures the required deceleration.

This curve, which has been plotted in Fig. 4.29, divides the M_{b_1}, M_{b_2} plane into two zones: The characteristics of the braking system must lie completely in the zone above the curve to comply with the standards, while it must lie below the curve of ideal braking to assure that the rear wheels do not lock.

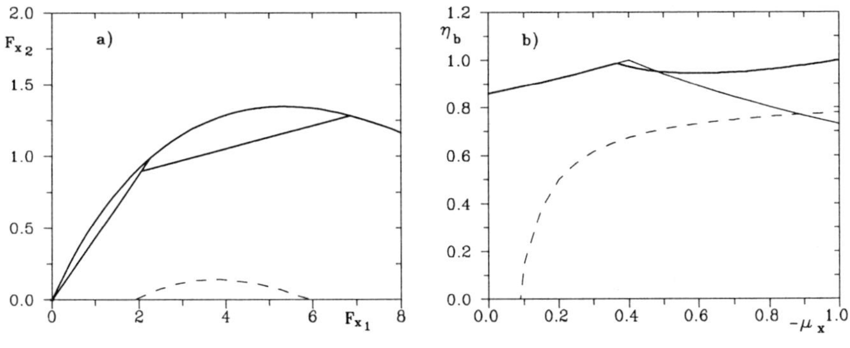

Fig. 4.35 Braking characteristics of the vehicle of Appendix 1.1. (a) Ideal braking conditions and characteristics of the braking system. (b) Braking efficiency with and without pressure proportioning valve. The dashed lines show the minimum conditions stated by CEE standards.

The zone in which the characteristics of the braking system must be included is larger or smaller depending on the allowed excursion of the centre of mass and many other construction parameters. From the graph the need of resorting to devices which modify the braking characteristics can be assessed.

Many other rules apply to vehicles with more than two axles, to trailers and all other devices as ABS systems.

Example 4.12

Plot the braking efficiency of the car of Appendix 1.1, assuming that the braking system is designed to reach the ideal conditions for a longitudinal force coefficient $\mu_x = -0.4$. Use a pressure proportioning valve in such a way that the front wheels lock before the rear ones up to a value of μ_x equal to unity. Neglect aerodynamic forces and rolling resistance.

The curve which characterizes the conditions for ideal braking in the plane F_{x_1}, F_{x_2} is plotted (Fig. 4.35a). In order to obtain the ideal conditions at a value of the longitudinal force coefficient $\mu_x^* = 0.4$, ratio K_B is immediately computed from Eq. (4.81): $K_B = 2.283$. The braking forces corresponding to the ideal conditions are $F_{x_1} = 2.265$ kN and $F_{x_2} = 0.992$ kN.

The pressure proportioning valve is assumed to start acting when values of the forces equal to 90% of those for ideal conditions are reached: $F_{x_1} = 2.038$ kN and $F_{x_2} = 0.893$ kN.

As the point at which the ideal conditions with $\mu_x = 1$ are reached is easily computed ($F_{x_1} = 6.861$ kN and $F_{x_2} = 1.282$ kN), the equation which expresses the force F_{x_2} as a function of F_{x_1} when the valve is operating is immediately found. From its slope, the value of constant $\rho_c = 0.184$ is obtained. The characteristics of the braking system is plotted in Fig. 4.35a.

At each point of the characteristic a pair of values F_{x_1} and F_{x_2} are obtained. From them the deceleration and the maximum value of the longitudinal force coefficient can be computed, obtaining finally the braking efficiency. The results are plotted in Fig. 4.35b. In the same figures the curves related to the CEE standards are also plotted (dashed lines). Note that the position of the centre of gravity results in a very low position of the dashed line in Fig. 4.35a.

Chapter 5

Handling of a rigid vehicle

5.1 Trajectory control in road vehicles

Two basic functions can be identified in all types of vehicles: Propulsion and trajectory control. As far as the second issue is concerned, all vehicles can be divided into two categories:

a) Guided vehicles, or better, kinematically guided vehicles, whose trajectory is fixed by a set of kinematic constraints;
b) piloted vehicles, in which the trajectory, a tri-dimensional or a planar curve, is determined by a guidance system, controlled by a human pilot or by a device, usually electro-mechanical. The guidance system acts by exerting forces on the vehicle which are able to change its trajectory.

In the first case the kinematic constraint exerts all forces needed to modify the trajectory without any deformation, i.e. is assumed to be infinitely stiff and infinitely strong. A perfect kinematic guidance is therefore an abstraction, although it is well approximated in many actual cases.

In the second case the forces are due to the changes of the attitude of the vehicle which in turn are caused by forces and moments due to the guidance devices. These vehicles can be said to be dynamically guided.

Apart from the cases in which the forces needed to change the trajectory are directly exerted by thrusters (usually rockets), there can be two cases: The attitude changes can be quite large, large enough to be directly felt by the pilot or driver, or small enough to be unnoticed. The first case is that of aerodynamically or hydrodynamically controlled vehicles, in which the pilot acts on a control surface, which causes the changes of attitude needed to generate the forces which modify the trajectory. There is also usually a certain delay between the changes of attitude and the actual generation of forces and consequently the drivers feels clearly that a dynamic control, i.e. a control through the application of forces, takes place.

In the case of road vehicles the situation is similar but the driver has a completely different impression: The driver operates the steering wheel causing some wheels to work with a sideslip and to generate lateral forces. These forces cause a change of attitude of the vehicle (change of angle β) and then a sideslip of all wheels: The resulting forces bend the trajectory. However the linearity of the behaviour of the tire and the very high value of the cornering stiffness give the driver the impression of a kinematic, not dynamic, driving. The wheels seem to be in pure rolling and the trajectory seems to be determined by the directions of the midplanes of the wheels.

This impression has influenced the study of the handling of motor vehicles for a long time, originating the very concept of kinematic steering and in a sense hiding the true meaning of the phenomena.

The impressions of the driver is in good accordance with this kinematic approach, at least for all the linear part of the behaviour of the tire. When high values of the sideslip angles are reached, the average driver has the impression of losing control of the vehicle, much more so if this occurs abruptly. This impression is confirmed by the fact that in normal road conditions, particularly if radial tires are used, the sideslip angles become large only when approaching the limit lateral forces.

These considerations are only an indication, as there are cases in between those considered here like kinematic guidance with deformable constraints or magnetic levitation vehicles. The difference turns out to be more quantitative than qualitative, and depends mostly on the greater or smaller "stiffness" with which the vehicle responds to the variations of attitude due to the guidance devices.

5.2 Low-speed or kinematic steering

5.2.1 *Vehicles without trailer*

Low speed or kinematic steering is defined as the motion of a wheeled vehicle determined by pure rolling of the wheels. The velocities of the centres of all the wheels lie in their midplane, i.e., the sideslip angles α are vanishingly small. In these conditions the wheels can exert no cornering force to balance the centrifugal force due to the curvature of the trajectory. Kinematic steering is possible only if the velocity is vanishingly small.

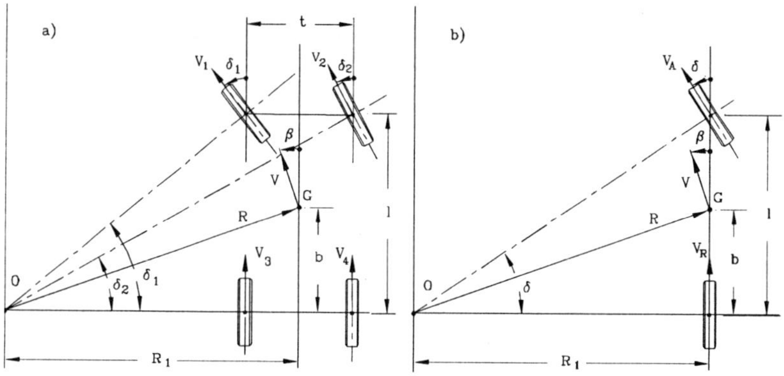

Fig. 5.1 Kinematic steering of a four-wheeled and a two-wheeled vehicle.

Consider a vehicle with 4 wheels, two of which can steer (Figure 5.1). The relationship which must be verified to allow kinematic steering is easily found by imposing that the perpendiculars to the midplanes of the front wheels meet the one of the rear wheels at the same point

$$\tan(\delta_1) = \frac{l}{R_1 - \frac{t}{2}}, \qquad \tan(\delta_2) = \frac{l}{R_1 + \frac{t}{2}} . \tag{5.1}$$

Instead of the track t, Eq. (5.1) should contain the distance between the kingpin axes of the wheels, or better, between their intersections with the ground. By eliminating R_1 between the two equations, a direct relationship between δ_1 and δ_2 is readily found

$$\cot(\delta_1) - \cot(\delta_2) = \frac{t}{l} . \tag{5.2}$$

A device which allows to steer the wheels complying exactly with Eq. (5.2) is usually referred to as *Ackerman steering* or *Ackerman geometry*. No actual steering mechanism allows to follow exactly such law and a steering error, defined as the difference between the actual value of δ_2 and that obtained from Eq. (5.2) can be obtained as a function of δ_1.

Consider for instance the device based on an articulated quadrilateral shown in Fig. 5.2a. The relationship linking angle δ_1 to angle δ_2 is[1]

[1]G. Pollone *Il veicolo*, Levrotto & Bella, Torino, 1970

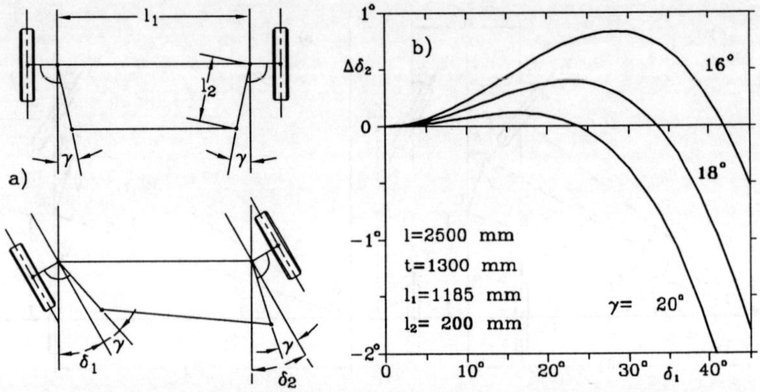

Fig. 5.2 Steering mechanism based on an articulated quadrilateral. (a) Sketch; (b) steering error $\Delta\delta_2 = \delta_2 - \delta_{2_c}$ as a function of δ_1.

$$\sin(\gamma - \delta_2) + \sin(\gamma + \delta_1) =$$

$$= \sqrt{\left[\frac{l_1}{l_2} - 2\sin(\gamma)\right]^2 - \left[\cos(\gamma - \delta_2) - \cos(\gamma - \delta_1)\right]^2} \,. \qquad (5.3)$$

A steering error $\Delta\delta_2 = \delta_2 - \delta_{2_c}$, i.e. the difference between the actual value of δ_2 and the kinematically correct one can be computed for each value of δ_1, as shown in Fig. 5.2b. Three values of angle γ have been considered: 16°, 18° and 20°; the higher the value of γ the lower the error is for small values of the steering angle. However, low values of the error at low steering angles are accompanied by large errors at large steering angles and a trade-off is needed: In the case of the figure a value of 18° can be a good starting point.

Much effort has been devoted to design devices allowing to minimize this error; the importance of this issue has however been overstated from the viewpoint of the directional response of the vehicle: The facts that (a) a sideslip angle is always present, (b) most suspension mechanisms allow a certain amount of roll steer, (c) in most cases the steering wheels are intentionally not exactly parallel but have a certain toe-in and (d) the deformations of the suspensions induce small angles depending on the forces exerted by the wheel on the road, reduce the importance of small steering errors.

The steering error has on the contrary a larger effect on the wear of the front tires and on the centring torques of the steering system, the latter affecting the feel the driver obtains from the steering wheel. It is important that the torque increases steadily with the steer angle, a feature which is obtained with a correct Ackerman geometry.

The radius of the trajectory of the centre of mass of the vehicle is

$$R = \sqrt{b^2 + R_1^2} = \sqrt{b^2 + l^2 \cot^2(\delta)}\ , \tag{5.4}$$

where δ is the steering angle of the "equivalent" two-wheeled vehicle (Fig. 5.1 b). Although it should be computed by averaging the cotangents of the angles of the two wheels,

$$\cot(\delta) = \frac{R_1}{l} = \frac{\cot(\delta_1) + \cot(\delta_2)}{2}\ , \tag{5.5}$$

it is very close to the direct average of the angles. Consider for example the same vehicle of Fig. 5.2 with centre of mass at midwheelbase on a curve with a radius $R = 10$ m. The correct values of the steering angles are $\delta_1 = 15.090°$, $\delta_2 = 13.305°$ and $\delta = 14.142°$. By direct averaging the steering angles of the wheels, it would follow that $\delta = 14.197°$, with an error of only 0.36%.

In case the radius of the trajectory is large if compared with the wheelbase of the vehicle, Eq. (5.4) reduces to

$$R \approx l \cot(\delta) \approx \frac{l}{\delta}\ . \tag{5.6}$$

Equation (5.6) can be rewritten in the form

$$\frac{1}{R\delta} = \frac{1}{l}\ . \tag{5.7}$$

The expression $1/R\delta$ has an important physical meaning: It is the ratio between the response of the vehicle, in terms of curvature $1/R$ of the trajectory, and the input which causes it. It is therefore a sort of transfer function for the directional control and can be referred to as *trajectory curvature gain*. In kinematic steering conditions it is equal to the reciprocal of the wheelbase.

Another important transfer function is the ratio β/δ. The sideslip angle of the vehicle at its centre of mass can be expressed as a function of the radius of the trajectory R as

$$\beta = \arctan\left(\frac{b}{\sqrt{R^2 + b^2}}\right). \tag{5.8}$$

By linearizing Eq. (5.8) and introducing the expression (5.7) which links R with δ, it follows

$$\frac{\beta}{\delta} = \frac{b}{l}. \tag{5.9}$$

Ratio β/δ can be referred to as *sideslip angle gain*.

If both axles can steer, the optimum condition for low-speed steering is that in which the steering angles are equal and opposite: In this case the radius of curvature is half of that of the same vehicle with the same steering angle but only front wheel steering.

In case of a vehicle with more than two axles, a true kinematic steering is possible only if the wheels of several axles (all except one) can steer and if the steering angles comply with conditions similar to equation (5.1). In order to avoid strong wear of the tires it is possible to lift one axle from the ground in certain conditions: In some countries it is legal to design the suspensions in such a way that not all axles are on the ground when the vehicle is unloaded while in others this is not allowed. Some axles can be lifted for low-speed manoeuvring while being on the ground in normal driving. Some axles can also be self-steering, i.e. the wheels are allowed to orient themselves in such a way to minimize sideslip. Clearly an axle of this type cannot exert side forces and reduce the overall cornering ability of the vehicle. Again different laws hold in different countries, sometimes allowing the use of self-steering axles in normal driving and sometimes stating that self-steering axles must be blocked except in low speed manoeuvres. In the case of a three-axle vehicle with nonsteering axles near to each other, an approximation as the one shown in Fig. 5.3 can be used to study low speed steering.

Particularly in the case of long vehicles, the off-tracking distance, i.e. the difference of the radii of the trajectories of the front and the rear wheels is an important parameter. If R_f is the radius of the trajectory of the front wheels, the off-tracking distance is

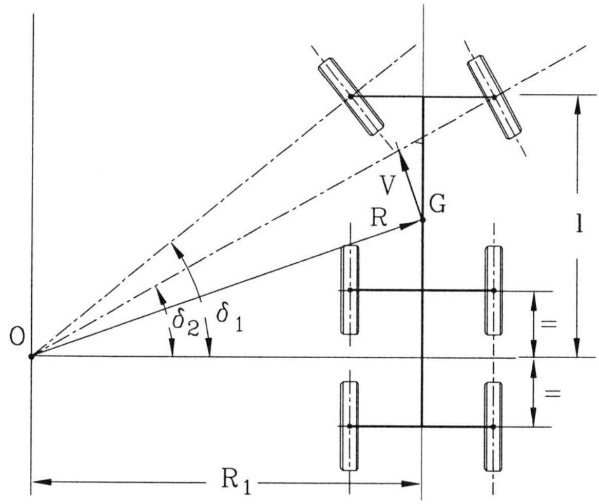

Fig. 5.3 Low speed steering of industrial vehicles; approximate kinematic condition for a truck with three axles.

$$R_f - R_1 = R_f \left\{ 1 - \cos \left[\arctan \left(\frac{l}{R_1} \right) \right] \right\} . \tag{5.10}$$

If the radius of the trajectory is large when compared to the wheelbase, Eq. (5.10) reduces to

$$R_f - R_1 \approx R \left[1 - \cos \left(\frac{l}{R} \right) \right] \approx \frac{l^2}{2R} . \tag{5.11}$$

5.2.2 *Vehicles with trailer*

If the vehicle has a trailer with one axle or a trailer with two axles, in which the front one is attached to the draw bar, kinematic steering is always possible if the tractor allows it.

Generally speaking, if the wheels of the trailer are fixed the trailer follows a trajectory which is internal to that of the tractor. In the case of the vehicle of Fig. 5.3a radius R_T is

$$R_T = \sqrt{R_1^2 + l_A^2 - l_T^2} . \tag{5.12}$$

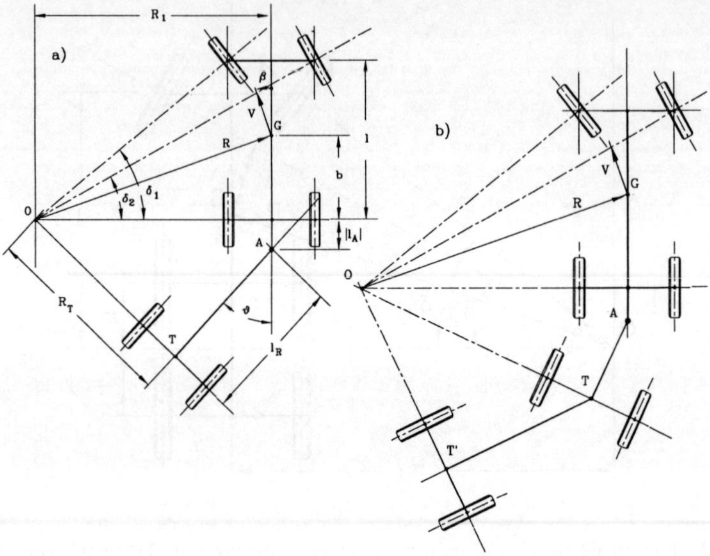

Fig. 5.4 Low speed steering of vehicles with trailer. (a) steering of a vehicle with a trailer with one axle or an articulated vehicle; (b) steering of a vehicle with a trailer with two axles.

The value of ratio θ/δ is, in linearized conditions,

$$\frac{\theta}{\delta} = \frac{l_A + l_T}{l}, \tag{5.13}$$

where l_A is positive if point A is outside the wheelbase.

In the case of Fig. 5.4b the radius of the trajectory of the trailer can be obtained by considering the latter as two subsequent trailers of the type already considered.

The only way to prevent the trailer from following a trajectory which is internal to the one of the tractor is to provide its wheels with a steering mechanism. The steering angle of the last axle must be opposite to the one of the tractor.

Note that the trajectory of the trailer is circular only after a certain time: When the tractor starts to follow a circular trajectory there is an initial transient in which the path of the trailer starts to bend and a long time is required in order to reach the steady state conditions. The trajectory of the trailer, or better of point T of Fig. 5.4a, can be computed as follows.

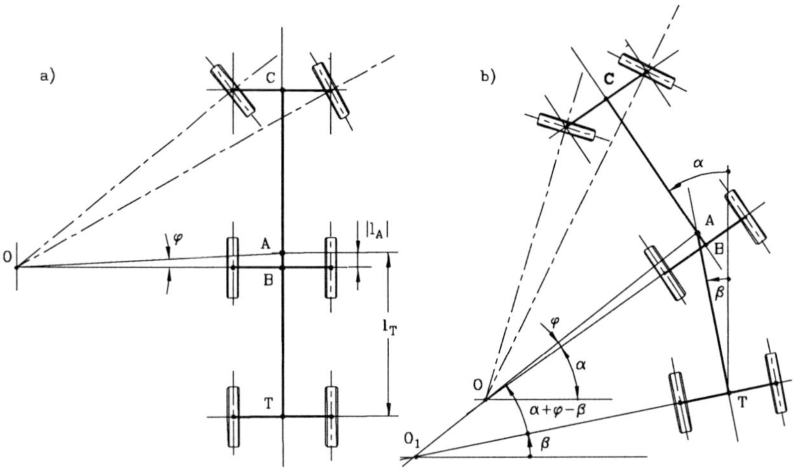

Fig. 5.5 Vehicle with two axles pulling a trailer with one axle. (a) Situation at time $t = 0$ with the vehicle in straight position; (b) Situation at time t.

In Fig. 5.5a the vehicle is sketched in its initial configuration with the trailer and tractor aligned; the generic configuration at time t is shown in Fig. 5.5b. In the second figure the tractor is rotated by an angle α and the trailer is rotated by an angle β. Note that angle ϕ is positive if A lies between B and C.

The positions of the centre of rotation of the tractor O and of the trailer O_1 at time t and $t + dt$ are shown in Fig. 5.6. Distances $\overline{\mathrm{RR}'}$, $\overline{\mathrm{AA}'}$ and $\overline{\mathrm{RR}''}$ are very small if compared with $\overline{\mathrm{AR}}$ and $\overline{\mathrm{A'R'}}$. Neglecting vanishingly small quantities, it follows that

$$\begin{array}{l} \overline{\mathrm{AA}'} = R_A d\alpha \\ \overline{\mathrm{A'A''}} = l_T d\beta = \overline{\mathrm{AA}'} \sin(\alpha + \phi - \beta)\ . \end{array} \tag{5.14}$$

Eqs. (5.14) yield

$$\frac{d\beta}{d\alpha} = \frac{R_A}{l_T} \sin(\alpha + \phi - \beta)\ . \tag{5.15}$$

Since $\alpha = \beta = 0$ at time $t = 0$, Eq. (5.15) can be easily integrated numerically. The radius of the trajectory of the trailer R_T is

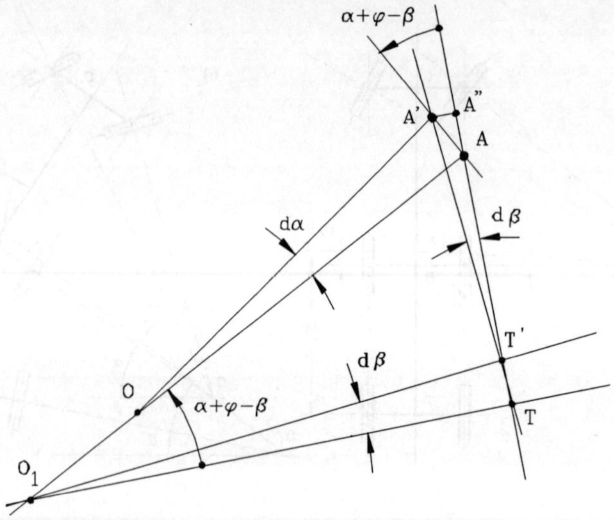

Fig. 5.6 Position of the vehicle of Fig. 5.5 at time t and $t + dt$.

$$R_T = \frac{l_T}{\tan(\alpha + \phi - \beta)} \ . \tag{5.16}$$

A long trailer on a narrow bend requires a change of direction of more than 90° before steady state conditions are reached and its trajectory becomes almost circular.

Many articulated vehicles now employ a steering rear axle. As already said, the difference between the radius of the path of the trailer and that of the tractor can be reduced and this allows the use of a narrower road but there are drawbacks too, as it becomes impossible for the driver to see the position of the external rear corner of the trailer which, at the beginning of the curve, seems to move laterally outside. The dynamic problems linked with steering rear axles of trailers will be dealt with later.

The low-speed steering of a vehicle with a trailer with two axles as the one shown in Fig. 5.3b can be dealt with the same equations seen above, applied to both the simple trailers which model the actual two-axle trailer. Note that the trajectory of the first one is initially not circular and this must be taken into account while integrating numerically Eq. (5.15).

Example 5.1

Study the conditions for kinematic steering of the articulated vehicle of Appendix 1.4. Assume a value of the radius of the centre mass of the tractor of 10 m and compute the trajectory of the trailer. Assume that the trailer has a single axle, coinciding with the third axle of the actual trailer.

The radius of the trajectories of the front and rear axles of the tractor are easily computed as 9.730 and 10.335 m; the off-tracking of the tractor is thus of 605 mm. The approximated expression (5.11) for the off-tracking yields 607 mm, very close to the correct value even if the radius of the trajectory is not actually very large if compared with the wheelbase (10 m versus 3.485 m).

The steering angles of the front wheels are 17.99° and 21.77° with an average value of 19.71°. This value is also very close to the correct value of 19.88° obtained without any linearization and to the linearized value of 19.77°.

The steady state radius of the trajectory of the trailer is of 5.446 m, yielding a value of 4.889 m for the total off-tracking distance.

The trajectory of the trailer has been computed by integrating numerically Eq. (5.15) for α spanning from 0 to 450°, with a step of 0.5°. The values of ϕ and R_A are respectively of 2.648° and 9.740 m. The trajectory and the locus of points O′ are reported in Fig. 5.7. Note that after a rotation of 90° the radius of the trajectory is still larger than that in steady state conditions.

5.3 Simplified model: Ideal steering

If the speed is not vanishingly small, the wheels must move with suitable sideslip angles to generate cornering forces. A simple evaluation of the steady state steering of a vehicle in high-speed or dynamic[2] steering conditions can be performed as follows. Consider a rigid vehicle moving on

[2]The term *dynamic steering* is used here to denote a condition in which the trajectory is determined by the balance of forces acting on the vehicle, as opposed to *kinematic steering* in which the trajectory is determined by the directions of the midplane of the wheels. Note that dynamic steering applies to both steady state and unstationary turning.

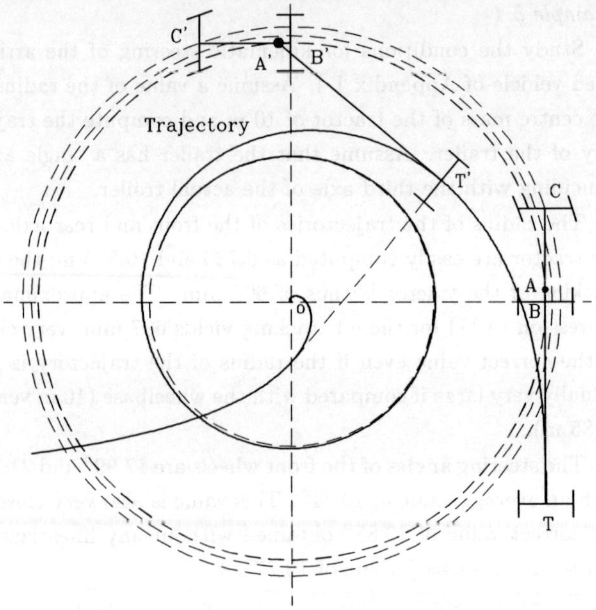

Fig. 5.7 Trajectory and locus of the centres of curvature of the trajectory of the trailer for an articulated vehicle. The positions of the vehicle before starting on the curved path and after a rotation of the trailer of 90° are reported.

level road with transversal slope angle α_t and neglect the aerodynamic side force. Define a η-axis parallel to the road surface, passing through the centre of mass of the vehicle and intersecting the vertical for the centre of the trajectory, which in steady-state condition is circular (Fig. 5.8). Axis η does not coincide with y axis, except at one particular speed.

The equilibrium equation in η direction is immediately written by equating the components of weight mg, of centrifugal force mV^2/R and of the forces P_η due to the tires

$$\frac{mV^2}{R}\cos(\alpha_t) - mg\sin(\alpha_t) = \sum_{\forall i} P_{\eta_i} . \tag{5.17}$$

For a first approximation study, forces P_η can be confused with the cornering forces F_y of the tires and all wheels can be assumed to work with the same side force coefficient μ_y. As the last assumption is similar to that seen for braking in ideal conditions, this approach will be referred to as

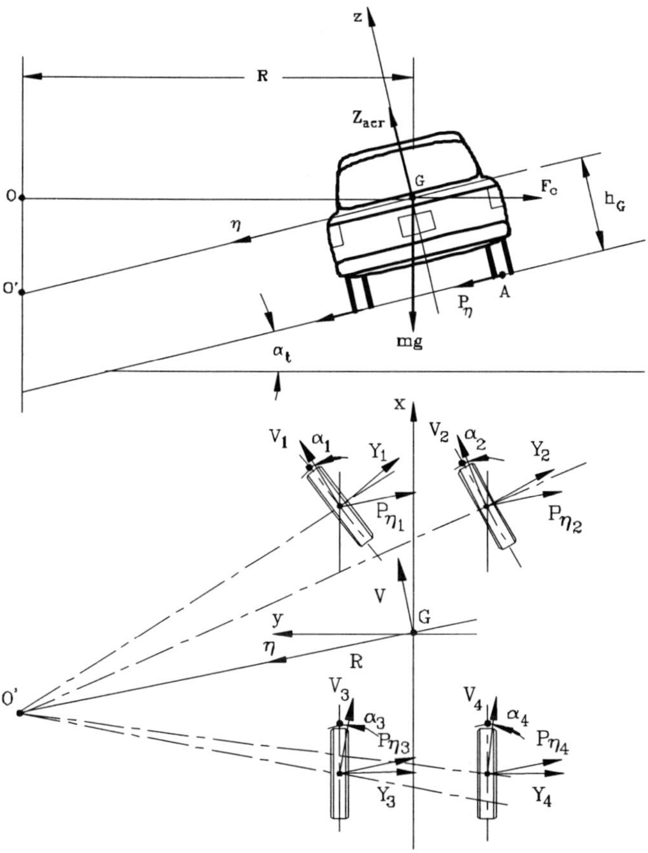

Fig. 5.8 Simplified model for dynamic steering.

"ideal steering". These two assumptions lead to substituting the expression $\sum_{\forall i} P_{\eta_i}$ with $\mu_y F_z$.

Force $F_z = \sum F_{z_i}$ exerted by the vehicle on the road is

$$F_z = mg\cos(\alpha_t) + \frac{mV^2}{R}\sin(\alpha_t) - \frac{1}{2}\rho V^2 S C_z \ . \tag{5.18}$$

By introducing Eq. (5.18) into Eq. (5.17) the ratio between the lateral acceleration and the gravitational acceleration g is

$$\frac{V^2}{Rg} = \frac{\tan(\alpha_t) + \mu_y(1 - MV^2)}{1 - \mu_y \tan(\alpha_t)}, \tag{5.19}$$

where

$$M = \frac{\rho S C_z}{2mg\cos(\alpha_t)}$$

in such a way that MV^2 is the ratio between the aerodynamic lift and the component of weight in direction perpendicular to the road surface. Note that M is negative if the lift is directed downwards.

By introducing the maximum value of the side force coefficient μ_{y_p} into Eq. (5.19) it is possible to obtain the maximum value of the lateral acceleration

$$\left(\frac{V^2}{R}\right)_{max} = g f_s , \tag{5.20}$$

where the so-called sliding factor f_s can be defined as[3]

$$f_s = \frac{\tan(\alpha_t) + \mu_{y_p}(1 - MV^2)}{1 - \mu_{y_p} \tan(\alpha_t)} \tag{5.21}$$

and is in general a function of the speed, if the aerodynamic lift is accounted for.

The sliding factor is reported as a function of μ_{y_p} for different values of the transversal slope of the road in Fig. 5.9a and for different values of ratio MV^2 in Fig. 5.9b. Note that if the road is flat and the aerodynamic lift is neglected it reduces to the maximum value of the side force coefficient μ_{y_p}.

The maximum speed at which a bend with radius R can be negotiated is

$$V_{max} = \sqrt{Rg}\sqrt{\frac{\tan(\alpha_t) + \mu_{y_p}}{1 - \mu_{y_p}\left[\tan(\alpha_t) - RgM\right]}} . \tag{5.22}$$

[3]The sliding factor is more commonly defined as the square root of the same quantity considered here. The present definition, which refers directly to the lateral acceleration instead of the speed at which a given radius can be obtained, is here preferred as in particular conditions it reduces to the side force coefficient.

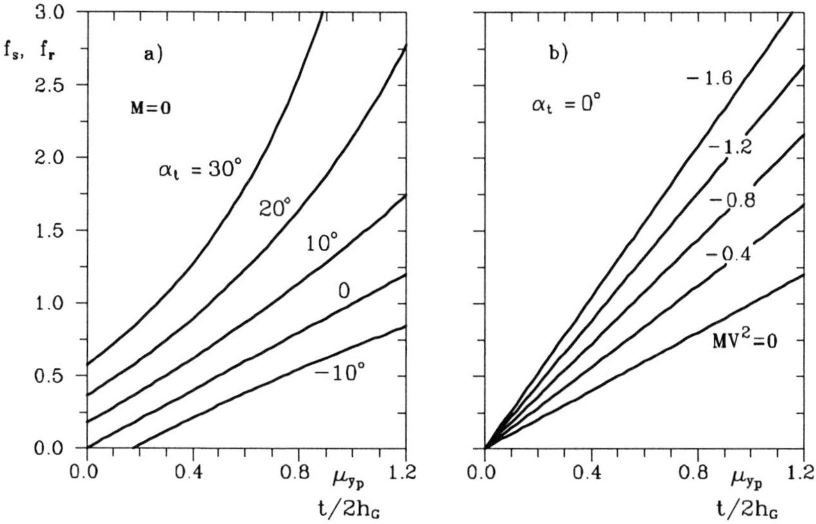

Fig. 5.9 Sliding and rollover factors as functions of μ_{y_p} and of $t/2h_G$ respectively for roads with different transversal slope (a) and for vehicle with different values of ratio MV^2 (b).

The limitation to the maximum lateral acceleration due to the cornering force the tires can exert is however not the only one, at least theoretically. A further one can come from the danger of rollover which occurs if the resultant of forces in yz plane crosses the road surface outside point A (Fig. 5.8).

The limit condition for rollover is then

$$\left(\frac{V^2}{R}\right)_{max} = g f_r \;, \tag{5.23}$$

where the rollover factor can be defined as

$$f_r = \frac{\tan(\alpha_t) + \frac{t}{2h_G}(1 - MV^2)}{1 - \frac{t}{2h_G}\tan(\alpha_t)} \;. \tag{5.24}$$

Its expression is identical to that of the sliding factor, once ratio $t/2h_G$ has been substituted for μ_{y_p} (Fig. 5.9). It depends on the speed owing to aerodynamic lift.

The maximum lateral acceleration is then

$$\left(\frac{V^2}{R}\right)_{max} = g \min\{f_s, f_r\} . \tag{5.25}$$

Whether the limit condition first reached is that related to sliding, with subsequent spin out of the vehicle, or related to rolling over depends on the relative magnitude of f_s and f_r. If $f_s < f_r$, as it often occurs, the vehicle spins out. This condition can be written in the form

$$\mu_{y_p} < \frac{t}{2h_G} .$$

The value of μ_{y_p} at which rollover can occur is as high as 1.2–1.7 for sports cars, 1.1–1.6 for saloon cars, 0.8–1.1 for pickup and passenger vans and 0.4–0.8 for heavy and medium trucks. Only in the latter case rollover seems to be an actual possibility.

The present model is only a rough approximation of the actual situation, as is based on the assumption that the side force coefficients μ_y of all wheels are equal, which implies that all wheels work with the same sideslip angle α. Also it ignores the effect of the different directions of the cornering forces of the various wheels which should be considered as perpendicular to the midplanes of the wheels and not directed along η axis. The load transfer between the wheels of the same axle and the presence of the suspensions have also been neglected, other assumptions which contribute to the lack of precision of this model.

If the maximum speed at which a circular path can be negotiated is measured during a steering pad test and the value of the lateral force coefficient is computed through Eq. (5.21), a value of μ_{y_p} which is well below that obtained from tests on the tires is obtained. The cornering force coefficient obtained in this way is that of the vehicle as a whole and the difference between its value and that referred to the tires gives a measure of how well the vehicle is able to exploit the cornering characteristics of its wheels.

The side force coefficient measured on the whole vehicle depends also on the radius of the trajectory, with a notable decrease on narrow bends. The majority of industrial and passenger vehicles are able to use only a fraction from 50% to 70% of the potential cornering force of the tires, with higher values found only in the case of sports cars. This reduction of the lateral forces makes the danger of rollover a more remote one. Actually rolling over in quasi-static condition is impossible for most vehicles; which is not in contrast with the observation that rollover actually occurs in many road accidents. Its occurrence can usually be ascribed to dynamic phenomena in nonstationary conditions or to lateral forces due to side contacts, e.g. of the wheels with the curb of the road, which rule out the possibility of side slipping and then cause far stronger lateral forces to be exerted on the wheels. Also the presence of the suspensions modifies this picture, making rollover a likely outcome of many accidents.

From the equations it is also clear that only the use of aerodynamic devices able to exert a strong negative lift allows to reach high values of lateral acceleration, which can be well above 1 g in the case of racers.

The cornering dynamics of a vehicle with two wheels is radically different from that of four wheeled vehicles (Fig. 5.10). If the gyroscopic moments of the wheels are neglected, the equation expressing the rolling equilibrium can be used to compute the roll angle the vehicle must maintain in order not to capsize, as a two-wheeled vehicle is a system underconstrained in roll. The limitation for the lateral acceleration and the speed on the trajectory comes from lateral sliding only, with a further geometrical limit on the maximum roll angle which can be reached before the vehicle or the driver touches the road on one side. Equation (5.20) yielding the maximum lateral acceleration still holds, with the difference that usually the global side force coefficient is higher.

The roll angle is easily computed

$$\phi = \arctan\left(\frac{V^2}{Rg}\right) , \tag{5.26}$$

and the geometrical limitation $\phi \le \pi/2 + \alpha_t - \gamma$ (Fig. 5.10) usually does not induce further limitations.

If gyroscopic moments of the wheels are considered further terms in the relevant equations must be introduced. When the vehicle runs on a circular path with radius R, the gyroscopic moment due to the i-th wheel with radius R_i and moment of inertia J_{p_i} about its spin axis is equal to

$$J_{p_i} V^2 \cos(\phi)/RR_i \; .$$

The equation expressing the equilibrium for rolling motions is then

$$mgh_G \sin(\phi) - \frac{V^2}{R}\cos(\phi)\left[mh_G - \sum_{\forall i}\left(\frac{J_{p_i}}{R_i}\right)\right] = 0 \; . \tag{5.27}$$

The roll angle is

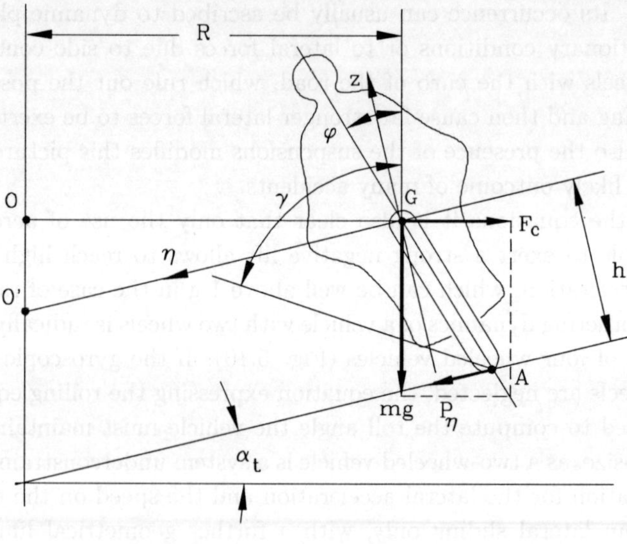

Fig. 5.10 High speed steering of a two wheeled vehicle. Point G is the centre of mass of the vehicle-driver system and can be displaced from the plane of symmetry of the former if the latter is displaced on a side as usually occurs in bends.

$$\phi = \arctan\left\{\frac{V^2}{Rg}\left[1 + \frac{1}{mh_G}\sum_{\forall i}\left(\frac{J_{p_i}}{R_i}\right)\right]\right\} . \tag{5.28}$$

Generally speaking, the effect of the gyroscopic moment of the wheels on the dynamic behaviour of the whole vehicle is small even in the case of vehicles with two wheels. Usually gyroscopic moments are important only in the dynamics of the steering device.

5.4 High-speed cornering of a rigid vehicle

5.4.1 *Equations of motion*

If a motor vehicle is considered as a rigid body moving on a surface, a model with three degrees of freedom is needed for the study of its motion. If the road is considered as a flat surface, the motion is planar. By using the inertial reference frame[4] XY shown in Fig. 5.11, it is possible to use

[4]Strictly speaking such reference frame is not inertial, as it is fixed to the road surface and hence it follows the motion of the earth. It is however "enough" inertial for the

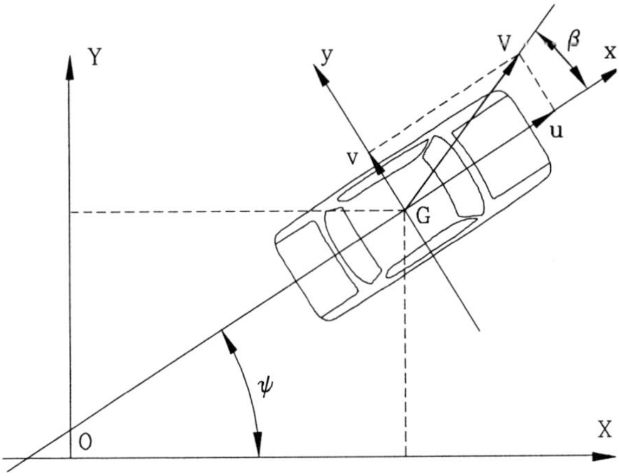

Fig. 5.11 Reference frame for the study of the motion of a rigid vehicle. The vehicle has three degrees of freedom and the coordinates X and Y of the centre of mass G and the yaw angle ψ can be used as generalized coordinates.

the coordinates X and Y of the centre of mass G of the vehicle and the yaw angle ψ between x and X axes as generalized coordinates.

The equations of motion of the vehicle are

$$\begin{cases} m\ddot{X} = F_X \\ m\ddot{Y} = F_Y \\ J_z\ddot{\psi} = M_z \ , \end{cases} \tag{5.29}$$

where F_X, F_Y and M_z are the total forces acting in X and Y directions and the total yawing moment.

Equations (5.29) are very simple but include the forces acting on the vehicle in the directions of the axes of the inertial frame. They are clearly linked with the forces acting in the directions of axes x and y of the vehicle by the obvious relationship

$$\begin{Bmatrix} F_X \\ F_Y \end{Bmatrix} = \begin{bmatrix} \cos(\psi) & -\sin(\psi) \\ \sin(\psi) & \cos(\psi) \end{bmatrix} \begin{Bmatrix} F_x \\ F_y \end{Bmatrix} . \tag{5.30}$$

If the model is used to perform a numerical integration in time they can be used directly without any difficulty. However, if the model has to be used to obtain linearized equations in order to gain a general insight on the

problems here studied and this issue will not be dealt with any further behaviour of the vehicle, it is better to write the equations of motion with reference to the non-inertial xy frame, to avoid dealing with the trigonometric functions of angle ψ, which is in general not a small angle and would make linearizations impossible.

Although there are many ways to obtain the mathematical model, a procedure based on Lagrange equations will be followed here, as it is consistent with what will be done later for more complex models. Note that in the present case other approaches would have been more straightforward.

The kinetic energy of the vehicle is

$$\mathcal{T} = \frac{1}{2}m\left(\dot{X}^2 + \dot{Y}^2\right) + \frac{1}{2}J_z\dot{\psi}^2 . \tag{5.31}$$

Note that the rotational kinetic energy of the wheels has been neglected: No gyroscopic effect of the wheels will be obtained in this way.

The equations of motion are

$$\frac{d}{dt}\left(\frac{\partial \mathcal{T}}{\partial \dot{q}_i}\right) - \frac{\partial \mathcal{T}}{\partial q_i} = Q_i , \tag{5.32}$$

where coordinates q_i are X, Y and ψ and forces Q_i are the corresponding generalized forces F_X, F_Y and M_z.

The derivatives needed to write the equations of motion are

$$\frac{\partial \mathcal{T}}{\partial \dot{X}} = m\dot{X} , \qquad \frac{\partial \mathcal{T}}{\partial \dot{Y}} = m\dot{Y} , \qquad \frac{\partial \mathcal{T}}{\partial \dot{\psi}} = J_z\dot{\psi} , \qquad \frac{\partial \mathcal{T}}{\partial X} = \frac{\partial \mathcal{T}}{\partial Y} = \frac{\partial \mathcal{T}}{\partial \psi} = 0 . \tag{5.33}$$

By remembering that

$$\begin{Bmatrix} \dot{X} \\ \dot{Y} \end{Bmatrix} = \begin{bmatrix} \cos(\psi) & -\sin(\psi) \\ \sin(\psi) & \cos(\psi) \end{bmatrix} \begin{Bmatrix} u \\ v \end{Bmatrix} , \tag{5.34}$$

and performing the relevant derivatives with respect to time, it follows that

$$\begin{cases} m\left[\left(\dot{u} - \dot{\psi}v\right)\cos(\psi) - \left(\dot{v} + \dot{\psi}u\right)\sin(\psi)\right] = F_X \\ m\left[\left(\dot{u} - \dot{\psi}v\right)\sin(\psi) + \left(\dot{v} + \dot{\psi}u\right)\cos(\psi)\right] = F_Y \\ J_z\ddot{\psi} = M_z . \end{cases} \tag{5.35}$$

By premultiplying the set of the first two equations (5.35) by the inverse of the rotation matrix and introducing forces F_x and F_y written with reference to xy frame, it follows

$$\begin{cases} m\left(\dot{u}-\dot{\psi}v\right)=F_x \\ m\left(\dot{v}+\dot{\psi}u\right)=F_y \\ J_z\ddot{\psi}=M_z \, . \end{cases} \tag{5.36}$$

Equations (5.36) are nonlinear in the velocities u, v and $\dot{\psi}$. Their linearization is however possible.

5.4.2 *Sideslip angles of the wheels*

The sideslip angles of the wheels can be expressed easily in terms of the generalized velocities. With reference to Fig. 5.12, the velocity of the centre P_i of the contact area of the i-th wheel, located in a point whose coordinates are x_i and y_i in the reference frame of the vehicle, is

$$\vec{V}_{P_i}=\vec{V}_G+\dot{\psi}\Lambda\overline{(\mathrm{P}_i-\mathrm{G})} \, . \tag{5.37}$$

Angle β_i between the direction of the velocity of point P_i and x-axis is

$$\beta_i=\arctan\left(\frac{v_i}{u_i}\right)=\arctan\left(\frac{v+\dot{\psi}x_i}{u-\dot{\psi}y_i}\right) \, . \tag{5.38}$$

If the i-th wheel has a steering angle δ_i, its sideslip angle is

$$\alpha_i=\beta_i-\delta_i=\arctan\left(\frac{v+\dot{\psi}x_i}{u-\dot{\psi}y_i}\right)-\delta_i \, . \tag{5.39}$$

5.4.3 *Forces acting on the vehicle*

As already stated, if the vehicle has more than three wheels, the forces it exerts on the road in normal direction are undetermined. This feature is strictly linked with the assumption of rigid vehicle and therefore is unavoidable within the present model. However it can be circumvented by taking into account the presence of the suspensions in the computation of the forces while neglecting them into the dynamic study of the vehicle.

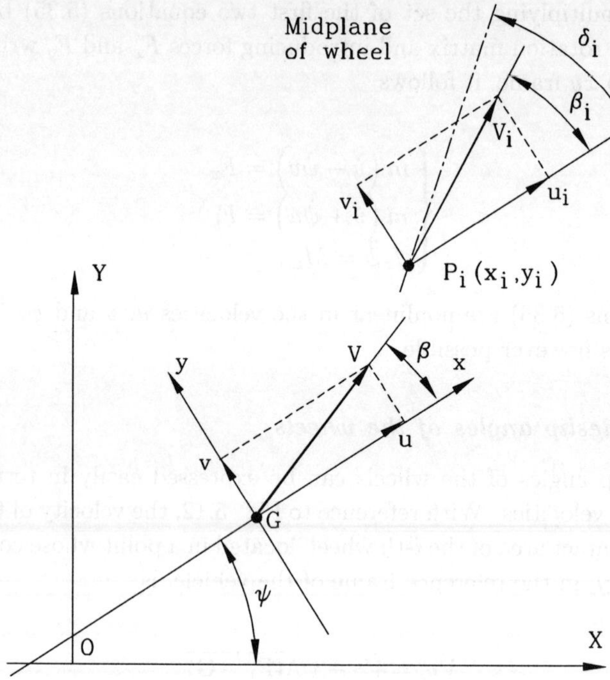

Fig. 5.12 Position and velocity of the centre P_i of the contact area of the i-th wheel.

The forces acting on the various axles in symmetrical conditions have been obtained in Sec. 4-1. By substituting the longitudinal acceleration $\dot{V}$ with $\dot{u} - \dot{\psi}v$ it is possible to use the same relationships for non-symmetrical conditions, as those occurring when moving on a curved path. In these conditions however the forces acting on each wheel cannot be simply obtained as half (or one quarter, for axles with four wheels) of the total force on the axle.

The forces F_{z_l} and F_{z_r} acting on the left and right wheels of the i-th axle can be expressed as

$$\begin{cases} F_{z_{i_l}} = \dfrac{F_{z_i}}{2} + \Delta F_{z_i} \\ F_{z_{i_r}} = \dfrac{F_{z_i}}{2} - \Delta F_{z_i}, \end{cases} \tag{5.40}$$

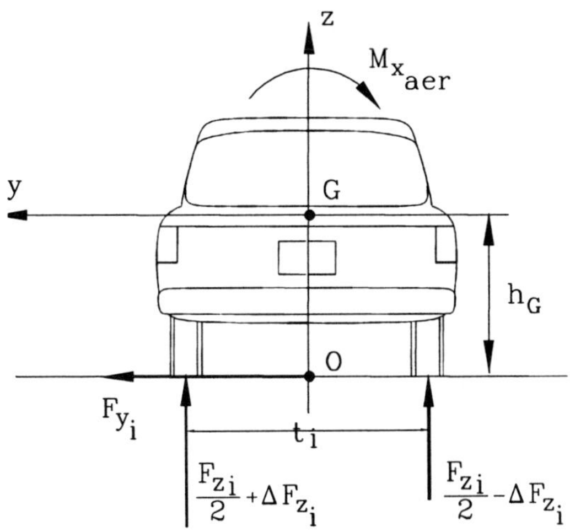

Fig. 5.13 Load shift of the i-th axle.

where F_{z_i} and ΔF_{z_i} are respectively the total load and the transversal load shift of the relevant axle (Fig. 5.13). The latter is positive if the wheels in the half-plane with positive y (left wheels) are more loaded.

With reference to Fig. 5.13 the following equation expressing equilibrium to rotations about x-axis can be written

$$\sum_{\forall i} F_{y_i} h_G + \sum_{\forall i} t_i \Delta F_{z_i} + M_{x_{aer}} = 0 . \tag{5.41}$$

Note that forces F_{y_i} are referred to the frame of the vehicle and not to that of the wheels: If some of them are steered the relevant steering angles must be accounted for. The rolling aerodynamic moment $M_{x_{aer}}$ can usually be neglected.

From Eq. (5.41) the total transversal load shift $\sum_{\forall i} t_i \Delta F_{z_i}$ can be immediately obtained.

On the contrary the assumption of rigid vehicle prevents to compute the load transfer acting on each single axis. However, it is possible to reintroduce the compliance of the suspensions only for the evaluation of the forces. The load shift on the i-th axle can be linked to the roll angle ϕ by the equation

$$t_i \Delta F_{z_i} = K_{t_i} \phi \,, \tag{5.42}$$

where K_{t_i} is the stiffness for a rolling displacement of the suspension. In many cases the force-displacement relationship of the suspensions is nonlinear and Eq. (5.42) relies on a linearization about the symmetrical condition ($\phi = 0$).

Note that Eq. (5.42) holds only in steady state conditions, as no allowance is taken for the presence of the shock absorbers. If the roll velocity $\dot{\phi}$ is not very small, a term $\Gamma_i \dot{\phi}$ should be added at the right hand side, where Γ_i is the damping coefficient for rolling rotations of the i-th axle.

A similar relationship holds for all suspensions and hence the global load transfer is given by

$$\sum_{\forall k} t_k \Delta F_{z_k} = \phi \sum_{\forall k} K_{t_k} \,. \tag{5.43}$$

By introducing Eq. (5.43) into Eq. (5.42) the load transfer on the i-th axle can be obtained from the global load transfer

$$\Delta F_{z_i} = \frac{K_{t_i}}{\sum_{\forall k} K_{t_k}} \frac{\sum_{\forall k} t_k \Delta F_{z_k}}{t_i} \,. \tag{5.44}$$

If the effect of the driving torque T_d on a driving rigid axle is accounted for, torque T_d must be added both on the axle and on the vehicle body as seen in Sec. 4-2. Its inclusion into the model does not increase its complexity in any substantial way.

The forces in xy plane acting at the road-wheel interface in a four-wheeled vehicle with steering on the front axle only are shown in Fig. 5.14.

The resultant of the forces acting on the vehicle in x direction, including aerodynamic drag and the component of weight due to a grade of the road are

$$F_x = \sum_{\forall i} F_{x_{i_t}} \cos(\delta_i) - \sum_{\forall i} F_{y_{i_t}} \sin(\delta_i) - \frac{1}{2}\rho v^2 S C_x - mg \sin(\alpha) \,. \tag{5.45}$$

Both driving, or braking, forces and rolling resistance of free wheels are included in forces Fx_{i_t}

In a similar way, the forces acting in y direction give a resultant equal to

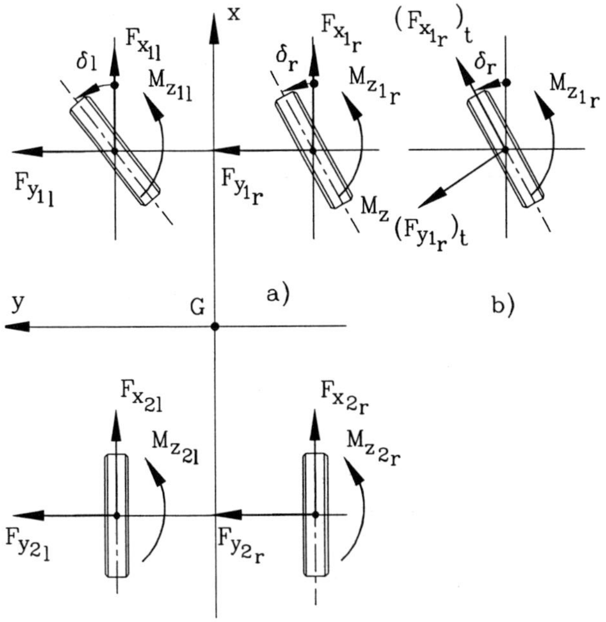

Fig. 5.14 Forces in xy plane acting at the road-wheel interface in a four-wheel vehicle with steering on the front axle only. (a) Forces in xy frame, (b) forces in the frame of the wheel.

$$F_y = \sum_{\forall i} F_{x_{i_t}} \sin(\delta_i) + \sum_{\forall i} F_{y_{i_t}} \cos(\delta_i) + \frac{1}{2}\rho v^2 S C_y + F_{y_e} , \tag{5.46}$$

where the external force F_{y_e} can be simply $mg\sin(\alpha_t)$ if it is due to the transversal slope of the road.

The total yawing moment about the centre of mass is

$$\begin{aligned} M_z = &\sum_{\forall i} F_{x_{i_t}} \sin(\delta_i) x_i + \sum_{\forall i} F_{y_{i_t}} \cos(\delta_i) x_i - \sum_{\forall i} F_{x_{i_t}} \cos(\delta_i) y_i + \\ &+ \sum_{\forall i} F_{y_{i_t}} \sin(\delta_i) y_i + \sum_{\forall i} M_{z_i} + \frac{1}{2}\rho v^2 S l C_{M_z} + M_{z_e} , \end{aligned} \tag{5.47}$$

where x_i and y_i are the coordinates of the centre of the contact areas of the wheels, M_{z_i} are the self-aligning torques and M_{z_e} is a generic external torque applied to the vehicle.

5.4.4 *Equations of motion*

The equations of motion of the vehicle are then

$$\begin{cases} m\left(\dot{u}-\dot{\psi}v\right)=\sum_{\forall i}F_{x_{i_t}}\cos(\delta_i)-\sum_{\forall i}F_{y_{i_t}}\sin(\delta_i)-\frac{1}{2}\rho v^2SC_x-mg\sin(\alpha) \\ m\left(\dot{v}+\dot{\psi}u\right)=\sum_{\forall i}F_{x_{i_t}}\sin(\delta_i)+\sum_{\forall i}F_{y_{i_t}}\cos(\delta_i)+\frac{1}{2}\rho v^2SC_y+F_{y_e} \\ J_z\ddot{\psi}=\sum_{\forall i}F_{x_{i_t}}\sin(\delta_i)x_i+\sum_{\forall i}F_{y_{i_t}}\cos(\delta_i)x_i-\sum_{\forall i}F_{x_{i_t}}\cos(\delta_i)y_i+ \\ \qquad +\sum_{\forall i}F_{y_{i_t}}\sin(\delta_i)y_i+\sum_{\forall i}M_{z_i}+\frac{1}{2}\rho V_r^2SlC_{M_z}+M_{z_e}\,. \end{cases} \tag{5.48}$$

Note that in the first equation the mass is essentially the mass of the vehicle, as the inertia of the engine and of the driving wheels are accelerated directly by the engine. The inertia of the free wheels should on the contrary be added to the mass when accelerating but not during braking; it is however usually neglected.

Equations (5.48), together with the equations yielding the sideslip angles of the wheels, those expressing the forces and moments of the tires as functions of the sideslip angles, the load, and the other relevant parameters, allow to study the handling of the vehicle.

The mathematical model so obtained is quite general and suffers only from the assumption of rigid vehicle. It is however strongly nonlinear and it is impossible to obtain closed form solutions for the handling of the vehicle. The numerical integration is straightforward and allows to perform numerical simulations of any manoeuvre.

5.5 Linearized handling model

Equations (5.48) can be easily linearized, to obtain closed form solutions allowing to perform a general study of the handling of the vehicle and particularly to study its stability in the small.

The sources of nonlinearities are mainly three: The presence of products of the variables of motion in the equation, the presence of trigonometric functions and the nonlinear nature of the forces due to the tires. All these nonlinearities can be neglected only if the sideslip angles of the wheels and of the vehicle, the steering angles etc. are very small. Such conditions usually hold for normal driving, i.e., when the vehicle is driven far from the limit performances.

When pushing the performances to the limit like in racing driving and, even more, when uncontrolled motion occur during an accident, no linearization is possible and a full nonlinear model must be used.

There are also intermediate conditions. Since the nonlinear dependence of tire forces from the sideslip angle occurs at values of the latter which are smaller than those at which the linearization of trigonometric functions becomes impossible, it is possible to build *semilinearized* models in which the linearized equations to be obtained below are used but the full nonlinear expressions of the forces due to the tires are introduced.

Although far simpler than fully nonlinear models, semilinearized ones must however resort to numerical integration and today it is questionable whether their use is expedient, owing to the ease with which the numerical integration of fully nonlinear models can be dealt with.

As a first consideration, angle β (Fig. 5.11) can be considered small and its trigonometric functions can be linearized

$$\begin{cases} u = V\cos(\beta) \approx V \\ v = V\sin(\beta) \approx V\beta \,. \end{cases} \tag{5.49}$$

Equation (5.36) reduces to

$$\begin{cases} m\left(\dot{V} - rv\right) = F_x \\ m\left(\dot{v} + rV\right) = F_y \\ J_z\dot{r} = M_z \,, \end{cases} \tag{5.50}$$

where $r = \dot{\psi}$ is the yaw velocity.

Alternatively it can be written as

$$\begin{cases} m\left(\dot{V} - rV\beta\right) = F_x \\ mV\left(\dot{\beta} + r\right) + m\beta\dot{V} = F_y \\ J_z\dot{r} = M_z \,. \end{cases} \tag{5.51}$$

The speed V can be considered a known function of time, which amounts to study the motion with a given law $V(t)$ (in many cases at constant speed) and to assume as an unknown the driving or the braking force needed to follow such law. In this case the unknown for the degree of freedom related to translation along x-axis is the force F_{x_d} exerted by the driving wheels. In case of braking, force F_{x_d} is the total braking force exerted by all wheels.

If the interaction between longitudinal and transversal forces due to the tires is neglected or accounted for in an approximate way, the first equation of motion uncouples from the other two and the latter are linear equations in the unknowns v and r, if Eqs. (5.50) are used, or β and r (Eqs. 5.51). This amounts to saying that the lateral behaviour is uncoupled from the longitudinal behaviour.

To study the motion the law $V(t)$ is stated first. The first Eq. (5.51) then reduces to

$$\dot{V} - Vr\beta = F_{x_d} + F_{x_{nd}} - \sum_{\forall i} Fy_{i_t}(\delta_i) - \frac{1}{2}\rho V_r^2 S C_x - mg\sin(\alpha)\,, \quad (5.52)$$

where $\sum_{\forall i} F_{x_{i_t}}$ has been substituted by the sum of force F_{x_d} defined above and force $F_{x_{nd}}$ exerted by all other non-driving wheels. If braking is performed on all wheels, the latter vanishes when studying braking.

The longitudinal behaviour is then studied using a linearized version of Eq. (5.52) in which the term in $r\beta$ and forces $F_{y_{i_t}}(\delta_i)$ are neglected: As the law $V(t)$ is known, force F_{x_d} is thus obtained. Note that operating in this way amounts to assuming that the longitudinal behaviour of the vehicle is the same as that occurring in straight running.

The handling of the vehicle can then be studied using the last two equations. At this stage the longitudinal forces on the tires are known and their interactions with lateral forces can be accounted for. The handling can thus be studied using a model with only two degrees of freedom.

If a better precision is required, once that laws $v(t)$ and $r(t)$ (or $\beta(t)$ and $r(t)$) are obtained, Eq. (5.52) can be studied again in its complete form (also lateral forces are now known) and the longitudinal behaviour on the curved path can be obtained. The handling of the vehicle can then be studied again.

5.5.1 *Linearized sideslip angles of the wheels*

Equation (5.39) can be easily linearized. By noting that $y_i\dot{\psi}$ is far smaller than the speed V, it follows

$$\alpha_i = \beta_i - \delta_i \approx \frac{v + rx_i}{V} - \delta_i = \beta + \frac{x_i}{V}r - \delta_i\,. \quad (5.53)$$

Note that the coordinate y_i of the centre of the contact area of the wheel does not appear in the expression for the sideslip angle α_i. If the differences between the steering angles δ_i of the wheels of the same axle are neglected, the values of their sideslip angles are then equal. This allows one to work in terms of axles instead of single wheels and to substitute a model of the type of that of Fig. 5.1b to that of Fig. 5.1a.

The explicit expressions of the sideslip angles of the front and rear axles of a vehicle with two axles are then

$$\begin{cases} \alpha_1 = \dfrac{v + ra}{V} - \delta_1 \\ \\ \alpha_2 = \dfrac{v - rb}{V} - \delta_2 \, , \end{cases} \tag{5.54}$$

or

$$\begin{cases} \alpha_1 = \beta + \dfrac{a}{V} r - \delta_1 \\ \\ \alpha_2 = \beta - \dfrac{b}{V} r - \delta_2 \, . \end{cases} \tag{5.55}$$

In the majority of cases only the front axle can steer and $\delta_2 = 0$. Note that the assumption of rigid vehicle prevents one from considering roll steering.

5.5.2 *Cornering forces*

Equation (5.46) reduces to

$$F_y = \sum_{\forall i} F_{x_{i_t}} \delta_i + \sum_{\forall i} F_{y_{i_t}} + \frac{1}{2} \rho V_r^2 S C_y + F_{y_e} \, , \tag{5.56}$$

where products $F_{x_{i_t}} \delta_i$ can usually be neglected, as they are far smaller than the other forces included in the equation.

Cornering forces can be expressed as the product of the cornering stiffness by the sideslip angle

$$F_{y_{i_t}} = -C_i \alpha_i = -C_i \left(\beta + \frac{x_i}{V} r - \delta_i \right) \, . \tag{5.57}$$

Equation (5.57) is written in terms of axles. The cornering stiffness is then that of the axle and not of the single wheel. In this way no allowance is taken for the camber force as, owing to the assumption of rigid vehicle, no roll is considered and the wheels of any axle have opposite camber. The camber forces then cancel each other.

No allowance is also taken for the transversal load transfer. If the dependence of the cornering stiffness were linear with the load F_z this would be correct as the increase of cornering stiffness of the more loaded wheel would exactly compensate for the decrease of the other one. As this is not exactly the case, the load transfer causes a decrease of cornering stiffness of each axle but this effect is usually considered as negligible at least for lateral accelerations lower than 0.5 g[5].

By linearizing also the value of the aerodynamic coefficient C_y and assuming that the steering angles of the various axles can be expressed as

$$\delta_i = K_i'\delta \,, \tag{5.58}$$

the expression of the total lateral force (5.56) can be reduced to the linear equation

$$F_y = Y_\beta \beta + Y_r r + Y_\delta \delta + F_{y_e} \,, \tag{5.59}$$

where

$$\begin{cases} Y_\beta = -\sum_{\forall i} C_i + \frac{1}{2}\rho V_r^2 S (C_y)_{,\beta} \\ Y_r = -\frac{1}{V}\sum_{\forall i} x_i C_i \\ Y_\delta = \sum_{\forall i} K_i' \left(C_i + F_{x_{i_t}}\right) \,. \end{cases} \tag{5.60}$$

Note that in the case of vehicles with only one steering axle, all K_i' vanish except $K_1' = 1$ while in other cases, they can be functions of many parameters. If also the variables of motion β or r enter such equation the model is no longer linear.

The first Eq. (5.60) has been obtained confusing the sideslip angle of the vehicle β with the aerodynamic sideslip angle β_a, as it happens when

[5] L. Segel, *Theoretical Prediction and Experimental Substantiation of the Response of the Automobile to Steering Control*, Cornell Aer. Lab., Buffalo, N.Y.

no side wind is present, and in the third one the terms in $F_{x_{i_t}}$ are usually neglected.

5.5.3 *Yawing moments*

Equation (5.47) can be linearized as

$$M_z = \sum_{\forall i} F_{x_{i_t}} \delta_i x_i + \sum_{\forall i} F_{y_{i_t}} x_i - \sum_{\forall i} F_{x_{i_t}} y_i + \sum_{\forall i} F_{y_{i_t}} \delta_i y_i + + \sum_{\forall i} M_{z_i} + \frac{1}{2}\rho V_r^2 S l C_{M_z} + M_{z_e} \; . \tag{5.61}$$

Once expression (5.59) is introduced into Eq. (5.61), the sums of the terms in $F_{y_{i_t}} \delta_i y_i$ for the wheels of the same axle are very small and are usually neglected. The moments $F_{x_{i_t}} y_i$ contain a part due to forces which are equal for the wheels of the same axis, (i.e. due to the driving or braking forces[6] and to the rolling resistance linked with forces $F_{z_i}/2$), which cancel each other, and a part due to the rolling resistance caused by the load transfer ΔF_{z_i} which, being opposite, do not cancel. It is possible to see that

$$\sum_{\forall i} F_{x_{i_t}} y_i = \sum_{\forall i} t_i \Delta F_{z_i} \left(f_0 + K V^2\right) \; . \tag{5.62}$$

As only the global load transfer is included into Eq. (5.62), at least if the rolling coefficients of all wheels are equal, from Eq. (5.41), in which the aerodynamic term has been neglected, it follows that

$$\sum_{\forall i} F_{x_{i_t}} y_i = -\sum_{\forall i} F_{y_i} h_G \left(f_0 + K V^2\right) \; . \tag{5.63}$$

The aligning torque can be expressed as a linear function of the sideslip angle,

$$M_z = (M_z)_{,\alpha} \alpha \; , \tag{5.64}$$

[6]The driving forces are assumed to be equally divided between the wheels of the same axle, which is not the case with limited slip differentials and similar devices. Also devices which control the braking forces could result in the generation of unsymmetrical longitudinal forces requiring modifications to the approach here followed. Also the use of brakes in a differential way to generate a yawing moment is not considered here.

which holds only in a range of α smaller than that for which the side force can be linearized. The same considerations seen for the cornering force hold here; moreover the aligning torque is far less important and the errors in its evaluation affect far less the global behaviour of the vehicle than those in the cornering force. In the following equations the values of $(M_z)_{,\alpha}$ are referred to the whole axle.

Acting similarly to what seen for the cornering forces, the linearized expression for the yawing moments is

$$M_z = N_\beta \beta + N_r r + N_\delta \delta + M_{z_e} , \tag{5.65}$$

where

$$\begin{cases} N_\beta = \sum_{\forall i} \left[-x_i C_i + (M_{z_i})_{,\alpha} + h_G \left(f_0 + K V^2 \right) C_i \right] + \dfrac{1}{2} \rho V_r^2 S l (C_{M_z})_{,\beta} \\ N_r = \dfrac{1}{V} \sum_{\forall i} \left[-x_i^2 C_i + (M_{z_i})_{,\alpha} x_i + h_G \left(f_0 + K V^2 \right) C_i x_i \right] \\ N_\delta = \sum_{\forall i} K_i' \left[C_i x_i - (M_{z_i})_{,\alpha} + F_{x_{i_t}} x_i - h_G \left(f_0 + K V^2 \right) C_i \right] . \end{cases} \tag{5.66}$$

Also in this case the terms in $F_{x_{i_t}}$ are usually neglected.

5.5.4 *Derivatives of stability*

The terms Y_β, Y_r, Y_δ, N_β, N_r and N_δ are nothing but the derivatives $\partial F_y / \partial \beta$, $\partial F_y / \partial r$, etc. They are usually referred to as derivatives of stability. N_r is sometimes referred to as yaw damping, as it is a factor which multiplied by an angular velocity yields a moment, like a damping coefficient.

In a simplified study of the handling of road vehicles, aerodynamic forces and the yawing moment due to load transfer are usually neglected as well as the interaction between longitudinal and transversal forces of the tires. In these conditions Y_β, Y_δ, N_β and N_δ are constant while Y_r and N_r are proportional to $1/V$. Note that at any rate they are strongly influenced by the load and road conditions through the cornering stiffness of the tires.

If aerodynamic forces are considered, the airspeed V_r is often substituted by the groundspeed V. They introduce a strong dependence with V^2 in Y_β and N_β and with V in N_r. A similar effect, but far less important is due to rolling resistance, through the term in KV^2. The latter slightly affects also N_δ.

Example 5.2

Compute the derivatives of stability at 100 km/h of the vehicle of Appendix 1.1, using the simplified and the complete formulations. Plot the derivatives of stability as functions of the speed for the same vehicle. In the whole computation neglect the longitudinal forces on the tires.

The normal forces on the ground are first computed. At 100 km/h, at constant velocity on level road they are 4.804 and 3.536 kN for the front and rear axles respectively. From these values the cornering and aligning stuffiness can be computed as $C_1 = 6.7369 \times 10^4$ N/rad, $C_2 = 6.3411 \times 10^4$ N/rad, $(M_{z_1})_{,\alpha} = 2010$ Nm/rad and $(M_{z_2})_{,\alpha} = 1366$ Nm/rad. Note that these values are referred to the axles; the normal load on each wheel must be first computed and introduced into the "magic formula"; the results are then multiplied by the number of wheels on the axles.

By taking into account only the cornering forces of the tires, the following values of the derivatives of stability at 100 km/h are obtained: $Y_\beta = -130,570$ N/rad; $Y_r = 824.62$ Ns/rad; $Y_\delta = 67,374$ N/rad; $N_\beta = 22,906$ Nm/rad; $Y_r = -5,622$ Nms/rad; $Y_\delta = 58,615$ Nm/rad.

If the complete expressions, including aligning torques, aerodynamic forces and load shift between the wheels of the same axle are used, the values of the derivatives of stability at 100 km/h are: $Y_\beta = -132,340$ N/rad; $Y_r = 824.62$ Ns/rad; $Y_\delta = 67,374$ N/rad; $N_\beta = 26,488$ Nm/rad; $Y_r = -5,630$ Nms/rad; $Y_\delta = 55,962$ Nm/rad.

The derivatives of stability are plotted as functions of the speed in Fig. 5.15. The values obtained from the complete expressions are reported as full lines while the dashed lines are the constant values (proportional to $1/V$ for Y_r and N_r) obtained considering only the cornering forces computed at 100 km/h. Note that the only value which is strongly affected by load shift, aligning torques and the other effects is that of N_β. Here an apparent strange result is obtained: From the formula a decrease with the speed seems to occur, as the aerodynamic

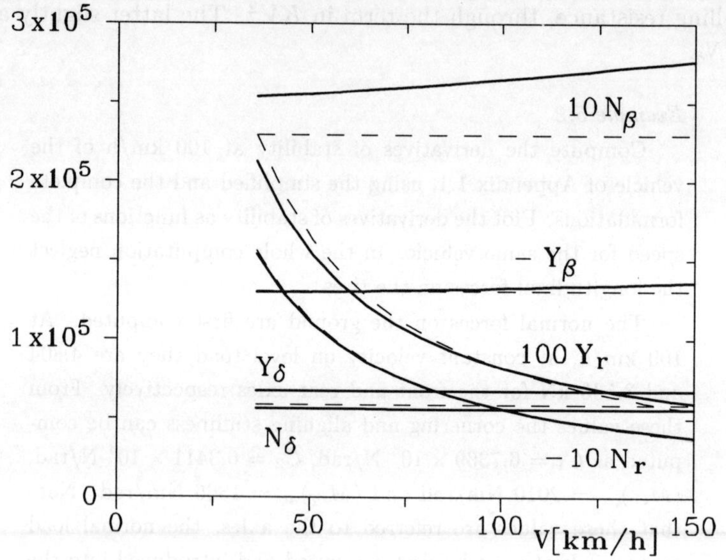

Fig. 5.15 Derivatives of stability as functions of the speed. Full lines: Values obtained from the complete expressions; dashed lines: Constant values (proportional to $1/V$ for Y_r and N_r) obtained considering only the cornering forces computed at 100 km/h.

term is negative while the plot shows an increase. The latter is due to the longitudinal load shift which, causing an increase of the load on the rear axle produces an increase of N_β which is larger than the decrease due to the aerodynamic moment M_z.

5.5.5 *Final expression of the equations of motion*

The final expression of the linearized equations of motion for the handling model is thus

$$\begin{cases} mV\left(\dot{\beta}+r\right)+\dot{m}V\beta = Y_\beta\beta + Y_r r + Y_\delta\delta + F_{y_e} \\ J_z\dot{r} = N_\beta\beta + N_r r + N_\delta\delta + M_{z_e}\,. \end{cases} \tag{5.67}$$

They are two first order differential equations in the two unknown β and r. Alternatively, a set of two first order differential equations in v and r could be written.

The steering angle δ can be considered as an input to the system, together with the external force and moment F_{y_e} and M_{z_e}. This way of proceeding is usually referred to as "locked controls" behaviour. Alternatively it is possible to study the "free controls" behaviour, in which the steering angle δ is one of the variables of the motion and a further equation expressing the dynamics of the steering system is added.

In the first case β and r can be considered as state variables and Eq. (5.67) can be directly written as a state equation

$$\{\dot{z}\} = [A]\{z\} + [B]\{u\} , \tag{5.68}$$

where the state and input vectors $\{z\}$ and $\{u\}$ are

$$\{z\} = \left\{ \begin{array}{c} \beta \\ r \end{array} \right\} , \qquad \{u\} = \left\{ \begin{array}{c} \delta \\ F_{y_e} \\ M_{z_e} \end{array} \right\} ,$$

the dynamic matrix is

$$[A] = \begin{bmatrix} \dfrac{Y_\beta}{mV} - \dfrac{\dot{V}}{V} & \dfrac{Y_r}{mV} - 1 \\ \dfrac{N_\beta}{J_z} & \dfrac{N_r}{J_z} \end{bmatrix}$$

and the input gain matrix is

$$[B] = \begin{bmatrix} \dfrac{Y_\delta}{mV} & \dfrac{1}{mV} & 0 \\ \dfrac{N_\delta}{J_z} & 0 & \dfrac{1}{J_z} \end{bmatrix} .$$

The study of the system is straightforward: The eigenvalues of the dynamic matrix allow one to see immediately whether the behaviour is stable or not and the study of the solution to given constant inputs yields the steady state response to a steering input or to external forces and moments.

There is however an interesting analogy which can be used. If the speed is kept constant in such a way that the derivatives of stability are constant in time, there is no difficulty in obtaining r from the first Eq. (5.67) and substituting it into the second, which becomes a second order differential equation in β. Similarly, solving the second in β and substituting it in the first one, an equation in r is obtained. The result is

$$P\ddot{\beta}+Q\dot{\beta}+U\beta=S'\delta+T'\dot{\delta}-N_rF_{y_e}+J_z\dot{F}_{y_e}-(mV-Y_r)M_{z_e} \qquad (5.69)$$

or

$$P\ddot{r}+Q\dot{r}+Ur=S''\delta+T''\dot{\delta}+N_\beta F_{y_e}-Y_\beta M_{z_e}+mV\dot{M}_{z_e}\,, \qquad (5.70)$$

where

$$\begin{cases} P=J_zmV \\ Q=-J_zY_b-mVN_r \\ U=N_\beta\left(mV-Y_r\right)+N_rY_\beta \end{cases} \qquad \begin{cases} S'=-N_\delta\left(mV-Y_r\right)+N_rY_\delta \\ S''=Y_\delta N_\beta-N_\delta Y_\beta \\ T'=J_zY_\delta \\ T''=mVN_\delta\,. \end{cases}$$

If the simplified expressions of the derivatives of stability are used, the expressions for P, Q, etc., for a vehicle with two axles reduce to

$$\begin{cases} P=J_zmV \\ Q=J_z(C_1+C_2)+m(a^2C_1+b^2C_2) \\ U=mV(-aC_1+bC_2)+C1C2\dfrac{l^2}{V} \end{cases} \qquad \begin{cases} S'=C_1\left(-amV+C2\dfrac{bl}{V}\right) \\ S''=lC_1C_2 \\ T'=J_zC_1 \\ T''=mVaC_1\,. \end{cases}$$

Each one of equations (5.69) and (5.70) is sufficient for the study of the dynamic behaviour of the vehicle. They are formally identical to the equation of motion of a spring-mass-damper system.

The linearized behaviour of a rigid motor vehicle at constant speed is then identical to that of a mass P suspended to a spring with stiffness U and a damper with damping Q and excited by the different forcing functions stated above.

The analogy here suggested is only a formal one: The state variables β and r are dimensionally an angular velocity (r) or are related to velocities (β has been introduced to express the lateral velocity v) and not displacements and thus P, Q and U are dimensionally far from being a mass, a damping coefficient and a stiffness.

5.6 Linearized steady state directional behaviour

5.6.1 *Response to a steering input*

In steady state driving the radius of the trajectory is constant, i.e. the path is circular. The relationship linking r to the radius R of the trajectory is thus

$$r = \frac{V}{R} \,. \tag{5.71}$$

To compute the steady state response is the same as computing the equilibrium position of the equivalent mass-spring-damper system under the effect of a constant force $S'\delta$ or $S''\delta$ since in steady state motion $\dot{\delta} = 0$

$$\begin{cases} \beta = \dfrac{S'}{U}\delta = \dfrac{-N_\delta\left(mV - Y_r\right) - N_r Y_\delta}{N_\beta\left(mV - Y_r\right) + N_r Y_\beta}\delta \\ r = \dfrac{S''}{U}\delta = \dfrac{Y_\delta N_\beta - N_\delta Y_\beta}{N_\beta\left(mV - Y_r\right) + N_r Y_\beta}\delta \,. \end{cases} \tag{5.72}$$

The transfer functions of the vehicle are thus the *trajectory curvature gain*

$$\frac{1}{R\delta} = \frac{Y_\delta N_\beta - N_\delta Y_\beta}{V\left[N_\beta\left(mV - Y_r\right) + N_r Y_\beta\right]} \,, \tag{5.73}$$

expressing the ratio between the curvature of the trajectory and the steering input, the *lateral acceleration gain*

$$\frac{V}{R\delta} = \frac{V^2\left[Y_\delta N_\beta - N_\delta Y_\beta\right]}{N_\beta\left(mV - Y_r\right) + N_r Y_\beta} \,, \tag{5.74}$$

expressing the ratio between the centrifugal acceleration and the steering input, the *sideslip angle gain*

$$\frac{\beta}{\delta} = \frac{-N_\delta\left(mV - Y_r\right) - N_r Y_\delta}{N_\beta\left(mV - Y_r\right) + N_r Y_\beta} \,, \tag{5.75}$$

expressing the ratio between the sideslip angle and the steering angle and the *yaw velocity gain*

$$\frac{r}{\delta} = \frac{Y_\delta N_\beta - N_\delta Y_\beta}{N_\beta (mV - Y_r) + N_r Y_\beta} , \tag{5.76}$$

expressing the ratio between the yaw velocity and the steering angle.

Often a simplified expression of the derivatives of stability in which only the lateral forces of the tires are accounted for and no dependence of the cornering stiffness of the tires with the speed due to longitudinal load shift is considered. In this case it is possible to define a *stability factor* K or an *undesteer gradient* K^*; under the above mentioned assumptions they are constant and, in the case of a vehicle with two axles, their expression is

$$K = \frac{m}{l^2}\left(\frac{b}{C_1} - \frac{a}{C_2}\right) , \qquad K^* = \frac{mg}{l}\left(\frac{b}{C_1} - \frac{a}{C_2}\right) . \tag{5.77}$$

The expressions of the above defined gains reduce to:

- trajectory curvature gain

$$\frac{1}{R\delta} = \frac{1}{l}\frac{1}{1 + KV^2} , \tag{5.78}$$

- lateral acceleration gain

$$\frac{V^2}{R\delta} = \frac{V^2}{l}\frac{1}{1 + KV^2} , \tag{5.79}$$

- sideslip angle gain

$$\frac{\beta}{\delta} = \frac{b}{l}\left(1 - \frac{maV^2}{blC_2}\right)\frac{1}{1 + KV^2} , \tag{5.80}$$

- yaw velocity gain

$$\frac{r}{\delta} = \frac{V}{l}\frac{1}{1 + KV^2} . \tag{5.81}$$

Note that in kinematic conditions $1/R\delta$ is equal to $1/l$. The expression $1+KV^2$ can thus be considered as a correction factor giving the response of the vehicle in dynamic conditions from that in kinematic conditions. This is true for all the mentioned gains except for the sideslip angle gain which depends on the speed even if factor $1 + KV^2 = 1$, i.e., if $K = 0$. If Eq.

(5.77) holds, i.e. if only the cornering forces of the tires are considered, K is a constant.

If $K = 0$ the value of $1/R\delta$ is constant and equal to the value which characterizes kinematic steering, i.e. the response of the vehicle to a steering input is, at any speed, equal to that in kinematic conditions. This however does not mean that the vehicle is in kinematic conditions, since the values of the sideslip angle is not equal to its kinematic value and those of the sideslip angles of the wheels are not equal to zero. A vehicle which behaves in this way is said to be *neutral-steer* (Fig. 5.16a).

If $K > 0$ the value of $1/R\delta$ decreases with increasing speed. The response of the vehicle is then smaller than that in kinematic conditions and, to maintain a constant radius of the trajectory, the steering angle must be increased at increasing speed. A vehicle which behaves in this way is said to be *understeer*. A quantitative measure of the understeering of a vehicle is given by the *characteristic speed*, defined as the speed at which the steering angle needed to negotiate a turn is equal to twice the Ackerman angle, i.e. the trajectory curvature gain is equal to $1/2l$.

Using the simplified approach outlined above, the characteristic speed is

$$V_{char} = \sqrt{\frac{1}{K}} \,. \tag{5.82}$$

If $K < 0$ the value of $1/R\delta$ increases with increasing speed until, for a speed

$$V_{crit} = \sqrt{-\frac{1}{K}} \tag{5.83}$$

the response tends to infinity, i.e., the system develops an unstable behaviour. A vehicle which behaves in this way is said to be *oversteer* and the speed given by Eq. (5.83) is said *critical speed*. The critical speed of any oversteer vehicle must be well above the maximum speed it can reach, at least in normal road conditions.

The value of β, or better, of β/δ, decreases with the speed from the kinematic value up to the speed

$$(V)_{\beta=0} = \sqrt{\frac{blC_2}{am}} \tag{5.84}$$

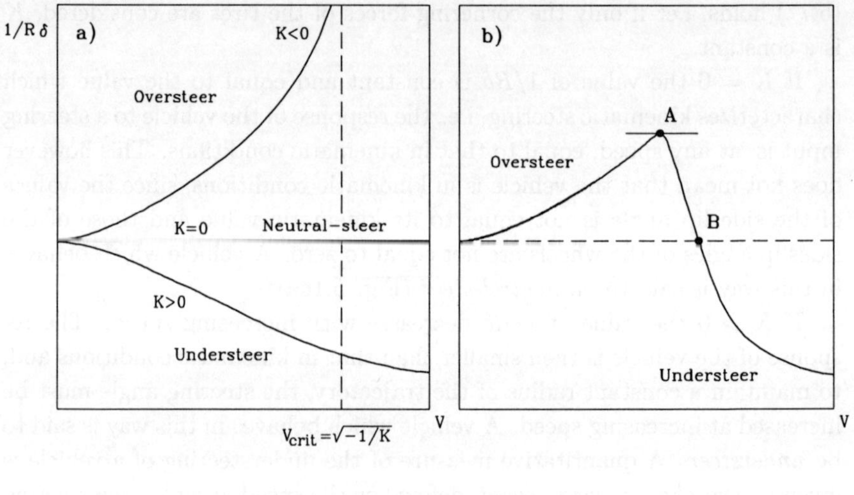

Fig. 5.16 Steady state response to a steering input.

at which it vanishes. At higher speed it becomes negative, tending to infinity when approaching the critical speed for oversteering vehicles and tending to $aC_1/(aC_1 - bC_2)$ when the speed tends to infinity in the case of understeering vehicles.

The sideslip angles of the tires are equal in the case of neutral-steer vehicles. In the case of oversteer vehicles the rear wheels have a greater sideslip angle (in absolute value, as the sideslip angles are negative when the radius of the trajectory is positive), while the opposite holds in the case of undesteer vehicles. It follows that oversteer vehicles can be expected to reach the limit conditions at the rear wheels and understeer vehicles at the front wheels, even if the present model cannot be applied in conditions approaching any limit.

The above mentioned considerations hold only in the case of vehicles with two axles and in which all effects which cause a dependence of the derivatives of stability from the speed (for Y_r and N_r a dependence different from that from $1/V$) are neglected. If this last assumption is dropped the stability factor K is not constant and the vehicle can have a different behaviour at different speeds.

A strong effect is due to the aerodynamic yawing moment. If $\partial C_{M_z}/\partial\beta$ is negative (the side force F_y acts forward of the centre of mass), the effect is to increase the oversteer behaviour or to the decrease the understeer one, at increasing speed. If a critical speed exists, such aerodynamic effect lowers it and an overall unstabilizing effect, which increases with the absolute value of $(C_{M_z})_{,\beta}$ is present. The opposite occurs if $(C_{M_z})_{,\beta}$ is positive.

Another important effect is due to the longitudinal load shift. If the load on the rear axle increases more, or decreases less, than that on the front axle the understeer increases with increasing speed.

The case of a vehicle which is oversteer at low speed and understeer at high speed, as it can be caused by a positive value of $(C_{M_z})_{,\beta}$, is shown in Fig. 5.16b. Following the definition seen above, the speed at which neutral-steer is obtained is that of point B.

If the simplified expressions for the derivatives of stability are not used, a new definition of neutral-steer, and hence under- and oversteer, vehicle can be introduced. Instead of referring to the condition $1/R\delta = 1/l$, neutral-steering can be defined by the relationship

$$\frac{d}{dV}\left(\frac{1}{R\delta}\right) = 0\,. \tag{5.85}$$

It is obvious that in case the derivatives of stability are constant (Y_r and N_r are proportional to $1/V$) the first definition, which can be said to be *absolute* and the second, which can be said to be *incremental*, coincide.

On the plot of Fig. 5.16b the speed at which neutral-steering is obtained is point A, in which the curve reaches its maximum. The incremental definition follows more closely the feeling of the driver, who feels the vehicle as oversteering if an increase of speed is accompanied by a decrease of radius of the trajectory and vice versa. The driver has clearly no reference to feel the kinematic value of the radius of the trajectory and hence the absolute definition has little meaning for him.

From the viewpoint of the equations of motion, on the contrary, the absolute definition is more significant.

5.6.2 *Neutral-steer point and static margin*

The neutral-steer point of the vehicle is usually defined as the point laying on the plane of symmetry in which is applied the resultant of the cornering forces due to the tires as a consequence of a sideslip angle β, obviously with $\delta = 0$ and $r = 0$. The cornering forces, computed through the linearized model, in these conditions are simply $-C_1\beta$ and $-C_2\beta$ and the x coordinate of the neutral point is

$$x_N = \frac{aC_1 - bC_2}{C_1 + C_2} \, . \tag{5.86}$$

A better definition of neutral-steer point can however be introduced. If all forces and moments due to a sideslip angle β, with $\delta = 0$ and $r = 0$ are considered, the resultant force and moment are simply $Y_\beta \beta$[7] and $N_\beta \beta$. The x coordinate of the neutral-steer point, defined as the point of applications of the resultant of all lateral forces is thus

$$x_N = \frac{N_\beta}{Y_\beta} \tag{5.87}$$

The static margin $\mathcal{M}_s$ is the ratio between the x coordinate of the neutral point and the wheelbase

$$\mathcal{M}_s = \frac{x_N}{l} \, . \tag{5.88}$$

As will be seen when dealing with the response to external forces and moments, if an external force is applied to the neutral-steer point it does not cause any steady-state yaw velocity. Owing to the mathematical model used in the present chapter, the position in height of the neutral-steer point cannot be defined.

Note that the condition to obtain a neutral-steer response is that the neutral-steer point coincides with the centre of mass, i.e. $x_N = 0$, $\mathcal{M}_s = 0$, $N_\beta = 0$. If they are positive the vehicle is oversteer[8] (centre of gravity behind the neutral point); the opposite applies to understeer vehicles.

The signs of parameters K, K^*, $\mathcal{M}_s$, x_N, $|\alpha_1| - |\alpha_2|$ and N_β which correspond to an oversteer, undesteer or neutral-steer behaviour are reported in Table 5.1.

As in case of neutral-steer $N_\beta = 0$, the second equation of motion (5.67) uncouples from the first one and simplify as

[7] Y_β can be considered as a sort of cornering stiffness of the vehicle.

[8] Sometimes the position of the neutral-steer point and the static margin are defined with different sign conventions: Instead of referring to the position of the neutral point with respect to the centre of mass, the position of the latter with respect to the former is given. In this case the signs of x_N and $\mathcal{M}_s$ are changed and an understeer vehicle has a positive static margin.

Behaviour	K	K^*	$\mathcal{M}_s$	x_N	$\|\alpha_1\| - \|\alpha_2\|$	N_β
Understeer	>0	>0	<0	<0	>0	>0
Neutral-steer	0	0	0	0	0	0
Oversteer	<0	<0	>0	>0	<0	<0

$$J_z\dot{r} = N_r r + N_\delta \delta + M_{z_e} \ . \tag{5.89}$$

The behaviour of the motor vehicle is then that of first order system instead of being that of a second order system.

Example 5.3

Study the directional behaviour of the vehicle of Appendix 1.1, using the simplified and the complete formulations.

The value of N_β is positive and hence the vehicle is understeer. Using the values of the derivatives of stability computed from the cornering stuffiness at 100 km/h the values of the coordinate of the neutral-steer point and of the static margin are $x_N = -175$ mm, $\mathcal{M}_s = -0.081$, while the values obtained, always at 100 km/h using a complete expression of the derivatives of stability are $x_N = -200$ mm, $\mathcal{M}_s = -0.093$.

The trajectory curvature gain, the lateral acceleration gain, the sideslip angle gain and the yaw velocity gain are plotted as functions of the speed in Fig. 5.17. The values obtained from the complete expressions of the derivatives of stability are reported as full lines while the dashed lines refer to the simplified expressions for the derivatives of stability (constant or proportional to $1/V$ for Y_r and N_r) obtained considering only the cornering forces computed at 100 km/h. The dotted lines refer to a neutral-steer vehicle.

The vehicle has a strong understeering behaviour, even more so if the complete expression of the derivatives of stability is considered. However, the simplified approach allows one to obtain a fair approximation of the directional behaviour of the vehicle.

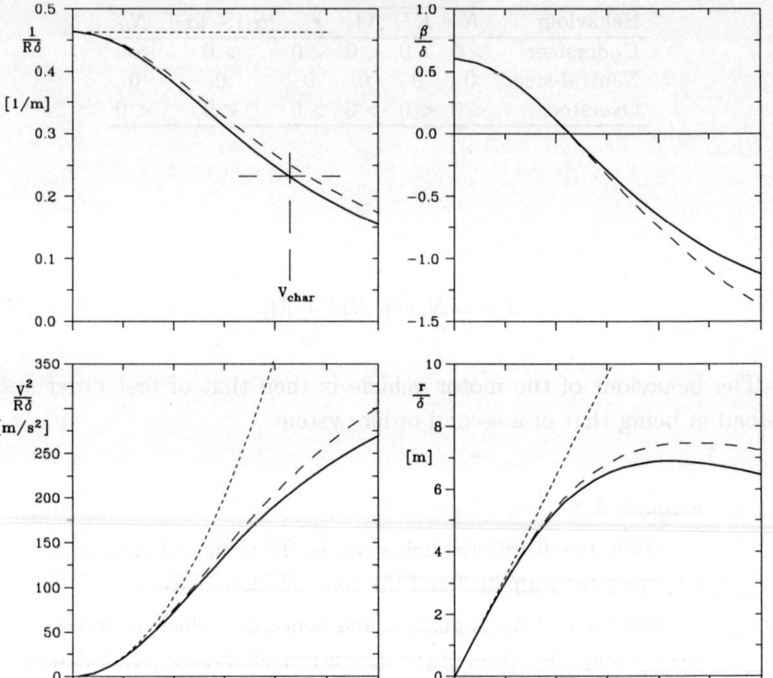

Fig. 5.17 Example 5.3: Trajectory curvature gain, lateral acceleration gain, sideslip angle gain and yaw velocity gain as functions of the speed. Full lines: Values obtained from the complete expressions of the derivatives of stability; dashed lines: Simplified approach (constant derivatives of stability, Y_r and N_r proportional to $1/V$, obtained considering only the cornering forces computed at 100 km/h); dotted lines: Neutral-steer vehicle.

5.6.3 *Response to external forces and moments*

From the equivalent mass-spring-damper model the steady state response to an external force F_{y_e} or an external moment M_{z_e} is immediately obtained. The relevant gains are

$$\left\{\begin{array}{l}\dfrac{1}{RF_{y_e}}=\dfrac{N_\beta}{VU}\\[2ex]\dfrac{V^2}{RF_{y_e}}=\dfrac{VN_\beta}{U}\\[2ex]\dfrac{\beta}{F_{y_e}}=\dfrac{-N_r}{U}\end{array}\right.\qquad\left\{\begin{array}{l}\dfrac{1}{RM_{z_e}}=\dfrac{-Y_\beta}{VU}\\[2ex]\dfrac{V^2}{RM_{z_e}}=\dfrac{-VY_\beta}{U}\\[2ex]\dfrac{\beta}{M_{z_e}}=\dfrac{-mV+Y_r}{U}\end{array}\right.\tag{5.90}$$

If the vehicle is neutral-steer, $N_\beta = 0$ and consequently $1/RF_{y_e} = 0$: The trajectory under the effect of an external force remains straight (Fig. 5.18Ia). The trajectory is however changed from the one preceding the application of force F_{y_e}: The deviation is equal to angle β, i.e. to $-F_{y_e}/Y_\beta$. The lateral velocity of the vehicle is simply $v = V\beta = -VF_{y_e}/Y_\beta$: Particularly in the case of fast vehicles it is important that Y_β is as large as possible in order to avoid large lateral velocities. The fact that the trajectory remains straight can be easily understood considering that the neutral-steer point lies in the centre of mass, i.e. in the point of application of the external force.

Actually this condition can be used to define the neutral-steer point as the point in which the application of an external force does not cause a yaw rotation of the vehicle. If the presence of the suspension is accounted for, instead of a neutral-steer point it is possible to define a neutral-steer line as the locus of the points in xz plane in which an external force applied in y direction does not cause any yaw rotation.

If the vehicle is understeer, the neutral-steer point is behind the centre of mass and the trajectory bends as in Fig. 5.18Ib. Opposite effect can be found in the case of oversteer vehicles. Note that the trajectories so computed are steady-state trajectories and when the force is applied an unstationary motion occurs (dashed lines in the figure). This first part of the trajectory cannot be computed with the above mentioned equations.

If the vehicle is oversteer all the gains expressed by Eq. (5.90) tend to infinity when approaching the critical speed while they decrease with the speed in case of understeer vehicles.

The effect of a crosswind can be considered as the combined effect of a force and a moment. If the relative velocity is changed by angle ψ_w with respect to the velocity in still air, the force and the moment acting on the vehicle due to crosswind are

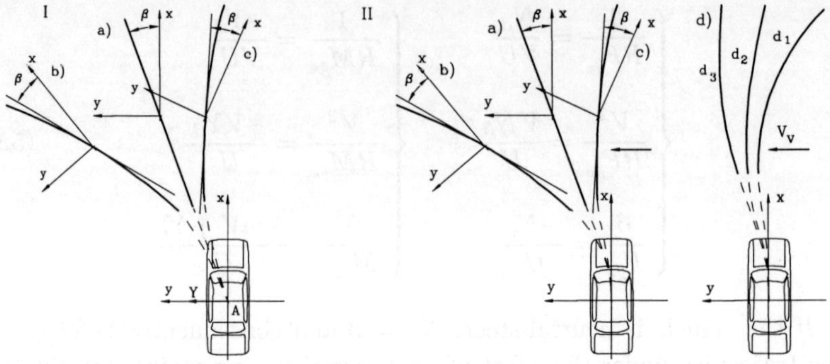

Fig. 5.18 I. Response to a force F_{y_e} applied to the centre of mass; (a) neutral-steer, (b) understeer and (c) oversteer vehicle. II. Response to a lateral wind; point of application of the side force in the neutral-steer point (a), forward (b) and after the neutral-steer point (c) and (d).

$$F_{y_w} = (F_{y_{aer}})_{,\beta}\psi_w \,, \qquad M_{z_w} = (M_{z_{aer}})_{,\beta}\psi_w \,. \tag{5.91}$$

Note that this way of operating, which is essentially a linearization of aerodynamic forces, holds only for small values of ψ_w, or, better, for values causing angle $\beta + \psi_w$ to remain within the field in which the sideforce and the yawing moment can be linearized. This occurs either for feeble crosswinds or for head- or tailwinds. If the wind velocity is not small, the aerodynamic terms of the derivatives of stability must be computed using V_r instead of V.

The response in terms of curvature of the trajectory, computed as the sum of the response to a force and to a moment, is

$$\frac{1}{R} = \frac{F_{y_w} N_\beta - M_{z_w} Y_\beta}{V\left[N_\beta\left(mV - Y_r\right) N_r Y_\beta\right]} = \frac{F_{y_w} Y_\beta}{VU}\left(\frac{N_\beta}{Y_\beta} - \frac{M_{z_w}}{F_{y_w}}\right). \tag{5.92}$$

Ratio M_{z_w}/F_{y_w} is nothing but the distance of the point of application of the aerodynamic side force from the centre of mass. If it is equal to N_β/Y_β the aerodynamic force is applied in the neutral steer point and a straight trajectory occurs. The deviation angle is

$$\beta = \frac{M_{z_w}}{Y_\beta} = -\frac{F_{y_w} x_N}{N_\beta} \,. \tag{5.93}$$

In general, the value of β is

$$\beta = \frac{F_{y_w}}{U}\left[-\frac{M_{z_w}}{F_{y_w}}(mV - T_r) - N_r\right] . \tag{5.94}$$

The trajectories are shown in Fig. 5.18II.

Usually the point of application of the aerodynamic force is in front of the centre of mass and also the neutral-steer point. In this case the trajectory bends downwind (curve b).

On the contrary, if the point of application of aerodynamic forces is behind the neutral-steer point the trajectory bends upwind (curves c and d). If the effect is not too strong (curve d_3) the effect is beneficial as very little correction is needed, but if the result is like that of curve d_1 a large correction can be required in a direction which is opposite to the instinctive reaction of the driver. Again it must be noted that the present steady-state model has limited application to the case of wind gusts, which involve mainly unsteady phenomena.

The application of a side force to the centre of mass is easy: It is sufficient to use a road with a transversal slope fashioned in a proper way. Wind gusts can be simulated using jet engines and suitable ducts to distribute the gust with the required profile.

5.6.4 *Influence of longitudinal forces on the handling*

The directional behaviour is strongly influenced by the presence of longitudinal forces between the tires and the road. Any longitudinal force causes a reduction of the cornering stiffness: If it is applied to the front axle it reduces the value of C_1 and consequently makes the vehicle more understeering or less oversteering. Opposite effect is due to a longitudinal force applied to the rear axle.

In the linearized model this can be easily accounted for by using the elliptical approximation (Eq. 2.25) which, if a complete linearization of the behaviour of the tires is assumed, can be applied directly to each axle

$$C_i = C_{0_i}\sqrt{1 - \left(\frac{F_{x_i}}{\mu_p F_{z_i}}\right)^2} . \tag{5.95}$$

Note that the forces and the cornering stiffness are referred to the whole axle.

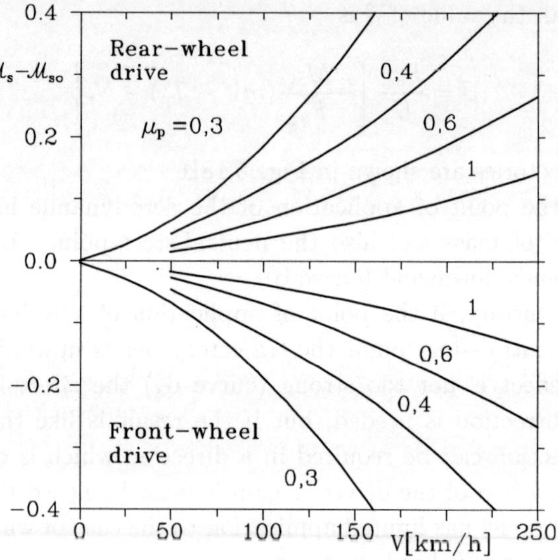

Fig. 5.19 Variation of the stability margin due to the longitudinal forces on the tires in the cases of front- and rear wheel-drive saloon cars. Various values of μ_p; a completely linearized model has been used.

The driving force needed to maintain a constant speed increases with the latter and as a consequence the cornering stiffness of the tires of the driving axle decreases. The effect is felt particularly if the conditions of the road are poor, as in Eq. (5.95) the ratio between the actual and the maximum value of the driving force is present.

The variation of static margin for a front wheel drive and a rear wheel drive saloon car with the speed due to the effect of the driving forces is shown in Fig. 5.19. It is clear that the effect is very small in the whole practical speed range of the car if the road conditions are good while if μ_p is low the change of the handling of the car due to traction is quite strong.

In the case of rear wheel drive vehicles the driving forces increase the oversteer behaviour or decrease the understeer one. The critical speed, if exist, lowers or a critical speed can appear. In bad road conditions a rear wheel drive vehicle may have a very low critical speed and the driver may be required to limit the speed for stability reasons, to avoid spinout. Starting and accelerating the vehicle may be difficult and the driver has to exert a great care in operating the accelerator control; antispin devices are very useful in these conditions.

Front wheel drive vehicles on the contrary become tendentially understeering and more stable with increasing speed or decreasing μ_p and an increasingly large steering angle is needed to maintain the vehicle on a given trajectory. The limit condition is that of a vehicle which is infinitely stable, i.e. can only move on a straight line.

In the case the vehicle has more than one driving axle and in the case of braking the effect on handling depends on how the longitudinal forces are shared between the axles. If the front axle is working with a larger longitudinal force coefficient μ_x than the rear axle, which does not necessarily imply that force F_x is larger but that the ratio F_x/F_z of the front wheels is larger than that of the rear wheels, the vehicle becomes more understeering and is in a sense more stable. When the limit conditions are reached and the front wheels slip (lock in braking or spin in traction) the vehicle cannot be steered and follows a straight trajectory.

A larger ratio F_x/F_z at the rear wheels makes the vehicle more oversteering and easily introduces a critical speed. When reaching the limit conditions a spinout occurs, unless the driver promptly reduces the longitudinal forces and countersteers, manoeuvre which can be expected only from very proficient drivers. To avoid this situation the braking system must prevent the working point on the X_2, X_1 plane from lying above the curve for ideal braking. Antispin and antilock devices are very important from this viewpoint.

In the case all values of μ_x are equal, theoretically the behaviour should not be affected by the longitudinal forces; however when the limit conditions occur the vehicle can spinout or go straight depending on small changes in many parameters, like the conditions of the individual wheels and brakes, the load transfer, etc.

Example 5.4

Study the directional behaviour of the vehicle of Appendix 1.1, taking into account the reduction of the cornering stiffness of the driving wheels caused by the longitudinal forces needed to move at constant speed. Repeat the computation for two values of μ_p, namely 1 and 0.2.

The study is performed by computing, at each speed, the values of the longitudinal and normal component of the tire forces, using the "magic formula" for the cornering stiffness and then reducing it through the elliptic expression (5.95). The results, in terms of trajectory curvature gain, lateral acceleration gain, sideslip angle gain and yaw velocity gain, are plotted as functions of the speed in Fig. 5.20 for both values of the maximum longitudinal force coefficient. The dashed lines refer to the simplified expressions for the derivatives of stability (constant or proportional to $1/V$ for Y_r and N_r) obtained considering only the cornering forces computed at 100 km/h; the dotted lines refer to a neutral-steer vehicle.

By comparing Fig. 5.20 with Fig. 5.17 it is clear that the effect of the driving force is almost negligible in the whole speed range if the road conditions are good ($\mu_t = 1$): The lines of the two figures are almost completely superimposed. However if μ_t is lowered to 0.2 the understeer behaviour becomes much more marked, particularly at high speed.

5.6.5 *Transversal load shift*

No allowance has yet been taken for the transversal load shift. If the dependence of the cornering stiffness of a single wheel from the load is of the type shown in Fig. 5.21 this does not introduce errors if the load transfer ΔZ is small, lower than $(\Delta Z)_{lim}$ in the figure (condition a).

But if the load transfer is larger, as in the case of ΔZ_b, the increase of the stiffness of the more loaded wheel cannot compensate for the decrease of the other one and the cornering stiffness of the axle is reduced. This effect introduces a nonlinearity in the behaviour of the vehicle.

The simultaneous presence of longitudinal forces and load transfer makes things more complicated. Even if the cornering stiffness is still in the linear part of the plot of Fig. 5.21, i.e. the load transfer is smaller than $(\Delta Z)_{lim}$, the combined effect yields a nonlinear behaviour. Assuming that the longitudinal force splits equally on its two wheels, the cornering stiffness of the axle, computed using the elliptical approximation, is

$$C = \frac{1}{2}\left(C_0 + \Delta F_z \frac{\partial C}{\partial F_z}\right)\sqrt{1 - \left[\frac{F_x}{\mu_p(F_z + 2\Delta F_z)}\right]^2} + \\ + \frac{1}{2}\left(C_0 - \Delta F_z \frac{\partial C}{\partial F_z}\right)\sqrt{1 - \left[\frac{F_x}{\mu_p(F_z - 2\Delta F_z)}\right]^2}, \tag{5.96}$$

where forces F_x and F_z are referred to the whole axle.

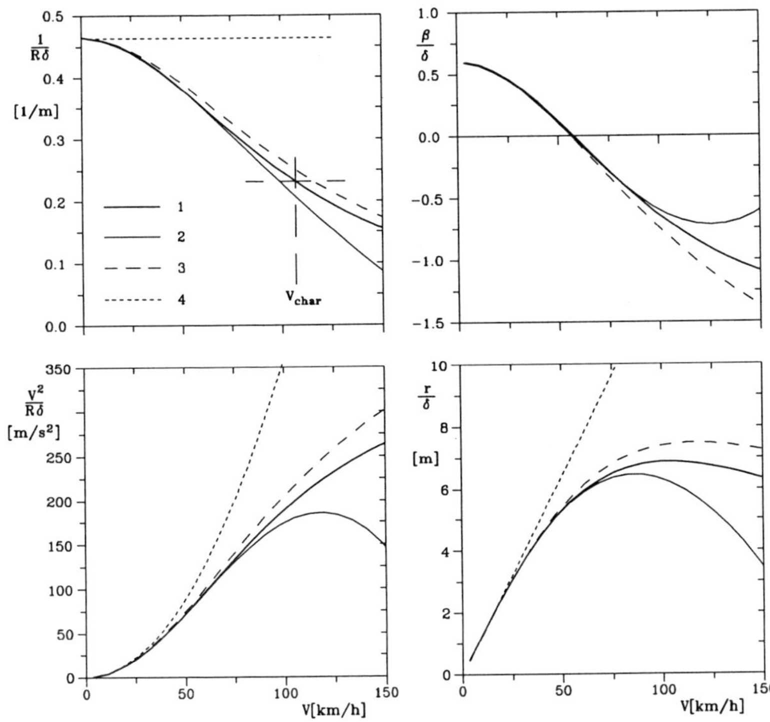

Fig. 5.20 Example 5.4: Trajectory curvature gain, lateral acceleration gain, sideslip angle gain and yaw velocity gain as functions of the speed. Values obtained from the complete expressions of the derivatives of stability, with the effect of the driving forces accounted for; (1): $\mu_t = 1$; (2): $\mu_t = 0.2$; (3): Simplified approach (constant derivatives of stability, Y_r and N_r proportional to $1/V$, obtained considering only the cornering forces computed at 100 km/h and no longitudinal force effects); (4): Neutral-steer vehicle.

Owing to the presence of the square root, the decrease of the cornering stiffness of the less loaded wheel is greater, particularly if μ_x is low, than the increase at the other wheel.

Load transfer on the driving axle then increases the effect of longitudinal forces; this combined action can be reduced by introducing an anti-roll bar on the other axle. Operating in this way, the increased load transfer on the non-driving axle reduces also its cornering stiffness, reducing the overall effect of longitudinal forces on handling.

Anti-roll bars affect the distribution of transversal load shift between the axles, increasing the load shift on the interested one, obviously decreasing that on the other axles. They can be used to correct the behaviour of the

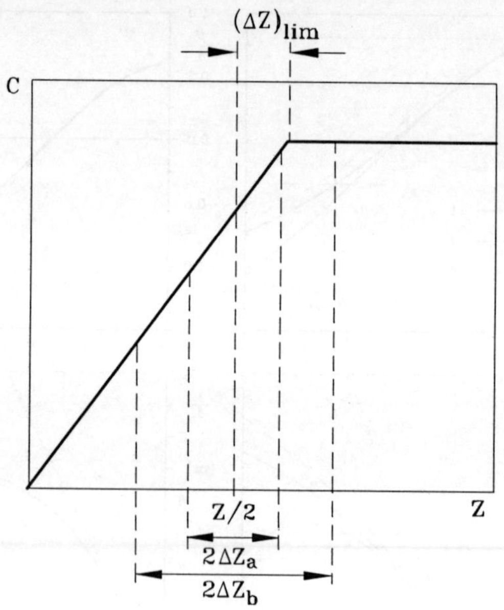

Fig. 5.21 Effect of load transfer on the cornering stiffness.

vehicle, particularly in conditions approaching the limit lateral acceleration as their effect on the cornering stiffness increases when the latter increases. As an example, a large rear-wheel drive saloon car can benefit from the application of an anti-roll bar at the front axle to correct the strong oversteering tendency when the rear wheels approach their traction limit, while a small front wheel car can use an anti-roll bar at the rear axle to reduce its understeering behaviour.

Note that it is impossible to state the effect of anti-roll bar on the gains defined in the previous sections as they introduce a strong nonlinearity into the mathematical model of the vehicle and the very definition of the gains is based on a complete linearization. It is only possible to study a number of specific cases, in which the lateral acceleration is defined, and to compute the response of the vehicle in such conditions.

5.6.6 *Toe in*

Consider an axle (e.g., the front axle), in which the midplanes of the wheels are not exactly parallel and assume that the x-axes of the reference frames of the wheels converge in a point laying forward with respect to the axle This effect is usually referred to as *toe in*[9].

Let α_c be the angle each wheel makes with the symmetry plane of the vehicle, positive when the toe-in is positive. With reference to Figure 5.1, the steering angle of the wheel on the right side of the vehicle is incremented by an angle equal to α_c, while the steering angle of the wheel on the left side is decreased by the same quantity. If the usual linearization assumptions are accepted, the sideslip angles of the two wheels of the axle are then

$$\begin{cases} \alpha_{i_R} = \beta + \dfrac{x_i}{V} r - \delta_i - \alpha_c = \alpha_i - \alpha_c \\ \\ \alpha_{i_L} = \beta + \dfrac{x_i}{V} r - \delta_i + \alpha_c = \alpha_i + \alpha_c \ , \end{cases} \tag{5.97}$$

where subscripts R and L refer to the right and left wheels respectively and i refers to the i-th axle. Toe in then produces an increase of the sideslip angle of the wheels external to the road bend and a decrease of the sideslip angle of the other wheels.

If transversal load shift is not taken into account, and the two wheels have the same cornering stiffness, toe in has no effect within the validity of the linearized model, since the increase of the cornering force of one wheel exactly compensates the decrease on the other one.

The situation is different if load shift is included into the model. Consider a vehicle negotiating a bend on the left; the sideslip angle α_i is negative while the side force is positive. The transversal load shift causes an increase of the load on the wheels on the right and the total side force the axle exerts is

$$F_y = C_R \alpha_R + C_l \alpha_l =$$

$$= -\frac{1}{2} \left[(\alpha_i - \alpha_c) \left(C + \Delta F_z \frac{\partial C}{\partial F_z} \right) + (\alpha_i + \alpha_c) \left(C - \Delta F_z \frac{\partial C}{\partial F_z} \right) \right] , \tag{5.98}$$

where C is the cornering stiffness of the whole axle and the load transfer has been assumed to be small enough to accept a linear dependence of C with the load.

[9]Toe in is usually defined as the difference between the distance of the front part and the rear part of the wheels of an axle, measured at the height of the hub, when the steering is in its central position. It is positive when the midplanes converge forward.

Equation (5.98) reduces to

$$F_y = -C\alpha_i + \alpha_c \Delta F_z \frac{\partial C}{\partial F_z} , \tag{5.99}$$

The wheel located on the outside of the curve is more loaded than that at the inside and hence the increase of side force due to the increase of its sideslip angle is larger than the decrease of cornering force of the other wheel: an increase of the total cornering power of the axle than results. This has the same effect of an increase of cornering stiffness of the axle: toe-in at front wheels or toe-out of the rear ones have than an oversteering effect.

The model shown above is just a rough approximation of a more complicated phenomenon: apart from the linearizations involved, toe-in depends on the steering angle, owing to steering errors, and on the geometry and the stiffness of the suspensions.

5.7 Stability of the vehicle

It is customary to define a static and a dynamic stability. A system is statically stable in a given equilibrium condition if, when its state is perturbed, it tends to return to the previous situation. If the motion which follows this tendency towards the previous state of equilibrium succeeds, at least asymptotically, to actually restore it, then the system is also dynamically stable. This motion can tend to the equilibrium condition monotonically or through a damped oscillation. If on the contrary the equilibrium conditions are not reached, usually because a divergent oscillation takes place, the system is dynamically unstable. If an undamped oscillation occurs, as in the case of an undamped spring-mass system, the dynamic stability is neutral.

If the system is linear this definition of stability holds in the whole field in which its state variables are defined. If the system is nonlinear it holds "in the small", i.e. for small variations of the values of the state variables about each equilibrium point in the state space. The linearized model of the motor vehicle studied here can be considered as a linearization well suited for this study in the small.

The above mentioned definition of stability refers to the state of the system and in the case of the handling model with two degrees of freedom the state variables are β and r (or v and r). A motor vehicle is then stable if, when in motion with given values β_0 and r_0 of β and r, after a small external perturbation, it follows that

$$\beta(t) \to \beta_0 \,, \qquad r(t) \to r_0 \,.$$

No reference is made to the trajectory: After a perturbation the vehicle cannot return to the previous trajectory and a correction by the driver or by an automatic control system is required in order to maintain the vehicle on the road.

5.7.1 *Locked controls*

If the steering wheel is kept in a given position, that allowing to maintain the required trajectory, the stability can be studied simply using the homogeneous equation of motion $\{\dot{z}\} = [A]\{z\}$. The eigenvalues of the dynamic matrix $[A]$ are readily found and the stability is assessed from the sign of their real part, which must be negative. If the imaginary part is nonzero the behaviour is oscillatory; which does not necessarily imply that the trajectory is oscillatory but only that the time histories $\beta(t)$ and $r(t)$ are such.

The analogy with the spring-mass-damper system allows a simpler approach to the study of the stability at constant speed: To ensure static stability the stiffness U must be positive, to ensure dynamic stability the damping coefficient Q must be positive and to allow an oscillatory free behaviour Q must be lower than the critical damping $2\sqrt{PU}$.

Using the simplified expression of the derivatives of stability, the following expression of the "stiffness" U can be readily obtained

$$U = \frac{C_1 C_2 l^2}{V}(1 + KV^2)\,, \tag{5.100}$$

where K is the stability factor defined by Eq. (5.77).

U is then always positive for understeer and neutral-steer vehicles, and in the latter case it tends to zero when the speed tends to infinity. In the case of oversteer vehicles it is positive up to the critical speed, where it vanishes to become negative at higher speed. The critical speed is thus the threshold of instability for oversteering vehicles. Similar results hold also if the complete expressions for the derivatives of stability are used.

It is also easy to verify that Q is always positive: If the vehicle is statically stable it is also dynamically stable. If the simplified expression for the derivatives of stability is accepted, the value of Q is independent of the speed

$$Q = J_z(C_1 + C_2) + m(a^2 C_1 + b^2 C_2) \, . \tag{5.101}$$

The "critical damping" of the equivalent system Q_{crit} is, under the same simplifying assumptions

$$Q_{crit} = 2\sqrt{PU} = 2\sqrt{C_1 C_2 J_z m l^2 (1 + KV^2)} \, . \tag{5.102}$$

It is a constant in the case of neutral-steer vehicles, increases with speed for understeer vehicles while it decreases, vanishing at the critical speed, in case of oversteer ones. By comparing the actual damping with the critical one, it follows that understeer vehicles tend to develop an oscillatory behaviour (in the same way as a spring-mass-damper system with constant damping and increasing stiffness) with a frequency which increases with the speed, while oversteer vehicles tend to return to the original state without oscillations but in a way which is slower and slower with increasing speed, in the same way as a spring-mass-damper system with constant damping and decreasing stiffness.

In the case of a neutral-steer vehicle, under the same assumptions seen above, when $K = 0$ and $C_1 a = C_2 b$, the values of Q_{crit} and Q are

$$Q_{crit} = 2\frac{C_1 l J_z}{b}\sqrt{\frac{mab}{J_z}} \, ,$$

$$Q = \frac{C_1 l J_z}{b}\left(1 + \frac{mab}{J_z}\right) . \tag{5.103}$$

In many cases ratio mab/J_z is not very far from unity. By writing $mab/J_z = 1 + \epsilon$, expanding the above expressions in power series in ϵ and truncating the series after the linear term, the actual value of the damping coefficient can be shown to coincide with the critical value. A neutral-steer vehicle is then critically damped, at least in an approximated way, while understeering and oversteering vehicles are respectively underdamped and overdamped: The free behaviour of the former can then be expected to be oscillatory. It must be however noted that the issue whether a given vehicle has an oscillatory behaviour or not cannot be satisfactorily solved

	1	2	3	4
P kg^{2}m^3/s	2.790×10^7	2.790×10^7	2.790×10^7	2.790×10^7
Q kg^{2}m^3/s^2	2.876×10^8	2.899×10^8	2.897×10^8	2.829×10^8
U kg^{2}m^3/s^3	1.243×10^9	1.334×10^9	1.335×10^9	1.369×10^9
Q_{crit} kg^{2}m^3/s^2	3.725×10^8	3.858×10^8	3.860×10^8	3.908×10^8
$\Re(s)$ 1/s	-5.155	-5.196	-5.192	-5.070
$\Im(s)$ 1/s	± 4.242	± 4.562	± 4.573	± 4.834

using the present rigid body model as the presence of rolling motions, which are neglected here and are almost always underdamped and then oscillatory, can induce an oscillatory behaviour also for what β and r are concerned. This is particularly true for vehicles whose suspensions exhibit roll steer.

Example 5.5

Study the stability with locked controls of the vehicle of Appendix 1.1, taking into account the reduction of the cornering stiffness of the driving wheels caused by the longitudinal forces needed to move at constant speed.

The parameters of the equivalent spring-mass-damper system are evaluated first and then the poles of the system are computed. The values obtained at 100 km/h (27.78 m/s) are reported in Table 5.2. It is clear that the effect of the driving forces on the stability at 100 km/h is not great, even if the available traction is quite low, and that the simplified formulae already yield satisfactory results.

The values of P, Q and U are reported, together with that of Q_{crit}, as functions of the speed in Fig. 5.22a. In the same figure the real and imaginary parts of s and the roots locus are also shown. The figure has been obtained using the complete expressions of the derivatives of stability, but neglecting the effect of driving forces. Note that the stiffness U reduces with the speed without tending to zero as in the case of neutral vehicles and that the vehicle is almost always underdamped, except for very low speed, when $Q > Q_{crit}$.

5.7.2 *Free controls*

If the steering wheel is not controlled, motion of the vehicle with free controls occurs. The steering angle δ in this case is not an input to the system

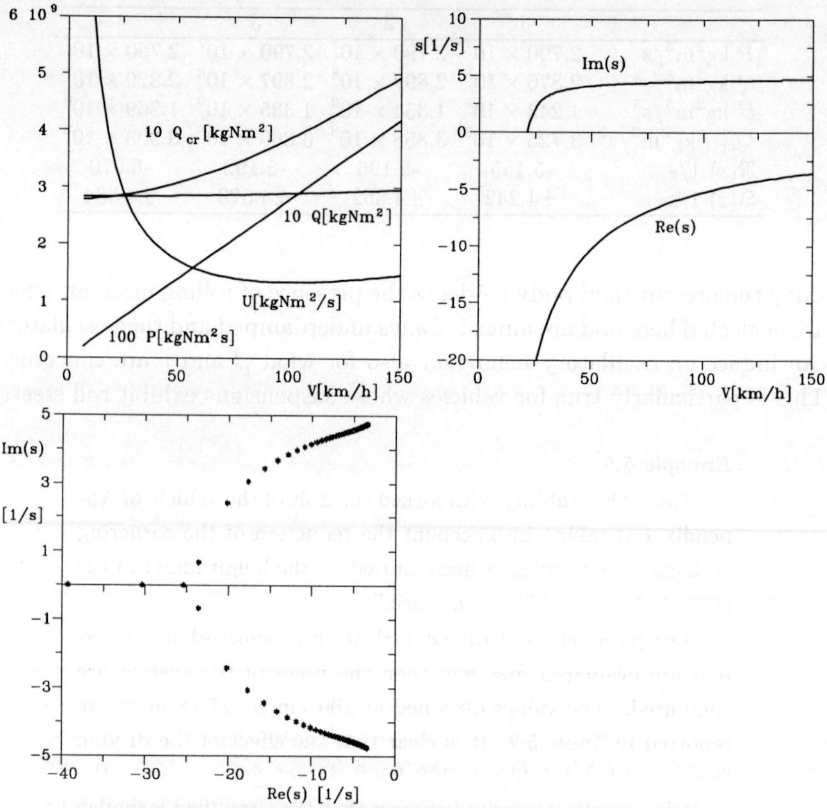

Fig. 5.22 Example 5.5: Study of the stability. (a) Parameters of the equivalent spring-mass-damper system as functions of the speed. (b) Real and imaginary parts of the eigenvalues as functions of the speed. (c) Roots locus at varying speed. Complete expressions of the derivatives of stability, with the effect of driving forces neglected.

but one of its state variables and a new equation stating the equilibrium of the steering system has to be included. Note that the same approach could be followed in the study of the motion with locked controls, as what is locked is actually not the steering angle δ but the position of the steering wheel and, if the compliance of the steering system is accounted for, the two things do not coincide. But if the compliance of the steering system is considered, oscillatory motions with high frequency can usually be found and it is unrealistic to consider the driver as a device which inputs a "position signal" δ to the vehicle. It is more realistic to consider the driver as a device supplying a driving torque on the steering wheel and consequently the motion occurs more in a free control situation than in a locked control one.

The actual situation is a mixed one: at low frequencies, like those typical of the motion of the vehicle as a whole, the locked control model is adequate while for high frequency modes the free control model is more suitable.

At any rate, as the motion of the vehicle includes high frequency components, the dynamic behaviour of the tires cannot be neglected. The simplest way to include it into a linearized model is to use relationships of the type

$$\begin{array}{l} F_y = -C\left(\alpha - B\dot{\alpha}\right), \\ M_z = \left(M_z\right)_{,\alpha}\left(\alpha - B'\dot{\alpha}\right) \end{array} \tag{5.104}$$

for the cornering force and the aligning torque.

The time derivatives of the sideslip angles are obviously

$$\dot{\alpha}_i = \dot{\beta} + \frac{x_i}{V}\dot{r} - \dot{\delta}_i \; . \tag{5.105}$$

The equations of motion (5.68) modify as

$$\left\{ \begin{array}{l} mV\left(\dot{\beta} + r\right) + m\dot{V}\beta = Y_\beta\beta + Y_r r + Y_\delta\delta + Y_{\dot{\beta}}\dot{\beta} + Y_{\dot{r}}\dot{r} + Y_{\dot{\delta}}\dot{\delta} + F_{y_e} \\ J_z\dot{r} = N_\beta\beta + N_r r + N_\delta\delta + N_{\dot{\beta}}\dot{\beta} + N_{\dot{r}}\dot{r} + N_{\dot{\delta}}\dot{\delta} + M_{z_e} \; , \end{array} \right. \tag{5.106}$$

where the expressions of the derivatives of stability already seen still hold while those of the other ones are

$$\left\{ \begin{array}{l} Y_{\dot{\beta}} = \sum_{\forall i} C_i B_i \\ \\ Y_{\dot{r}} = \frac{1}{V}\sum_{\forall i} x_i C_i B_i \\ \\ Y_{\dot{\delta}} = -\sum_{\forall i} K_i' C_i B_i \end{array} \right. \tag{5.107}$$

$$\begin{cases} N_{\beta} = \sum_{\forall i} \left[x_i C_i B_i - (M_{z_i})_{,\alpha} B_i' - h_G \left(f_0 + K V^2 \right) C_i B_i \right] \\ N_{\dot{r}} = \frac{1}{V} \sum_{\forall i} \left[x_i^2 C_i B_i - (M_{z_i})_{,\alpha} x_i B_i' - h_G \left(f_0 + K V^2 \right) C_i x_i B_i \right] \\ N_{\dot{\delta}} = \sum_{\forall i} \left[- K_i' C_i x_i B_i + K_i' (M_{z_i})_{,\alpha} B_i' + h_G \left(f_0 + K V^2 \right) K_i' C_i B_i \right] . \end{cases}$$

Note that in the equations above the moment due to rolling resistance and load transfer has been computed neglecting the presence of the shock absorbers, which cannot be accounted for in a rigid body model; in high frequency motion this assumption is more rough than in the previous model.

The equation which must be added to Eqs. (5.106) states the equilibrium to rotation of the steering system, assumed to be a rigid system. The geometry of the steering system is sketched in Fig. 5.23. The wheel rotates about an axis, the kingpin axis, which is neither perpendicular to the ground nor passing through the centre of the contact area: The caster angle ν, the lateral inclination angle λ and the lateral offset at the ground d are reported in the figure. In the figure the kingpin axis intersects the rotation axis of the wheel; this is a very common situation and the case in which the two axes are skew will not be dealt with here. However, if this is the case, it is also necessary to introduce a longitudinal offset at the ground.

If the kingpin axis were perpendicular to the ground and no offset were present, the torque acting on the wheel as a consequence of the road-tire interaction forces would just have been the aligning torque. The actual situation is however different and the torque about the kingpin axis contains all forces and moments acting on the wheel.

With geometrical reasoning, assuming that all angles are small, the total moment M_k about the kingpin axis of both wheels of a steering axle can be approximated as[10]

$$M_k = -(F_{z_l} + F_{z_r}) d \sin(\lambda) \sin(\delta) + (F_{z_l} - F_{z_r}) d \sin(\nu) \cos(\delta) + \quad (5.108)$$

$$+ (F_{y_l} + F_{y_r}) r_l \tan(\nu) + (F_{x_l} - F_{x_r}) d + (M_{z_l} + M_{z_r}) \cos\left(\sqrt{\lambda^2 + \nu^2} \right) ,$$

10 T. D. Gillespie, *Fundamentals of Vehicle Dynamics*, SAE, Warrendale, 1992.

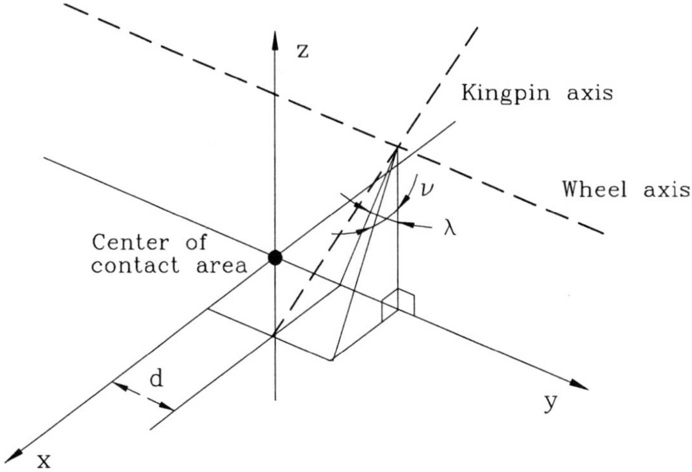

Fig. 5.23 Simplified geometry of the steering system and definition of the caster angle ν, the lateral inclination angle λ and the lateral offset at the ground d. The right wheel is sketched and ν, λ and d are positive. The kingpin axis is assumed to intersect the rotation axis of the wheel and consequently no longitudinal offset at the ground (other than that due to the caster angle) is present.

where r and l indicate the right and left wheels respectively.

In symmetrical conditions, the forces on ground at the two wheels are equal. By assuming that the steering angle is small, Eq. (5.108) reduces to

$$M_k = -F_z d \sin(\lambda)\delta + F_y r_l \tan(\nu) + M_z \cos\left(\sqrt{\lambda^2 + \nu^2}\right) , \qquad (5.109)$$

where forces and moments are referred to the whole axle.

By introducing expressions (5.104) into Eq. (5.109) the following linearized expression of the moment about the kingpin is obtained

$$M_k = M_{\dot{\beta}}\dot{\beta} + M_{\dot{r}}\dot{r} + M_{\dot{\delta}}\dot{\delta} + M_\beta \beta + M_r r + M_\delta \delta , \qquad (5.110)$$

where

$$\begin{aligned} &M_{\dot\beta} = CBr_l\tan(\nu) - (M_z)_{,\alpha}B'\cos\left(\sqrt{\lambda^2+\nu^2}\right) , \\ &M_{\dot r} = M_{\dot\beta}\frac{a}{V} , \qquad M_{\dot\delta} = -M_{\dot\beta} , \\ &M_\beta = -Cr_l\tan(\nu) + (M_z)_{,\alpha}\cos\left(\sqrt{\lambda^2+\nu^2}\right) , \\ &M_r = M_\beta\frac{a}{V} , \qquad M_\delta = -M_\beta - F_z d\sin(\lambda) . \end{aligned} \tag{5.111}$$

The linearized equation of motion of the steering system is then

$$J_s\ddot\delta + c_s\dot\delta = M_{\dot\beta}\dot\beta + N_{\dot r}\dot r + M_{\dot\delta}\dot\delta + M_\beta\beta + M_r r + M_\delta\delta + M_s\tau_s , \tag{5.112}$$

where M_s, τ_s, c_s and J_s are respectively the torque exerted by the driver on the steering wheel, the steering ratio (the ratio between the rotation angle of the wheel and that of the kingpin), the damping coefficient of the steering damper and the moment of inertia of the whole system, the latter two reduced to the kingpin. Note that the steering ratio is often not constant and that the compliance of the mechanism, here neglected, may have a large effect on it.

No gyroscopic effect of the wheels has been accounted for, which is consistent with the assumption of rigid vehicle, even if a weak gyroscopic effect should be present if the kingpin axis is not perpendicular to the road.

Equation (5.112) holds also when more complicated geometries are accounted for, provided that a linearization about a reference position is performed. In this case the expressions of the derivatives of stability M_β, M_r etc. also contain the longitudinal offset at the ground.

As the second derivative of the state variable δ enters the equations of motion, a further state variable $v_\delta = \dot\delta$ must be introduced and a further equation stating the mentioned identity must be added. The state equation is still Eq. (5.68)

$$\{\dot z\} = [A]\{z\} + [B]\{u\} ,$$

where the state and input vectors $\{z\}$ and $\{u\}$ are

$$\{z\} = \begin{Bmatrix} \beta \\ r \\ v_\delta \\ \delta \end{Bmatrix} , \qquad \{u\} = \begin{Bmatrix} F_{y_e} \\ M_{z_e} \\ M_s\tau_s \end{Bmatrix} ,$$

the dynamic matrix is

$$[A] = \begin{bmatrix} mV - Y_{\dot{\beta}} & -Y_{\dot{r}} & -Y_{\dot{\delta}} & 0 \\ -N_{\dot{\beta}} & J_z - N_{\dot{r}} & -N_{\dot{\delta}} & 0 \\ -M_{\dot{\beta}} & -M_{\dot{r}} & J_s & 0 \\ 0 & 0 & 0 & 1 \end{bmatrix}^{-1}$$

$$\times \begin{bmatrix} Y_\beta & -mV + Y_r & 0 & Y_\delta \\ N_\beta & N_r & 0 & N_\delta \\ M_\beta & M_r & (M_{\dot{\delta}} - c_s) & M_\delta \\ 0 & 0 & 1 & 0 \end{bmatrix}$$

and the input gain matrix is

$$[B] = \begin{bmatrix} mV - Y_{\dot{\beta}} & -Y_{\dot{r}} & -Y_{\dot{\delta}} & 0 \\ -N_{\dot{\beta}} & J_z - N_{\dot{r}} & -N_{\dot{\delta}} & 0 \\ -M_{\dot{\beta}} & -M_{\dot{r}} & J_s & 0 \\ 0 & 0 & 0 & 1 \end{bmatrix}^{-1} \begin{bmatrix} 1 & 0 & 0 \\ 0 & 1 & 0 \\ 0 & 0 & 1 \\ 0 & 0 & 0 \end{bmatrix} .$$

This equation can be used to study the stability of the vehicle and the response to any given law $M_s(t)$. In a similar way it is possible to study the steady-state performance simply by assuming that all derivatives are vanishingly small (the last state equation can then be dropped, as it reduces to the identity $0 = 0$)

$$\begin{bmatrix} -Y_\beta & mV - Y_r & -Y_\delta \\ -N_\beta & -N_r & -N_\delta \\ -M_\beta & -M_r & -M_\delta \end{bmatrix} \begin{Bmatrix} \beta \\ r \\ \delta \end{Bmatrix} = \begin{Bmatrix} F_{y_e} \\ M_{z_e} \\ M_s \tau_s \end{Bmatrix} . \qquad (5.113)$$

The *steering wheel torque gain* M_s/δ referred to the steering angle and that referred to the curvature of the trajectory $M_s R$, can be easily computed.

The eigenproblem

$$\det([A] - s[I]) = 0 \qquad (5.114)$$

allows one to study the stability in a straightforward way. Since the size of the dynamic matrix $[A]$ is just four, it is possible to write the characteristic equation and to solve it using the formula for 4-th degree algebraic equations. However, no closed form solution from which to draw general conclusions is available. The eigenvalues are either a pair of complex conjugate solutions, yielding damped oscillations (if both real parts are negative), one usually at low frequency and the other at high frequency, or two nonoscillatory solutions and one high frequency oscillation. The high frequency solution is usually linked with the dynamics of the steering device while the other ones are mainly linked with the behaviour of the vehicle.

The vibrations of the steering system were an actual concern in the past, particularly in the thirties, and were referred to as the "steering shimmy". They were also present in the tailwheel of aircraft undercarriage. The use of tires with lower pneumatic trail and above all the introduction of damping in the steering mechanism has completely rectified the problem. Both viscous damping and dry friction have been used with success, but the latter decreases the reversibility of the steering system and thus decreases its precision and its centring characteristics.

However the present model is too rough to be used to study in detail this phenomenon, as the compliance of the steering system and the lateral compliance of the suspension are important factors in originating this type of vibrations which can become self-excited.

If only the low-frequency overall behaviour of the vehicle is studied, it is possible to neglect the dependence of the tire forces with the time derivative of the sideslip angle. In this case the expressions of the dynamic matrix and of the input gain matrix simplify as follows

$$[A]=\begin{bmatrix} \dfrac{Y_\beta}{mV} & \dfrac{Y_r}{mV}-1 & 0 & \dfrac{Y_\delta}{mV} \\ \dfrac{N_\beta}{J_z} & \dfrac{N_r}{J_z} & 0 & \dfrac{N_\delta}{J_z} \\ \dfrac{M_\beta}{J_s} & \dfrac{M_r}{J_s} & \dfrac{-c_s}{J_s} & \dfrac{-M_\delta}{J_s} \\ 0 & 0 & 1 & 0 \end{bmatrix}$$

$$[B] = \begin{bmatrix} \frac{1}{mV} & 0 & 0 \\ 0 & \frac{1}{J_z} & 0 \\ 0 & 0 & \frac{1}{J_s} \\ 0 & 0 & 0 \end{bmatrix} .$$

If the inertia and the damping of the steering system can also be neglected, Eq. (5.112) can be solved in δ. By introducing it into the equations of motion of the vehicle an approximate model for the behaviour of the vehicle with free controls is obtained.

By assuming that the speed V is constant, the homogeneous state equation for a vehicle with steering on the front axle only is then

$$\left\{ \begin{matrix} \dot{\beta} \\ \dot{r} \end{matrix} \right\} = \begin{bmatrix} \frac{Y_\beta + Y_\delta}{mV} & \frac{Y_r + Y_\delta \frac{a}{V}}{mV} - 1 \\ \frac{N_\beta + N_\delta}{J_z} & \frac{N_r + N_\delta \frac{a}{V}}{J_z} \end{bmatrix} \left\{ \begin{matrix} \beta \\ r \end{matrix} \right\} . \tag{5.115}$$

The equation is formally identical to the homogeneous Eq. (5.67) and in this case it is also possible to resort to a spring-mass-damper analogy and to study the constant speed stability in a very simple way. It is possible to show that in this case both the "stiffness" and the "damping coefficient" are always positive, denoting both static and dynamic stability.

By introducing only the cornering forces due to the tires, the vehicle is overdamped at low speed, up to

$$V = \frac{1}{2}\left(b^2 + \frac{J_z}{m}\right)\sqrt{\frac{C_2}{J_z b}} ;$$

above that speed the behaviour becomes more and more underdamped, with a more and more marked oscillatory behaviour.

Note however that usually the last simplification is too rough: The high value of the steering ratio τ_s makes the inertia of the steering wheel reduced to the kingpin axis not negligible in most cases and the use of equation (5.115) can lead to errors which are not negligible. Also to neglect the steering damping cannot be justified, as a certain amount is at any rate present in the system and the effect of neglecting it can be a dynamic instability of the free behaviour.

Example 5.6

Compute the torque which must be exerted on the steering wheel needed to maintain the vehicle of Appendix 1.1 in a circular trajectory with a radius of 100 m and to counteract a transversal slope of 1° at constant speed. The additional data for the steering system are: $\lambda = 11°$, $\nu = 3°$, $d = 5$ mm and $\tau_s = 16$.

The steering wheel torque gain $M_s R$ can be computed from Eq. (5.113). By stating $F_{y_e} = 0$, $M_{z_e} = 0$ and $M_s = 1$ it is possible to obtain the yaw velocity r which follows the application of a unit torque to the steering wheel.

As $R = V/r$ it is immediate to compute the gain $M_s R$ and then the value of the torque needed to maintain any given circular trajectory. The results for $R = 100$ m are reported in Fig. 5.24a.

To obtain the steering torque needed to counteract a transversal road slope, Eq. (5.112) needs to be rearranged. The slope α_t is felt by the vehicle as a side force $F_{y_e} = mg\sin(\alpha_t)$. If the trajectory is straight $r = 0$ and also M_{z_e} is equal to zero as no external moment acts on the vehicle. The unknowns are β, δ and M_s. The equation is rearranged as

$$\begin{bmatrix} -Y_\beta & -Y_\delta & 0 \\ -N_\beta & -N_\delta & 0 \\ -M_\beta & -M_\delta & \tau_s \end{bmatrix} \begin{Bmatrix} \beta \\ \delta \\ M_s \end{Bmatrix} = \begin{Bmatrix} mg\sin(\alpha_t) \\ 0 \\ 0 \end{Bmatrix} .$$

The results obtained for a slope of 1° are reported in Fig. 5.24b.

5.8 Unstationary motion

There is no difficulty in integrating numerically the equation of motion (5.68) or even the complete nonlinear equation (5.35) once laws $\delta(t)$, $F_{y_e}(t)$, $M_{z_e}(t)$ and $V(t)$ are stated. If the motion occurs at constant speed and the manoeuvre is very simple, as a step input, it is even possible to perform a closed form integration of the linear model. However nowadays the numerical integration is so straightforward that it turns out that closed form solutions which are complicated and do not give a good insight into the rel-

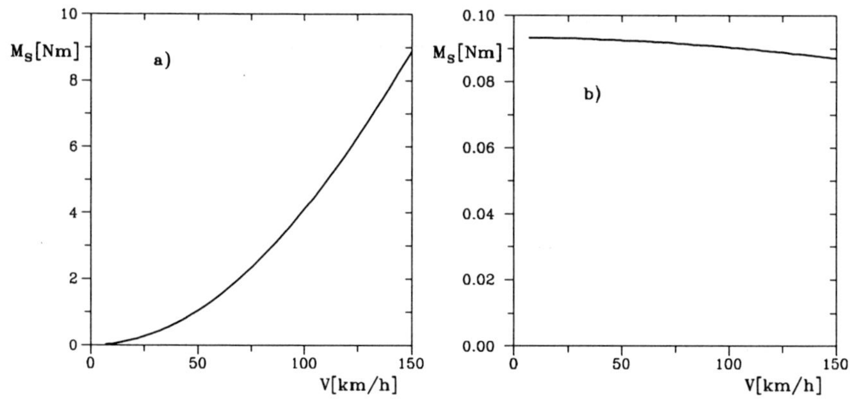

Fig. 5.24 Example 5.6: Steering wheel torque needed to maintain the vehicle on a circular trajectory with a radius of 100 m (a) and to counteract a transversal slope of 1° at constant speed (b).

evant phenomena are often of little use. In the present case the trajectory has at any rate to be computed through numerical integration.

Once the law $r(t)$ has been obtained, it is possible to integrate it to yield the yaw angle

$$\psi(t) = \int_0^t r(u)du \; . \tag{5.116}$$

The trajectory can then be obtained directly in the inertial coordinates X,Y. The velocities $\dot{X}$ and $\dot{Y}$ can be expressed in terms of angles β and ψ

$$\left\{ \begin{array}{c} \dot{X} \\ \dot{Y} \end{array} \right\} = V \begin{bmatrix} \cos(\psi) & -\sin(\psi) \\ \sin(\psi) & \cos(\psi) \end{bmatrix} \left\{ \begin{array}{c} \cos(\beta) \\ \sin(\beta) \end{array} \right\} . \tag{5.117}$$

By integrating equations (5.117) the trajectory is readily obtained

$$\left\{ \begin{array}{l} X = \int_0^t V\left[\cos(\beta)\cos(\psi) - \sin(\beta)\sin(\psi)\right] du \\ Y = \int_0^t V\left[\cos(\beta)\sin(\psi) + \sin(\beta)\cos(\psi)\right] du \end{array} \right. . \tag{5.118}$$

Note that the integration must be performed numerically even in the case of the linearized model as angle ψ can be too large to linearize its

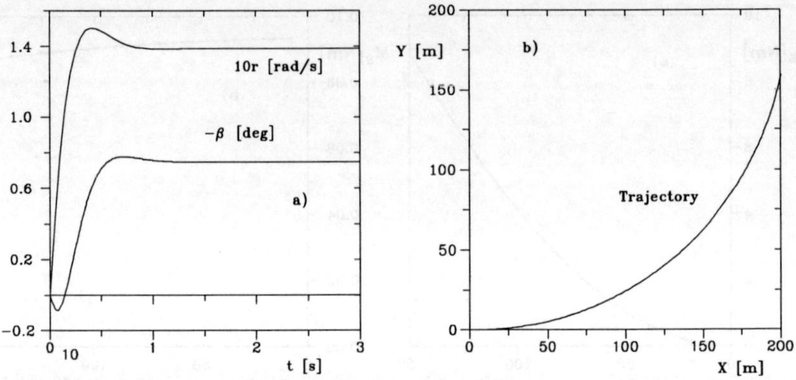

Fig. 5.25 Example 5.7: Response to a step steering input. (a) Time histories of the yaw velocity and sideslip angle and (b) trajectory.

trigonometric functions.

Example 5.7

Study the motion with locked controls of the vehicle of Appendix 1.1 following a step steering input. Assume that the value of the steering angle is that needed to obtain a circular trajectory with a radius of 200 m at a speed of 100 km/h.

At 100 km/h the trajectory curvature gain $1/R\delta$ is equal to 0.2472 1/m. To perform a curve with a radius of 200 m a steering angle $\delta = 0.0202$ rad $= 1.159°$ is needed. Note that in kinematic conditions the radius of the trajectory corresponding to the same value of δ is 106.8 m. The fact that it is almost half was easily predictable, as 100 km/h is only slightly less than the characteristic speed.

The steady state values of r and β are respectively of 0.1389 rad/s and -0.0131 rad $= -0.749°$.

The equation of motion of the vehicle has been integrated numerically for a duration of 30 s. The results are plotted in Fig. 5.25. The time histories of the yaw velocity and sideslip angle are reported together with the trajectory. Note that the steady-state conditions are reached after a few seconds, with a slightly underdamped behaviour.

Example 5.8

Study the motion with locked controls of the vehicle of Appendix 1.1 following a wind gust. Assume a step lateral gust, as the one encountered when exiting a tunnel. Assume an ambient wind velocity $v_a = 10$ m/s and a vehicle speed of 100 km/h. The driver does not react to the gust and the steering angle is kept equal to zero.

The presence of a cross-wind is accounted for by adding a side force F_{y_e} and a yawing moment M_{z_e} equal to

$$\begin{cases} F_{y_e} = \frac{1}{2}\rho V^2 S(C_y)_{,\beta}\psi_w \\ M_{y_e} = \frac{1}{2}\rho V^2 Sl(C_{M_z})_{,\beta}\psi_w \ , \end{cases}$$

where ψ_w is the angle between the direction of the relative velocity and the tangent to the trajectory. This is clearly an approximation as it relies on the linearity of the aerodynamic forces and moments with the aerodynamic sideslip angle and holds only if angle $\beta + \psi_w$ remains small.

As the trajectory of the vehicle bends after the manoeuvre, the components of the relative velocity along the trajectory and in a direction perpendicular to it are

$$\begin{cases} V_{\parallel} = V - v_a \sin(\psi + \beta) \\ V_{\perp} = -v_w \cos(\psi + \beta) \ , \end{cases}$$

yielding

$$\psi_w = \arctan\left(\frac{-v_w \cos(\psi + \beta)}{V - v_w \cos(\psi + \beta)}\right) \ .$$

The above relationships can be approximated by neglecting angle β. Another approximation is that of neglecting the contribution of the wind velocity to the airspeed, which is always considered at 100 km/h.

The equation of motion of the vehicle has been integrated numerically for a duration of 10 s. The results are plotted in Fig. 5.26. The time histories of the yaw velocity and sideslip angle are reported together with the trajectory. Note that quasi steady-state conditions are reached after a few seconds, with a slightly underdamped behaviour. Actually, there are no steady-state conditions as the direction of the wind is fixed, while the

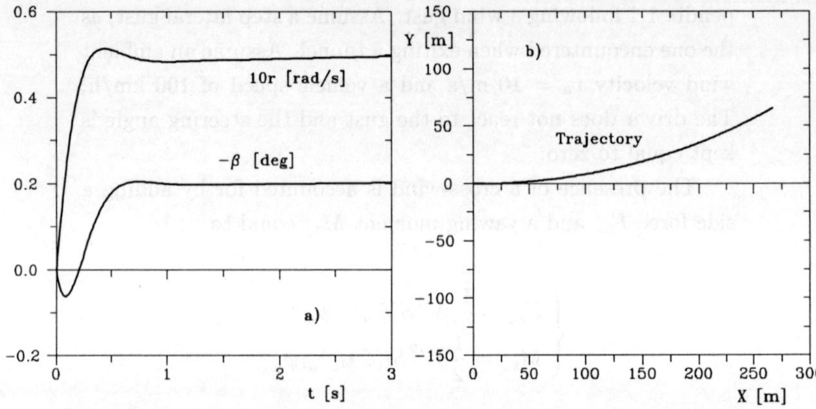

Fig. 5.26 Example 5.8: Response to a cross-wind gust. (a) Time histories of the yaw velocity and sideslip angle and (b) trajectory.

directions of the vehicle axes change. However, this effect is barely felt for the whole duration of the manoeuvre, and a very good approximation could have been obtained by assuming a constant value for angle ψ_w (ψ_w increases from 19.8° to 20.9° for $t = 0$ to $t = 10$ s).

At the end of the manoeuvre the values of r and β are respectively of 0.0505 rad/s and −0.0036 rad = −0.2073°. The errors linked to neglecting β in the above expression are then negligible.

Example 5.9

A manoeuvre which is often performed by test drivers to assess the handling and stability of a car is the following: A step steering input is supplied and the steering wheel is kept in position for a short time. The driver releases the wheel and the vehicle returns to a straight trajectory. The whole manoeuvre is performed at constant speed.

Study the motion of the vehicle of Appendix 1.1 following a manoeuvre of this kind with a 45° steering wheel input held for 1.5 s at 100 km/h.

The data of the steering system are $J_s = 15$ kg m^2, $c_s =$ 150 Nms/rad, $\lambda = 11°$, $\nu = 3°$, $d = 5$ mm and $\tau_s = 16$.

The first part of the manoeuvre is the same as in Example 5.6, only with a greater value of δ: 2.81°.

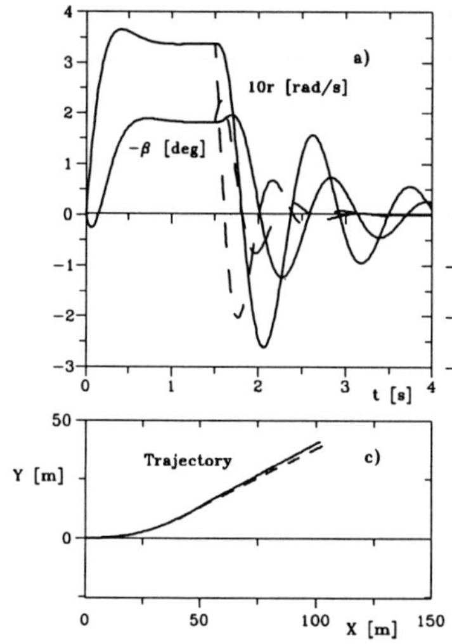

Fig. 5.27 Example 5.9: Response to a step steering input and a subsequent recovery of the straight trajectory with free controls. (a) Time histories of the yaw velocity and sideslip angle and (b) of the steering angle; (c) trajectory. The inertia and damping of the steering system are considered (full lines) and then neglected (dashed lines).

The integration in time is performed in two parts: Firstly a locked controls model is used for the first 1.5 s; a free control model is used after the driver releases the wheel. This second part of the simulation is performed alternatively using two models: One in which the dependence of tire forces from the derivative $\dot{\alpha}$ is neglected and a second one in which the inertia and damping of the steering system are also not considered.

The time histories of the yaw velocity, sideslip angle and steering angle are reported together with the trajectory in Fig. 5.27.

Note that the inertia of steering system plays an important role in the response, as it slows down the recovery of the vehicle, thus affecting also the trajectory. It also increases the oscillatory behaviour of the vehicle, and if no damping is considered, an unstable behaviour is found.

The effect of neglecting the inertia of the steering system can be verified by comparing the poles of the system: If no inertia and damping is accounted for the two eigenvalues are

$-3.011 \pm 7.709i$, while the more complete model yields four eigenvalues $-9.129 \pm 8.2921i$ and $-1.065 \pm 5.563i$. The first one is quite damped and is not important in the motion, but the second one is clearly different from the one obtained from the simpler model. The high value of the steering ratio, which enters as a square in the computation of the equivalent inertia of the steering wheel, is responsible for this effect.

5.9 Vehicles with two steering axles (4WS)

In the majority of the vehicles with two axles only the front wheels are provided with a steering system. However, starting from the eighties, an increasing number of cars with steering on all four wheels (4WS) appeared on the market, at the beginning mainly due to Japanese industries. The main aim has been an increase of manoeuvrability and in general of the handling characteristics both in low-speed and high-speed steering.

A very simple four-wheel steering can be implemented by supplying the rear axle with a purposely designed compliance to provide the required steering action under the effect of the road loads without adding an actual steering device. This approach is defined as passive steering. Active steering occurs when the rear axle is provided of a second steering device, operated by the driver together with that of the front axle, through adequate actuators.

To reduce the radius of the trajectory in low-speed (kinematic) conditions the rear axle must steer in opposite direction from the front one; if the absolute values of the steering angles are equal, the radius is halved and the off-tracking of the rear axle is reduced to zero. Using the notation introduced in the preceding sections, this situation is characterized by $K_1' = 1$, $K_2' = -1$ (in the following it will always be assumed that $K_1' = 1$). Practically this value is too high as, if starting the motion with the wheels in a steered position, the rear axle would initially be displaced too much outwards the line connecting the centres of the wheels in the initial position and, for example, would be very difficult to move a vehicle parked near the curb or, worse, near a wall.

Assuming that $K_1' = 1$ and K_2' is constant, the trajectory curvature gain and the off-tracking distance are

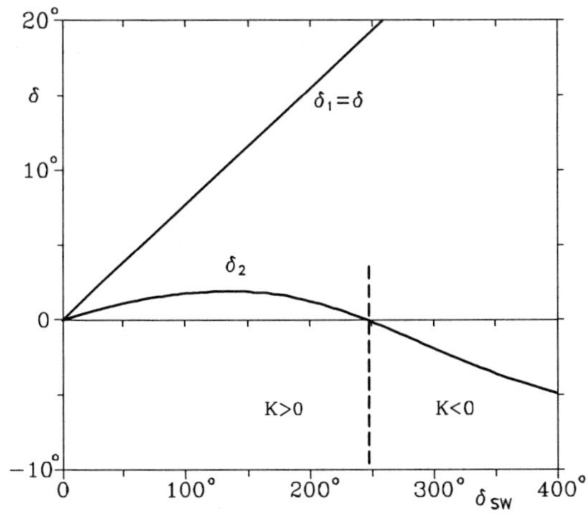

Fig. 5.28 Steering angles of the two axles as a function of angle δ_{sw} of the steering wheel in a mechanical device with variable gear ratio introduced by Honda.

$$\frac{1}{R\delta} \approx \frac{1+K_2'}{l}, \qquad R_f - R_1 \approx \frac{l^2(1-K_2')}{2R(1+K_2')} . \tag{5.119}$$

In high-speed cornering the situation is different: The possibility of putting all wheels with a sideslip angle without waiting for a rotation of the whole vehicle makes the response to a steering input far quicker. In this case rear wheels must exert cornering forces with the same direction as front ones and consequently the steering angles must be in the same direction. The limiting case, for a vehicle with neutral-steer point at the centre of the wheelbase, will be that of equal steering angles $K_1' = K_2' = 1$. This is again an unpractical result, as the vehicle would react very quickly in a lane change, simply moving sideways, but would never be able to negotiate a road bend: Instead of turning it would accelerate transversely.

It is then clear that the steering mechanism must adapt the value of K_2' to the external conditions and the requests of the driver. The simplest strategy is that of using a device, possibly mechanical, linking the two steering boxes with a variable gear ratio: When angle δ is small, as typically occurs in high speed driving, K_2' is positive and the steering angles have the same directions while when δ is large, as occurs when manoeuvring at low speed, K_2' is negative (Fig. 5.28).

However, to exploit fully the potential advantages of 4WS more complicated control laws for the steering of the rear axle must be implemented. The parameters which can enter such law are plenty, e.g. the speed V, the lateral acceleration, the sideslip angles α_i, etc. Such devices must be based on electronic controllers and actuators of different types and their implementation enters in that important but still not yet completely clear field of autronics. The reliability and cost requirements for autronic applications, particularly when vital functions like the steering are involved, are the key issues.

From the viewpoint of the mathematical modelling, the situation is, at least in principle, simple. There is no difficulty in introducing a suitable function $K_2'(V, \delta, \dots)$ into the equations (actually it would appear only in the derivatives of stability Y_δ and N_δ) and to modify accordingly the equations of the rigid-body model seen above. Also the more advanced models of next sections can be modified along the same lines. If function K_2' includes some of the state variables, the modifications can be larger but no conceptual difficulty arises.

Note that, except in the latter case, the locked control stability is not affected by the introduction of 4WS, while the stability with free controls can be affected by it.

Generally speaking the advantages are mainly linked with an increase of the quickness of the response of the vehicle to a steering input, but this cannot be true for all types of manoeuvres: Steering all axles in the same direction can make the vehicle very quick in lane change manoeuvres but can make it more slow in acquiring a given yaw velocity. The feeling of the driver can be strange and, at least at the beginning, unpleasant. A solution can be a device which initially steers the rear wheels in opposite direction for a very short time, to initiate a yaw rotation, and then steers them in the same direction as the front wheels, to generate cornering forces. This requires a more complicated control logic, possibly based on microprocessors.

As a last consideration, most applications are based on vehicles already designed for conventional steering to which 4WS is then added, normally as an option. In this case the steering of the rear wheel is limited to $1° - 2°$ or even less as there is not the required space in the rear wheel wells for a larger movement. Even if the car is designed from the beginning for 4WS a trade-off between the advantages and the loss of available space in the trunk due to 4WS must be performed. If the car is designed to fully exploit the advantages of all wheel steering, it can be equipped with integral drive as well (4WDS): In this case the standardization of the components of all the suspensions can be an added advantage, also from the viewpoint of costs.

5.10 Vehicle dynamics control (VDC)

Up to this point the control of the vehicle trajectory has been shown to be performed using the capability of the wheels to exert side forces, while the longitudinal forces are used to control longitudinal dynamics. The fact that longitudinal force generation interacts with the generation of cornering forces does not change much this picture.

Actually, in dynamic steering the trajectory must be controlled by forces which are perpendicular to the trajectory and then, if the sideslip angle of the vehicle β is small, almost perpendicular to the xz plane. However, it has been stated that dynamic driving occurs in two phases: initially the driver, operating some control device, exerts forces (mainly moments) changing the attitude of the vehicle on the trajectory and then, owing to this change of attitude, the required lateral forces are produced. In both two-wheel and four-wheel steering the forces which change the attitude of the vehicle are lateral forces due to the steering of some wheels. It is however possible to imagine different possibilities and also longitudinal forces can be used to modify the attitude of the vehicle, mainly by generating yaw moments.

Conceptually, it is even possible to imagine a vehicle with no steering wheels, in which the trajectory is controlled only by differential braking and driving of right and left wheels. However, while this practice is well consolidated in sea (e.g., differential thrust of the propellers) or air (e.g. differential braking of the main wheels during takeoff and landing) transportation, the trajectory of motor vehicles is universally controlled by steering some wheels.

Steering control has however its own limitations, particularly when limit conditions are approached. In particular, the average driver has difficulties in realizing that tire adhesion limits are approached, and in reacting in the proper way when these limits are met. An automatic device able to recognizing the imminent danger and acting to avoid the loss of control of the vehicle would be an important asset to increase road safety.

Such vehicle dynamics control (VDC) systems are presently under study and several versions of the relevant devices are already on the road, applied both to experimental and production vehicles. The input to these devices need to be some state variables of the vehicle, like the yaw velocity r and the sideslip angle β. A reasonable control philosophy would be to keep the former at the value required by the driver by stating a steering wheel angle (yaw-rate control); as a consequence a third variable which must be acquired is the steering wheel angle i.e., neglecting the steering gear ratio, angle δ. The quantities r, β and δ are the parameters entering the rigid-body handling models described in the previous sections, so it is fairly easy to introduce VDC in the models seen above.

While the yaw velocity can be easily acquired using a rate gyro, the sideslip angle can be difficult to measure. Instead, the lateral acceleration $\dot{v}$ can be easily measured and can be even more directly linked with the reaching of critical conditions for handling: while in steady state cornering the lateral acceleration and the yaw velocity are linked by the obvious relationship $\dot{v} = Vr$ (confusing the component of the acceleration perpendicular to the trajectory with that along the y-axis of the vehicle), in a spinout the yaw rate grows without a matching increase of the lateral acceleration.

The use of the longitudinal (braking or driving) forces for implementing VDC systems is particularly interesting both for the quickness of the control action and for the possibility of integrating the relevant hardware with that already present for anti-lock and anti-spin systems. The main cornering control is still performed by steering some of the wheels and is operated directly by the driver, while quick corrections can be performed by the VDC system using differential application of longitudinal forces.

5.11 Model with 4 degrees of freedom for articulated vehicles

5.11.1 *Equations of motion*

An articulated vehicle modelled as two rigid bodies hinged to each other has, in its motion on the road surface, four degrees of freedom (Fig. 5.29). The assumption of rigid bodies implies that the hinge is cylindrical and its axis is perpendicular to the road: In practice different setups are used, but if rolling is prevented the present one is the only possible layout. There is no difficulty in writing the six equations of motion of the two rigid bodies (each one has three degrees of freedom in the planar motion on the road) and then in introducing the two equations for the constraints due to the hinge to eliminate two of them. The forces exchanged between the two bodies are so explicitly introduced.

Here a different approach is followed and the equations of motion are obtained through Lagrange equations. To this end a set of four generalized

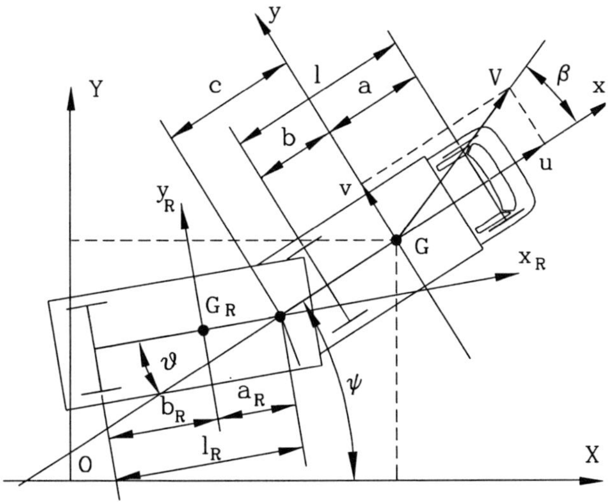

Fig. 5.29 Articulated vehicle. Reference frames and generalized coordinates.

coordinated is first stated: X and Y are the inertial coordinates of the centre of mass of the tractor and ψ is its yaw angle. They are the same used in the study of the insulated vehicle. The added coordinate is angle θ between the longitudinal axes x of the tractor and x_R of the trailer. Positive angles are shown in Fig. 5.29.

As a damper with damping coefficient Γ is associated to the hinge between tractor and trailer, a Rayleigh dissipation function must be written together with the kinetic energy. No conservative forces act in the plane of the road, at least if the hinge has no elastic restoring force as usual, and hence no potential energy needs to be computed.

The positions of the centre of mass of the trailer is

$$\overline{(G_R - O)} = \left\{ \begin{array}{l} X - c\cos(\psi) - a_R\cos(\psi - \theta) \\ Y - c\sin(\psi) - a_R\sin(\psi - \theta) \end{array} \right\} . \qquad (5.120)$$

The velocity of the centre of mass of the tractor is simply $V_G = [\dot{X}, \dot{Y}]^T$ while that of point G_R is

$$V_{G_R} = \left\{ \begin{array}{c} \dot{X} + \dot{\psi} c \sin(\psi) + \left(\dot{\psi} - \dot{\theta}\right) a_R \sin(\psi - \theta) \\ \dot{Y} - \dot{\psi} c \cos(\psi) - \left(\dot{\psi} - \dot{\theta}\right) a_R \cos(\psi - \theta) \end{array} \right\} . \tag{5.121}$$

The kinetic energy of the system is then

$$\mathcal{T} = \frac{1}{2} m_T V_G^2 + \frac{1}{2} m_R V_{G_R}^2 + \frac{1}{2} J_T \dot{\psi}^2 + \frac{1}{2} J_R \left(\dot{\psi} - \dot{\theta}\right)^2 , \tag{5.122}$$

where m_T, m_R, J_T and J_R are respectively the masses and the baricentric moments of inertia about an axis perpendicular to the road of the tractor and the trailer.

By introducing the expressions for the velocities into Eq. (5.122), it follows

$$\begin{aligned} \mathcal{T} = {} & \tfrac{1}{2} m \left(\dot{X}^2 + \dot{Y}^2\right) + \tfrac{1}{2} J_1(\theta) \dot{\psi}^2 + \tfrac{1}{2} J_3 \dot{\theta}^2 - J_2(\theta) \dot{\psi} \dot{\theta} + \\ & + m_R \left[c\dot{\psi} + a_R \left(\dot{\psi} - \dot{\theta}\right) \cos(\theta) \right] \left[\dot{X} \sin(\psi) - \dot{Y} \cos(\psi) \right] + \\ & - m_R a_R \left(\dot{\psi} - \dot{\theta}\right) \sin(\theta) \left[\dot{X} \cos(\psi) + \dot{Y} \sin(\psi) \right] , \end{aligned} \tag{5.123}$$

where

$$\left\{ \begin{array}{l} m = m_T + m_R , \\ J_1(\theta) = J_T + J_R + m_R \left[a_R^2 + c^2 + 2 a_R c \cos(\theta) \right] , \\ J_2(\theta) = J_R + m_R \left[a_R^2 + a_R c \cos(\theta) \right] , \\ J_3 = J_R + m_R a_R^2 . \end{array} \right.$$

Note that again the rotation kinetic energy of the wheels has been neglected: No gyroscopic effect of the wheels will be obtained in this way.

The Rayleigh dissipation function due to the above mentioned viscous damper is simply

$$\mathcal{F} = \frac{1}{2} \Gamma \dot{\theta}^2 . \tag{5.124}$$

The equations of motion obtained in the form of Lagrange equations are

$$\frac{d}{dt} \left(\frac{\partial \mathcal{T}}{\partial \dot{q}_i} \right) - \frac{\partial \mathcal{T}}{\partial q_i} + \frac{\partial \mathcal{F}}{\partial \dot{q}_i} = Q_i , \tag{5.125}$$

where the coordinates q_i are X, Y, ψ and θ and Q_i are the corresponding generalized forces F_X, F_Y and the moments related to rotations ψ and θ.

The derivatives needed to write the first two equations of motion, those related to the displacement degrees of freedom, are

$$\begin{cases} \dfrac{\partial \mathcal{T}}{\partial \dot{X}} = m\dot{X} + m_R \left\{ \dot{\psi} \left[c + a_R \cos(\theta) \right] - a_R \dot{\theta} \cos(\theta) \right\} \sin(\psi) + \\ \qquad -m_R a_R \left(\dot{\psi} - \dot{\theta} \right) \sin(\theta) \cos(\psi) \\ \dfrac{\partial \mathcal{T}}{\partial \dot{Y}} = m\dot{Y} - m_R \left\{ \dot{\psi} \left[c + a_R \cos(\theta) \right] - a_R \dot{\theta} \cos(\theta) \right\} \cos(\psi) + \\ \qquad -m_R a_R \left(\dot{\psi} - \dot{\theta} \right) \sin(\theta) \sin(\psi) \\ \dfrac{\partial \mathcal{T}}{\partial X} = \dfrac{\partial \mathcal{T}}{\partial Y} = \dfrac{\partial \mathcal{F}}{\partial \dot{X}} = \dfrac{\partial \mathcal{F}}{\partial \dot{Y}} = 0 \,. \end{cases} \tag{5.126}$$

Operating in the same way as for the insulated vehicle, and remembering expressions (5.34) for the absolute velocities as functions of components u and v in the reference frame xy of the tractor, the first two derivatives can be written in the form

$$\begin{cases} \dfrac{\partial \mathcal{T}}{\partial \dot{X}} = A \cos(\psi) - B \sin(\psi) \\ \\ \dfrac{\partial \mathcal{T}}{\partial \dot{Y}} = A \sin(\psi) + B \cos(\psi) \,, \end{cases} \tag{5.127}$$

where

$$A = mu - m_R a_R \left(\dot{\psi} - \dot{\theta} \right) \sin(\theta) \,,$$

$$B = mv - m_R \left\{ \dot{\psi} \left[c + a_R \cos(\theta) \right] - a_R \dot{\theta} \cos(\theta) \right\} \,.$$

By performing the derivatives with respect to time, the first two equations of motion are

$$\begin{cases} \left(\dot{A} - B\dot{\psi} \right) \cos(\psi) - \left(\dot{B} + A\dot{\psi} \right) \sin(\psi) = Q_X \\ \left(\dot{A} - B\dot{\psi} \right) \sin(\psi) + \left(\dot{B} + A\dot{\psi} \right) \cos(\psi) = Q_Y \,. \end{cases} \tag{5.128}$$

By premultiplying equations (5.128) by the inverse of the yaw rotation matrix (see Eq. (5.34)) and introducing the generalized forces Q_x and Q_y written with reference to xy frame, it follows

$$\begin{cases} \dot{A} - B\dot{\psi} = Q_x \\ \dot{B} + A\dot{\psi} = Q_y \end{cases} \tag{5.129}$$

i.e.

$$\begin{cases} m(\dot{u} - vr) - m_R a_R \left(\dot{r} - \ddot{\theta}\right)\sin(\theta) + 2m_R a_R r\dot{\theta}\cos(\theta) + \\ \qquad + m_R a_R \dot{\theta}^2 \cos(\theta) + m_R\left[c + a_R\cos(\theta)\right] r^2 = Q_x \\ m(\dot{v} + ur) - m_R\left[c + a_R\cos(\theta)\right]\dot{r} + m_R a_R r\ddot{\theta}\cos(\theta) + \\ \qquad - m_R a_R \sin(\theta)\left(r - \dot{\theta}\right)^2 = Q_y \,. \end{cases} \tag{5.130}$$

The third and fourth equations refer to the degrees of freedom ψ and θ. By performing all relevant derivatives, they are

$$\begin{cases} J_1(\theta)\ddot{\psi} - J_2(\theta)\ddot{\theta} + m_R a_R c\left(\dot{\theta}^2 - 2\dot{\theta}\dot{\psi}\right)\sin(\theta) + \\ \qquad - m_R\left[c + a_R\cos(\theta)\right]\left(\dot{v} + u\dot{\psi}\right) - m_R a_R \sin(\theta)\left(\dot{u} - v\dot{\psi}\right) = Q_\psi \\ J_3\ddot{\theta} - J_2(\theta)\ddot{\psi} + m_R a_R\cos(\theta)\left(\dot{v} + u\dot{\psi}\right) + \\ \qquad + m_R a_R \sin(\theta)\left[\dot{u} - \dot{\psi}\left(v - c\dot{\psi}\right)\right] = -\Gamma\dot{\theta} + Q_\theta \,. \end{cases} \tag{5.131}$$

5.11.2 *Sideslip angles of the wheels*

The sideslip angles of the wheels of the tractor are the same as for the insulated vehicle. In a similar way it is possible to write the sideslip angles of the wheels of the trailer.

With reference to Fig. 5.30, the coordinates of point P_i, centre of the contact zone of the i-th wheel of the trailer, are

$$\begin{cases} X_{P_i} = X - c\cos(\psi) - l_i\cos(\psi - \theta) - y_{R_i}\sin(\psi - \theta) \\ Y_{P_i} = Y - c\sin(\psi) - l_i\sin(\psi - \theta) + y_{R_i}\cos(\psi - \theta) \,. \end{cases} \tag{5.132}$$

The velocity of the same point can be obtained by differentiating the expressions of the coordinates. For the computation of the sideslip angle the velocity of point P_i must be expressed in the reference frame $G_R x_R y_R$ of the trailer

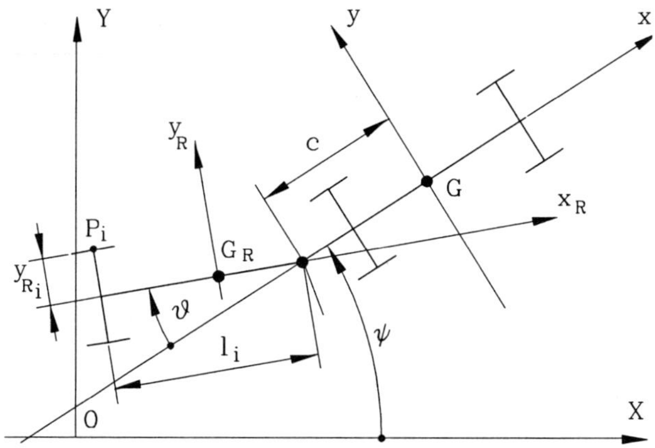

Fig. 5.30 Position of the centre P_i of the contact area of the i-th wheel of the trailer.

$$\left\{\begin{array}{c}\dot{X}_{P_i}\\ \dot{Y}_{P_i}\end{array}\right\}_R=\left[\begin{array}{cc}\cos\left(\psi-\theta\right) & \sin\left(\psi-\theta\right)\\ -\sin\left(\psi-\theta\right) & \cos\left(\psi-\theta\right)\end{array}\right]\left\{\begin{array}{c}\dot{X}_{P_i}\\ \dot{Y}_{P_i}\end{array}\right\}. \tag{5.133}$$

The velocity of the centre of the contact area can thus be expressed in the reference frame of the trailer as

$$\left\{\begin{array}{l}\dot{V}_{x_R}(P_i)=u\cos(\theta)-v\sin(\theta)+c\dot{\psi}\sin(\theta)-y_{R_i}\left(\dot{\psi}-\dot{\theta}\right)\\ \dot{V}_{y_R}(P_i)=u\sin(\theta)+v\cos(\theta)-c\dot{\psi}\cos(\theta)-l_i\left(\dot{\psi}-\dot{\theta}\right).\end{array}\right. \tag{5.134}$$

Since the sideslip angle of a steering wheel can be obtained as the arctangent of the ratio of y and x components of the velocity minus the steering angle δ, it follows that

$$\alpha_i=\arctan\left[\frac{u\sin(\theta)+v\cos(\theta)-c\dot{\psi}\cos(\theta)-l_i\left(\dot{\psi}-\dot{\theta}\right)}{u\cos(\theta)-v\sin(\theta)+c\dot{\psi}\sin(\theta)-y_{R_i}\left(\dot{\psi}-\dot{\theta}\right)}\right]-\delta_i\,. \tag{5.135}$$

5.11.3 *Generalized forces*

The contributions to the generalized forces Q_x, Q_y and Q_ψ due to the tractor are the same as for the insulated vehicle. The tractor does not give any contribution to force Q_θ. To compute the contributions due to the i-th wheel of the trailer and to the aerodynamic forces of the latter the easiest way is to write their virtual work $\delta\mathcal{L}$ due to a virtual displacement $\{\delta s\} = [\delta x, \delta y, \delta\psi, \delta\theta]^T$.

$$\begin{cases} \delta x_R(P_i) = \delta x\cos(\theta) - \delta y\sin(\theta) + c\delta\psi\sin(\theta) - y_{R_i}(\delta\psi - \delta\theta) \\ \delta y_R(P_i) = \delta x\sin(\theta) + \delta y\cos(\theta) - c\delta\psi\cos(\theta) - l_i(\delta\psi - \delta\theta)\ . \end{cases} \tag{5.136}$$

If the i-th wheel has a steering angle δ_i, the forces it exerts in the reference frame $G_R x_R y_R z_R$, the same in which the virtual displacement has been written, are simply

$$\begin{cases} F_{x_{i_R}} = F_{x_{i_t}}\cos(\delta_i) - F_{y_{i_t}}\sin(\delta_i) \\ F_{y_{i_R}} = F_{x_{i_t}}\sin(\delta_i) + F_{y_{i_t}}\cos(\delta_i)\ . \end{cases} \tag{5.137}$$

By multiplying the forces and the moment (the aligning torque M_{z_i}) by the corresponding virtual displacements (for the latter the rotation $\delta\psi - \delta\theta$), the virtual work is immediately obtained

$$\begin{aligned} \delta\mathcal{L} = &\left[F_{x_{i_t}}\cos(\theta-\delta_i) + F_{y_{i_t}}\sin(\theta-\delta_i)\right]\delta x + \\ &+\left[-F_{x_{i_t}}\sin(\theta-\delta_i) + F_{y_{i_t}}\cos(\theta-\delta_i)\right]\delta y + \\ &+\left\{F_{x_{i_t}}\left[c\sin(\theta-\delta_i) - y_{R_i}\cos(\delta_i) - l_i\sin(\delta_i)\right] + \right. \\ &\left. +F_{y_{i_t}}\left[-c\cos(\theta-\delta_i) + y_{R_i}\sin(\delta_i) - l_i\cos(\delta_i)\right] + M_{z_i}\right\}\delta\psi + \\ &+\left\{F_{x_{i_t}}\left[y_{R_i}\cos(\delta_i) + l_i\sin(\delta_i)\right] + \right. \\ &\left. +F_{y_{i_t}}\left[-y_{R_i}\sin(\delta_i) + l_i\cos(\delta_i)\right] - M_{z_i}\right\}\delta\theta. \end{aligned} \tag{5.138}$$

The generalized forces due to the i-th wheel of the trailer are then obtained by differentiating the virtual work $\delta\mathcal{L}$ with respect to the virtual displacements δx, δy, $\delta\psi$ and $\delta\theta$

$$
\begin{cases}
Q_{x_i} = \dfrac{\partial \delta \mathcal{L}}{\partial \delta x} = F_{x_{i_t}} \cos(\theta - \delta_i) + F_{y_{i_t}} \sin(\theta - \delta_i) \\
Q_{y_i} = \dfrac{\partial \delta \mathcal{L}}{\partial \delta y} = -F_{x_{i_t}} \sin(\theta - \delta_i) + F_{y_{i_t}} \cos(\theta - \delta_i) \\
Q_{\psi_i} = \dfrac{\partial \delta \mathcal{L}}{\partial \delta \psi} = F_{x_{i_t}} \left[c \sin(\theta - \delta_i) - y_{R_i} \cos(\delta_i) - l_i \sin(\delta_i) \right] + \\
\qquad + F_{y_{i_t}} \left[-c \cos(\theta - \delta_i) + y_{R_i} \sin(\delta_i) - l_i \cos(\delta_i) \right] + M_{z_i} \\
Q_{\theta_i} = \dfrac{\partial \delta \mathcal{L}}{\partial \delta \theta} = F_{x_{i_t}} \left[y_{R_i} \cos(\delta_i) + l_i \sin(\delta_i) \right] + \\
\qquad + F_{y_{i_t}} \left[-y_{R_i} \sin(\delta_i) + l_i \cos(\delta_i) \right] - M_{z_i} \; .
\end{cases}
\tag{5.139}
$$

In a similar way the generalized forces due to the aerodynamic forces and moments acting on the trailer can be accounted for. Here the problem is that it is usually very difficult to distinguish between the forces acting on the tractor and on the trailer, as the forces acting on the whole vehicle are measured in the wind tunnel. In the following equations it will be assumed that the forces acting on the tractor are measured separately from those acting on the trailer and that they are applied on the centre of mass of the relevant rigid body and decomposed along the axes fixed to it. Those acting on the trailer are so decomposed along axes $x_R y_R z_R$.

The aerodynamic forces acting on the tractor give forces Q_x, Q_y and Q_ψ which are the same as for the insulated vehicle while the expression of the generalized aerodynamic forces applied on the trailer can then be obtained from equations (5.139), by substituting $F_{x_{R_{aer}}}$, $F_{y_{R_{aer}}}$, $M_{z_{R_{aer}}}$ and a_R to $F_{x_{i_t}}$, $F_{y_{i_t}}$, M_{z_i} and l_i and by setting to zero both y_{R_i} and δ_i.

The external force $F_{y_{e_R}}$ acting in the centre of mass or the trailer and the component of the weight $m_R g \sin(\alpha)$ due to a longitudinal grade α of the road will be assumed to act in the directions of axes x and y of the tractor and consequently the relevant equations must be modified accordingly.

5.11.4 *Final expression of the equations of motion*

The final expression of the equation of motion involving coordinate x, i.e. the equation for longitudinal motion of the vehicle, is

$$m(\dot{u}-vr)-m_R a_R\left(\dot{r}-\ddot{\theta}\right)\sin(\theta)+2m_R a_R r\dot{\theta}\cos(\theta)+$$
$$+m_R a_R\dot{\theta}^2\cos(\theta)+m_R\left[c+a_R\cos(\theta)\right]r^2=$$
$$=\sum_{i=1}^{n_T}\left[F_{x_{i_t}}\cos(\delta_i)-F_{y_{i_t}}\sin(\delta_i)\right]+\sum_{i=1}^{n_R}\left[F_{x_{i_t}}\cos(\theta-\delta_i)+F_{y_{i_t}}\sin(\theta-\delta_i)\right]+$$
$$+\tfrac{1}{2}\rho V^2SC_x+\tfrac{1}{2}\rho V^2S_R\left[C_{X_R}\cos(\theta)+C_{Y_R}\sin(\theta)\right]-mg\sin(\alpha)+F_{y_{e_R}}\sin(\theta)\,,\tag{5.140}$$

where n_T and n_R are respectively the number of wheels of the tractor and of the trailer.

The equation of motion involving coordinate y, i.e. the equation for transversal motion of the vehicle, is

$$m(\dot{v}+ur)-m_R\left[c+a_R\cos(\theta)\right]\dot{r}+m_R a_R\ddot{\theta}\cos(\theta)-m_R a_R\sin(\theta)\left(r-\dot{\theta}\right)^2=$$
$$=\sum_{i=1}^{n_T}\left[F_{x_{i_t}}\sin(\delta_i)+F_{y_{i_t}}\cos(\delta_i)\right]+\sum_{i=1}^{n_R}\left[-F_{x_{i_t}}\sin(\theta-\delta_i)+F_{y_{i_t}}\cos(\theta-\delta_i)\right]+$$
$$+\tfrac{1}{2}\rho V^2SC_y+\tfrac{1}{2}\rho V^2S_R\left[-C_{X_R}\sin(\theta)+C_{Y_R}\cos(\theta)\right]+F_{y_{e_T}}+F_{y_{e_R}}\cos(\theta)\,.\tag{5.141}$$

The third equations for degree of freedom ψ, i.e. the equation for yaw rotation of the whole vehicle, is

$$J_1(\theta)\ddot{\psi}-J_2(\theta)\ddot{\theta}+m_R a_R c\left(\dot{\theta}^2-2\dot{\theta}\dot{\psi}\right)\sin(\theta)+$$
$$-m_R\left[c+a_R\cos(\theta)\right]\left(\dot{v}+u\dot{\psi}\right)-m_R a_R\sin(\theta)\left(\dot{u}-v\dot{\psi}\right)=$$
$$=\sum_{i=1}^{n_T}\left\{F_{x_{i_t}}\left[x_i\sin(\delta_i)-y_i\cos(\delta_i)\right]+F_{y_{i_t}}\left[x_i\cos(\delta_i)+y_i\sin(\delta_i)\right]+M_{z_i}\right\}+$$
$$+\sum_{i=1}^{n_R}\left\{F_{x_{i_t}}\left[c\sin(\theta-\delta_i)-y_{R_i}\cos(\delta_i)-l_i\sin(\delta_i)\right]+\right.$$
$$\left.+F_{y_{i_t}}\left[-c\cos(\theta-\delta_i)+y_{R_i}\sin(\delta_i)-l_i\cos(\delta_i)\right]+M_{z_i}\right\}+$$
$$+\tfrac{1}{2}\rho V^2SC_{M_z}+\tfrac{1}{2}\rho V^2S_R\left\{C_{xR}c\sin(\theta)+C_{yR}\left[-c\cos(\theta)-a_R\right]+l_RC_{M_{zR}}\right\}+$$
$$+M_{z_{e_T}}+M_{z_{e_R}}+F_{y_{e_R}}\left[-a_R\cos(\theta)-c\right]\,.\tag{5.142}$$

The fourth equation for the degrees of freedom θ, i.e. the equation for yaw rotation of the trailer about the hinge, is

$$
\begin{gathered}
J_3\ddot{\theta} - J_2(\theta)\ddot{\psi} + m_R a_R \cos(\theta)\left(\dot{v} + u\dot{\psi}\right) + \\
+ m_R a_R \sin(\theta)\left[\dot{u} - \dot{\psi}\left(v - c\dot{\psi}\right)\right] = -\Gamma\dot{\theta} + \\
+ \sum_{i=1}^{n_R}\left\{F_{x_{i_t}}\left[y_{R_i}\cos(\delta_i) + l_i \sin(\delta_i)\right] + F_{y_{i_t}}\left[-y_{R_i}\sin(\delta_i) + l_i\cos(\delta_i)\right] - M_{z_i}\right\} + \\
+ \frac{1}{2}\rho V^2 S_R\left(C_{y_R} a_R - l_R C_{M_{z_R}}\right) - M_{z_{e_R}} - F_{y_{e_R}} - a_R\cos(\theta)\ .
\end{gathered}
\tag{5.143}
$$

The four equations (5.140) – (5.143), together with the equations yielding the sideslip angles of the wheels, those expressing the forces and moments of the tires as functions of the sideslip angles, the load, and the other relevant parameters, allow to study the handling of the vehicle. The numerical integration is straightforward and allows to perform numerical simulations of the motion of the vehicle.

5.12 Linearized model for articulated vehicles

5.12.1 *Linearization of the equations of motion*

Equations (5.140) – (5.143) are a set of four nonlinear differential equations. The nonlinearities are present owing to the trigonometric functions of angles θ and δ_i, to some products of generalized coordinates and, in the external forces, to the nonlinearities which are linked with road-vehicle interaction forces and aerodynamic forces.

It is possible to perform a linearization of the equations of motion yielding a simplified set of equations well suited for the study of the motion of the vehicle in normal handling conditions. Actually, angles θ and δ_i are usually well within the range for which the linearization of the trigonometric functions is possible and the products of such angles by two of the variables v and r are negligible, as their order of magnitude is the same as that of the neglected terms in the series for the trigonometric functions of the angles.

As it was the case for the equations of the insulated vehicle, the linearization of the equations allows to uncouple the longitudinal behaviour (first equation of motion) from the lateral, or handling, behaviour, which can be studied using only the remaining three equations. This occurs if the law $u(t)$, which can be confused with $V(t)$, is considered as a stated law, while the unknowns are the driving or braking forces F_x for the longitudinal behaviour and β, r and θ for handling.

The linearized equation for the longitudinal behaviour is simply

$$m\dot{V} = Q_x \, , \tag{5.144}$$

while those for the lateral behaviour are

$$\begin{cases} m\left(\dot{v}+Vr\right)-m_R\left(c+a_R\right)\dot{r}+m_Ra_R\ddot{\theta}=Q_y \\ J_1\dot{r}-J_2\ddot{\theta}-m_R\left(c+a_R\right)\left(\dot{v}+Vr\right)-m_Ra_R\dot{V}\theta=Q_\psi \\ J_3\ddot{\theta}-J_2\dot{r}+m_Ra_R\left(\dot{v}+Vr\right)+m_Ra_R\theta\dot{V}=Q_\theta \, , \end{cases} \tag{5.145}$$

where now J_1 and J_2 are constant and the term in $\Gamma\dot{\theta}$ has been inserted into Q_θ.

5.12.2 *Linearized expressions of the sideslip angles*

By linearizing Eq. (5.135), the following expressions for the sideslip angles of the wheels of the trailer are easily obtained

$$\alpha_i = \theta + \beta - \frac{r}{V}(c+l_i) + \frac{\dot{\theta}}{V}l_i - \delta_i \, . \tag{5.146}$$

Note that the term in y_{R_i} does not enter the expression of the sideslip angle: The wheels of the same axle have the same sideslip angle and it is possible to work in terms of axle instead of in terms of single wheels also for the trailer.

The steering angle δ_i is either 0 or, if the axle can steer, is usually not directly controlled by the driver but is linked with the variables of the motion, e.g. with angle θ. If the law $\delta_i(\theta)$ is simply $\delta_i = -K_i'\theta$, the expression for the sideslip angle is

$$\alpha_i = \theta(1+K_i') + \beta - \frac{r}{V}(c+l_i) + \frac{\dot{\theta}}{V}l_i - \delta_i \, . \tag{5.147}$$

If some of the axles of the trailer are free to pivot about their kingpin, an equilibrium equation of the relevant parts of the steering system of those axles must be written, in a way which is similar to what was done for the study of the motion with free controls.

5.12.3 *Linearized expressions of the forces*

The expressions of the generalized forces Q_x, Q_y, Q_ψ and Q_θ can be linearized in the same way as seen for the insulated vehicle. The linearization can be performed by introducing the cornering stiffness C_i and the aligning stiffness $(M_{z_i})_{,\alpha}$ of the various axle (the subscript i now refers to the i-th axle and not to the i-th wheel). Similarly the slopes of the aerodynamic coefficients $(C_y)_{,\beta}$, etc. are introduced.

Operating in this way a very simple formulation for Q_x is obtained

$$\begin{aligned} Q_x = X_m - \left(f_0 + KV^2\right)\left[mg\cos(\alpha) - \tfrac{1}{2}\rho V^2\left(SC_z + S_R C_{z_R}\right)\right] + \\ -\tfrac{1}{2}\rho V^2\left(SC_x + S_R C_{x_R}\right) - mg\sin(\alpha), \end{aligned} \tag{5.148}$$

where, as usual, X_m is the tractive force due to the engine but can also be the total braking force.

By substituting the global sideslip angle β for the ratio v/V, the expressions of the forces entering the lateral behaviour are

$$\begin{cases} Q_y = (Q_y)_{,\beta}\beta + (Q_y)_{,r}r + (Q_y)_{,\dot\theta}\dot\theta + (Q_y)_{,\theta}\theta + (Q_y)_{,\delta}\delta + F_{y_e} + F_{y_{e_R}} \\ Q_\psi = (Q_\psi)_{,\beta}\beta + (Q_\psi)_{,r}r + (Q_\psi)_{,\dot\theta}\dot\theta + (Q_\psi)_{,\theta}\theta + (Q_\psi)_{,\delta}\delta + M_{z_e} + \\ \qquad + M_{z_{e_R}} - (c + a_R)F_{y_{e_R}} \\ Q_\theta = (Q_\theta)_{,\beta}\beta + (Q_\theta)_{,r}r + (Q_\theta)_{,\dot\theta}\dot\theta + (Q_\theta)_{,\theta}\theta - M_{z_{e_R}} + a_R F_{y_{e_R}} \,. \end{cases} \tag{5.149}$$

The derivatives of stability entering the expression for Q_y are

$$\begin{cases} (Q_y)_{,\beta} = Y_\beta - \sum_{\forall i_R} C_i + \dfrac{1}{2}\rho V_r^2 S_R (C_{Y_R})_{,\beta} \\ (Q_y)_{,r} = Y_r + \dfrac{1}{V}\left[\sum_{\forall i_R}(c + l_i)C_i - \dfrac{1}{2}\rho V_r^2 S_R (c + a_R)(C_{Y_R})_{,\beta}\right] \\ (Q_y)_{,\dot\theta} = -\dfrac{1}{V}\left[\sum_{\forall i_R} l_i C_i - \dfrac{1}{2}\rho V_r^2 S_R a_R (C_{Y_R})_{,\beta}\right] \\ (Q_y)_{,\theta} = -\sum_{\forall i_R} C_i + \dfrac{1}{2}\rho V_r^2 S_R (C_{Y_R})_{,\beta} \\ (Q_y)_{,\delta} = Y_\delta \,, \end{cases} \tag{5.150}$$

where Y_β, Y_r and Y_δ are the derivatives of stability of the tractor expressed by equations (5.60) and all axles of the trailer have been assumed as nonsteering.

The derivatives of stability entering the expression for Q_ψ and Q_θ are respectively

$$\begin{cases} (Q_\psi)_{,\beta} = N_\beta + \sum_{\forall i_R} M_{1_i} + C_1 \\ (Q_\psi)_{,r} = N_r - \frac{1}{V}\left[\sum_{\forall i_R}(c+l_i)M_{1_i} + (c+a_R)C_1\right] \\ (Q_\psi)_{,\dot{\theta}} = \frac{1}{V}\left[\sum_{\forall i_R} l_i M_{1_i} + a_R C_1\right] \\ (Q_\psi)_{,\theta} = \sum_{\forall i_R} M_{1_i} + C_1 \\ (Q_\psi)_{,\delta} = N_\delta \end{cases} \tag{5.151}$$

$$\begin{cases} (Q_\theta)_{,\beta} = -\sum_{\forall i_R} M_{2_i} - C_2 \\ (Q_\theta)_{,r} = \frac{1}{V}\left[\sum_{\forall i_R}(c+l_i)M_{2_i} + (c+a_R)C_2\right] \\ (Q_\theta)_{,\dot{\theta}} = -\frac{1}{V}\left[\sum_{\forall i_R} l_i M_{2_i} + a_R C_2\right] - \Gamma \\ (Q_\theta)_{,\theta} = (Q_\theta)_{,\beta} \\ (Q_\theta)_{,\delta} = 0 \,, \end{cases} \tag{5.152}$$

where

$$\begin{aligned} M_{1_i} &= (c+l_i)C_i + (M_{z_i})_{,\alpha} + h_{G_R}\left(f_0 + KV^2\right)C_i \,, \\ M_{2_i} &= l_i C_i + (M_{z_i})_{,\alpha} + h_{G_R}\left(f_0 + KV^2\right)C_i \,, \\ C_1 &= \frac{1}{2}\rho V_r^2 S_R\left[l_R (C_{N_R})_{,\beta} - (c+a_R)(C_{Y_R})_{,\beta}\right] , \\ C_2 &= \frac{1}{2}\rho V_r^2 S_R\left[l_R (C_{N_R})_{,\beta} - a_R (C_{Y_R})_{,\beta}\right] , \end{aligned}$$

N_β, N_r and N_δ are the derivatives of stability of the tractor expressed by equations (5.66), all axles of the trailer have been assumed as non-steering and the moment due to rolling resistance and load transfer on the trailer has been computed assuming that the tractor-trailer connection does not exert any rolling moment on the latter.

If the axles of the trailer can steer and their steering angles δ_i are linked with angle θ by the law $\delta_i = -K_i'\theta$, the expressions of the derivatives of stability reported above still hold, except for $(Q_y)_{,\theta}$, $(Q_\psi)_{,\theta}$ and $(Q_\theta)_{,\theta}$ in which all terms in C_i and $(M_{z_i})_{,\alpha}$ must be multiplied by $(1+K_i')$.

5.12.4 *Final expressions of the equations of motion*

The linearized equations for the lateral behaviour of the articulated vehicle can be expressed in the space of the configurations as

$$[M]\{\ddot{x}\} + [C]\{\dot{x}\} + [K]\{x\} = \{F\} , \tag{5.153}$$

where the vectors of the generalized coordinates and of the forces are

$$\{x\} = \left\{\begin{array}{c} y \\ \psi \\ \theta \end{array}\right\} , \qquad \{F\} = \left\{\begin{array}{c} (Q_y)_{,\delta}\delta + F_{y_e} + F_{y_{e_R}} \\ (Q_\psi)_{,\delta}\delta + M_{z_e} + M_{z_{e_R}} - (c+a_R)F_{y_{e_R}} \\ -M_{z_{e_R}} + a_R F_{y_{e_R}} \end{array}\right\} \tag{5.154}$$

and the matrices are

$$[M] = \begin{bmatrix} m & -m_R(c+a_R) & m_R a_R \\ -m_R(c+a_R) & J_1 & -J_2 \\ m_R a_R & -J_2 & J_3 \end{bmatrix}$$

$$[C] = \begin{bmatrix} -\dfrac{(Q_y)_{,\beta}}{V} & mV - (Q_y)_{,r} & -(Q_y)_{,\dot{\theta}} \\ -\dfrac{(Q_\psi)_{,\beta}}{V} & -m_R V(c+a_R) - (Q_\psi)_{,r} & -(Q_\psi)_{,\dot{\theta}} \\ -\dfrac{(Q_\theta)_{,\beta}}{V} & m_R V a_R - (Q_\theta)_{,r} & -(Q_\theta)_{,\dot{\theta}} \end{bmatrix} \tag{5.155}$$

$$[K] = \begin{bmatrix} 0 & 0 & -(Q_y)_{,\theta} \\ 0 & 0 & -(Q_\psi)_{,\theta} \\ 0 & 0 & -(Q_\theta)_{,\theta} \end{bmatrix} .$$

The set of differential equations (5.153) is actually of the fourth order and not of the sixth, as variables y and ψ appear in the equation only as first and second derivatives (the first two columns of matrix $[K]$ vanish). The equation can thus be written in the state space in form of a set of four first order differential equations by introducing a fourth state variable $v_\theta = \dot{\theta}$

$$\{\dot{z}\} = [A]\{z\} + [B]\{u\} .$$

The state vector $\{z\}$ is simply

$$\{z\} = \left[v \; r \; v_\theta \; \theta \right]^T ,$$

the dynamic matrix is

$$[A] = \begin{bmatrix} -[M]^{-1}[C] & [M]^{-1}\begin{Bmatrix} (Q_y)_{,\theta} \\ (Q_\psi)_{,\theta} \\ (Q_\theta)_{,\theta} \end{Bmatrix} \\ [0\,0\,1] & 0 \end{bmatrix} ,$$

the input gain matrix is

$$[B] = \begin{bmatrix} [M]^{-1} \begin{bmatrix} (Q_y)_{,\delta} & 1 & 1 & 0 & 0 \\ (Q_\psi)_{,\delta} & 0 & -(c+a_R) & 1 & 1 \\ 0 & 0 & a_R & 0 & -1 \end{bmatrix} \\ [0\,0\,0\,0\,0] \end{bmatrix} ,$$

and the input vector is

$$\{u\} = \left[\delta \; F_{y_e} \; F_{y_{e_R}} \; M_{z_e} \; M_{z_{e_R}} \right]^T .$$

5.12.5 *Steady-state motion*

To study the steady-state behaviour of the vehicle, Eq. (5.153) can be used together with the assumption that $\dot{v} = \dot{r} = \dot{\theta} = \ddot{\theta} = 0$. The following equation is thus obtained

$$\begin{bmatrix} -(Q_y)_{,\beta} & mV - (Q_y)_{,r} & -(Q_y)_{,\theta} \\ -(Q_\psi)_{,\beta} & -m_R V(c+a_R) - (Q_\psi)_{,r} & -(Q_\psi)_{,\theta} \\ -(Q_\theta)_{,\beta} & m_R V a_R - (Q_\theta)_{,r} & -(Q_\theta)_{,\theta} \end{bmatrix} \begin{Bmatrix} \beta \\ r \\ \theta \end{Bmatrix} = \tag{5.156}$$

$$= \begin{Bmatrix} (Q_y)_{,\delta}\delta + F_{y_e} + F_{y_{e_R}} \\ (Q_\psi)_{,\delta}\delta + M_{z_e} + M_{z_{e_R}} - (c+a_R)F_{y_{e_R}} \\ -M_{z_{e_R}} + a_R F_{y_{e_R}} \end{Bmatrix} .$$

There is no difficulty in solving such a set of equations e.g., after having stated that $\delta = 1$ and set all other inputs to zero, to compute the gains $1/R\delta$, β/δ etc.

A particularly simple solution is obtained for a two-axle vehicle with a one-axle trailer if only the cornering forces of the wheels are accounted for

$$\begin{cases} \dfrac{1}{R\delta} = \dfrac{1}{l}\dfrac{1}{1+KV^2} \\ \\ \dfrac{\theta}{\delta} = \dfrac{a+c+K'V^2}{l(1+KV^2)} \end{cases}, \tag{5.157}$$

where the stability factor K and K' are

$$\begin{cases} K = \dfrac{1}{l^2}\left[\left(m_T + m_R\dfrac{l_R - a_R}{l_R}\right)\left(\dfrac{b}{C_1} - \dfrac{a}{C_2}\right) - m_R\dfrac{c(l_R - a_R)}{l_R}\left(\dfrac{1}{C_1} + \dfrac{1}{C_2}\right)\right] \\ K' = \dfrac{1}{l}\left\{m\dfrac{a}{C_2} + \dfrac{m_R}{l_R}\left[\dfrac{(a+c)(l_R - a_R)}{C_2} - \dfrac{l_R a_R}{C_1}\right]\right\}. \end{cases} \tag{5.158}$$

The same definitions for the insulated vehicle hold also in this case and, if the derivatives of stability are constant or proportional to $1/V$ the sign of the stability factor allows one to state immediately whether the vehicle is oversteer, neutral-steer or understeer.

The simplified expression of the stability factor (5.158) is made of two terms: The first one usually has the same sign of $bC_1 - aC_2$, i.e. of the factor which decides the behaviour of the tractor alone. The second term is negative, unless the product $c(l_R - a_R)$ is negative, i.e. the centre of mass of the trailer is behind its axle.

If $c(l_R - a_R) > 0$, the trailer increases the understeering character of the vehicle, more so if the hinge is far from the centre of mass of the tractor and the centre of mass of the trailer is near the hinge. In the case of trailers with a single axle, as caravans, this effect can be reduced by reducing the distance between its centre of mass and the axle. If the centre of mass is exactly on the axle the trailer has no effect on the steady state behaviour of the tractor; it does however affect its dynamic behaviour and stability.

If the centre of mass of the trailer is behind its axle ($l_R - a_R < 0$), the trailer increases the oversteering behaviour of the tractor. If the vehicle is oversteering, the presence of a critical speed can be expected.

However, this way of comparing the behaviour of the tractor alone with that of the complete vehicle is not correct: The presence of the trailer can change the loads on the wheels of the former thus affecting their cornering stiffness.

Example 5.10

Study the steady state directional behaviour of the articulated truck of Appendix 1.4. Compare the results obtained using the complete expressions of the derivatives of stability with those computed considering only the cornering forces of the tires.

The computation is straightforward. At each value of the speed the normal forces on the ground must be computed, although they do not change much with the speed. From the normal forces the cornering stiffness and the aligning stiffness of the axles are readily obtained. At 100 km/h, for instance, the normal forces on the axles are 57.25, 107.28, 79.83, 83.56 and 56.14 kN, yielding the following values for the cornering stiffness and the aligning stiffness: 422.05, 806.64, 641.34, 665.89, 416.42 kN/rad and 22.724, 41.472, 26.102, 28.175, 22.116 kNm/rad.

The trajectory curvature gain, the sideslip angle gain and the trailer angle gain θ/δ are plotted as functions of the speed in Fig. 5.31. The values obtained from the complete expressions of the derivatives of stability are reported as full lines while the dashed lines refer to the simplified expressions for the derivatives of stability obtained by considering only the cornering forces.

Note that when the speed tends to zero the trajectory curvature gain does not tend to the kinematic value $1/l$ of the tractor: The trailer has a number of axles greater than one and correct kinematic steering is impossible. The vehicle is understeer, even if not in a very pronounced way. The simplified approach allows one to obtain a fair approximation of the directional behaviour of the vehicle, the differences between the two results being due mostly to the aligning torques of the tires and only marginally to aerodynamic forces and moments.

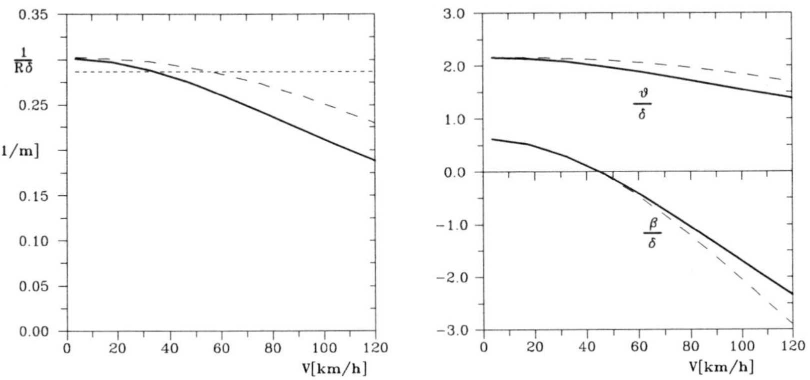

Fig. 5.31 Example 5.10: Trajectory curvature gain, sideslip angle gain and trailer angle gain as functions of the speed. Full lines: Values obtained from the complete expressions of the derivatives of stability; dashed lines: Simplified approach obtained considering only the cornering forces.

5.12.6 *Stability and nonstationary motion*

The study of the stability in the small, i.e., for small changes of the state of the system around the equilibrium conditions, can be performed by computing the eigenvalues of the dynamic matrix. The plot of the eigenvalues (their real and imaginary parts) as functions of the speed and that of the roots locus give a picture of the stability of the system which can be very easily interpreted.

The eigenvalues of the system are four, two of which are usually complex conjugate showing an oscillatory behaviour; the corresponding eigenvector shows that the motion of the trailer is mainly involved. These oscillations are usually lightly damped, and can become, mainly at high speed, self excited leading to a global instability of the vehicle. It must be however stated that the presence of an eigenvalue with positive real part, and hence of an instability in the mathematical sense, can be felt by the driver more as a source of discomfort than as an actual instability. If the value of both the imaginary and the real parts of the eigenvalue are low enough, i.e. if the frequency is low and the amplitude grows slowly, the driver is forced to introduce continuously steering corrections without actually realizing the instability of the vehicle.

The introduction of a damper at the trailer-tractor connection can solve this problem, while the use of steering axles on the trailer makes things worse. A steering axle, with a law causing the wheels to steer in the opposite direction than those of the tractor with a magnitude proportional to angle θ, provides a sort of restoring force trying to keep the trailer aligned with the tractor. The effect is similar to that of increasing the stiffness of a system: If the damping is not increased the underdamped character is magnified, while the natural frequency is also increased.

For the study of the motion in nonstationary conditions, the same considerations seen for the insulated vehicle still hold. The more complicated nature of the equations of motion compels however to resort to numerical integration in a larger number of cases.

Example 5.11

Study the stability with locked controls of the articulated truck of Appendix 1.4.

The plot of the real and imaginary parts of s and the roots locus are reported in Fig. 5.32.

The figure has been obtained using the complete expressions of the derivatives of stability, but neglecting the effect of driving forces. At 100 km/h the eigenvalues are $-2.3364 \pm 1.5896i$ and $-2.2698 \pm 3.4037i$; the corresponding eigenvectors are

$$\begin{Bmatrix} -0.8723 \pm 0.4849i \\ 0.0305 \pm 0.0424i \\ -0.0037 \mp 0.0346i \\ 0.0058 \pm 0.0109i \end{Bmatrix} , \quad \begin{Bmatrix} -0.6448 \mp 0.6533i \\ -0.0521 \pm 0.0862i \\ -0.1322 \pm 0.3429i \\ 0.0518 \mp 0.0734i \end{Bmatrix} .$$

Note that the vehicle has a strong oscillatory behaviour, even if both modes are well damped and no dynamic instability occurs; both modes interest the tractor as well as the trailer.

Example 5.12

Study the directional response and the stability with locked controls of the car of Appendix 1.2 with a caravan with a single axle. Assume the following data for the caravan: Mass $m_R = 600$ kg, moment of inertia $J_R = 800$ kg m^2, $c = 2.87$ m, $a_R = l_3 = 2.5$ m, $h_R = 1$ m, $S_R = 2.5$ m^2; $(C_{Y_R})_{,\beta} = -1.5$, $(C_{N_R})_{,\beta} = -0.6$. Assume that the trailer has the same tires used on the tractor.

The trajectory curvature gain, sideslip angle gain and trailer angle gain are plotted against the speed in Fig. 5.33. Both the

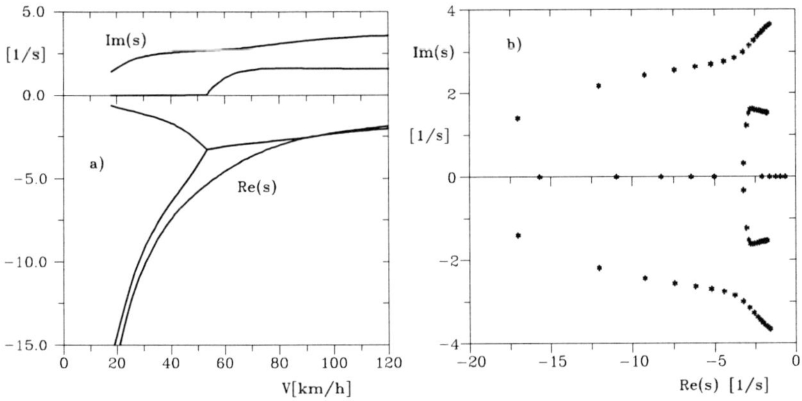

Fig. 5.32 Example 5.11: Study of the stability. (a) Real and imaginary parts of s as functions of the speed. (b) Roots locus at varying speed. Complete expressions of the derivatives of stability, with the effect of driving forces neglected.

complete and simplified expressions of the derivatives of stability have been used while the effect of driving forces has been neglected. Note that the curve obtained from the simplified expressions of the derivatives of stability is completely superimposed on that describing the behaviour of the insulated vehicle, as it was predictable as $a_R = l_3$. Note also that the trajectory curvature gain tends to the kinematic value for a speed tending to zero, as the trailer has a single axle and correct kinematic steering is possible.

The plot of the real and imaginary parts of s and the roots locus are reported in Figs. 5.32a and b. Here only the complete expressions of the derivatives of stability have been used. The vehicle is stable, but the absolute value of the real part of the Laplace variable s is quite low at high speed, denouncing a strong and little damped oscillatory motion, which occurs at low frequency.

To compare the behaviour of the vehicle with and without trailer the computation has been repeated without the latter and the results have been reported in Figs. 5.32c and d.

The comparison shows that the modes affecting mainly the vehicle are fairly uncoupled from the ones affecting mainly the trailer, although for a correct analysis of such coupling a

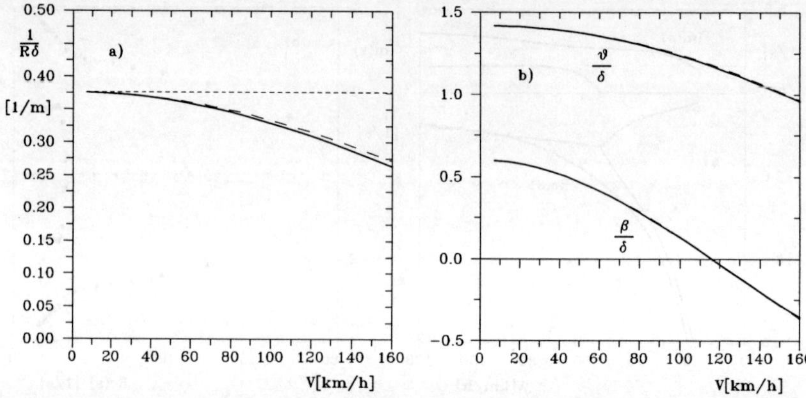

Fig. 5.33 Example 5.12: Trajectory curvature gain, sideslip angle gain and trailer angle gain as functions of the speed. Full lines: Values obtained from the complete expressions of the derivatives of stability; dashed lines: Simplified approach obtained considering only the cornering forces.

through analysis of the eigenvectors should be performed. The "trailer mode" with low frequency and low dynamic stability is superimposed to the more stable "vehicle mode", which is not strongly affected by the presence of the trailer. Note that the motion of the tractor in the trailer mode can also be quite large, as this mode affects the whole system.

Example 5.13

Study the stability with locked controls of the car of Appendix 1.1 with the same caravan of Example 5.12. Assume that the tires of the caravan are the same as those used on the tractor. Then study the motion with locked controls of the same vehicle following a step steering input at 80 and 140 km/h. Assume that the value of the steering angle is that needed to obtain a circular trajectory with a radius of 200 m, computed neglecting the presence of the trailer.

The plot of the real and imaginary parts of s and the roots locus computed using the complete expressions of the derivatives of stability are reported in Figs. 5.35a and b. The vehicle is stable only up to a speed of about 120 km/h, where the real part of the Laplace variable s related to one of the two modes vanishes to become positive at higher speed. The absolute value of the real part of s is always quite low, denoting a marginal

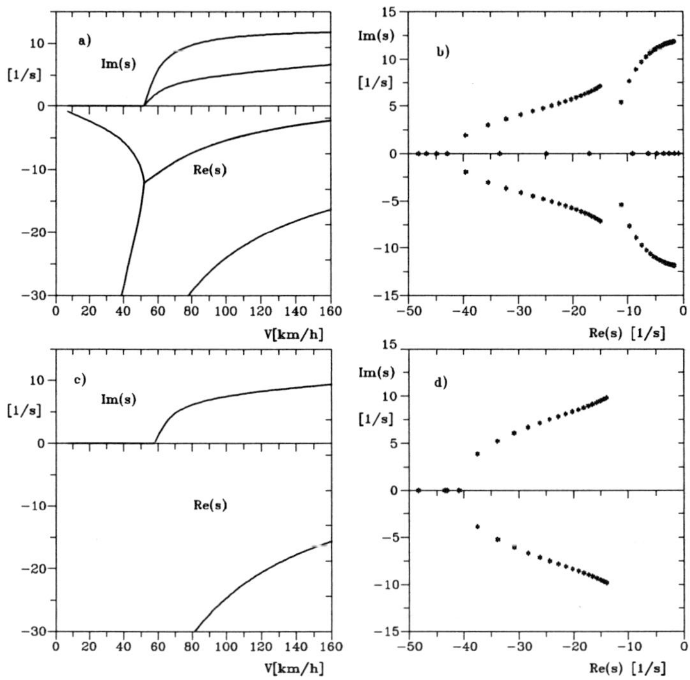

Fig. 5.34 Example 5.12: Study of the stability. (a) Real and imaginary parts of s as functions of the speed. (b) Roots locus at varying speed. (c), (d): Same as (a), (b) but for the vehicle without trailer. Complete expressions of the derivatives of stability, with the effect of driving forces neglected.

dynamic stability at low speed and a marginal instability at higher speed.

This type of behaviour is quite evident in the response to a step steering input. The steering angle needed to obtain a radius of the trajectory of 200 m is 0.9659° at 80 km/h and 1.7271° at 140 km/h. The integration of the equation of motion was performed numerically. At 80 km/h the response is stable but the step input excites a strong oscillatory behaviour which is slowly damped (Fig. 5.36a). The strong oscillatory behaviour is mainly due to the trailer and the time history which shows more pronounced oscillations is that of the trailer angle θ. After 6 s the values of $r/V\delta$, $\beta\delta$ and $\theta\delta$ are almost stabilized at the values of 0.3018, −0.4056 and 0.3098 which characterize the steady state behaviour (the former two are almost the same as

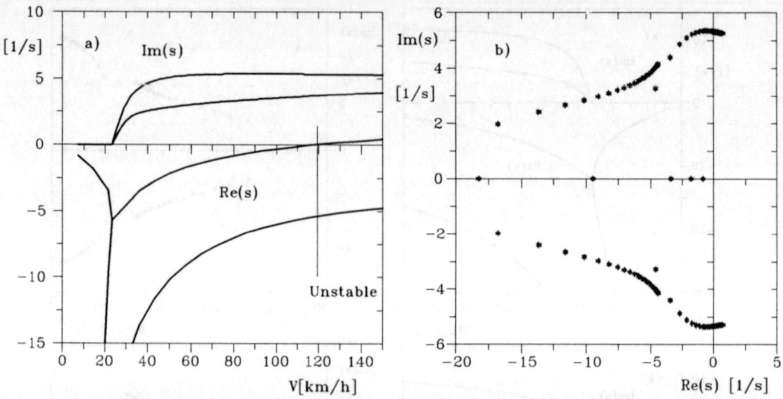

Fig. 5.35 Example 5.13: Study of the stability. (a) Real and imaginary parts of s as functions of the speed. (b) Roots locus at varying speed. Note the instability threshold at about 120 km/h. Complete expressions of the derivatives of stability, with the effect of driving forces neglected.

those obtained for the vehicle without trailer, except for a small difference due to the difference in aerodynamic drag, which influences the road loads and hence the cornering stiffness). The trajectory is however not oscillatory.

At 140 km/h the vehicle is unstable and the oscillations of r, β and θ quickly diverge. However the trajectory, reported in Fig. 5.36b, is not strongly oscillatory.

Note that this example is a limiting case as the trailer is not correctly matched to the vehicle and also the tires are not correct for the trailer. An example of unstable behaviour which can occur in incorrectly designed vehicles with trailer has been shown. Note that a step input is prone to excite very strongly an unstable behaviour and is the worst thing to do with a marginally stable vehicle. The oscillations have a low frequency and it is possible that the driver is able to stabilize the vehicle even at speeds at which the real part of s is positive: A test driver would probably state that the handling and the comfort of the vehicle are very poor rather than stating that the vehicle is unstable, owing to the need of continuous steering corrections.

On the other hand it is possible that a vehicle with a low negative real part of s ends to be unstable owing to the action

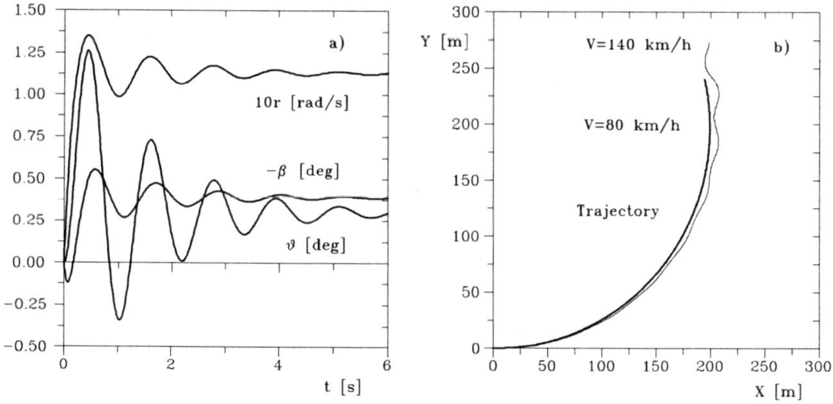

Fig. 5.36 Example 5.13: Response to a step steering input. (a) Time histories of the yaw velocity, sideslip angle β and trailer angle θ at 80 km/h and (b) trajectory at 80 and 140 km/h.

of the driver. The stability of the vehicle-driver system is what counts at the end, but intrinsic stability of the vehicle is needed, to avoid that the driver is forced to act as a "stabilizer" for a system which is unstable by itself.

5.13 Multibody articulated vehicles

5.13.1 *Equations of motion*

Consider a vehicle with a trailer with two axles, one connected to its body and the other connected to the draw bar (Fig. 5.37a). Its dynamic behaviour can be studied using a model of the same type seen in the previous section, in which the trailer is modelled as two simple trailers connected one after the other. The model has five degrees of freedom and the five generalized coordinates can be X Y, ψ, θ_1 and θ_2. The first two coordinates can be substituted by displacements x and y referred to the frame of the tractor and the first equation for longitudinal motion can be decoupled from the other ones, if the equations of motion are linearized. The transversal behaviour can be studied using a set of four differential equations, which can be linearized under the usual conditions yielding a set of linear differential equation whose order is six.

This procedure can be generalized to a generic multibody vehicle, made

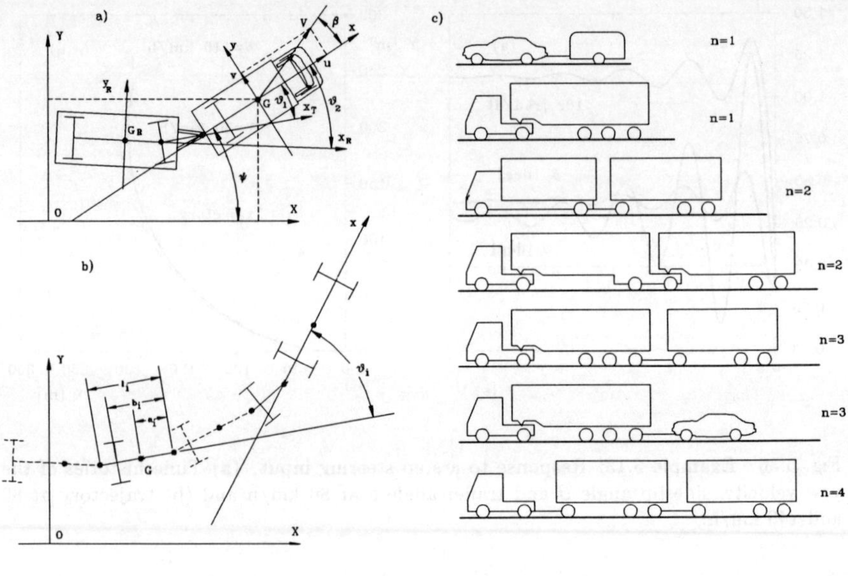

Fig. 5.37 (a) Vehicle with a trailer with two axles. (b) model of a multibody articulated vehicle; parameters for the i-th trailer. (c) examples of multibody vehicles; note that only the first three are road legal in Europe.

of a tractor and a set of n trailers (Fig. 5.37b). Note that while in Europe no vehicle with multiple trailers is road-legal, in America and Australia such vehicles can be used, although they are subjected to some restrictions. The model here described, which leads to a set of $n+3$ differential equations ($n+2$ for the lateral behaviour if the first equation is uncoupled) allows one to study the behaviour of any vehicle of this type.

With reference to Fig. 5.37b, the position of the centre of mass of the i-th trailer is

$$\overline{(\mathrm{G}_i - \mathrm{O})} = \left\{ \begin{array}{c} X - c\cos(\psi) - \sum_{k=1}^{i-1} l_k \cos(\psi - \theta_k) - a_i \cos(\psi - \theta_i) \\ Y - c\sin(\psi) - \sum_{k=1}^{i-1} l_k \sin(\psi - \theta_k) - a_i \sin(\psi - \theta_i) \end{array} \right\} . \tag{5.159}$$

The velocity of the same point G_i is

$$V_{G_i} = \left\{ \begin{array}{c} \dot{X} + \dot{\psi} c \sin(\psi) + \sum_{k=1}^{i-1} \left(\dot{\psi} - \dot{\theta}_k \right) l_k \sin(\psi - \theta_k) + \left(\dot{\psi} - \dot{\theta}_i \right) a_i \sin(\psi - \theta_i) \\ \dot{Y} - \dot{\psi} c \cos(\psi) - \sum_{k=1}^{i-1} \left(\dot{\psi} - \dot{\theta}_k \right) l_k \cos(\psi - \theta_k) - \left(\dot{\psi} - \dot{\theta}_i \right) a_i \cos(\psi - \theta_i) \end{array} \right\} . \tag{5.160}$$

The contribution to the kinetic energy due to the i-th trailer with mass m_i and moment of inertia J_i about a baricentric axis parallel to z-axis is then

$$\mathcal{T}_i = \frac{1}{2} m_i V_i^2 + \frac{1}{2} J_i \left(\dot{\psi} - \dot{\theta}_i \right)^2 , \tag{5.161}$$

i.e.

$$\begin{array}{l} \mathcal{T}_i = \frac{1}{2} m \left[\dot{X}^2 + \dot{Y}^2 - 2 \left(\dot{X} \alpha_i + \dot{Y} \beta_i \right) \cos(\psi) + \right. \\ \left. \quad +2 \left(\dot{X} \beta_i - \dot{Y} \alpha_i \right) \sin(\psi) \right] + \frac{1}{2} J_i \left(\dot{\psi} - \dot{\theta}_i \right)^2 , \end{array} \tag{5.162}$$

where

$$\alpha_i = \sum_{j=1}^{i} l_{ij} \left(\dot{\psi} - \dot{\theta}_j \right) \sin(\theta_j) ,$$

$$\beta_i = c\dot{\psi} + \sum_{j=1}^{i} l_{ij} \left(\dot{\psi} - \dot{\theta}_j \right) \cos(\theta_j)$$

and constants l_{ij} are the elements of the matrix

$$[l] = \begin{bmatrix} a_1 & 0 & 0 & 0 \\ l_1 & a_2 & 0 & 0 \\ l_1 & l_2 & a_3 & 0 \\ \cdots & \cdots & \cdots & \cdots \\ l_1 & l_2 & l_3 & a_n \end{bmatrix} .$$

Here again the rotation kinetic energy of the wheels has been neglected and no gyroscopic effect of the wheels can be obtained.

The Rayleigh dissipation function due to a generic viscous damper located between the $(i-1)$-th and the i-th trailer is simply

$$\mathcal{F} = \frac{1}{2}\Gamma\left(\dot{\theta}_i - \dot{\theta}_{i-1}\right)^2 . \tag{5.163}$$

Operating in the same way seen for the insulated vehicle, the first equation of motion, related to the displacement in x direction, is

$$m\left(\dot{u} - v\dot{\psi}\right) + \sum_{i=1}^{n}\left[-\sum_{j=1}^{i} l_{ij}\left(\ddot{\psi} - \ddot{\theta}_j\right)\sin(\theta_j) - 2\dot{\psi}\sum_{j=1}^{i} l_{ij}\dot{\theta}_j\cos(\theta_j) + \right.$$
$$\left. + \sum_{j=1}^{i} l_{ij}\dot{\theta}_j^2\cos(\theta_j) + \dot{\psi}^2\left(c + \sum_{j=1}^{i} l_{ij}\cos(\theta_j)\right)\right] = Q_x , \tag{5.164}$$

where $m = m_T + \sum_{i=1}^{n} m_i$ is the total mass of the vehicle.

The second equation of motion, related to the displacement in y direction, is

$$m\left(\dot{v} + u\dot{\psi}\right) + \sum_{i=1}^{n}\left\{-\ddot{\psi}\left[c + \sum_{j=1}^{i} l_{ij}\cos(\theta_j)\right] + \right.$$
$$\left. + \sum_{j=1}^{i} l_{ij}\ddot{\theta}_j\cos(\theta_j) - \sum_{j=1}^{i} l_{ij}\left(\dot{\psi} - \dot{\theta}_j\right)^2\sin(\theta_j)\right\} = Q_y , \tag{5.165}$$

The third equation refers to the degree of freedom ψ

$$\left\{J_T + \sum_{i=1}^{n}\left[m_i\left(S_i^2 + C_i^2\right) + J_i\right]\right\}\ddot{\psi} + \sum_{i=1}^{n} m_i\left\{\left(-\dot{u} + v\dot{\psi}\right)S_i + \right.$$
$$-\left(\dot{v} + u\dot{\psi}\right)C_i - S_i\sum_{j=1}^{i} l_{ij}\left[\ddot{\theta}_j\sin(\theta_j) - \dot{\theta}_j\left(\dot{\psi} - \dot{\theta}_j\right)^2\cos(\theta_j)\right] +$$
$$-C_i\sum_{j=1}^{i} l_{ij}\left[\ddot{\theta}_j\cos(\theta_j) + \dot{\theta}_j\left(\dot{\psi} - \dot{\theta}_j\right)^2\sin(\theta_j)\right] + \tag{5.166}$$
$$\left. + \dot{\psi}S_i\sum_{j=1}^{i} l_{ij}\dot{\theta}_j\cos(\theta_j) - \left(\dot{\psi}C_i + v\right)\sum_{j=1}^{i} l_{ij}\dot{\theta}_j\sin(\theta_j)\right\} = Q_\psi ,$$

where

$$S_i = \sum_{j=1}^{i} l_{ij} \sin(\theta_j) \ ,$$

$$C_i = c + \sum_{j=1}^{i} l_{ij} \cos(\theta_j) \ .$$

The following n equations refer to the rotational generalized coordinates θ_j (for $j = 1, 2, \ldots, n$). The generic equation for θ_k, i.e. the $(3+k)$-th equation, is

$$\sum_{i=k}^{n} m_i l_{ik} \left\{ \sin(\theta_k) \left[\dot{u} - v\dot{\psi} + \dot{\psi}^2 C_i - \ddot{\psi} S_i - \sum_{j=1}^{i} l_{ij} \ddot{\theta}_j \sin(\theta_j) + \right.\right.$$
$$\left. -2\dot{\psi} \sum_{j=1}^{i} l_{ij} \dot{\theta}_j \cos(\theta_j) + \sum_{j=1}^{i} l_{ij} \dot{\theta}_j^2 \cos(\theta_j) \right] +$$
$$+ \cos(\theta_k) \left[\dot{v} + u\dot{\psi} - \ddot{\psi} C_i + -\dot{\psi}^2 S_i + \sum_{j=1}^{i} l_{ij} \ddot{\theta}_j \cos(\theta_j) + 2\dot{\psi} \sum_{j=1}^{i} l_{ij} \dot{\theta}_j \sin(\theta_j) + \right.$$
$$\left.\left. - \sum_{j=1}^{i} l_{ij} \dot{\theta}_j^2 \sin(\theta_j) \right] \right\} + J_k \left(\ddot{\theta}_k - \ddot{\psi} \right) = Q_{\theta_k} \ . \tag{5.167}$$

Note that the derivatives of the Rayleigh dissipation function have not been included into the equations: The generalized forces due to the dampers, if they exist at all, will be included in the forces Q_{θ_k}.

5.13.2 *Sideslip angles of the wheels and generalized forces*

The sideslip angles of the wheels of the trailer can be computed as seen for the articulated vehicle. If the r-th wheel of the i-th trailer has a steering angle δ_{i_r}, its sideslip angle is

$$\alpha_{i_r} = \arctan\left(\mathcal{A}\right) - \delta_{i_r} \ , \tag{5.168}$$

where

$$\mathcal{A} = \frac{u \sin(\theta_i) + v \cos(\theta_i) - \alpha_{i_r}^* \sin(\theta_i) - \beta_{i_r}^* \cos(\theta_i)}{u \cos(\theta_i) - v \sin(\theta_i) - \alpha_{i_r}^* \cos(\theta_i) + \beta_{i_r}^* \sin(\theta_i) - y_{i_r} \left(\dot{\psi} - \dot{\theta}_i \right)} \ ,$$

$\alpha^*_{i_r}$ and $\beta^*_{i_r}$ are the same as α_i and β_i but are computed using the distance b_{i_r} of the axle instead of a_i.

The contributions to the generalized forces Q_x, Q_y and Q_ψ due to the tractor are the same as for the insulated vehicle. As usual, the tractor does not give any contribution to the forces Q_{θ_k}. To compute the contributions due to the r-th wheel of the i-th trailer and to the aerodynamic forces of the latter it is possible to proceed as seen for the previous models, by writing their virtual work and then by differentiating with respect of the virtual displacements.

The results obtained for the wheels are

$$\begin{cases}
Q_{x_{i_r}} = F_{x_{i_{rt}}} \cos(\theta_i - \delta_{i_r}) + F_{y_{i_{rt}}} \sin(\theta_i - \delta_{i_r}) \\
Q_{y_{i_r}} = -F_{x_{i_{rt}}} \sin(\theta_i - \delta_{i_r}) + F_{y_{i_{rt}}} \cos(\theta_i - \delta_{i_r}) \\
Q_{\psi_{i_r}} = F_{x_{i_{rt}}} \left[c\sin(\theta_i - \delta_{r_i}) + \sum_{j=1}^{i} l^*_{ij} \sin(\theta_i - \theta_j - \delta_{r_i}) - y_{i_r} \cos(\delta_{r_i}) \right] + \\
\qquad +F_{y_{i_{rt}}} \left[-c\cos(\theta_i - \delta_{r_i}) - \sum_{j=1}^{i} l^*_{ij} \cos(\theta_i - \theta_j - \delta_i) + y_{i_r} \sin(\delta_{r_i}) \right] + M_{z_{r_i}} \\
Q_{\theta_{k_{i_r}}} = F_{x_{i_{rt}}} l^*_{ik} \sin(\theta_i - \theta_k - \delta_{r_i}) + F_{y_{i_{rt}}} l^*_{ik} \cos(\theta_i - \theta_k - \delta_{r_i}) \qquad \text{if } k < i \\
Q_{\theta_{k_{i_r}}} = F_{x_{i_{rt}}} [y_{i_r} \cos(\delta_{r_i}) + l^*_{ik} \sin(\delta_{r_i})] + \\
\qquad +F_{y_{i_{rt}}} [-y_{i_r} \sin(\delta_{r_i}) + l^*_{ik} \cos(\delta_{r_i})] - M_{z_{r_i}} \qquad \text{if } k = i \\
Q_{\theta_{k_{i_r}}} = 0 \qquad \text{if } k < i\,,
\end{cases} \tag{5.169}$$

where l^*_{ij} are the same as l_{ij} but are defined using the distance b_{i_r} of the axle instead of a_i.

In a similar way the generalized forces due to the aerodynamic forces and moments acting on the trailers can be accounted for. Assuming that it is possible to distinguish between the forces acting on the various rigid bodies, the generalized forces can be immediately computed from equations (5.169), using l_{ij} instead of l^*_{ij} and the aerodynamic forces and moments instead of the forces acting between road and wheels. Obviously y_{i_r} and δ_{i_r} vanish.

The generalized forces due to dampers located between the various bodies are

$$\begin{cases}
Q_x = Q_y = 0 \\
Q_\psi = -\Gamma_1 \dot\psi + \Gamma_1 \dot\theta_1 \\
Q_{\theta_1} = \Gamma_1 \dot\psi - (\Gamma_1 + \Gamma_2)\, \dot\theta_1 + \Gamma_2 \dot\theta_2 \\
Q_{\theta_k} = \Gamma_k \dot\theta_{k-1} - (\Gamma_k + \Gamma_{k+1})\, \dot\theta_k + \Gamma_{k+1} \dot\theta_{k+1} \qquad k = 2, ..., n\,.
\end{cases} \tag{5.170}$$

The external forces $F_{y_{e_i}}$ acting in the centres of mass or the trailers and the components of the weight $m_i g \sin(\alpha)$ are assumed to act in the directions of axes x and y of the tractor; the expressions of the generalized forces must therefore be modified accordingly.

The equations of motion are $n+3$; together with the equations yielding the sideslip angles of the wheels, those expressing the forces and moments of the tires as functions of the sideslip angles, the load, and the other relevant parameters, they allow one to study the handling of the vehicle.

5.13.3 *Linearization of the equations of motion*

As usual there is no difficulty in obtaining a linearized model, since in the normal usage of the vehicle, angles θ_i are all small enough to linearize their trigonometric functions.

As it was the case for all the previous models, the linearization of the equations allows one to uncouple the longitudinal behaviour (first equation of motion) from the lateral behaviour, which can be studied using the remaining $n+2$ equations. The linearized equation for the longitudinal behaviour is the usual one

$$m\dot{V} = Q_x$$

while those for the lateral behaviour are

$$\begin{cases} m\left(\dot{v} + u\dot{\psi}\right) - \ddot{\psi}\sum_{i=1}^{n} m_i d_i + \sum_{i=1}^{n} \ddot{\theta}_i \left(\sum_{j=i}^{i} m_j l_{ji}\right) = \\ \quad = (Q_y)_\beta \beta + (Q_y)_\psi \psi + \sum_{i=1}^{n} \left[(Q_y)_{\theta_i}\theta_i\right] + (Q_y)_{\dot{\theta}_i}\dot{\theta}_i + (Q_y)_\delta \delta + F_{y_e} + \sum_{i=1}^{n} F_{y_{e_i}} \\ J'\ddot{\psi} - \sum_{i=1}^{n} J_i'\ddot{\theta}_i + \sum_{i=1}^{n} m_i \left\{ -\dot{V}\sum_{j=1}^{i} l_{ij}\theta_j - \left(\dot{v} + V\dot{\psi}\right) d_i \right\} = \\ \quad = (Q_\psi)_\beta \beta + (Q_\psi)_\psi \psi + \sum_{i=1}^{n} \left[(Q_\psi)_{\theta_i}\theta_i + (Q_\psi)_{\dot{\theta}_i}\dot{\theta}_i\right] + (Q_\psi)_\delta \delta + \\ \quad + M_{z_e} + \sum_{i=1}^{n} M_{z_{e_i}} - \sum_{i=1}^{n} F_{y_{e_i}} d_j \\ \sum_{i=k}^{n} m_i l_{ik} \left(\theta_k \dot{V} + \dot{v} + V\dot{\psi} - d_i\ddot{\psi} + \sum_{j=1}^{i} l_{ij}\ddot{\theta}_j\right) = (Q_{\theta_k})_\beta \beta + (Q_{\theta_k})_\psi \psi + \\ \quad + \sum_{i=1}^{n} \left[(Q_{\theta_k})_{\theta_i}\theta_i + (Q_{\theta_k})_{\dot{\theta}_i}\dot{\theta}_i\right] + (Q_{\theta_k})_\delta \delta - M_{z_{e_i}} + \sum_{i=1}^{k} F_{y_{e_i}} l_{ik} , \end{cases} \tag{5.171}$$

where

$$d_i = \sum_{j=1}^{i} l_{ij}, J' = J_T + \sum_{i=1}^{n} (J_i + m_i d_i^2)$$

$$J_i' = \sum_{j=1}^{i} (J_i + m_i d_j l_{ji}).$$

The sideslip angles of the wheels of the trailers are linearized in the usual way

$$\alpha_{i_r} = \theta_i + \beta - \dot{\psi}\frac{d_i^*}{V} + \sum_{j=1}^{i} l_{ij}^* \frac{\dot{\theta}_i}{V} - \delta_{i_r} \,. \tag{5.172}$$

Again the wheels of the same axle have the same sideslip angle and it is possible to work in terms of axles rather than in terms of single wheels.

By linearizing the generalized forces Q_x, Q_y, Q_ψ and Q_{θ_k} in the same way as for the previous models the derivatives of stability entering Eq. (5.171) are readily computed.

The set of $(n+2)$ differential equations (5.171) is of the $(2n+2)$-th order, since variables y and ψ appear in the equation only as first and second derivatives. The equation can thus be written in the state space in the form of a set of $2n+2$ first order differential equations by introducing the state variables $v_{\theta_i} = \dot{\theta}_i$.

5.14 Limits of linearized models

Linearized models have some features which make them particularly useful. Namely

- The great simplification of the equations of motion and then the possibility of obtaining closed form solutions which, when simple enough, allow to gain a general insight into the dynamic behaviour of the vehicle, particularly for what the effect of the changes of the parameters of the vehicle is concerned.
- The possibility of studying the stability with the usual methods of linear dynamics.

The disadvantages are also clear: They can be applied only within a limited range of sideslip angles and lateral acceleration and for trajectories whose radius is large with respect to the dimensions of the vehicle. They can thus be applied with confidence to the conditions corresponding to a normal use of the vehicle, while they fail for sport driving and above all for the motions which take place during road accidents.

Another consideration for the models seen in the present chapter is that they are based on rigid body dynamics and that the presence of the suspensions is neglected. Also this assumption is well suited for the description of the behaviour of a vehicle driven in a relaxed way: Although depending on the stiffness of the suspensions, in these conditions the roll and pitch angles are very small and can be assumed to have little effect on the dynamic behaviour.

However it must be stated that a linearization carried too far will lead to results which contradict the experimental evidence. If the cornering stiffness is assumed to be proportional to the load Z acting on the wheel not only for the small load variations acting on each wheel but also for the differences of load between front and rear axle, in the case of a vehicle with two axles with equal tires it follows

$$\frac{C_1}{C_2} = \frac{F_{z_1}}{F_{z_2}} = \frac{b}{a} . \tag{5.173}$$

If only the cornering forces of the tires are included in the formula for the neutral-steer point, it follows that it always coincides with the centre of mass leading to the conclusion, which is clearly incorrect, that all vehicles with four equal wheels are neutral-steer.

5.15 Semilinearized models

The assumptions leading to linearized models are of two different kinds: Some of them deal with geometrical linearization, i.e. with the possibility of truncating the series for the trigonometric functions of the relevant angles after the first term, and others with the linearization of the behaviour of the tires, i.e. essentially with the use of the cornering and aligning stiffness. While the assumptions of the first type hold up to values of the angles larger than $10^\circ - 20^\circ$, those of the second type hold only up to values of the sideslip angle of about $4^\circ - 5^\circ$ for the cornering stiffness and even less for the aligning torque.

There is a large field in which the first ones still apply while the behaviour of tires can no longer be considered linear. In this field it is possible to write equations of motion which are basically the same as those already seen but in which the tire forces are expressed using more complex models, e.g. the "magic formula". In this way some of the featurcs of linearized models still apply, as the possibility of uncoupling the equation for the longitudinal behaviour from those dealing with handling and that of working in terms of axles instead of single wheels. However the very fact that the equations are nonlinear does not allow one to study the stability or to obtain general expressions of the gains, compelling to resort to a numerical study of selected manoeuvres of the vehicle, exactly in the same way as in the case of fully nonlinear models.

In the opinion of the author nowadays the advantage of semilinearized models, namely the use of simpler equations, does not compensate for the drawbacks of lack of generality and lower accuracy. The large computational power of even personal computers make the use of full nonlinear model straightforward.

In the present section only one model which can be considered as semilinearized will be dealt with, mostly because it was, and still is, quite common for the evaluation of steady-state directional behaviour.

Consider a vehicle with two axles and model it as a vehicle with two wheels, each one having the characteristics of the whole axle. If the vehicle is moving in a circular path with radius R the situation is that sketched in Fig. 5.38.

In low speed steering point H coincides with B and the various angles are

$$\delta_c = \arctan\left(\frac{l}{\sqrt{R^2-b^2}}\right) \, , \quad \beta_c = \arctan\left(\frac{b}{\sqrt{R^2-b^2}}\right) \, , \quad \alpha_1 = \alpha_2 = 0 \, . \tag{5.174}$$

In Fig. 5.38 the cornering forces are positive and the sideslip angles are negative. From triangle ORB the following relationship can be written

$$\frac{b}{\sin(\beta-\alpha_2)} = \frac{R}{\sin(90^\circ-\alpha_2)} = \frac{R}{\cos(\alpha_2)} \, . \tag{5.175}$$

The relationship between the sideslip angle of the vehicle and that of the rear wheels is then

$$\beta = \arcsin\left[\frac{b\cos(\alpha_2)}{R}\right] + \alpha_2 \, . \tag{5.176}$$

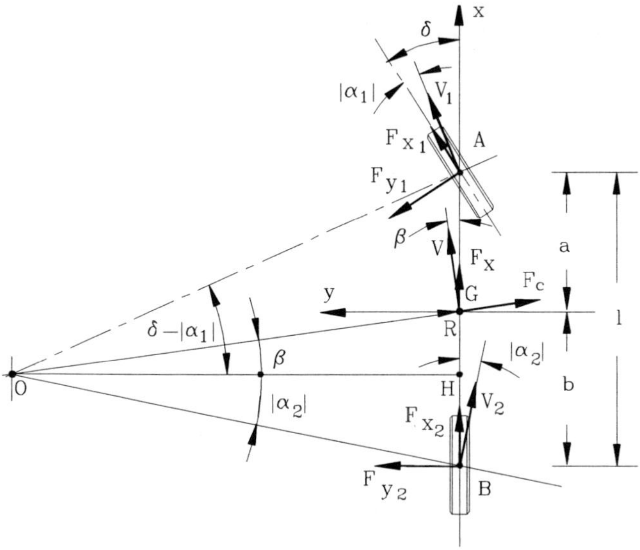

Fig. 5.38 Simplified semilinearized model for the study of steady state cornering of a vehicle with two axles.

A similar relationship between the sideslip angles of the vehicle and the steering angle can be obtained from triangles OHA and OHG

$$\tan(\delta + \alpha_1) = \frac{\overline{\mathrm{AH}}}{\overline{\mathrm{OH}}} \qquad \overline{\mathrm{OH}} = R\cos(\beta) \qquad \overline{\mathrm{AH}} = a + R\sin(\beta) \qquad (5.177)$$

and then

$$\delta = \arctan\left(\frac{\alpha + R\sin(\beta)}{R\cos(\beta)}\right) - \alpha_1 \, . \qquad (5.178)$$

Equations (5.176) and (5.178) are similar to equations (5.174) but take into account also the sideslip angles of the wheels which are present when dealing with high speed cornering.

If only the cornering forces or the tires are accounted for, neglecting aerodynamic forces, aligning moments and the effects of transversal load shift, the equilibrium equations of the vehicle are, with reference to Fig. 5.38

$$\begin{cases} F_{x_1}\cos(\delta)+F_{x_2}+F_x-F_{y_1}\sin(\delta)+F_c\cos(\beta)=0 \\ F_{x_1}\sin(\delta)+F_{y_1}\cos(\delta)+F_{y_2}-F_c\sin(\beta)=0 \\ F_{x_1}a\sin(\delta)+F_{y_1}a\cos(\delta)-F_{y_2}b=0\ . \end{cases} \tag{5.179}$$

Some of the forces are known: The centrifugal force is obviously

$$F_c = mV^2/R\ ,$$

force F_x is aerodynamic drag, or better aerodynamic x-force, plus the drag due to the longitudinal grade of the road, force F_{x_i} on the non-driving axle is the rolling resistance of that axle. The aerodynamic drag coefficient and the rolling resistance coefficient (or better, the coefficient of the x-component of the force) should be computed taking into account the sideslip angles of the vehicle and of the relevant wheels, which are unknown, but this would introduce additional complications and is usually neglected.

The unknown forces are then three: F_{y_1}, F_{y_2} and force F_{x_i} related to the driving axle. In case of front wheel drive vehicles the latter unknown is F_{x_1} and the set of equations can be written in the form

$$\begin{bmatrix} \cos(\delta) & -\sin(\delta) & 0 \\ \sin(\delta) & \cos(\delta) & 1 \\ a\sin(\delta) & a\cos(\delta) & -b \end{bmatrix} \begin{Bmatrix} F_{x_1} \\ F_{y_1} \\ F_{y_2} \end{Bmatrix} = \begin{Bmatrix} -F_{x_2}-F_x-F_c\sin(\beta) \\ F_c\cos(\beta) \\ 0 \end{Bmatrix} . \tag{5.180}$$

Equation (5.180) can be readily solved, yielding

$$\begin{cases} F_{x_1} = \dfrac{mV^2}{Rl}\Big[b\sin(\delta)\cos(\beta)-l\cos(\delta)\sin(\beta)\Big]-(F_{x_2}+F_x)\cos(\delta) \\ F_{y_1} = \dfrac{mV^2}{Rl}\Big[b\cos(\delta)\cos(\beta)+l\sin(\delta)\sin(\beta)\Big]+(F_{x_2}+F_x)\sin(\delta) \\ F_{y_2} = \dfrac{mV^2}{Rl}a\cos(\beta)\ . \end{cases} \tag{5.181}$$

In a similar way, in case of rear wheel drive vehicles, the unknown forces are

$$\begin{cases} F_{x_2} = \dfrac{mV^2}{Rl\cos(\delta)}\left[b\sin(\delta)\cos(\beta) - l\cos(\delta)\sin(\beta)\right] - F_x - \dfrac{F_{x_1}}{\cos(\delta)} \\ F_{y_1} = \dfrac{mV^2 b\cos(\beta)}{Rl\cos(\delta)} - F_{x_1}\tan(\delta) \\ F_{y_2} = \dfrac{mV^2}{Rl}a\cos(\beta)\ . \end{cases} \tag{5.182}$$

The unknowns of the problem are seven: Forces F_{y_1}, F_{y_2} and F_{x_1} (or F_{x_2}), angles β, α_1 and α_2 and either angle δ or the radius R depending on which one of the latter is considered as an input, i.e. whether the problem is that of finding the radius obtained with a given steering angle or the steering angle needed to obtain a given trajectory.

Also the equations are seven: Three equilibrium equations, either (5.181) or (5.182), two geometrical relationships (5.176) and (5.178) and the characteristics $F_y(\alpha)$ of the tires, all of them nonlinear. An iterative solution scheme is usually followed. The computation starts from the values of β_c and δ_c related to low speed steering. The values of the forces are then computed through either Eq. (5.181) or Eq. (5.182). Once the forces are known, the sideslip angles of the wheels can be obtained from the characteristics of the tires, possibly taking into account the interaction between longitudinal and transversal forces. This part of the computation is usually performed by resorting to the elliptical approximation but other approaches are possible. Finally Eqs. (5.176) and (5.178) allow one to compute new values of β and δ, which are obviously different from the kinematic values.

The computation can then proceed using the new values of β and δ as starting values for a new iteration cycle, until the difference between the values obtained at the end of the computation are close to the starting ones within a given tolerance. Convergence is usually quite fast, few iterations being needed to obtain errors of about 1%.

The model here described is semilinearized in the sense that, although dealing with nonlinear tire characteristics and also using the complete expressions of the trigonometric functions of angles β, δ and α_i, retains one of the essential features of linear models, namely the assumptions that the sideslip angles of the wheels of each axle are equal. Clearly this is inconsistent with the use of the trigonometric functions of the angles.

Similar models can be built for vehicles with more than two axles or vehicles with trailer, with the only difference that in some cases the iteration scheme must be different because it is impossible to obtain directly the cornering and driving forces from the equilibrium equations.

Example 5.14

Compute the steering angle needed to put the car of Appendix 1.1 on trajectories with radii of 20, 40, 60, 100, 200 and 500 m on level road.

The computation is performed using the iterative scheme outlined above. The dependence of the cornering forces is obtained through the "magic formula", by computing various values of F_y and α and interpolating the table so obtained to compute sideslip angles. Note that this computation must be repeated for each axle and for each speed, to account for the variations of normal forces F_z; normal forces are however computed neglecting centrifugal forces. The interaction between longitudinal and transversal forces has been accounted for using the elliptical approximation as the use of the equations included in the model of the magic formula leads to added complexities in the iteration scheme.

The results are reported in the form of a plot of the curvature gain $1/R\delta$ as a function of the speed in Fig. 5.39. The gain has been used even if it depends on the radius to compare the results with those of the linearized model (dashed line).

The curve obtained for a large value of the radius is not superimposed with that related to the linearized model, as the latter takes into account also aerodynamic forces and others effects, which are not included in the present model.

Note that the understeering behaviour of the vehicle is increased by the effect of driving forces, particularly on low radius trajectories. At a certain speed the gain goes to zero, i.e. the vehicle is unable to maintain the curved path as the limit cornering forces are reached at the front wheels.

The vehicle had been oversteer, particularly in case of a rear drive vehicle, the gain would have gone to infinity as a spinout would have occurred.

The results are however affected by large errors when the limit conditions are approached: on a radius of 20 m the limit speed is about 52 km/h, corresponding to a value of μ_y of 1.06, which is clearly too high for a vehicle of this type.

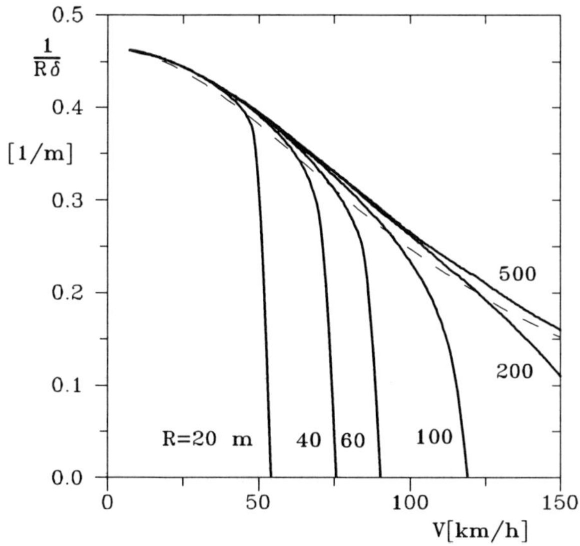

Fig. 5.39 Example 5.14: Trajectory curvature gain, obtained with the simplified nonlinear model, on circular trajectories with various radii. The dashed line has been obtained from the linear model.

5.16 Vehicle-driver interaction

5.16.1 *Vehicle-driver system*

As already stated, a road vehicle on pneumatic tires cannot maintain a given trajectory under the effect of external perturbations unless it is controlled by some external driving device, which is usually a human driver. Its stability concerns only such state variables as the sideslip angle β and the yaw velocity r.

As it is not said that a system made of two subsystems which are stable is globally stable, the stability assessment of the vehicle should be performed taking into account also the behaviour of the driver or of the automatic trajectory controller if present.

A tentative scheme of the vehicle-driver system is shown in Fig. 5.40. The driver has been assumed to detect the yaw angle ψ, the angular and linear accelerations $\dot{\beta}$, $\dot{r}$, dV/dt, V^2/R and to be able to assess his position on the road (X and Y). Moreover, the driver receives a number of other informations from the vehicle, as forces, moments, noise, vibrations, etc. which allow him to assess, mostly without being really conscious of this process,

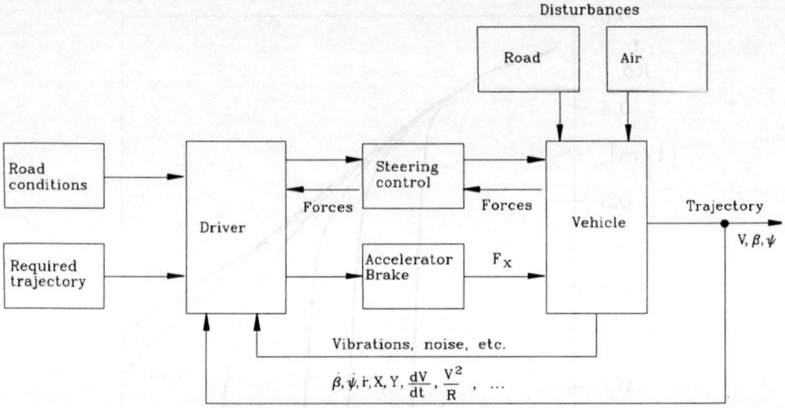

Fig. 5.40 Tentative scheme of the vehicle-driver system.

the conditions of the vehicle and those of the road-wheel interactions.

The stability of the vehicle-driver system is mandatory but is not sufficient to assess the required handling and comfort characteristics of the vehicle. The greater the stability with free and locked controls of the vehicle alone, the lower the number of corrections the driver has to introduce to obtain the required trajectory. A vehicle which is stable for what β and r are concerned requires from the driver only those inputs needed to follow the required trajectory but not those needed to stabilize the motion on it.

On the other hand, a vehicle which is too stable may lack the manoeuvrability needed to cope with emergency conditions or simply to allow sport driving. The amount of stability must be assessed in each case, taking into account the type of vehicle, the market target, the traditions and the image of the manufacturer.

Usually stability, handling and comfort characteristics of a vehicle are assessed on the basis of prolonged road testing performed by skilled test drivers. Such approach has the drawback of being in a way subjective and above all of focusing on global characteristics of the vehicle, without giving detailed suggestions on causal relationships between the construction parameters of the vehicle and its behaviour. It also requires to perform long and costly road tests and, above all, forces to postpone the evaluation of the performances of the vehicle to a stage in which the prototypes are ready.

The availability of mathematical models for the driver-vehicle interaction has a number of advantages which are too obvious for a detailed discussion. The difficulty of translation into mathematical functions concepts like comfort and user friendliness is a serious obstacle in this way and the experimental and numerical approaches are bound to remain complementary.

For the study of the man-machine interactions a model able to simulate the behaviour of the driver must be built. The difficulties encountered in such task are so large that many different approaches have been attempted and up to now there is no standard driver model which can be used in general.

The first systematic studies were performed in the aeronautical field[2], but, mainly starting from the seventies, a large number of models specialized for the vehicular field have been published. A quick bibliographic scan allowed to identify more than sixty models, published in less than 25 years. They span from very simple constant-parameters single-input single-output linear models to multivariable, nonlinear, adaptive models or models based on fuzzy logic and/or neural networks.

As usual the complexity of the model must be chosen in a way which is consistent with the aims of the study and with the availability of significant input data.

5.16.2 *Simple linearized driver*

As already stated, the driver can be thought as a controller receiving a number of inputs from the vehicle and the environment and outputting a few control signals to the vehicle. As usual with manual control, the driver performs the tasks of the sensors, the controller, the actuators and the source of the control power, even if the control action due to the driver can be assisted by devices as power steering or braking.

In building a simple driver model a small number of the inputs the driver receives is selected and very simple control algorithms are chosen to link them with the outputs. The latter are usually only the steering angle δ and the position of the accelerator/brake pedals. Only the former is considered if the driver model is used in connection with a constant speed handling model.

The controller is assumed to be a tracking system affected by a delay, which is usually assumed as the sum of three distinct delays: A reaction time delay, due to the time needed to elaborate the informations from the

[2]See, for instance, D.T. McRuer, E.S.Crendel, *Dynamic response of human operators*, WADC T.R. 56-524, Oct. 1957.

vehicle and the environment, a neuromuscular delay due to the time needed for the command to reach the relevant muscles, and an execution delay, due to the time needed to perform the control action.

In some models the three forms of delay are accounted for separately and also a lead time is considered, as the human operator performs also a predictive task. Practical experience shows that this function, which clearly is greatly enhanced by training, is crucial in actual driving conditions. A simple open loop transfer function for the linearized driver is

$$\frac{y(s)}{u(s)} = K_d \frac{(1+T_L s)e^{-\tau s}}{1+T_D s} , \tag{5.183}$$

where y, u, K_d, T_L, τ, T_D are respectively the output and the input to the driver, the gain, the lead time, the reaction time delay and the neuromuscular delay.

In many cases the lead time is neglected and all delays are summarized as a single delay τ, yielding the simpler open loop transfer function

$$\frac{y(s)}{u(s)} = K_d e^{-\tau s} .$$

By expressing the exponential as a power series and truncating it after three or two terms, it reduces to

$$\frac{y(s)}{u(s)} \approx K_d \frac{1}{1+\tau s+\frac{1}{2}\tau^2 s^2} \approx K_d \frac{1}{1+\tau s} . \tag{5.184}$$

The choice of the inputs is critical and there is a large difference between the use of a quantity linked with the position, as coordinates X and Y or, better, the lateral deviation from the required trajectory and the yaw angle ψ.

It is a common experience that, when a driver takes as reference a point very close to the front of the vehicle, like when driving in the fog using as a reference the side of the road, an oscillatory trajectory is usually obtained while the oscillations disappear if the reference point is far ahead.

The simplest driver model is a linear tracking system whose input is the yaw angle of the vehicle and applies a control action in terms of angle δ proportional to the error between the actual value of ψ and that corresponding to the required path[3]. If Eq. (5.184) is used to account for a delay and the

[3]P.G. Perotto, *Sistemi di automazione, Vol.I, Servosistemi,* UTET, Torino, 1970. Actually the driver-vehicle model there reported is even simpler as the vehicle is assumed

series is truncated after two terms, the time domain equation is

$$\tau\dot{\delta}(t) + \delta(t) = -K_d[\psi(t) - \psi_0(t)] \,, \tag{5.185}$$

where ψ_0 is the required yaw angle.

By introducing Eq. (5.185) into the simplest mathematical model of a vehicle with two axles (Eq. 5.67), the following state equation for the vehicle-driver system is obtained

$$\left\{\begin{array}{c}\dot{\beta}\\ \dot{r}\\ \dot{\delta}\\ \dot{\psi}\end{array}\right\} = \left[\begin{array}{cccc}\dfrac{Y_\beta}{mV} & \dfrac{Y_r}{mV}-1 & \dfrac{Y_\delta}{mV} & 0\\ \dfrac{N_\beta}{J_z} & \dfrac{N_r}{J_z} & \dfrac{N_\delta}{J_z} & 0\\ 0 & 0 & -\dfrac{1}{\tau} & -\dfrac{K_d}{\tau}\\ 0 & 1 & 0 & 0\end{array}\right]\left\{\begin{array}{c}\beta\\ r\\ \delta\\ \psi\end{array}\right\} + \left\{\begin{array}{c}0\\ 0\\ \dfrac{K_d}{\tau}\\ 0\end{array}\right\}\psi_0 \,. \tag{5.186}$$

Note that the steering ratio must be included into the gain K_d, as δ is the angle of the wheels about the kingpin and not that of the steering wheel.

This simple model is actually too simple to yield useful results. The lack of predictive behaviour and the unrealistic assumption that the driver reacts only to the yaw angle makes such a driver quite unstable. The only useful result, though obvious, is that the action of the driver must be quick, i.e. with a short delay, and soft, i.e., with a small gain, not to induce instability.

Example 5.15

Consider the vehicle described in Appendix 1.1. A driver modelled by Eq. (5.185) steers it along a standard ISO lane change manoeuvre. Plot the root locus of the vehicle-driver system for various values of the gain at 80 km/h and compute the trajectory obtained.

The ISO lane change manoeuvre, aimed to simulate overtaking, requires the vehicle to travel for 15 m in the original lane, to change lane with a lateral displacement of 3.5 m in 30 m, to stay in this lane for 25 m and to return to the original lane

to be neutral steer and a first order system is used to model it.

in 25 m. The manoeuvre must be performed at 80 km/h. The lane changes can be performed using any trajectory, provided that none of the cones delimiting the three straight lanes are touched. The width of these lanes are, in meters, $1.1\,B + 0.25$ for the first lane, $1.2\,B + 0.25$ for the second and $1.3\,B + 0.25$ for the third, where B is the width of the vehicle. The three lanes are then 1.966, 2.122 and 2.278 m wide, leaving a margin of 0.203, 0.281 and 0.359 m at both sides of the theoretical trajectory.

The actual lane changes are left to the driver. In the present simulation a cosine function is used, which has the advantage of being very simple and the drawback of yielding a discontinuity of curvature at each transition with a straight path.

The trajectory and angle ψ_0 are then

$$\begin{array}{ll} Y = 0 & \text{for } X < 15 \\ Y = \dfrac{3.5}{2}\left\{1 - \cos\left[\dfrac{\pi}{30}(X-15)\right]\right\} & \text{for } 15 \le X < 45 \\ Y = 3.5 & \text{for } 45 \le X < 70 \\ Y = \dfrac{3.5}{2}\left\{1 + \cos\left[\dfrac{\pi}{25}(X-70)\right]\right\} & \text{for } 70 \le X < 95 \\ Y = 0 & \text{for } 95 \le X < 125 \end{array}$$

$$\begin{array}{ll} \psi_0 = 0 & \text{for } X < 15 \\ \psi_0 = \arctan\left\{\dfrac{3.5\pi}{60}\sin\left[\dfrac{\pi}{30}(X-15)\right]\right\} & \text{for } 15 \le X < 45 \\ \psi_0 = 0 & \text{for } 45 \le X < 70 \\ \psi_0 = -\arctan\left\{\dfrac{3.5\pi}{50}\sin\left[\dfrac{\pi}{25}(X-70)\right]\right\} & \text{for } 70 \le X < 95 \\ \psi_0 = 0 & \text{for } 95 \le X < 125\,. \end{array}$$

The values of the time delay span from 0.08 s for a professional sport driver to more than 0.25 s for an occasional driver. A value of 0.20 s is assumed here.

The roots locus, obtained for values of K_d spanning from 0.1 to 2 with increments of 0.1 is shown in Fig. 5.41a. The system becomes unstable if the gain becomes greater than 1.6 and a strong oscillatory behaviour is always expected. A value of the gain of 0.5 has been chosen for the simulation as a compromise

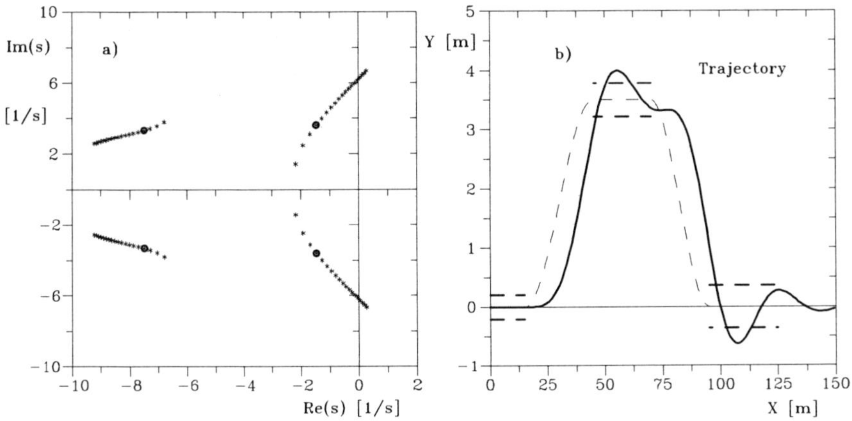

Fig. 5.41 Example 5.15. (a): Root locus of the vehicle-driver system at 80 km/h. (b): Trajectory obtained during an ISO lane change test with a gain $K_d = 0.5$. Note that the scales of the axes are quite different and that the trajectory is strongly distorted.

between low values leading to a large delay in the response and high values yielding to strong trajectory oscillations.

The trajectory is shown in Fig. 5.41b. Note that the trajectory follows the wanted path with a strong delay and the oscillations are very strong. The driver fails the task of performing the manoeuvre without touching the cones.

A simple way to incorporate a sort of predictive behaviour into the model is that of using as error not the difference between the desired and the actual value of the yaw angle, but the distance between a point on vehicle x-axis at a given distance L in front of the vehicle and the required trajectory (distance d in Fig. 5.41a).

With simple computations and assuming that angle $\psi - \psi_1$ is small, such distance can be approximated as

$$d = L\left(\psi - \psi_1 + \frac{y}{L}\right) , \tag{5.187}$$

where y is the lateral displacement of the vehicle, i.e. the integral of the lateral velocity v. If the speed of the vehicle is constant, with the usual linearization, it coincides with the integral of β, multiplied by V. Angle ψ_1 is the angle between X-axis and a line passing through two points of the trajectory at a distance L; it can be easily computed from the shape of the

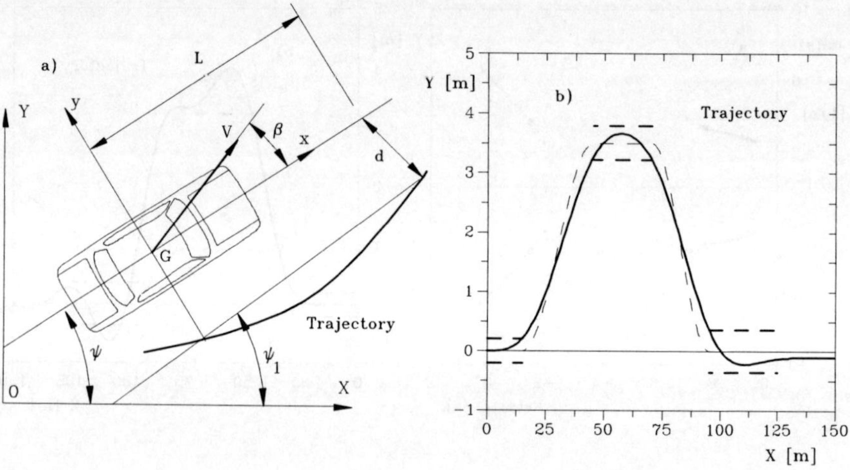

Fig. 5.42 (a). Definition of distance d. (b): Example 5.16. Trajectory obtained during an ISO lane change test. Note that the scales of the axes are quite different.

trajectory.

The equation expressing the time domain model of the driver is then

$$\tau\dot{\delta}(t) + \delta(t) = -K_d\left[\psi(t) - \psi_1(t) + \frac{y(t)}{L}\right] , \tag{5.188}$$

where now the gain K_d contains also the forward distance L.

By introducing Eq. (5.188) into the simplest mathematical open-loop model of linearized vehicle and operating in the same way as for the previous model, the state equation for the vehicle-driver system is

$$\left\{\begin{matrix}\dot{\beta}\\ \dot{r}\\ \dot{\delta}\\ \dot{\psi}\\ \dot{y}\end{matrix}\right\} = \begin{bmatrix} \frac{Y_\beta}{mV} & \frac{Y_r}{mV}-1 & \frac{Y_\delta}{mV} & 0 & 0\\ \frac{N_\beta}{J_z} & \frac{N_r}{J_z} & \frac{N_\delta}{J_z} & 0 & 0\\ 0 & 0 & -\frac{1}{\tau} & -\frac{K_d}{\tau} & -\frac{K_d}{L\tau}\\ 0 & 1 & 0 & 0 & 0\\ 0 & V & 0 & 0 & 0\end{bmatrix}\left\{\begin{matrix}\beta\\ r\\ \delta\\ \psi\\ y\end{matrix}\right\} + \left\{\begin{matrix}0\\ 0\\ \frac{K_d}{\tau}\\ 0\\ 0\end{matrix}\right\}\psi_0 \, . \tag{5.189}$$

Both the models here described are drastic oversimplification of the actual human behaviour, but the second one already performs in a satisfactory way in many instances. As already stated, more complex and realistic models can be found in the literature. The use of a much simplified vehicle model in connection with these driver models is adequate and very little improvement can be expected from the use of more detailed vehicle models.

Example 5.16

Repeat the simulation of Example 5.15 using the driver model of Eq. (5.189).

The values of the time delay, the gain and the prediction distance L are assumed to be respectively 0.20 s, 0.30 and 25 m. The trajectory is shown in Fig. 5.42b. The driver model is now successful in performing the required manoeuvre.

5.16.3 *Simple model for longitudinal control*

Apart from acting on the steering wheel to maintain the trajectory, the driver needs to regulate the vehicle speed acting on the accelerator pedal and, occasionally, on the brakes. A common task is that of maintaining a given distance from another vehicle proceeding at roughly the same speed on the same lane; in heavy traffic conditions on motorways long lines of vehicles are formed, each driver trying to regulate its speed to maintain a given distance. Let V_i and d_i be the speed of the i-th vehicle and the distance between the i-th and the $(i-1)$-th vehicle.

The derivative of the distance with respect to time is obviously linked to the speed by the relationship

$$\frac{d}{dt}(d_i) = V_{i-1} - V_i \, . \tag{5.190}$$

The driver of the i-th vehicle tries to maintain the distance at a fixed value by accelerating when the distance increases and decelerating while it decreases, but this action is applied with a certain delay. A linear model for this action is the following

$$\frac{dV_i(t+\tau)}{dt} = K\frac{d}{dt}(d_i) = K(V_{i-1} - V_i) \, , \tag{5.191}$$

where K is the gain and τ the delay time.

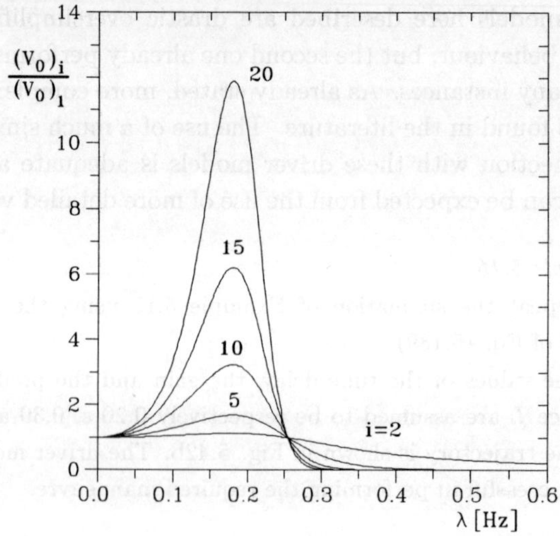

Fig. 5.43 Frequency response of the velocity of the i-th vehicle in a line using as input the velocity variations of the first one. $K = 1.6$ 1/s; $\tau = 0.6$ s.

By using the series for the function $V_i(t+\tau)$ truncated after the second term, Eq. (5.191) reduces to

$$\tau \dot{V}_i + \dot{V}_i + KV_i = KV_{i-1} , \tag{5.192}$$

which is formally identical to the equation of motion of a second-order system, with mass τ, unit damping and stiffness K. If $4K\tau > 1$ the system has an oscillatory behaviour with natural frequency $\lambda_n = \sqrt{K/\tau}$ and damping ratio $\zeta = 1/2\sqrt{K\tau}$.

A line of vehicles is then formally identical to a system made of a set of masses linked to each other by springs and dampers. If all drivers have the same gain and delay, the frequency response of the speed of the i-th vehicle with respect of that of the first one is

$$\frac{(V_0)_i}{(V_0)_1} = \frac{K^{i-1}}{\left[(K-\tau\lambda^2)^2 + \lambda^2\right]^{(i-1)/2}} . \tag{5.193}$$

The frequency response is reported for various values of i in Figure 5.43, for $K = 1.6$ 1/s and $\tau = 0.6$ s. A high value of τ is realistic in this case, as it is not easy to detect immediately the variations of speed of a preceding vehicle. The line of vehicles has then a resonance and can be expected to oscillate occasionally.

This linearized model is obviously too rough to give quantitative indications but qualitatively explains the oscillatory behaviour of long lines of vehicles and also the fact that in the case of very dense motorway traffic periodic stops can be experienced without any obvious explanation. Clearly in this case when the oscillations become too large the linearized model loses validity: The acceleration required for some vehicles can exceed its possibility or the distance becomes too small and a vehicle is forced to stop.

vehicle. The line of vehicles has then a resonance and can be expected to oscillate occasionally.

This linearized model is obviously too rough to give quantitative indications but qualitatively explains the oscillatory behaviour of long lines of vehicles and also the fact that in the case of very dense motorway traffic periodic stops can be experienced without any obvious explanation. Clearly in this case when the oscillations become too large the linearized model loses validity. The acceleration required for some vehicles can exceed its possibility or the distance becomes too small and a vehicle is forced to stop.

Chapter 6

Motor vehicle on elastic suspensions

If the number of wheels is greater than three a rigid vehicle is a statically undetermined system, as already stated in the previous chapters. The structure of the vehicle must then either be compliant enough or include a compliant connection between the wheels and the body. In all motor vehicles the second solution is followed, while the first one, which was widely used in the past for carts with animal traction, is now restricted to particularly slow vehicles. However there are cases in which the compliance of the structure acts together with the presence of the suspensions and has an important effect on the characteristics of the vehicle.

The suspensions have two main functions:

- Distributing the load on the ground in a way corresponding to the design goals and allowing the vehicle to sit in the proper attitude, under the effect of the static and quasi-static loads in stationary conditions;
- allowing the wheels to follow an uneven road profile without transferring excessive loads and accelerations to the vehicle body.

For the first task the suspensions must be an elastic system, linear or nonlinear, while for the second they must also incorporate damping, at least to avoid the onset of resonant oscillations. The second task is so important that suspensions are also used in the case of vehicles with two wheels, which do not need them in order to distribute the loads to the ground in a predictable way.

Note that theoretically both tasks could be performed by the tires also in the case of rigid vehicles, but their compliance is not sufficient and their damping is too low to do it in a satisfactory way.

Ideally, the suspensions should allow the wheels to move with respect to the body of the vehicle in a direction perpendicular to the road, maintaining the plane of the wheel parallel to itself and constraining all motions in the xy: A suspension of a single wheel should be a system with a single degree of freedom, the displacement in z direction.

However, none of the systems which are used at present are able to perform in that way and each one of them has a peculiar behaviour; the approximations with which the suspensions perform their task of constraining the five other degrees of freedom of the wheel hub are very important in giving any particular vehicle its own character. As when the body of the vehicle is moved in vertical direction (displacement z) or rotates about its x axis (roll angle ϕ) the position of the wheel changes, it is possible to plot the camber angle γ, the track t, the characteristic angles of the steering system, the steering angle δ, etc. as functions of z and ϕ. These functions are generally strongly nonlinear but can be linearized about any equilibrium position and the derivatives $\partial t/\partial z$, $\partial t/\partial \phi$, $\partial \gamma/\partial z$, $\partial \gamma/\partial \phi$, $\partial \delta/\partial z$, $\partial \delta/\partial \phi$, etc.[1] can be easily defined. They can be considered as constants in the small motions about an equilibrium position and define the behaviour of the suspension.

Apart from these kinematic characteristics of the suspension, the position of the wheel with respect to both the road and the body can be influenced by the compliance of the joints, which are often not spherical or cylindrical hinges but compliant links and that of other parts of the suspension. The displacements due to deformations are not univocally determined by the position of the body and for them it is impossible to introduce the relevant derivatives. It must be stated that the trend in the design of suspensions is towards a replacement of joints working in a kinematically correct way by elastic hinges or compliant elements and towards an integration in a single element of the guiding functions of the linkages and the elastic functions of the springs. It is increasingly more difficult to define the kinematic parameters of the suspension.

A vehicle on elastic suspensions can be modelled as a lumped parameters system with a rigid body, the "sprung mass", connected to a number of masses which include the wheels, the "unsprung masses", thorough massless springs and dampers simulating the suspensions. The unsprung masses are connected to the ground through massless springs and dampers simulating the tires. This model is clearly an approximation, as the suspensions and the tires have their own mass and hence also their own natural frequencies and the body and the linkages are not rigid bodies. Models of this type,

[1] In the following the notation $(t)_{,z}$, $(t)_{,\phi}$, etc. will be used.

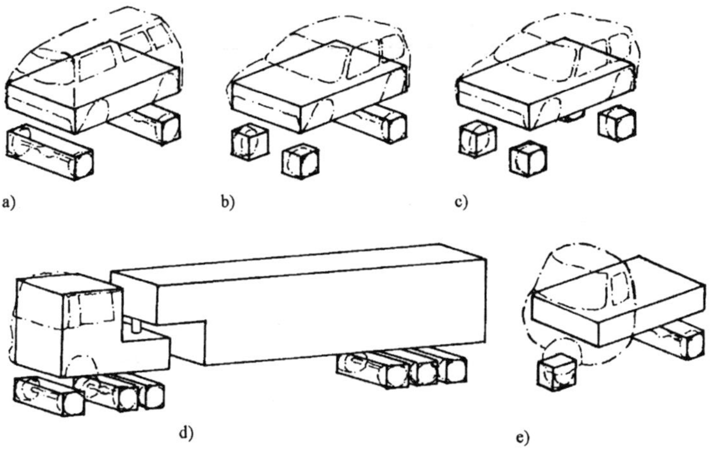

Fig. 6.1 Example of models for the dynamic study of road vehicles. (a), (b) and (c): Vehicle with two axles, 10 d.o.f.; (d): Articulated truck with 6 axles, 21 d.o.f.; (e): Vehicle with 3 wheels; 9 d.o.f.

whose complexity and precision can be increased by modelling the body as an elastic body, are however very useful in the study of the dynamic behaviour of road vehicles.

A vehicle with four wheels can be modelled as a system with 10 degrees of freedom, six for the body and one for each wheel. This holds for any type of suspension, as the wheels of each axle can be suspended separately (independent suspensions) or together (solid axle suspensions) but the total number of degrees of freedom is the same (Fig. 6.1). Additional degrees of freedom as the rotation of the wheels about their axis or about the kingpin can be inserted into the model to allow one to take into account the longitudinal slip or the compliance of the steering system.

Each suspension is thus characterized by its mass, the stiffness and damping parameters of the suspension and of the tire. The latter can be strongly nonlinear but can often be linearized, particularly for what the stiffness is concerned, if the small motion about an equilibrium position is studied.

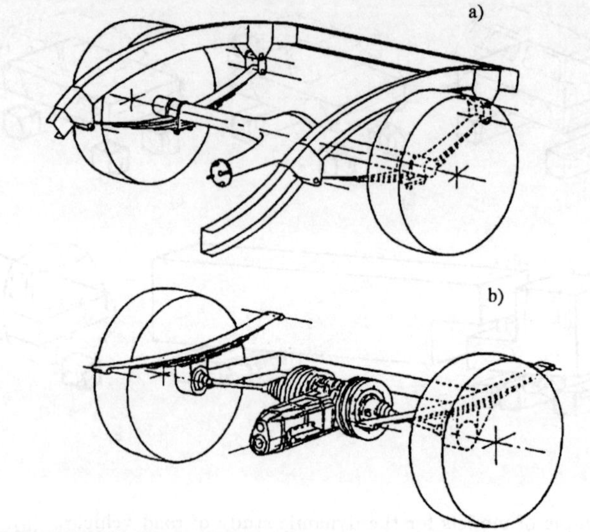

Fig. 6.2 Rear axles with suspensions in which the springs also act as constraints to guide the axle in the suspension motions.

6.1 An overview of the different types of suspensions

6.1.1 *Solid-axle*

If both wheels of the same axle are connected by a rigid beam, the latter must have two degrees of freedom, namely a translation in vertical direction and a roll rotation. The rigid beam, which can include also the final drive (solid drives or live axles, Fig. 6.2a), can be guided in its motion by some linkages or by the springs themselves, like in the solutions based on semi-elliptic leaf springs shown in Fig. 6.2, in which the shock absorbers are not represented.

The solution of Fig. 6.2a, often referred to as Hotchkiss axle, has the disadvantages of approximating the correct kinematic behaviour in which the axle can move only along z coordinate and rotate about the roll axis, in a poor way. The stiffness in x and y directions, although far higher than that in z direction, is not high enough, as is the stiffness for rotations about the y axis and, above all, any rolling motion is linked to a steering of the whole axis (roll steer). In other words, the derivative $\partial\delta/\partial\phi$ can be quite high (Fig. 6.3). The latter characteristic is due to the fact that the motion

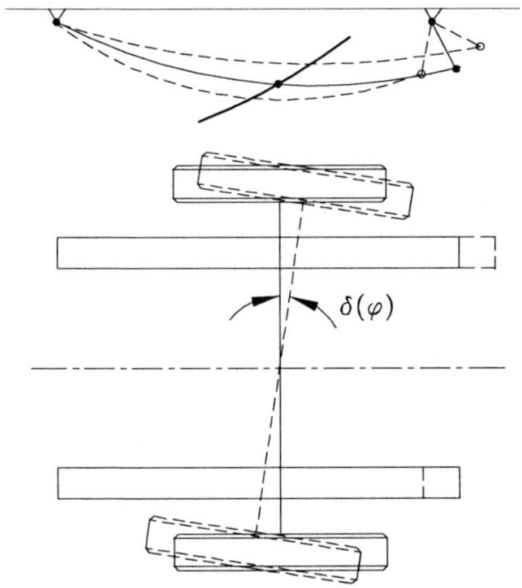

Fig. 6.3 Roll steer of a Hotchkiss axle.

of the points in which the axle is connected to the springs is not exactly vertical, although the deviation from the vertical direction of the trajectory shown in the figure is exaggerated, to show the phenomenon.

The axle shown in Fig. 6.2b is usually referred to as a De Dion axle. It has been widely used in a slightly different version, in which guiding elements are present and helical spring are used instead of leaf springs.

Some solutions in which different types of linkages are used to control the motion of the axle are shown in Fig. 6.4. The transversal guide can be supplied by a Watt quadrilateral (a), by a reaction bar (g), by a triangle hinged with two joints to the body (c) or to the axle (e), or by a straight guide (f). In other cases (b, c) the bending stiffness in xy plane of the leaf springs or of the longitudinal links act as transversal constraints. Only in the cases (a) and (f) the axle is guided in transversal direction in an almost kinematically correct way.

The longitudinal guide is supplied by an articulated quadrilateral (a, f) which constrains also rotations about y-axis, to a Watt's quadrilateral (c), which links rotations about y-axis with translations along z-axis, or by the longitudinal stiffness of the springs. Solutions (b), which contains also a compliant element, (d) and (e), kinematically similar to (f), can be assimilated to a quadrilateral. Only solution (c) uncouples exactly displacements

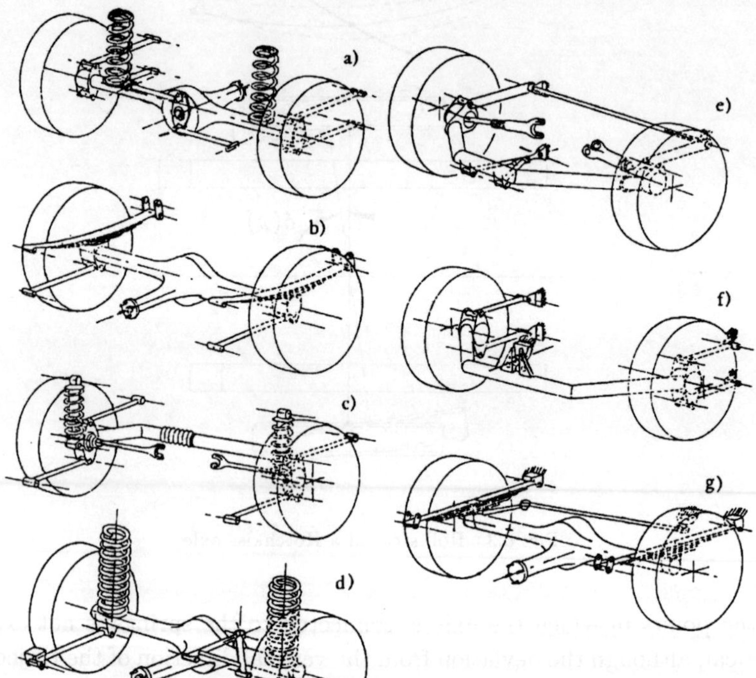

Fig. 6.4 Rigid-axle suspensions with different geometry of the various linkages.

z and roll rotations with displacements in x direction and rotations about z-axis, avoiding completely roll steering.

Note that in (c) the torsional stiffness of the axle must be low, or better, a cylindrical hinge must uncouple the rotations of the two parts of the axle which are different from each other in any rolling motion. If the deformation of the axle is used to uncouple them a correct kinematic working of the suspension is impossible, as this layout relies on the deformation of some elements to allow the required displacements.

Devices which rely on the compliance of some elements are increasingly popular and allow one to reduce the number of parts and the cost of the system. The same can be said for the traditional suspensions based on leaf springs which act as guide elements, but the use of spring of a different type allows one to obtain better performances, particularly if the friction between metal parts (as occurs between the leaves of the spring) is avoided.

The choice of the type of guide elements is dictated by the stressing and deformation of the various elements, the contact forces which can lock the motion in certain cases, as can occur in solution (f), and also by the tradition of the manufacturer and by the market for which the vehicle is intended.

In all cases the track is constant as is the camber angle, at least if the compliance of the tires is neglected. It follows that $(t)_{,z} = (t)_{,\phi} = (\gamma)_{,z} = (\gamma)_{,\phi} = 0$.

An important parameter in the study of motor vehicle suspensions is the position of the roll centre RC of the suspension. The roll axis of the vehicle is the instantaneous axis for roll rotations of the vehicle when it is in symmetrical conditions (i.e., with the roll angle $\phi = 0$). Note that it is an instantaneous axis of rotation as the rolling motion with a large angle is not a pure rotation about a well defined axis and can be defined only for small rotations about a given position, namely the symmetrical equilibrium position.

The point in which the roll axis crosses the plane perpendicular to the ground through the centres of the wheels of a given suspension is the roll centre of that suspension. For symmetry reasons, the roll axis must lie in the symmetry plane of the vehicle (xz plane) and therefore also the roll centres of the suspensions must lie in it. Note that in case of a two-axle vehicle the roll centre of each suspension can be determined from the characteristics of the relevant suspension only and that the roll axis can be defined as a line connecting the roll centres of the two suspensions. If the vehicle has more than two axles, the roll centres of the suspensions need not be aligned: A roll axis still exist, but it does not pass through the roll centres of the single suspensions, considered as insulated.

The roll centre of each suspension can also be defined as the point on a plane perpendicular to the ground and to the symmetry plane in which the application of a lateral force F_y to the vehicle body does not cause any roll. The two definitions obviously coincide.

Other important points are the centres of rotation of the body with respect to the wheels BW and of the wheels with respect to the ground WG.

As in a solid axle the two wheels are rigidly connected, the two points are located in the symmetry plane. Moreover, if the compliance of the tires is neglected, the wheels cannot rotate with respect to the ground: Points

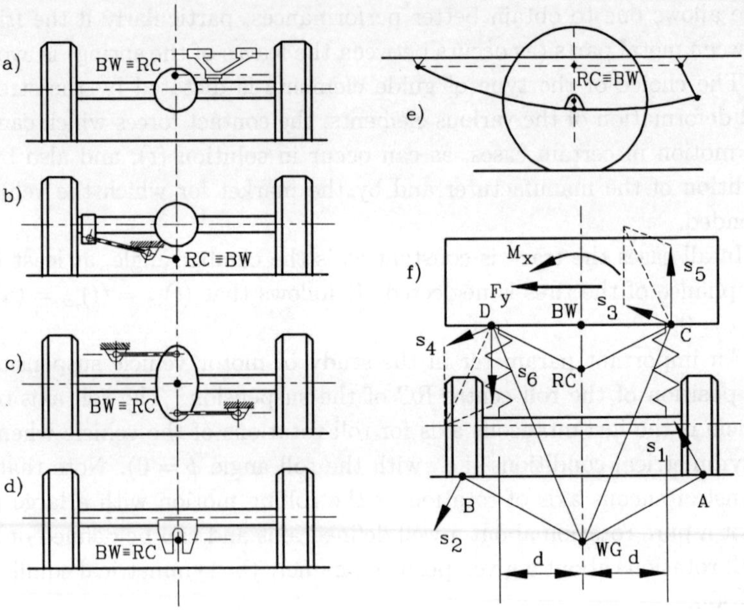

Fig. 6.5 From (a) to (e): Position of point BW in some rigid-axle suspensions; the tires are considered as rigid. (f): Position of points WG, RC and BW, taking into account the compliance of the tires.

WG are located at infinity on the intersection between the ground and the plane parallel to yz plane passing through the centres of the wheels and the two points BW coincide with the roll centre RC.

If the compliance of the tires is accounted for points WG lie on the symmetry plane slightly below the ground, but their position is not exactly defined as they depend on the deflection of the tires and hence on the forces applied to them; it is however possible to define a zone in which they lie. Also points BW coincide and are located in the symmetry plane of the vehicle below the roll centre RC.

In some cases points BW are physically defined as there is a material hinge between the axle and the body (Fig. 6.5a-d). If the axle is guided laterally by the leaf springs, BW is on the symmetry plane, at the level of the attachment of the springs to the body (Fig. 6.5e). The lateral deflection of the springs causes BW to be located at a lower level and the inflection of the tires causes the roll centre RC to be located below BW.

A four-link suspension is shown in Fig. 6.6. To obtain the position of

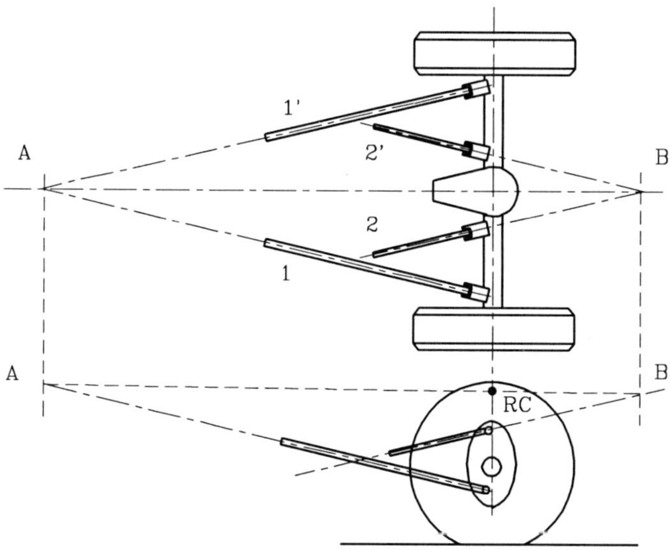

Fig. 6.6 Four-link solid-axle suspension. Position of point RC.

the roll centre, the intersections A and B of the axes of links 1-1′ and 2-2′ must be found first. They lie in the midplane of the vehicle. The roll centre is found as the intersection of line AB with the plane perpendicular to the ground containing the centres of the wheels. If two links are parallel (say links 1 and 1′) the intersection is at infinity and line AB is parallel to the projection of the relevant links on the symmetry plane.

The situation for a three-link suspension is similar (Fig. 6.7). The only difference is that point B is the intersection of the axis of the transversal link with the plane of symmetry.

In the case of a Hotchkiss axle the roll centre is located at the intersection between the projection on the symmetry plane of a line connecting the points in which the springs are connected to the body and the perpendicular to the ground through the centre of the wheel (Fig. 6.5e). If the compliance of the tires is not neglected, to find an approximate location of BW, RC and WG it is possible to proceed as in Fig. 6.5f: A force F_y and a moment M_x are applied to the body: Displacements s_1, s_2, s_3 and s_4 of points A, B, C and D are due to the inflection of the tires; while the compliance of the springs causes the displacements s_5 and s_6 of points C and D. The intersection of the lines perpendicular to s_1 and s_2 locate point

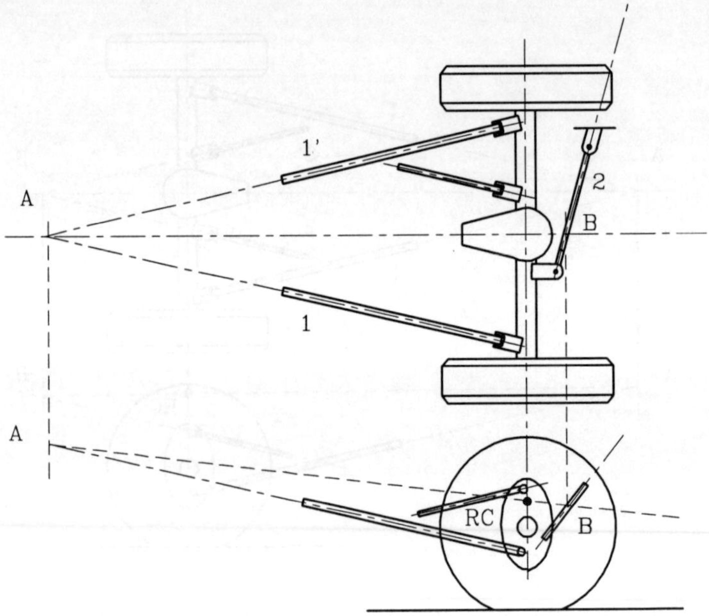

Fig. 6.7 Three-link solid-axle suspension. Position of point RC.

WG; BW is the intersection of the lines perpendicular to s_5 and s_6 and that of the perpendiculars to the resultants of s_3 and s_5 and of s_4 and s_6 locate point RC.

The derivative $(\gamma)_{,\phi}$ can be approximately computed as

$$\frac{\partial \gamma}{\partial \phi} = \frac{\chi}{\chi + \Pi}, \tag{6.1}$$

where χ and Π are respectively the rolling stiffness of the suspension ($\chi = 2kd^2 + \chi_t$, where k is the stiffness of the springs and χ_t that of the antiroll bar) and of the tires.

6.1.2 *Independent-wheels suspensions*

If the wheels are suspended independently, the linkages must constrain five out of the six degrees of freedom of the wheel (or better, of the wheel hub, because the wheel is then free to rotate about its axis). The unconstrained degree of freedom should be the translation in a direction perpendicular

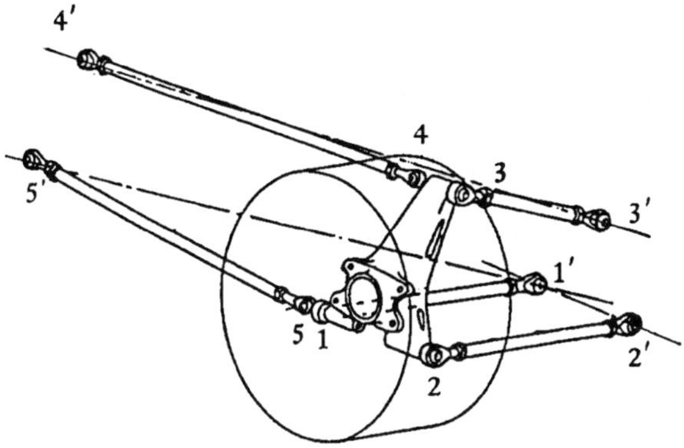

Fig. 6.8 General suspension based on five linkages to constrain the five degrees of freedom of the wheel.

to the ground. None of the many devices which are currently used fulfills exactly this requirement.

As the suspension must restrain five degrees of freedom, it can be materialized as a system made of five bars with spherical hinges at the ends (Fig. 6.8). This layout, often referred to as multilink suspensions, which has the advantage of allowing a very large freedom of adjustment by changing the length of the bars by screwing in or out the joints, has little application outside the field of racing cars for its complexity, even if simpler multilink suspensions are now more widespread. From it almost all configurations can be obtained by grouping the bars in different ways.

Note that in general the motion of the wheel is not planar and as a consequence the study of the kinematic behaviour is not easy. Nowadays it is however easy to obtain the exact kinematics of any suspension by using computer generated trajectories.

If points 1 and 2 and points 3 and 4 coincide with each other the corresponding bars become triangular elements: The suspension obtained is a transversal quadrilaterals suspension, often referred to as SLA (short-long arm) or A-arm suspension (Fig. 6.9). If lines $1'2'$ and $3'4'$ are parallel the motion of the wheel is contained in a plane perpendicular to the line $1'2'$ and the projection of the mechanism in such plane is an articulated quadrilateral, whose side $1'3'$ is made by the vehicle body. A front suspension of this type is shown in Fig. 6.10, in which the engine is also sketched to show

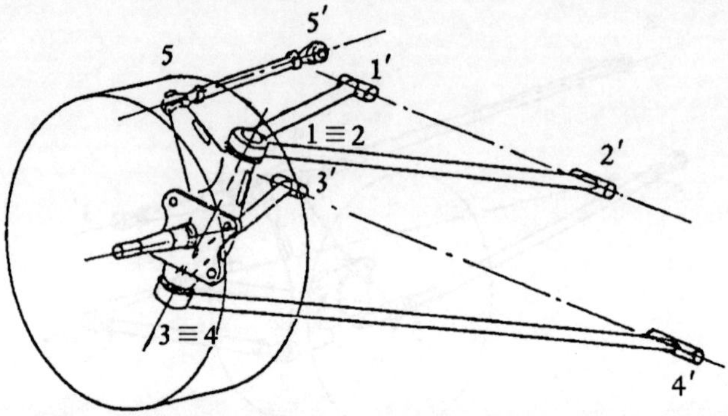

Fig. 6.9 Suspension based on transversal articulated quadrilaterals (SLA or A-arm suspension).

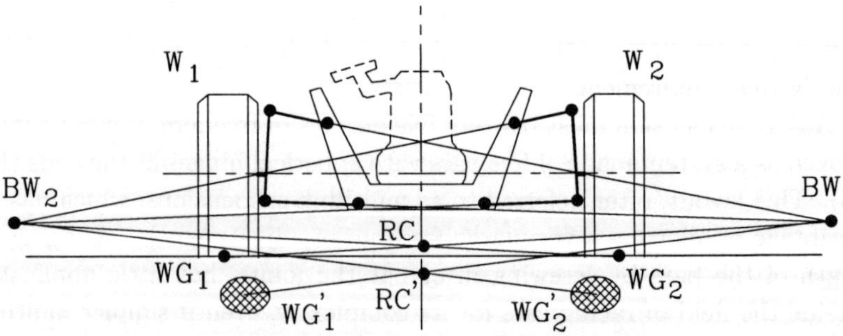

Fig. 6.10 Position of the roll centre RC for a front suspension based on transversal articulated quadrilaterals with axes $1'2'$ and $3'4'$ parallel to x-axis. RC: Position with rigid tires; RC′: Position obtained taking into account the compliance of the tires.

that this solution allows to locate the mechanical parts of the vehicle with a greater freedom than solutions based on rigid axles.

Points BW1 and BW2 of the wheels are at the intersection of the directions of the upper and lower links, which can converge towards the outside of the vehicle (Fig. 6.11a, negative swing arm suspension) or towards the midplane (Fig. 6.10, positive swing arm suspension). It is possible to obtain $\partial t/\partial z = 0$ by stating that points BW1 and BW2 lie on the ground (Fig. 6.11b), but this condition can be obtained only for one or two values of the load. If $\partial \gamma/\partial \phi$ must also vanish points BW1, BW2 and RC must be

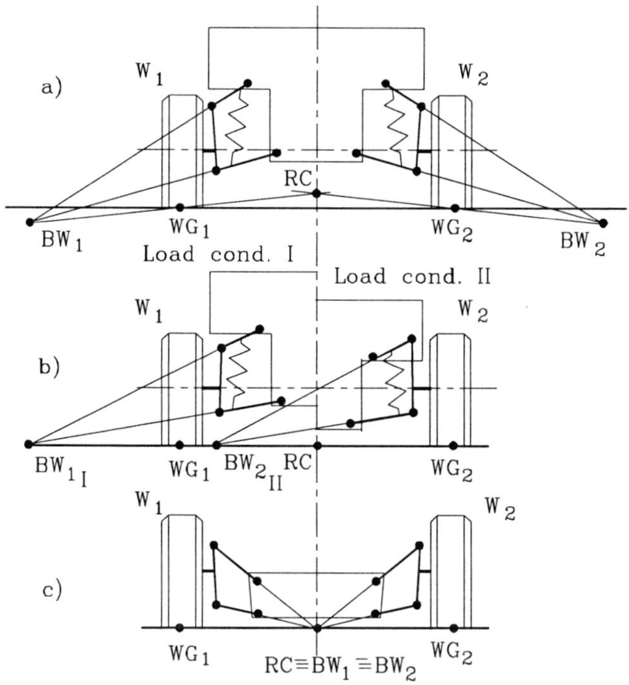

Fig. 6.11 Different schemes for transversal articulated quadrilateral suspensions.

located on the ground in the symmetry plane (Fig. 6.11c).

If the wheels were thin rigid discs, points WG1 and WG2 would coincide with the centres of the contact areas. If the compliance of the tires is accounted for their approximated position can be located under the ground, slightly inboard of the centres of contact. By connecting points BW1 and WG1 and points BW2 and WG2 and intersecting such lines the roll centre BW, which lies in the symmetry plane, can be located. In the case of transversal articulated quadrilaterals it is usually close to the ground or, if the deformation of the tires is considered, even below it. If the axes of the hinges of the two triangular linkages are not horizontal or not parallel (Fig. 6.12) the determination of the roll centre and of the motion of the latter is far more complicated.

If the upper triangle is substituted by a prismatic guide a MacPherson suspension is obtained (Fig. 6.13). Its simplicity and the fact that it leaves much free space for the engine has made it a very common solution for the

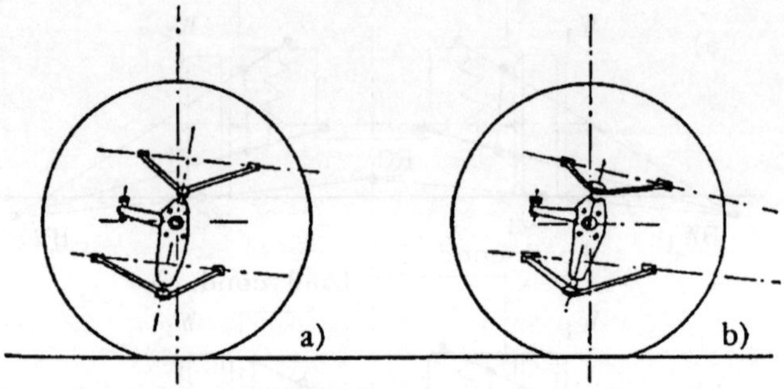

Fig. 6.12 Suspension based on articulated quadrilaterals with hinge axes not horizontal (a) and not parallel (b).

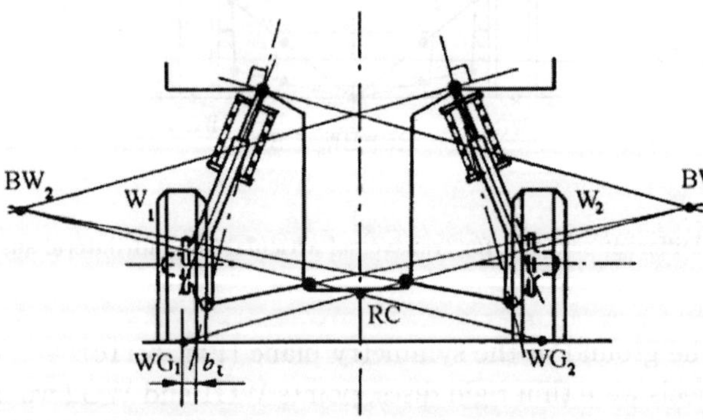

Fig. 6.13 MacPherson suspension.

front axle of cars, particularly small ones.

A different approach is that of using trailing arms (Fig. 6.14). The arms can be hinged to an axis which is perpendicular to the symmetry plane of the vehicle but this is not always the case. In the first case the track remains constant ($\partial t/\partial z = 0$, $\partial t/\partial \phi = 0$) and the camber angle does not change in the vertical motions ($\partial \gamma/\partial z = 0$) and is equal to the roll angle ($\partial \gamma/\partial \phi = 1$). If the compliance of the tires is neglected the roll centre is on the ground (Fig. 6.14a) or slightly below (Fig. 6.14b).

When a suspension of this type is used for a steering axle, as in the case

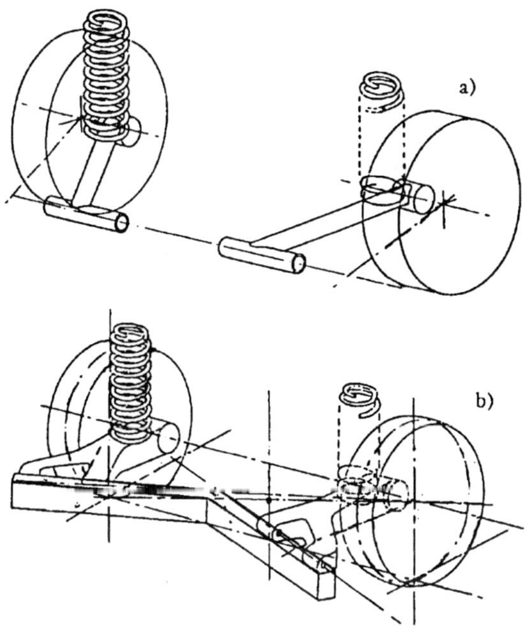

Fig. 6.14 Suspension based on trailing arms, with hinge axis parallel to y-axis (a) and inclined (b).

of Fig. 6.15 the orientation of the kingpin axis changes in both vertical and rolling motions.

Another solution is that based on swing arms. The hinges of the arms can be located at different points (Fig. 6.15a) or be coincident (Fig. 6.16b). The roll centre can be quite high on the road and the values of $\partial t/\partial z$ and of $\partial t/\partial \phi$ cannot be small. The swing arms can be connected to the engine block instead of being hinged to the body, as it is common in small cars with rear engine. The fact that the engine is suspended through rubber blocks decreases the precision of the suspension.

The main kinematic characteristics of the various type of suspensions are summarized in Table 6.1. This overview does not cover all the solutions which have been used or have been suggested like, as an example, the multi-link solution sketched in Fig. 6.17, which is common on racers. It is similar to a suspension with transversal quadrilaterals but the greater number of distinct elements allows a more detailed adjustment.

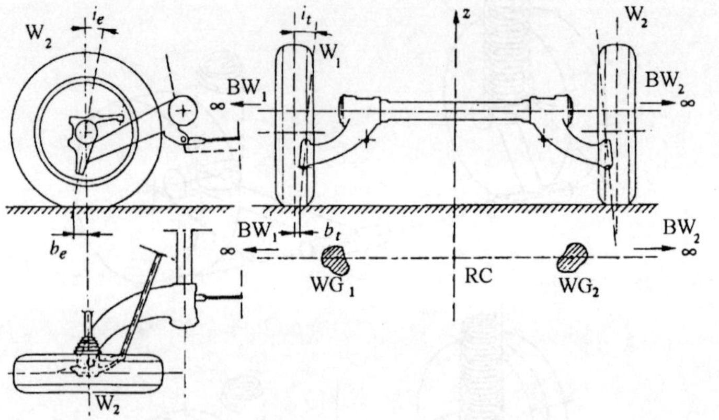

Fig. 6.15 Front axle with trailing arm suspension (the springs, which act through the pull rods, are not represented). The position of the roll centre has been obtained taking into account the compliance of the tires.

Table 6.1 Main kinematic characteristics of some types of suspensions. TA: Trailing arms; SAL: Swing arms (limit); SA: Swing arms; SLP: SLA with parallel A-arms; SLL: SLA (limit); SLA: SLA; MP: MacPherson; A, possible also high on the ground; B, on the ground or below, C, as high as the centre of the wheel; D, higher than the centre of the wheels; E, on the ground. (a): 0 in one configuration; (b): 0 in two configurations; (c): i: Inclination of kingpin axis.

	Rigid axle	Independent wheels						
		TA	SAL	SA	SLP	SLL	SLA	MP
$\partial t/\partial z$	0	0	#0	#0	#0 [a]	0	#0 [b]	#0
$\partial t/\partial \phi$	0	0	0	#0	#0	#0	#0	#0
$\partial \gamma/\partial z$	0	0	#0	#0	#0	#0	#0	#0
$\partial \gamma/\partial \phi$	0	1	0	#0	1	#0	$>0<1$	$<0>1$
Position RC	A	B	C	D	B	E	B	B
Variations RC	0 (≈ 0)	0	0	0	#0	0	#0	#0
$\partial i/\partial z$ [c]	0	#0	#0	#0	0	0	#0	#0

If the compliance of the suspension is also accounted for, other characteristics can be obtained, as $\partial x/\partial F_x$. Usually a number of plots are obtained by measuring experimentally some outputs, as camber, horizontal displacement, normal load on the ground, etc. when inputs as vertical displacement, longitudinal forces etc. are applied to the suspension. A

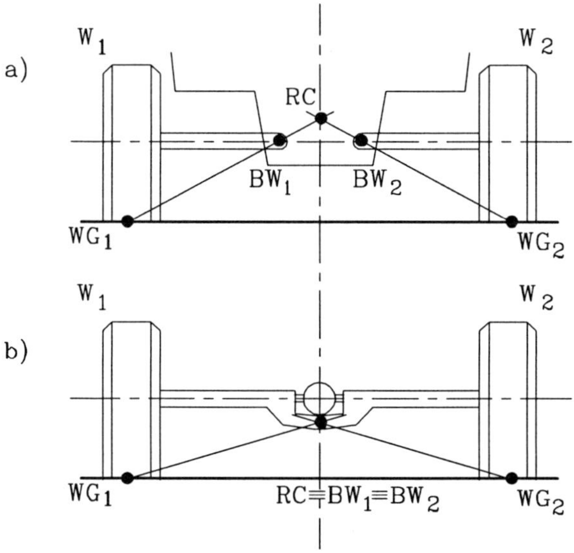

Fig. 6.16 Suspension based on swing arms, with hinges located in two different points (a) and in the plane of symmetry (b).

complete set of the mentioned plots is summarized in Table 6.2.

Table 6.2 Plots characterizing the behaviour of suspensions. X: Plot commonly used; – useless in most cases or meaningless; δ_{sw}: Rotation angle of the steering wheel.

Inputs	Outputs					
	Toe in	Camber $\Delta\gamma$	Wheelbase Δx	Track Δt	Normal load F_z	Wheel steering δ
Δz	X	X	X	X	X	–
F_x	X	–	X	–	–	–
F_y	X	X	–	X	–	–
M_z	X	–	–	–	–	–
δ_{sw}	X	–	X	X	–	X

The experimental input-output curves can be approximated by polynomials to be entered into a full nonlinear analysis; the slope in the static equilibrium position supplies the data required for the linearized study.

Another possible alternative, very seldom considered, is the use of vertical prismatic guides. Though it seems to be the simplest and more kinematically correct solution, it suffers from sticking problems of the guide due

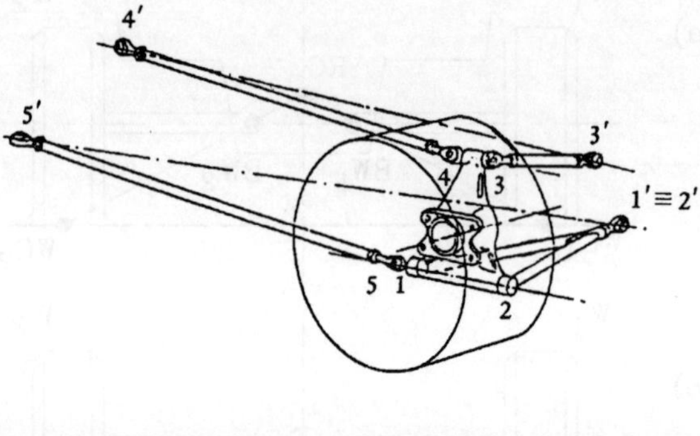

Fig. 6.17 Multi-link suspension used on some racing cars.

to transversal forces, which in some cases are large. All solutions based on prismatic guides have this problem, which in MacPherson suspensions is often solved by setting the spring in the direction of the load and not of the strut, in order to relieve from it all transversal forces.

The two suspensions of the same axle can be interconnected using mechanical springs (as the antiroll bars), or pneumatic or hydraulic devices. Apart from this "transversal" interconnection, sometimes the two wheels of the same side are connected to each other, leading to a "longitudinal" interconnection.

6.1.3 *Anti-dive and anti-squat designs*

When the vehicle accelerates or brakes a load transfer between front and rear wheels occurs (see Eq. (4.4)). This causes the body to pitch up (squat) or down (dive). Apparently, as forces F_{z_1} and F_{z_2} can be approximated as

$$\begin{cases} F_{z_1} = F^*_{z_1} - m\dfrac{h_G}{l}\dot{V} \\ F_{z_2} = F^*_{z_2} + m\dfrac{h_G}{l}\dot{V} \, , \end{cases} \tag{6.2}$$

where forces $F^*_{z_i}$ are those occurring when the vehicle does not accelerate, the lift of the front and the rear of the body are respectively

$$\Delta z_1 = m\frac{h_G}{lK_f}\dot{V} \, ; \qquad \Delta z_2 = -m\frac{h_G}{lK_r}\dot{V} \, ,$$

where K_f and K_r are the vertical stiffness of the front and rear suspensions. The pitch angle due to an acceleration is then

$$\theta = \frac{1}{l}\left(-\Delta z_1 + \Delta z_2\right) = -m\frac{h_G}{l^2}\dot{V}\left(\frac{1}{K_f} + \frac{1}{K_r}\right) . \tag{6.3}$$

A positive value of θ occurs when the vehicle dives (pitches down) as it occurs with a negative acceleration, hence the minus sign in the formula.

This expression is however an oversimplification, for two reasons: Firstly the longitudinal forces due to the driving or braking wheels can cause themselves a pitching moment due to the coupling of the suspensions and, secondly, the driving and braking torque reactions can be applied, at least partly, to the suspensions instead of the body, inducing further pitching. Both effects cause pitching even in constant speed driving.

If the suspension allows the wheels to move also in x direction, i.e. if the characteristic $\partial x/\partial z$ is not vanishingly small, a fraction $F_x \partial x/\partial z$ of force F_x acting between the road and the wheel acts on the suspension and causes pitching. Equation (6.2) then becomes

$$\begin{cases} F_{z_1} = F^*_{z_1} - m\frac{h_G}{l}\dot{V} - \left(\frac{\partial x}{\partial z}\right)_1 F_{x1} \\ F_{z_2} = F^*_{z_2} + m\frac{h_G}{l}\dot{V} - \left(\frac{\partial x}{\partial z}\right)_2 F_{x2} \end{cases} \tag{6.4}$$

and Eq. (6.3) transforms into

$$\theta = -m\frac{h_G}{l^2}\dot{V}\left(\frac{1}{K_f} + \frac{1}{K_r}\right) - \left(\frac{\partial x}{\partial z}\right)_1 \frac{F_{x1}}{lK_f} + \left(\frac{\partial x}{\partial z}\right)_2 \frac{F_{x2}}{lK_r} . \tag{6.5}$$

If only longitudinal forces needed to accelerate or to brake the vehicle are considered and the percentage of the longitudinal force assigned to the front axle is k_l, it follows

$$F_{x_1} = k_l m\dot{V} , \qquad F_{x_2} = (1 - k_l)m\dot{V} .$$

Equation (6.3) yields

$$\theta = -m\frac{\dot{V}}{l}\left[\frac{h_G}{lK_f} + \frac{h_G}{lK_r} + \frac{k_l}{K_f}\left(\frac{\partial x}{\partial z}\right)_1 - \frac{(1 - k_l)}{K_r}\left(\frac{\partial x}{\partial z}\right)_2\right] . \tag{6.6}$$

Obviously Eq. (6.3) holds in the case of acceleration and braking alike, provided that the sign of $\dot{V}$ is correct and a suitable value for k_l is used.

Consider for example the trailing arm suspension of Fig. 6.18a. With simple geometrical reasoning it is easy to assess that

$$\left(\frac{\partial x}{\partial z}\right) = \frac{e}{d} \,. \tag{6.7}$$

A similar equation holds also for the suspension of Fig. 6.18b. If a torque M_y is applied to sprung mass it causes an increase of the force acting on the spring equal to M_y/d. As the torque linked to a generation of driving or braking forces is equal to $-F_x R_l$, the result is that the application of the braking torque to the suspension can be accounted for by substituting

$$\left(\frac{\partial x}{\partial z}\right)_i + \left(\frac{R_l}{d}\right)_i \qquad \text{for} \qquad \left(\frac{\partial x}{\partial z}\right)_i \,.$$

Note that d is positive when point A is in front of the wheel and negative otherwise.

The driving torque is applied to the unsprung mass in the case of live axles, while in De Dion axles and independent suspensions it is applied directly to the vehicle body and this correction does not apply. Braking torques are on the contrary applied usually to the unsprung masses, so the term in R_l/d must always be accounted for. However, if the torque transmission between the sprung and the unsprung masses is supplied by linkages which prevent any relative rotation about y axis, as d tends to infinity, these effects are minimized.

The above relationships allow one to design the suspensions to compensate, usually partially, for squat or dive. A total compensation occurs when Eq. (6.5) yields $\theta = 0$. If

$$\frac{h_G}{lK_f} + \frac{k_l}{K_f}\left(\frac{\partial x}{\partial z}\right)_1 = 0 \tag{6.8}$$

the front of the car does not lift in acceleration or dive in braking, while if

$$\frac{h_G}{lK_r} - \frac{(1-k_l)}{K_r}\left(\frac{\partial x}{\partial z}\right)_2 = 0 \tag{6.9}$$

the rear does not squat in acceleration or lift in braking.

Note that in case of a single driving axle either $k_l = 0$ or $k_l = 1$ and both front and rear compensations cannot be performed together. To obtain a complete compensation the term in square brackets in Eq. (6.5) must vanish and the front of the car must dive to compensate for the squat of the rear axle in front drives or the rear must lift in rear drives.

In case of braking a total compensation of the front axle leads to the condition

$$\frac{k_l}{K_f}\left[\left(\frac{\partial x}{\partial z}\right)_1 + \left(\frac{R_l}{d}\right)_1\right] = -\frac{h_G}{lK_f}\,, \text{ i.e. } \quad \frac{k_l}{K_f}\left(\frac{e+R_l}{d}\right)_1 = -\frac{h_G}{lK_f} \tag{6.10}$$

and that of the rear axle to

$$\frac{(1-k_l)}{K_r}\left[\left(\frac{\partial x}{\partial z}\right)_2 + \left(\frac{R_l}{d}\right)_2\right] = \frac{h_G}{lK_r}\,, \text{ i.e. } \quad \frac{(1-k_l)}{K_r}\left(\frac{e+R_l}{d}\right)_2 = \frac{h_G}{lK_r}\,. \tag{6.11}$$

A simple geometrical construction is shown in Fig. 6.18c. If the pivots are on the dotted lines a complete compensation in braking is obtained while if they lie below the lines dive compensation is only partial. If they are above the line the front will lift and the rear will squat in braking.

Usually no complete anti-dive compensation is obtained for many reasons, psychological (a flat braking is not desirable) and objective (complete antidive compensation can lead to overcompensation of acceleration squat and to geometries which are poor both for comfort and performance).

6.1.4 *Active suspensions*

The strong vertical aerodynamic loads which could be obtained by the aerodynamic devices used on the Formula 1 racers of the recent past compelled designers to use suspensions which were so stiff that the riding comfort was very poor. Almost unbearable vibrations were present in all conditions except very smooth road surfaces. This pushed toward the development of active suspension devices, in which the conventional spring-damper setup is substituted by an electrohydraulic system which is able to control the motion of the wheels with respect to the vehicle body following predetermined laws. The use of one or more microprocessors allows to easily program the device and to follow optimal control laws, which can be tailored to optimize

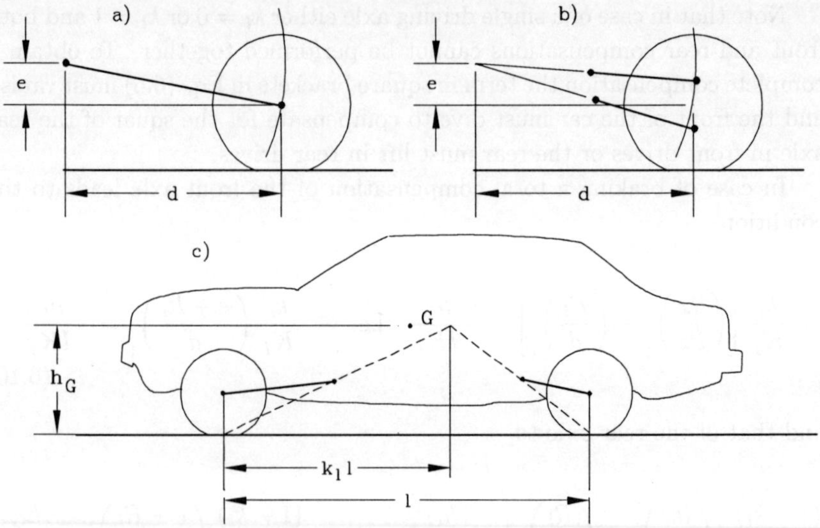

Fig. 6.18 Relationship between $\partial x/\partial z$ and geometry in suspensions with a single (a) and two (b) trailing arms. Note that if the arms are parallel $d \to \infty$. Geometrical construction for anti-dive compensation (c).

the handling, or the comfort performances or a suitable mix of them.

Clearly these devices give a wider freedom to the designer and results which were impossible with conventional systems can be obtained. It is possible for instance to have no steady state rolling under the effect of centrifugal forces or even to roll the suspended mass towards the inside of the curve to maintain the z-axis along the local vertical. A vanishing roll angle allows to avoid roll steer and any change of the camber angle of the wheels, both effects increasing the handling performances of the vehicle. It is also possible to avoid diving when braking, maintaining the required attitude of the sprung mass in all driving conditions.

The introduction of active control of the suspensions allow one to simplify the mechanical layout, transferring partly the complexities from the mechanical system to the controller. Actually the general definition of active suspensions covers a number of different approaches. The term "active" should be used only for a suspension provided of sensors and actuators and of a control system with an external power source which provides the actuation. A device is usually termed as "intelligent" if the control and the mechanical device are strictly linked and integrated in such a way that the working in purely mechanical mode is no more possible.

Solutions in which the control system only changes the stiffness and damping parameters of an essentially mechanical suspension are better referred to as "adaptive" suspensions. The simplest and oldest types are the so-called *self-levelling* suspensions, in which the spring system adapts the height of ride in such a way to maintain it constant in spite of changes of load. The most common types are based on air springs and used on industrial vehicles: A pneumatic system controls the pressure in the air springs, increasing it when the load increases for a sufficient length of time, typically several seconds. This approach has the added advantage of increasing also the stiffness of the suspension maintaining the natural frequency of the suspension almost constant at changing load. Self-levelling suspensions should strictly speaking be considered active devices, as some energy is required for their operation, but as very small quantities of energy are involved, the term "active suspensions" is not used.

Semi-active suspensions are also adaptive devices. Depending on the characteristic time of their operation they are often subdivided into *slow-active*, *low-bandwidth* and *high-bandwidth* suspensions. The first ones are suspensions in which the stiffness or the damping can be adjusted between discrete levels. The input signal which forces the suspension to adapt its parameters can be brake pressure, steering angle or suspension motion. The suspension adapts quickly to the new conditions while the switch back to the previous conditions takes a certain time, greater than a single oscillation of the vehicle.

In low- and high-bandwidth suspensions the stiffness or the damping are modulated continuously following only the low frequency motions of the sprung mass or also the high frequency motions of the unsprung masses.

In *fully-active* suspensions the system incorporates actuators, usually hydraulic cylinders, which exert forces between sprung and unsprung masses. Also active suspensions can be classified as low- or high-bandwidth suspensions following the criterion described above.

As it is far easier to control quasi-static load conditions, as rolling in bends or diving when brakes are applied than to control the motions occurring at low and particularly at high frequency, the complexity and the cost of active or adaptive devices increase if they are called to work at high frequencies and also the power requirements of the former become more demanding.

The potential performances of various types of adaptive and active suspensions are summarized in Table 6.3. It must also be noted that with active suspensions trade-offs are necessary as when optimizing the performance of a certain type, other performances can be negatively affected.

Table 6.3 Potential performances of the various types of adaptive and active suspensions.

Type	Performance mode					
	Ride	Height	Roll	Dive	Squat	Handling
Passive	Performance is a compromise between all modes					
Self-levelling	High	High	–	–	–	–
Semi-active	Medium	–	Low	Low	Low	Medium
Full-active	High	High	High	High	High	High

6.2 Generalized coordinates for rotations

The sprung mass is a rigid body and consequently it has six degrees of freedom. Three of them can be considered as translational and the corresponding generalized coordinates can be the coordinates of its centre of mass in any suitable inertial reference frame. The choice for the other three coordinates is less obvious. A possible choice is to resort to two coordinates of a second point and to one coordinate of a third point not aligned with the other two, but this choice is generally not advisable. An obvious choice can be the use of Euler angles[2]. They have however the drawbacks of being indeterminate if the xy plane of the rigid body is parallel to the xy plane of the inertial reference frame and of not giving results which are intuitively understood. Another alternative which is sometimes used also in road vehicle dynamics is the use of quaternions.

Here a different way will be followed. Consider an inertial reference frame XYZ with XY plane parallel to the road surface. In this frame the xyz axes which are fixed to the vehicle body are located. In order to define the angular position of the second frame with respect to the first one the relative position of the origins is immaterial; the two frames will then be represented as centred at the same point (Fig. 6.19).

Rotate the XYZ frame about Z-axis until axis X coincides with the projection of x-axis on XY plane (Fig. 6.19a). Indicate this position of axis X as x^*; the rotation angle between axes X and x^* is the yaw angle ψ. The rotation matrix allowing to pass from the frame x^*y^*Z to the inertial

[2]In the following the term Euler angles will be used to designate the precession, rotation and nutation angles as used in the study of gyroscopes. If the more comprehensive definition for which any set of three angles taken in a given order is a set of Euler anlges, also the yaw, pitch and roll angles defined below can be considered as such.

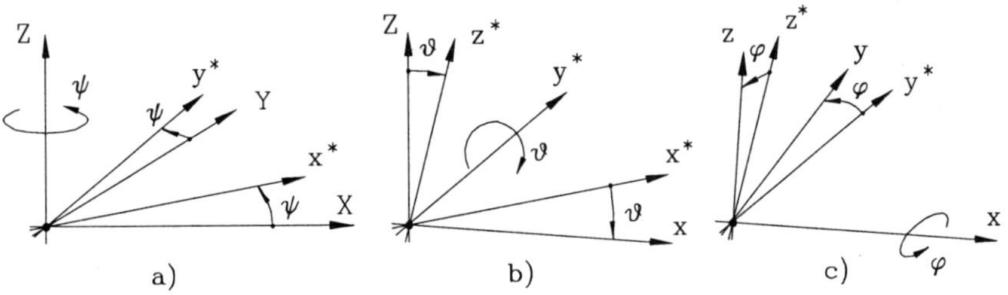

Fig. 6.19 Definition of the yaw ψ (a), pitch θ (b) and roll ϕ (c) angles.

frame XYZ is

$$[R_1] = \begin{bmatrix} \cos(\psi) & -\sin(\psi) & 0 \\ \sin(\psi) & \cos(\psi) & 0 \\ 0 & 0 & 1 \end{bmatrix} . \tag{6.12}$$

The second rotation is the pitch rotation θ about axis y^* which leads axis x^* in the position of x-axis (Fig. 6.19b). θ is positive if x-axis points downwards. The rotation matrix is

$$[R_2] = \begin{bmatrix} \cos(\theta) & 0 & \sin(\theta) \\ 0 & 1 & 0 \\ -\sin(\theta) & 0 & \cos(\theta) \end{bmatrix} . \tag{6.13}$$

The third rotation is the roll rotation ϕ about x-axis. It leads axes y^* and z^* in the position of axes y and z (Fig. 6.19c). ϕ is positive if y-axis points upwards. The rotation matrix is

$$[R_3] = \begin{bmatrix} 1 & 0 & 0 \\ 0 & \cos(\phi) & -\sin(\phi) \\ 0 & \sin(\phi) & \cos(\phi) \end{bmatrix} . \tag{6.14}$$

The global rotation matrix allowing to rotate any vector from frame xyz to the inertial frame XYZ is obviously the product of the three matrices: $[R] = [R_1][R_2][R_3]$.

The product of the last two matrices can be written in the form

$$[R_2][R_3] = \left[\begin{array}{cc|c} \cos(\theta) & \sin(\theta)\sin(\phi) & \sin(\theta)\cos(\phi) \\ 0 & \cos(\phi) & -\sin(\phi) \\ \hline -\sin(\theta) & \cos(\theta)\sin(\phi) & \cos(\theta)\cos(\phi) \end{array}\right] = \begin{bmatrix} [R_{11}] & [R_{12}] \\ [R_{21}] & [R_{22}] \end{bmatrix} . \tag{6.15}$$

By introducing the planar rotation matrix for yaw rotations in XY plane

$$[R_1^*] = \begin{bmatrix} \cos(\psi) & -\sin(\psi) \\ \sin(\psi) & \cos(\psi) \end{bmatrix} ,$$

the global rotation matrix can be written in the form

$$[R] = \begin{bmatrix} [R_1^*][R_{11}] & [R_1^*][R_{12}] \\ [R_{21}] & [R_{22}] \end{bmatrix} , \tag{6.16}$$

which has the advantage of keeping separate the roll and pitch rotations which are usually small and can often be linearized, from the yaw rotations, which usually cannot be linearized.

The angular velocities $\dot{\psi}$, $\dot{\theta}$ and $\dot{\phi}$ are not directed along axes x, y and z and consequently do not coincide with the components p, q and r of the angular velocity in the reference frame fixed to the vehicle. Their directions are those of axes Z, y^* and x. The relationship between them is

$$\begin{Bmatrix} p \\ q \\ r \end{Bmatrix} = \begin{Bmatrix} \dot{\phi} \\ 0 \\ 0 \end{Bmatrix} + [R_3]^T \begin{Bmatrix} 0 \\ \dot{\theta} \\ 0 \end{Bmatrix} + [[R_2][R_3]]^T \begin{Bmatrix} 0 \\ 0 \\ \dot{\psi} \end{Bmatrix} , \tag{6.17}$$

i.e.

$$\begin{cases} p = \dot{\phi} - \dot{\psi}\sin(\theta) \\ q = \dot{\theta} + \dot{\psi}\sin(\phi)\cos(\theta) \\ r = \dot{\psi}\cos(\theta)\cos(\phi) - \dot{\theta}\sin(\phi) . \end{cases} \tag{6.18}$$

If the pitch and roll angles are small enough to allow to linearize their trigonometric functions, the four submatrices in Eq. (6.15) reduce to

$$[R_{11}] = [I] \, , \qquad [R_{12}] = \left\{ \begin{array}{c} \theta \\ -\phi \end{array} \right\} \, , \qquad (6.19)$$
$$[R_{21}] = -[R_{12}]^T = \left[-\theta \; \phi \right] \, , \qquad [R_{22}] = 1$$

and the components of the angular velocity can be approximated as

$$\left\{ \begin{array}{l} p = \dot{\phi} - \theta\dot{\psi} \\ q = \dot{\theta} + \phi\dot{\psi} \\ r = \dot{\psi} - \phi\dot{\theta} \, . \end{array} \right. \qquad (6.20)$$

6.3 Model for an insulated vehicle

A vehicle made of a single rigid suspended mass and a number n of wheels can be modelled as a system with $6+n$ degrees of freedom. A possible choice for the suspended mass is to choose as generalized coordinates the X, Y and Z coordinates of its centre of mass and the yaw, pitch and roll angles ψ, θ and ϕ defined in the previous section. For the independent suspensions it is possible to choose a particular parameter which allows one to define its position; the choice of this parameter depends on the particular type of suspension and cannot be specified in general. In the case of the suspensions of Fig. 6.16 it can be the angle between the swing arm and y-axis of the body or, alternatively, XY plane. In the case of solid axle suspensions, each suspension has two degrees of freedom, which can in general be the Z coordinate of its centre and the roll angle ϕ.

The kinetic energy of the suspended mass can be written in an explicit way

$$\mathcal{T} = \tfrac{1}{2} m_S \left(\dot{X}^2 + \dot{Y}^2 + \dot{Z}^2 \right) + \tfrac{1}{2} \left\{ \begin{array}{c} \dot{\phi} - \dot{\psi}\sin(\theta) \\ \dot{\theta} + \dot{\psi}\sin(\phi)\cos(\theta) \\ \dot{\psi}\cos(\theta)\cos(\phi) - \dot{\theta}\sin(\phi) \end{array} \right\}^T$$
$$\left[\begin{array}{ccc} J_x & 0 & -J_{xz} \\ 0 & J_y & 0 \\ -J_{xz} & 0 & J_z \end{array} \right] \left\{ \begin{array}{c} \dot{\phi} - \dot{\psi}\sin(\theta) \\ \dot{\theta} + \dot{\psi}\sin(\phi)\cos(\theta) \\ \dot{\psi}\cos(\theta)\cos(\phi) - \dot{\theta}\sin(\phi) \end{array} \right\} . \qquad (6.21)$$

Note that, owing to symmetry, all terms outside the main diagonal of the inertia matrix of the sprung mass are equal to zero except for J_{xz}. It is impossible to write a similar expression for the unsprung masses until the exact geometrical configuration of the suspensions is defined. However, once it has been stated, the position of their centres of mass can be easily obtained. By differentiating, also the velocities of such points are known, as are the angular velocities; the kinetic energy of the i-th unsprung mass is then

$$\mathcal{T}_i = f\left(\dot{X}, \dot{Y}, Z, \dot{Z}, \theta, \dot{\theta}, \phi, \dot{\phi}, \dot{\psi}, \gamma_i, \dot{\gamma}_i\right) , \tag{6.22}$$

where γ_i is the generalized coordinate used for the i-th independent suspension.

If a suspension is of the solid axle type the expression of the kinetic energy is of the same type, with the only difference that now the coordinates related to the suspension included in the expression of the kinetic energy are two instead of one. Note that coordinates X, Y and ψ do not appear directly in the equation but only with their derivatives; this is a feature which the model of the vehicle on elastic suspensions shares with that of the rigid vehicle.

From the position of the centres of mass of all the rigid bodies the gravitational potential energy is readily obtained

$$\mathcal{U}_g = m_S g Z + \sum_{\forall i} m_i g Z_i , \tag{6.23}$$

where the Z coordinate of the i-th suspension is $Z_i = f(Z, \theta, \phi, \gamma_i)$. Again in case of solid axle suspensions the coordinates related to the suspension are two.

There is no difficulty in writing also the position of the points of attachment of the springs to both the body and the suspensions and hence to write the potential energy due to the springs. As no assumption of linearity has been made, the function expressing the potential energy due to the springs of the i-th suspension can be different from a quadratic function of the generalized coordinates

$$\mathcal{U}_{s_i} = f(Z, \theta, \phi, \gamma_i) . \tag{6.24}$$

In a similar way the potential energy due to the deformation of the tires is again a function of the coordinates

$$\mathcal{U}_{t_i} = f(Z, \theta, \phi, \gamma_i) \; . \tag{6.25}$$

In this case the coordinates X, Y and ψ can also enter the equation if the road is not a horizontal flat surface.

The Rayleigh dissipation function due to the shock absorbers is computed from the velocities of the points in which they are connected to the body and to the suspensions

$$\mathcal{F}_{a_i} = f\left(\dot{X}, \dot{Y}, Z, \dot{Z}, \theta, \dot{\theta}, \phi, \dot{\phi}, \dot{\psi}, \gamma_i, \dot{\gamma}_i\right) \; . \tag{6.26}$$

There is no difficulty in obtaining the Rayleigh dissipation function due to tires, although tire damping is usually neglected.

The equation of motion is then obtained from the Lagrange equation in its complete form

$$\frac{d}{dt}\left(\frac{\partial \mathcal{T}}{\partial \dot{q}_i}\right) - \frac{\partial \mathcal{T}}{\partial q_i} + \frac{\partial \mathcal{U}}{\partial q_i} + \frac{\partial \mathcal{F}}{\partial \dot{q}_i} = Q_i \; . \tag{6.27}$$

Actually the procedure is simpler than it seems: The expressions of the energies can be written separately for each suspension and also the contributions of each one to the various equations can be computed separately. It is so possible to build a sort of library of different types of suspensions, from which the vehicle model can be obtained simply by assembling the relevant parts of the equations of motion.

The generalized forces can be obtained in the usual way: The virtual displacement of the centre of the contact area of each tire and of the centre of mass of the vehicle (which does not coincide with that of the suspended mass[3]) can be computed.

The forces due to the tires, applied in the former, and the aerodynamic forces, applied in the latter, are computed and then the virtual work is obtained. By differentiating the virtual work with respect to the virtual displacements δX, δY, etc. the generalized forces are computed. Note that all forces are present in general, as the kinematics of the suspension can

[3]This leads to an approximation: The aerodynamic forces are referred to a certain configuration, while here they are applied in the centre of mass of the vehicle in any configuration. This is unavoidable, since usually no data is available about how they change with different relative position of the unsprung masses with respect to the sprung mass. Sometimes some data on the variation of the aerodynamic forces with pitch and roll are available: In such case they can be introduced into the model.

couple all motions (e.g., a virtual displacement δZ of the body can cause the centre of the contact area to be displaced in all directions).

Although the complexity of this approach can be mitigated by the modular modelling of the system, the resulting equations of motion are highly nonlinear and can be attacked only by numerical integration. As a last remark, the use of symbolic computer codes can allow to write automatically the equations of motion and the codes for their numerical integration.

Note that no allowance for the gyroscopic moments due to the wheels and to the other rotating elements has been taken. Also the effects of torques acting between the sprung and unsprung masses have not been included, but can be easily introduced into the model.

6.4 Linearized model for an insulated vehicle

As already stated, some of the variables of the motion are angles which in normal driving remain quite small: Many simplifications can be performed if their trigonometric functions are linearized. Also, the concept of roll axis can be used if the roll angle is small. Consider a vehicle sitting on the road in a certain equilibrium position: If it has only two axles the positions of the roll centres of the suspensions are readily identified and the roll axis runs through these points. If the axles are more than two the roll centres of the suspensions cannot be aligned but the roll axis can nevertheless be defined as stated in the previous sections. The roll axis, taken as x-axis for the whole vehicle (Fig. 6.20), will be assumed to be fixed to the sprung mass and to follow the latter in all vertical and pitching motions.

6.4.1 *Sprung mass*

Let the reference frame of the sprung mass be $x_s y_s z_s$, with axis x_s parallel to the roll axis and axis z_s laying in the plane of symmetry. Axis x^* coincides with the projection on the ground of the roll axis. A plane perpendicular to the road and to axis x^* containing the centre of mass G of the vehicle in the reference position is defined (section B-B in Fig. 6.20). The roll axis intersects such plane in H; O is the point on the ground vertically under H (it can be located above H, as the roll axis can lie below the ground, in particular cases). Two further reference frames are defined. They are:
— xyz, fixed to the sprung mass, with origin in point H, x-axis coinciding with the roll axis, z axis laying in the symmetry plane of the sprung mass;

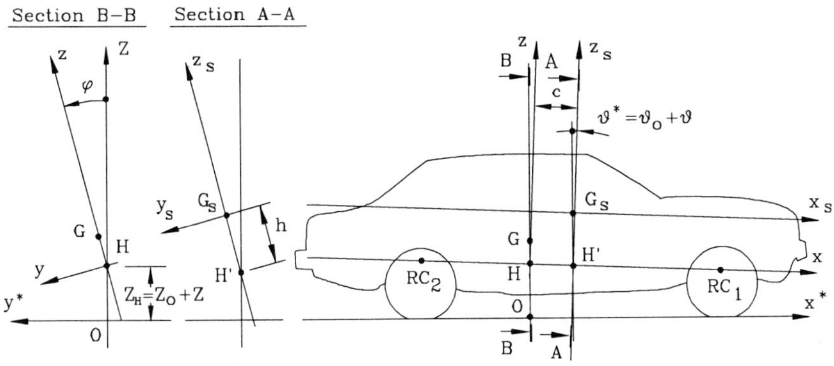

Fig. 6.20 Reference frames for the sprung mass and definition of points H and O.

— $x^*y^*z^*$ with origin in point O; x^*-axis coincides with the projection on the ground of the roll axis and z^* axis is perpendicular to the road.

Instead of using the coordinates of the centre of mass G_s of the sprung mass to define the generalized coordinates for the translational degrees of freedom, the coordinates X_H, Y_H and Z_H of point H will be used. In the following to simplify the notation it will be $X = X_H$ and $Y = Y_H$. Operating in this way, if the roll and pitch motions are blocked, the frame $x^*y^*z^*$ coincides with frame xyz defined in Fig. 5.11 and the model reduces to that of the rigid vehicle.

Coordinate Z_H can be considered as the sum of a constant value Z_0 corresponding to a reference position and a displacement Z. The generalized coordinates for the translations of the sprung mass will be X, Y and Z.

The generalized coordinates for rotations are the yaw angle ψ, the pitch angle, which will be considered as the sum of a value θ_0 related to the reference position plus a displacement θ, and the roll angle ϕ. Angle θ_0 is nothing but the inclination of the roll axis with respect to the horizontal and will be considered as a small angle.

The coordinates of the centre of mass of the sprung mass are c, 0 and h in the frame xyz. The corresponding values in the inertial frame are

$$\left\{\begin{matrix} X \\ Y \\ Z \end{matrix}\right\}_{G_s} = \left\{\begin{matrix} X \\ Y \\ Z_0 + Z \end{matrix}\right\} + [R]\left\{\begin{matrix} c \\ 0 \\ h \end{matrix}\right\} . \tag{6.28}$$

Differentiating the expressions of the coordinates the velocity is readily obtained

$$\begin{Bmatrix} \dot{X} \\ \dot{Y} \end{Bmatrix}_{G_s} = \begin{Bmatrix} \dot{X} \\ \dot{Y} \end{Bmatrix} + [\dot{R}_1^*]\left([R_{11}]\begin{Bmatrix} c \\ 0 \end{Bmatrix} + [R_{12}]\,h\right) +$$

$$+[R_1^*]\left(\left[\dot{R}_{11}\right]\begin{Bmatrix} c \\ 0 \end{Bmatrix} + \left[\dot{R}_{12}\right]h\right) \tag{6.29}$$

$$\dot{Z}_{G_s} = \dot{Z} + \left[\dot{R}_{21}\right]\begin{Bmatrix} c \\ 0 \end{Bmatrix} + \left[\dot{R}_{22}\right]h\,.$$

All the terms containing powers higher than the second of the generalized coordinates and their derivatives generate nonlinear terms in the equations of motion and must be dropped in order to obtain a linearized model. The trigonometric functions of angles θ^* and ϕ can be substituted by their series truncated at the terms of second degree. The only exception are the terms in $\dot{X}$ and $\dot{Y}$: They can be large since contain the speed V, that cannot be considered as small like the other velocity components. Dropping all terms leading to powers higher than the second one in the expression of the kinetic energy, equation (6.29) reduces to

$$\begin{Bmatrix} \dot{X} \\ \dot{Y} \end{Bmatrix}_{G_s} = \begin{Bmatrix} \dot{X} \\ \dot{Y} \end{Bmatrix} + [\dot{R}_1^*]\begin{Bmatrix} c + h\theta_0 + h\theta \\ -h\phi \end{Bmatrix} + [R_1^*]\begin{Bmatrix} h\dot{\theta} \\ -h\dot{\phi} \end{Bmatrix} \tag{6.30}$$

$$\dot{Z}_{G_s} = \dot{Z} - c\dot{\theta}\,.$$

The kinetic energy of the sprung mass is then

$$\mathcal{T}_s = \tfrac{1}{2}m_S\left(\begin{Bmatrix} \dot{X} \\ \dot{Y} \end{Bmatrix}_{G_s}^T \begin{Bmatrix} \dot{X} \\ \dot{Y} \end{Bmatrix}_{G_s} + \dot{Z}^2\right) +$$

$$+\frac{1}{2}\begin{Bmatrix} p \\ q \\ r \end{Bmatrix}^T \begin{bmatrix} J_{x_s} & 0 & -J_{xz_s} \\ 0 & J_{y_s} & 0 \\ -J_{xz_s} & 0 & J_{z_s} \end{bmatrix} \begin{Bmatrix} p \\ q \\ r \end{Bmatrix}. \tag{6.31}$$

By introducing the mentioned simplifications, it follows

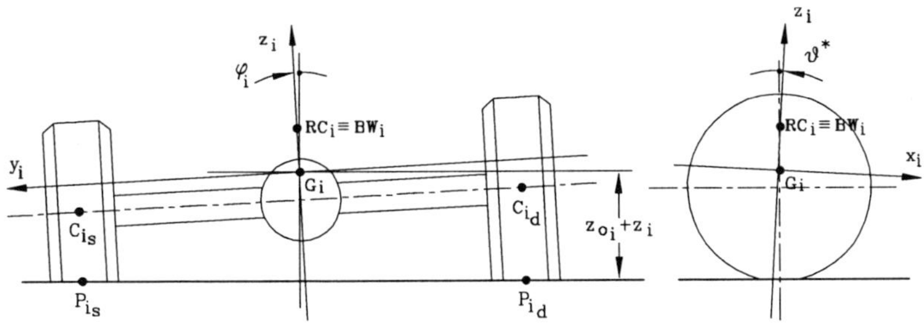

Fig. 6.21 Generalized coordinates for a solid axle suspension.

$$\mathcal{T}_s = \frac{1}{2} m s \left[\dot{X}^2 + \dot{Y}^2 + \dot{Z}^2 + c^2 \dot{\psi}^2 + h^2 \dot{\phi}^2 + \left(c^2 + h^2\right) \dot{\theta}^2 - 2ch\dot{\psi}\dot{\phi} + \right.$$

$$\left. -2c\dot{Z}\dot{\theta} + 2h \left\{ \begin{array}{c} \dot{X} \\ \dot{Y} \end{array} \right\}^T [R_1^*] \left\{ \begin{array}{c} \dot{\theta} \\ -\dot{\phi} \end{array} \right\} + 2\dot{\psi} \left\{ \begin{array}{c} \dot{X} \\ \dot{Y} \end{array} \right\}^T \frac{\partial}{\partial \psi} [R_1^*] \left\{ \begin{array}{c} c + h\theta_0 + h\theta \\ -h\phi \end{array} \right\} \right] +$$

$$+ \frac{1}{2} \left(J_{x_s} \dot{\phi}^2 + J_{y_s} \dot{\theta}^2 + J_{z_s} \dot{\psi}^2 \right) - J_{xz_s} \dot{\phi} \dot{\psi} \, . \tag{6.32}$$

6.4.2 *Solid axle suspensions*

A sketch of a solid axle is shown in Fig. 6.21. The simplest choice is that of using the coordinate $Z_{G_i} = Z_{i_0} + Z_i$ of its centre of mass, which can be assumed to be the sum of a constant term related to the static equilibrium position at standstill plus a displacement which is assumed to be small, and a roll angle ϕ_i.

The exact kinematics of the axle depends on the linkage used; however as a first approximation point G_i can be thought to remain in a plane parallel to yz plane containing the roll centre RC_i and the yaw and pitch angles of the unsprung mass can be assumed to be the same as those of the sprung mass. The relevant rotation matrices are the same as those of the sprung mass, with the difference that they now contain angle ϕ_i instead of ϕ. Although there is no difficulty to substitute $\psi + (\delta_i)_{,\phi}\phi$ to ψ to account for roll steer, the latter and the other motions of the same type of the axle give negligible contributions to the kinetic energy and then to the inertia forces.

Assuming that the roll centre RC coincides with point BW, i.e. that the axle rotates with respect to the sprung mass about point RC, the coordinates of point G_i in the reference frame of the sprung mass xyz are x_i, 0 and q_i, the latter being the distance between G_i and RC_i, positive when the former is above the latter (in Fig. 6.21 it is then negative).

The absolute position of the centre of mass of the i-th solid axle is then expressed by a relationship similar to Eq. (6.28)

$$\begin{Bmatrix} X \\ Y \\ Z \end{Bmatrix}_{G_i} = \begin{Bmatrix} X \\ Y \\ Z_0 + Z \end{Bmatrix} + [R_i] \begin{Bmatrix} x_i \\ 0 \\ q_i \end{Bmatrix} , \tag{6.33}$$

where x_i is the x coordinate of the i-th suspension and angle ϕ_i has been substituted for ϕ to obtain the rotation matrix $[R_i]$.

The value of coordinate Z_{G_i} computed through the third Eq. (6.33) can be equated to $Z_{i_0} + Z_i$, obtaining the following expression for q_i

$$q_i = Z_{i_0} - Z_0 + \theta_0 x_i + Z_i - Z + \theta x_i \; . \tag{6.34}$$

By resorting to the same type of linearization seen for the sprung mass and noting that also angle θ_0 can be considered as a small angle, the final expression of the coordinates of point G_i is

$$\begin{gathered} \begin{Bmatrix} X \\ Y \end{Bmatrix}_{G_i} = \begin{Bmatrix} X \\ Y \end{Bmatrix} + [R_1^*] \begin{Bmatrix} x_i + \theta_0 q_{i_0} + \theta q_{i_0} \\ -q_{i_0}\phi_i \end{Bmatrix} , \\ Z_{G_i} = Z_{i_0} + Z_i \; , \end{gathered} \tag{6.35}$$

where $q_{i_0} = Z_{i_0} - Z_0 + \theta_0 x_i$ is the value at rest of q_i.

By differentiating the expressions of the position of the centre of mass the components of the velocity are readily obtained. From them the kinetic energy can be computed

$$\begin{aligned} \mathcal{T}_i = \tfrac{1}{2} m_i \Bigg[& \dot{X}^2 + \dot{Y}^2 + \dot{Z}_i^2 + x_i^2 \dot{\psi}^2 + q_{i_0}^2 \dot{\phi}_i^2 + q_{i_0}^2 \dot{\theta}^2 - 2 x_i q_{i_0} \dot{\psi} \dot{\phi}_i + \\ & -2 q_{i_0} \begin{Bmatrix} \dot{X} \\ \dot{Y} \end{Bmatrix}^T [R_1^*] \begin{Bmatrix} \dot{\theta} \\ -\dot{\phi}_i \end{Bmatrix} + 2\dot{\psi} \begin{Bmatrix} \dot{X} \\ \dot{Y} \end{Bmatrix}^T \frac{\partial}{\partial \psi} [R_1^*] \begin{Bmatrix} x_i + q_{i_0}\theta_0 + q_{i_0}\theta \\ -q_{i_0}\phi_i \end{Bmatrix} \Bigg] + \\ & + \tfrac{1}{2} \left(J_{x_i} \dot{\phi}_i^2 + J_{y_i} \dot{\theta}^2 + J_{z_i} \dot{\psi}^2 \right) - J_{xz_i} \dot{\phi}_i \dot{\psi} \; . \end{aligned} \tag{6.36}$$

The coordinates of the centre of the left (right) wheel $C_{i_{l(r)}}$ in the reference frame of the axle are x_{i_C}, $\pm t/2$, z_{i_C}; in the majority of cases x_{i_C} and z_{i_C} are very small or even equal to zero (in the figure the first vanishes and the second is negative). Following the same procedure seen for the centre of mass of the axle, the linearized expression for the coordinates of point $C_{i_{l(r)}}$ is

$$\begin{aligned} \left\{ \begin{matrix} X \\ Y \end{matrix} \right\}_{C_{i_{l(r)}}} &= \left\{ \begin{matrix} X \\ Y \end{matrix} \right\} + [R_1^*] \left\{ \begin{matrix} x_i + x_{i_C} + (\theta_0 + \theta)(q_{i_0} + z_{i_C}) \\ \pm\frac{t}{2} - \phi_i(q_{i_0} + z_{i_C}) \end{matrix} \right\}, \\ Z_{C_{i_{l(r)}}} &= Z_{i_0} + z_{i_C} + Z_i \pm \phi_i \frac{t}{2} - x_{i_C}(\theta_0 + \theta) . \end{aligned} \tag{6.37}$$

The centre of the contact area $P_{i_{l(r)}}$ has the same coordinate x^* of point $C_{i_{l(r)}}$ while its coordinate y^* is slightly different from that of the centre of the wheel and coordinate Z vanishes, if the vehicle is on flat road. In the following analysis the coordinates x^* and y^* of the two points will be assumed as coinciding, as the differences due to the roll angle of the solid axle are for sure smaller than those due to the lateral compliance of the tire, and to the exact kinematics of the suspension. The position of the points of the unsprung mass can be obtained exactly once that the exact kinematic of the linkages of the suspension is known.

6.4.3 *Independent suspensions*

Owing to the large variety of solutions a general model will be used, based on the concept of roll centre. A swing arm in which points BW_i are actually materialized by hinges is sketched in Fig. 6.22 as an example. The same model can be used for any suspension if points BW_i can be considered as fixed to the sprung mass, as occurs in the case of linearized models.

The simplest choice for the generalized coordinates is that of using the angles between the arms and the horizontal plane γ_l and γ_r, which can be considered as the sum of a value γ_0 related to the static equilibrium position and a displacement γ. Note that for symmetry the two values of γ_0 have the same absolute value and opposite sign. Assuming that the linkage lies in a plane parallel to plane yz fixed to the sprung mass and linearizing as usual, the positions of the centres of mass of the left (right) unsprung mass in the xyz frame are

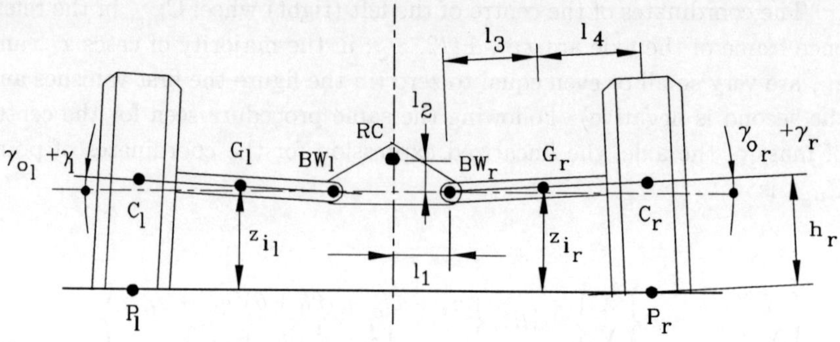

Fig. 6.22 Independent suspension: Geometrical definitions.

$$\left\{\begin{array}{c} x \\ y \\ z \end{array}\right\}_{G_{i_{l(r)}}} = \left\{\begin{array}{c} x_i \\ \pm(l_1 + l_3) \\ -l_2 \pm l_3\left(\gamma_{0i_{l(r)}} + \gamma_{i_{l(r)}}\right) \end{array}\right\}, \tag{6.38}$$

where l_2 is positive if RC lies above BW.

Instead of using angles γ as generalized coordinates, it is possible to use the coordinates Z_{i_l} and Z_{i_r} of the centres of mass, or better their sum and difference

$$\left\{\begin{array}{l} Z_{i_0} + Z_i = \dfrac{Z_{i_l} + Z_{i_r}}{2} \\ \phi_i = \dfrac{Z_{i_r} - Z_{i_l}}{d_0}, \end{array}\right. \tag{6.39}$$

where d_0 is the distance between the two centres of mass in the reference position. The linearized expressions of the coordinates Z of the centres of mass are

$$Z_{G_{i_{l(r)}}} = Z_0 + Z - \theta_0 x_i - \theta x_i \pm \phi_i(l_1 + l_3) - l_2 \pm l_3\left(\gamma_{0i_{l(r)}} + \gamma_{i_{l(r)}}\right). \tag{6.40}$$

If γ_0 is the value of angle γ of the right suspension in the reference position, by equating the expressions of the Z_i coordinate of the centre of mass given by Eqs. (6.39) and (6.40), it follows

$$\begin{aligned} \left\{ \begin{matrix} X \\ Y \end{matrix} \right\}_{G_{i_{l(r)}}} &= \left\{ \begin{matrix} X \\ Y \end{matrix} \right\} + [R_1^*] \left\{ \begin{matrix} x_i - \theta_0 l_2 - \theta l_2 \\ l_2\phi \pm \frac{d_0}{2} \end{matrix} \right\}, \\ Z_{G_{i_{l(r)}}} &= Z_{i_0} + Z_i \pm \tfrac{d_0}{2}\phi_i \,. \end{aligned} \tag{6.41}$$

The translational kinetic energy of the two suspensions can be computed by substituting a single mass located at the centre of the straight line which connects the two centres of mass, as the two masses are equal. With simple computations it is possible to show that the coordinates of this point are the same as for the solid axle, if ϕ and $-l_2$ are substituted for ϕ_i and q_{i_0} respectively. With this substitution the translational kinetic energy is also the same. For the rotational kinetic energy it is necessary to refer to the whole axle, which is made of two distinct unsprung masses. The resulting expression is

$$\begin{aligned} \mathcal{T}_i = \tfrac{1}{2}(m_{i_l} + m_{i_r}) \Bigg[& \dot{X}^2 + \dot{Y}^2 + \dot{Z}_i^2 + x_i^2\dot{\psi}^2 + l_2^2\dot{\phi}^2 + l_2^2\dot{\theta}^2 + 2x_i l_2 \dot{\psi}\dot{\phi} + \\ & -2l_2 \left\{ \begin{matrix} \dot{X} \\ \dot{Y} \end{matrix} \right\}^T [R_1^*] \left\{ \begin{matrix} \dot{\theta} \\ -\dot{\phi} \end{matrix} \right\} + 2\dot{\psi} \left\{ \begin{matrix} \dot{X} \\ \dot{Y} \end{matrix} \right\}^T \frac{\partial}{\partial \psi}[R_1^*] \left\{ \begin{matrix} x_i - l_2\theta_0 - l_2\theta \\ l_2\phi \end{matrix} \right\} \Bigg] + \\ & + \tfrac{1}{2} \Big[(m_{i_l} + m_{i_r})\tfrac{l_2^2}{8}\dot{\phi}^2 + (J_{xi_l} + J_{xi_r})\dot{\phi}_i^2 + (J_{yi_l} + J_{yi_r})\dot{\theta}^2 + \\ & + (J_{zi_l} + J_{zi_r}) + (m_{i_l} + m_{i_r})\tfrac{l_2^2}{8}\dot{\psi}^2 \Big] \,. \end{aligned} \tag{6.42}$$

The moments of inertia of the single suspensions are usually quite small and, as a consequence, the approximations here introduced have a small effect on the overall results. The terms containing the moments of inertia could be neglected.

The coordinates of the centre of the left (right) wheel $C_{i_{l(r)}}$ can be obtained from Eq. (6.38) in which $d_o/2 + l_4$ is substituted for $d_o/2$. The same holds for the centres of the contact area, even if in the present case to consider its coordinates x^* and y^* coinciding with those of the centre of the wheel leads to a more rough approximation. A partial correction can be done by substituting half of the track in the reference position $t_o/2$ for $d_o/2 + l_4$.

The equations here obtained for the swing-arm suspension hold, with a few little changes, even for the case in which points BW are at infinity.

6.4.4 *Expression of the Lagrangian function*

The Lagrangian function $\mathcal{L} = \mathcal{T} - \mathcal{U}$ of the whole vehicle is easily obtained. The total kinetic energy is

$$\begin{aligned}\mathcal{T} = &\tfrac{1}{2}m\left(\dot{X}^2 + \dot{Y}^2\right) + \tfrac{1}{2}m_S\dot{Z}^2 + \tfrac{1}{2}\textstyle\sum_{\forall i} m_i\dot{Z}_i^2 + \tfrac{1}{2}J_x\dot{\phi}^2 + \tfrac{1}{2}J_y\dot{\theta}^2 + \\ &+\tfrac{1}{2}J_z\dot{\psi}^2 + \tfrac{1}{2}\textstyle\sum_{\forall i} J^*_{x_i}\dot{\phi}_i^2 - J_{xz}\dot{\psi}\dot{\phi} - \dot{\psi}\sum_{\forall i} J^*_{xz_i}\dot{\phi}_i + \\ &+\left[\dot{X}\cos(\psi) + \dot{Y}\sin(\psi))\right]\left[J_s\dot{\theta} + \dot{\psi}\left(hm_S\phi + \textstyle\sum_{\forall i} m_i q_{i_0}\phi_i\right)\right] + \\ &+\left[\dot{X}\sin(\psi) - \dot{Y}\cos(\psi))\right]\left(m_S h\dot{\phi} + \textstyle\sum_{\forall i} m_i q_{i_0}\dot{\phi}_i\right) - m_S c\dot{Z}\dot{\theta}\,,\end{aligned} \tag{6.43}$$

where

$$\begin{aligned}m &= m_S + \textstyle\sum_{\forall i} m_i \\ J_x &= J_{x_S} + m_S h^2 + \textstyle\sum_{\forall i_I} J_{x_i} \\ J_y &= J_{y_S} + m_S\left(h^2 + c^2\right) + \textstyle\sum_{\forall i}\left(J_{y_i} + m_i q_{i_0}^2\right) \\ J_z &= J_{z_S} + m_S c^2 + \textstyle\sum_{\forall i}\left(J_{z_i} + m_i x_{i_0}^{*2}\right) \\ J_{xz} &= J_{xz_S} + m_S ch \\ J^*_{x_i} &= J_{x_i} + m_i q_{i_0}^2 \\ J^*_{xz_i} &= J_{xz_i} + m_i q_{i_0} x_i \\ J_s &= m_S h + \textstyle\sum_{\forall i} m_i q_{i_0}\end{aligned}$$

and the sums with suffix i_I are extended only to the axles with independent suspensions.

The gravitational potential energy on level road is readily computed once the Z coordinates of the centres of mass are known

$$\mathcal{U}_g = gm_S\left[Z + h\cos\left(\theta_0 + \theta\right)\cos(\phi) - c\sin\left(\theta_0 + \theta\right)\right] + g\sum_{\forall i} m_i Z_i\,. \tag{6.44}$$

Note that it is not possible to linearize the cosines in the expression of the potential energy: To include all terms leading to linear terms in the equations of motion the quadratic term must also be retained. As all constant terms can be dropped, the gravitational potential energy can be written in the form

$$\mathcal{U}_g = gm_S \left\{ Z - \frac{h}{2} \left[(\theta_0 + \theta)^2 + \phi^2 \right] - c(\theta_0 + \theta) \right\} + g \sum_{\forall i} m_i Z_i \,. \tag{6.45}$$

The potential energy due to the springs of the suspensions can easily be computed since the position of the points in which each spring is attached to the sprung and the unsprung masses are known. However it is easier and more consistent with the linearized nature of the present model to define a vertical and torsional stiffness of each suspension, K_i and χ_i, which in general are functions of the static equilibrium position of the system. The elastic elements of each suspension can thus be substituted by a linear spring connecting point RC, which is fixed to the sprung mass, and a point attached to the axle which in the static equilibrium condition coincides with RC and by a torsional spring which reacts to the relative rotations of the sprung and the unsprung masses. The former has a length at rest L_i which enables it to exert a force equal to the static equilibrium force F_{0_i} ($L_i = F_{0_i}/K_i$).

The potential energy due to the i-th suspension is thus

$$\mathcal{U}_{s_i} = \frac{1}{2} K_i \left(Z_i - Z + x_{s_i}\theta + L_i \right)^2 + \frac{1}{2} \chi_i \left(\phi - \phi_i \right)^2 \,, \tag{6.46}$$

where x_{s_i} is the x coordinate at which the spring system is located. It can be different from that of the centre of mass of the unsprung mass.

This expression of the potential energy is immediate for solid axle suspensions but holds also, within the limits of a linearized model, for independent suspensions provided that the values of the stiffness are correctly computed. Note that length L_i has a physical meaning only in the case of a true linear spring directed in vertical direction.

The potential energy due to the deformation of the tires can be computed exactly in the same way seen for the springs. For the left (right) tire of the i-th suspension of a vehicle moving on a flat road it follows

$$\mathcal{U}_{t_{il(r)}} = \frac{1}{2} P_i \left(Z_i \pm \frac{t}{2} \phi_i - L_{t_i} \right)^2 \,, \tag{6.47}$$

where P_i and L_{t_i} are respectively the vertical stiffness about the static equilibrium position and the length corresponding to the static load Z_i ($L_{t_i} = Z_i/P_i$). Note that in the expressions above the length x_{i_C} and z_{i_C} have been neglected, i.e. the centre of the wheel has been assumed to lie on the y axis of the unsprung mass. If this is not the case a term containing θ should be present, but its smallness allows one to neglect it.

The total potential energy of the tires of a suspension with two wheels is then

$$\mathcal{U}_{t_i} = P_i\left(Z_i^2 + \frac{t^2}{4}\phi_i^2 + L_{t_i}^2 - 2Z_iL_{t_i}\right). \tag{6.48}$$

From the expressions of the kinetic and potential energies of the various types the Lagrangian function is immediately computed.

6.4.5 *Rayleigh dissipation function*

The suspensions are provided with shock absorbers which can be modelled as linear viscous dampers, at least as a first approximation. Operating in the same way as for the springs, it is possible to linearize their behaviour and to substitute the shock absorbers of each axle with a linear damper with damping coefficient c_i and a damper for rotations with coefficient Γ_i. The Rayleigh dissipation function can be readily computed from the velocities of the same points whose displacements have been used in the computation of the potential energies

$$\mathcal{F}_{s_i} = \frac{1}{2}c_i\left(\dot{Z}_i - \dot{Z} + x_{d_i}\dot{\theta}\right)^2 + \frac{1}{2}\Gamma_i\left(\dot{\phi} - \dot{\phi}_i\right)^2, \tag{6.49}$$

where the position of the dampers x_{d_i} can be different from that of the springs x_{s_i}.

Note that the linearization of the shock absorber is less correct than that of the springs when the damping coefficient in the downward stroke (jounce) is different from that in the upward stroke (rebound): Even during the small oscillations the damping coefficient varies continuously. It will be shown later that it is nevertheless possible to find an average linear damping also for this case.

Following the same lines the Rayleigh dissipation function of the tires can be obtained

$$\mathcal{F}_{t_i} = c_{t_i}\left(\dot{Z}_i^2 + \frac{t^2}{4}\dot{\phi}_i^2\right), \tag{6.50}$$

where c_{t_i} is the equivalent damping coefficient of the tires.

6.4.6 *Sideslip angles of the wheels*

The sideslip angles of the wheels are readily obtained by computing the components of the velocity of the centres of the contact areas in the x^*y^*z reference frame

$$\{V_{P_{i_{l(r)}}}\}_{x^*y^*z} = \left\{ \begin{array}{c} u + \dot{\theta}\left(q_{i_0} + z_{i_C}\right) - \dot{\psi}\left[\pm\frac{t}{2} - \phi_i\left(q_{i_0} + z_{i_C}\right)\right] \\ v + \dot{\phi}_i\left(q_{i_0} + z_{i_C}\right) + \dot{\psi}\left[x_{w_i} + \left(\theta + \theta_0\right)\left(q_{i_0} + z_{i_C}\right)\right] \\ 0 \end{array} \right\}, \tag{6.51}$$

where u and v are the components of the velocity of the vehicle in x^* and y^* directions and $x_{w_i} = x_i + x_{i_C}$ is the x coordinate of the i-th wheel.

As the mean plane of the wheel forms with x^*z plane an angle equal to the steering angle δ_i, possibly added to the roll-steer angle $(\delta_i)_{,\phi}\phi$, with the usual linearizations it follows

$$\alpha_{i_{l(r)}} = \frac{v}{V} + \dot{\psi}\frac{x_{w_i}}{V} - \dot{\phi}_i\frac{\left(q_{i_0} + z_{i_C}\right)}{V} - \delta_i - (\delta_i)_{,\phi}\phi\,. \tag{6.52}$$

As it was the case for the rigid vehicle, the two wheels of the same axle have the same sideslip angle: Again linearization allows working in terms of axles instead of wheels. Note also that the term in $\dot{\phi}_i$ is usually very small; in the following equations it will be neglected and the sideslip angle will be expressed as

$$\alpha_{i_{l(r)}} = \frac{v}{V} + \dot{\psi}\frac{x_{w_i}}{V} - \delta_i - (\delta_i)_{,\phi}\phi\,. \tag{6.53}$$

Apart from roll steer it coincides with the expression obtained for the rigid vehicle.

Equation (6.53) has been obtained with reference to a solid axle suspension but, within the simplifications used here, holds also in the case of independent suspensions.

6.4.7 *Generalized forces*

The generalized forces Q_i which must be introduced into the equations of motion only include the forces due to the tires and aerodynamic forces, besides external forces which can be applied to the vehicle.

The virtual displacement of the centre of the left (right) wheel of the i-th axle has an expression close to Eq. (6.51)

$$\{\delta s_{Pi_{l(r)}}\}_{x^*y^*z} = \left\{ \begin{array}{c} \delta x^* + \delta\theta\,(q_{i_0} + z_{iC}) - \delta\psi\left[\pm\frac{t}{2} - \phi_i\,(q_{i_0} + z_{iC})\right] \\ \delta y^* + \delta\phi_i\,(q_{i_0} + z_{iC}) + \delta\psi\left[x_{w_i} + (\theta + \theta_0)\,(q_{i_0} + z_{iC})\right] \\ 0 \end{array} \right\} . \tag{6.54}$$

If the coupling between the vertical and longitudinal motion of the suspension must be taken into account, a term

$$\left(\frac{\partial x}{\partial z}\right)_i (\delta Z - x_{s_i}\delta\theta) = (x_i)_{,z}(\delta Z - x_{s_i}\delta\theta)$$

must be added to the x^* component of the virtual displacement. If

$$F_x{}^* = F_{xt}\cos\left[\delta_i - (\delta_i)_{,\phi}\phi\right] - F_{y_t}\sin\left[\delta_i - (\delta_i)_{,\phi}\phi\right]$$

$$F_y{}^* = F_{xt}\sin\left[\delta_i - (\delta_i)_{,\phi}\phi\right] + F_{y_t}\cos\left[\delta_i - (\delta_i)_{,\phi}\phi\right]$$

are the forces exerted by the tire in direction of the axes x^* and y^*, the expression of the virtual work is

$$\begin{array}{c} \delta\mathcal{L}_i = \delta x^* F_x{}^* + \delta y^* F_y{}^* + \delta Z (x_i)_{,z} F_x{}^* + \delta\theta F_x{}^*\,(q_{i_0} + z_{iC} - (x_i)_{,z} x_{s_i} F_x{}^*) + \\ -\delta\phi_i F_y{}^*\,(q_{i_0} + z_{iC}) + \\ +\delta\psi\left\{-F_x{}^*\left[\pm\frac{t}{2} - \phi_i\,(q_{i_0} + z_{iC})\right] + F_y{}^*\left[x_{w_i} + (\theta + \theta_0)\,(q_{i_0} + z_{iC})\right] + M_z\right\} . \end{array} \tag{6.55}$$

The generalized forces are then obtained by differentiating the virtual work with respect to the virtual displacements δx^*, δy^*, $\delta\theta$, etc. Note that the first two generalized forces are the actual forces directed along axes x^* and y^*; if the components along the inertial axes are needed a suitable rotation matrix can be used.

In the case of independent suspensions instead of $\delta\phi_i$ the expression of the virtual work contains $\delta\phi$, q_{i_0} and t must be substituted by $-l_2$ and the track in the reference conditions t_0.

The force $F_{y_{t_i}}$ on the i-th tire can be expressed by a linearized function of the sideslip and the camber angles α_i and $\gamma_{0_i} + (\gamma_i)_{,\phi}\phi + (\gamma_i)_{,\phi_i}\phi_i + (\gamma_i)_{,z}(Z - Z_i)$. As in a linearized model the terms γ_{0_i} and $(\gamma_i)_{,z}(Z - Z_i)$

vanish as they cause equal forces with opposite directions on the wheels of the same axle, the cornering force acting on each axle can be expressed as

$$F_{y_{t_i}} = -C_i\alpha_i + C_{\gamma_i}\left[(\gamma_i)_{,\phi}\phi + (\gamma_i)_{,\phi_i}\phi_i\right] \ , \tag{6.56}$$

where both C_i and C_{γ_i} are referred to the whole axle.

The aerodynamic forces are considered as applied to the centre of mass of the sprung mass. The virtual displacement of such point in x^*y^*z reference frame is

$$\{\delta s_{G_S}\}_{x^*y^*z} = \left\{\begin{array}{c} \delta x^* + h\delta\theta + h\phi\delta\psi \\ \delta y^* - h\delta\phi + (c + h\theta_0 + h\theta)\,\delta\psi \\ \delta Z - c\delta\theta \end{array}\right\} . \tag{6.57}$$

The virtual work due to aerodynamic forces and moments is thus

$$\begin{aligned} \delta\mathcal{L}_{aer} = {} & F_{x_{aer}}\delta x^* + F_{y_{aer}}\delta y^* + (F_{x_{aer}}h - F_{z_{aer}}c)\,\delta\theta + \\ & + \left[F_{x_{aer}}h\phi + F_{y_{aer}}(c + h\theta_0 + h\theta)\right]\delta\psi + \\ & - F_{y_{aer}}h\delta\phi + F_{z_{aer}}\delta Z + M_{y_{aer}}\left[\delta\phi - (\theta_0 + \theta)\delta\psi\right] + \\ & + M_{y_{aer}}(\delta\theta + \phi\delta\psi) + M_{z_{aer}}(\delta\psi + \phi\delta\theta) \ . \end{aligned} \tag{6.58}$$

Again the generalized forces are readily obtained by differentiating the virtual work with respect to the virtual displacements.

Note that, owing to the linearization of the model, force $F_{x_{aer}}$ can be assumed to be constant while $F_{y_{aer}}$, $M_{x_{aer}}$ and $M_{z_{aer}}$ can be considered as linear in angle β_a (if there is no crosswind, in β), while $F_{z_{aer}}$ and $M_{y_{aer}}$ can be considered as linear functions of angle θ.

As the aerodynamic forces have been applied in the centre of mass of the sprung mass instead of in that of the vehicle, the terms $(h\theta_0 + c)C_y + lC_{M_z}$ and $lC_{M_y} - cC_z$ included in the expressions of the aerodynamic forces are better substituted by lC'_{M_z} and lC'_{M_y}, where prime indicates the coefficients defined is the usual way.

6.4.8 *Equations of motion*

The equations of motion are then

$$\frac{d}{dt}\left(\frac{\partial\mathcal{L}}{\partial\dot{q}_i}\right) - \frac{\partial\mathcal{L}}{\partial q_i} + \frac{\partial\mathcal{F}}{\partial\dot{q}_i} = Q_i \ . \tag{6.59}$$

The derivatives needed to write the first two equations of motion, those for degrees of freedom X and Y, are

$$\frac{\partial \mathcal{L}}{\partial \dot{X}} = m\dot{X} + A\cos(\psi) - B\sin(\psi)$$

$$\frac{\partial \mathcal{L}}{\partial \dot{Y}} = m\dot{Y} + A\sin(\psi) + B\cos(\psi) \qquad (6.60)$$

$$\frac{\partial \mathcal{L}}{\partial X} = \frac{\partial \mathcal{L}}{\partial Y} = \frac{\partial \mathcal{F}}{\partial \dot{X}} = \frac{\partial \mathcal{F}}{\partial \dot{Y}} = 0\,,$$

where

$$\begin{array}{l} A = J_s\dot{\theta} + \dot{\psi}\left(hm_S\phi + \sum_{\forall i} m_i q_{i_0}\phi_i\right) \\ B = -m_S h\dot{\phi} - \sum_{\forall i} m_i q_{i_0}\dot{\phi}_i\,. \end{array}$$

By operating in the same way as seen for the rigid vehicles, and expressing the equations in x^*y^*Z instead of in XYZ frame and linearizing, the following equations of motion are readily obtained

$$\begin{cases} m\dot{V} + \ddot{\theta}J_s = Q_{x^*} \\ m\left(\dot{v} + V\dot{\psi}\right) - m_S h\ddot{\phi} - \sum_{\forall i} m_i q_{i_0}\ddot{\phi}_i = Q_{y^*}\,. \end{cases} \qquad (6.61)$$

The expressions of the generalized forces are then

$$Q_{x^*} = \sum_{\forall i} F_{xi} - \frac{1}{2}\rho V^2 S C_x \qquad (6.62)$$

and

$$Q_{y^*} = Y_\beta\beta + Y_{\dot{\psi}}\dot{\psi} + Y_\phi\phi + \sum_{\forall i} Y_{\phi_i}\phi_i + Y_\delta\delta + F_{y_e}\,, \qquad (6.63)$$

where

$$\begin{cases} Y_{\beta} = -\sum_{\forall i} C_i + \frac{1}{2}\rho V_a^2 S(C_y)_{,\beta} \\ Y_r = -\frac{1}{V}\sum_{\forall i} x_i C_i \\ Y_{\phi} = \sum_{\forall i} C_i(\delta_i)_{,\phi} + \sum_{\forall i} C_{\gamma_i}(\gamma_i)_{,\phi} \\ Y_{\phi_i} = C_{\gamma_i}(\gamma_i)_{,\phi} \\ Y_{\delta} = \sum_{\forall i} K_i' C_i + \sum_{\forall i} K_i' F_{x i_t} \, . \end{cases} \tag{6.64}$$

The equation of motion for Z degree of freedom is also readily obtained by differentiating the expression of the Lagrangian function. With simple computations it follows

$$m_S\ddot{Z} - m_S c\ddot{\theta} + \sum_{\forall i} c_i\dot{Z} - \sum_{\forall i} c_i\dot{Z}_i - \sum_{\forall i} c_i x_{d_i}\dot{\theta} + \sum_{\forall i} K_i Z - \sum_{\forall i} K_i Z_i + \\ -\sum_{\forall i} K_i x_{s_i}\theta = Z_{\theta}\theta + \frac{1}{2}\rho V^2 S(C_Z)_{,\theta}\theta_0 + \sum_{\forall i}\left(K_i L_i + (x_i)_{,z}F_{x_i}\right) - m_S g \, , \tag{6.65}$$

where $Z_{\theta} = \frac{1}{2}\rho V^2 S(C_z)_{,\theta}$.

The fourth equation of motion, that for ψ degree of freedom, is obtained in the same way. Its linearized expression is

$$J_z\ddot{\psi} - J_{xz}\ddot{\phi} + \dot{V}\left(m_S h\phi + \sum_{\forall i} m_i q_{io}\phi_i\right) - \sum_{\forall i} J_{xz_i}\ddot{\phi}_i = \\ = N_{\beta}\beta + N_{\dot{\psi}}\dot{\psi} + N_{\phi}\phi + \sum_{\forall i} N_{\phi_i}\phi_i + N_{\delta}\delta + M_{ze} \, , \tag{6.66}$$

where

$$\begin{cases} N_\beta = -\sum_{\forall i} x_{w_i} C_i + \sum_{\forall i} (M_{z_i})_{,\alpha} + \frac{1}{2}\rho V_a^2 S l (C'_{M_z})_{,\beta} \\ N_{\dot{\psi}} = \frac{1}{V} \sum_{\forall i} \left[x_{w_i}^2 C_i + x_{w_i} (M_{z_i})_{,\alpha} \right] \\ N_\phi = \left[\sum_{\forall i} x_{w_i} C_i (\delta_i)_{,\phi} + x_{w_i} C_{\gamma_i} (\gamma_i)_{,\phi} - (M_{z_i})_{,\alpha} (\delta_i)_{,\phi} \right] - \frac{1}{2}\rho V_a^2 S h C_x \\ N_{\phi_i} = x_{w_i} C_{\gamma_i} (\gamma_i)_{,\phi} + \left(f_0 + K V^2\right) P_i \frac{t_i^2}{2} \\ N_\delta = \sum_{\forall i} \left[x_{w_i} K'_i C_i + x_{w_i} K'_i F_{x i_t} - (M_{z_i})_{,\alpha} \right] . \end{cases} \tag{6.67}$$

The fifth equation of motion, that for θ degree of freedom, is

$$\begin{aligned} & J_y \ddot{\theta} + J_s \dot{V} - m_S c \ddot{Z} - \sum_{\forall i} c_i x_{d_i} \dot{Z} + \sum_{\forall i} c_i^2 x_{d_i} \dot{\theta} + \sum_{\forall i} c_i x_{d_i} \dot{Z}_i - \sum_{\forall i} K_i x_{d_i} Z + \\ & + \sum_{\forall i} K_i x_{s_i}^2 \theta + \sum_{\forall i} K_i x_{d_i} Z_i - m_S g h \theta = - \sum_{\forall i} \left[K_i x_{d_i} L_i + (x_i)_{,z} x_{s_i} F_{x_i} \right] + \\ & + M_\theta \theta - \frac{1}{2}\rho V^2 S \left[h C_x - (C'_{M_y})_{,\theta} \theta_0 \right] + \sum_{\forall i} F_{xi} (q_{i_0} + z_{iC}) + m_S g \left(h \theta_0 + c\right) , \end{aligned} \tag{6.68}$$

where $M_\theta = \frac{1}{2}\rho V_a^2 S (C'_{M_y})_{,\theta}$.

The linearized expression of the sixth equation of motion, that for ϕ degree of freedom, is

$$\begin{aligned} J_x \ddot{\phi} - J_{xz} \ddot{\psi} - m_S h \left(\dot{v} + V \dot{\psi} \right) = L_\beta \beta + L_\phi \phi + L_{\dot{\phi}} \dot{\phi} + \\ + \textstyle\sum_{\forall i} L_{\phi_i} \phi_i + \sum_{\forall i} L_{\dot{\phi}_i} \dot{\phi}_i , \end{aligned} \tag{6.69}$$

where

$$\begin{cases} L_\beta = \frac{1}{2}\rho V_a^2 S \left[h (C_y)_{,\beta} + t (C_{M_x})_{,\beta} \right] \\ L_{\dot{\phi}} = -\sum_{\forall i} \Gamma_i \\ L_\phi = -\sum_{\forall i} \chi_i + m_S g h \\ L_{\dot{\phi}_i} = \Gamma_i \\ L_\phi = \chi_i . \end{cases} \tag{6.70}$$

The n equations for the degrees of freedom z_i' are

$$\begin{aligned}m_i\ddot{Z}_i+\left(c_i+2c_{P_i}\right)\dot{Z}_i-c_i\dot{Z}+c_ix_{d_i}\dot{\theta}-K_iZ+\\+\left(K_i+2P_i\right)Z_i+K_ix_{d_i}\theta=-K_ix_{d_i}\theta_0-K_iL_i+2P_iL_{t_i}-m_ig\,.\end{aligned}\tag{6.71}$$

Finally, the n equations for the degrees of freedom ϕ_i are

$$\begin{aligned}J^*_{x_i}\ddot{\phi}_i-J^*_{xz_i}\ddot{\psi}-m_iq_{i_0}\left(\dot{v}+V\dot{\psi}\right)=Li_\beta\beta+L_{i_{\dot{\psi}}}\dot{\psi}+L_{i_\phi}\phi+L_{i_{\dot{\phi}}}\dot{\phi}+\\+L_{i_{\phi_i}}\phi_i+L_{i_{\dot{\phi}_i}}\dot{\phi}_i\,,\end{aligned}\tag{6.72}$$

where

$$\begin{cases}L_{i_\beta}=\left(q_{i_0}+z_{i_C}\right)C_i\\ L_{i_{\dot{\psi}}}=\left(q_{i_0}+z_{i_C}\right)C_i\dfrac{x_{w_i}}{V}\\ L_{i_{\dot{\psi}}}=L_{\dot{\psi}_i}\\ L_{i_\phi}=\chi_i-q_i'\left[C_i(\delta_i)_{,\phi}+C_{\gamma_i}(\gamma_i)_{,\phi}\right]\\ L_{i_{\dot{\phi}_i}}\dot{\phi}_i=-\Gamma_i-c_{P_i}\dfrac{t_i^2}{2}\\ L_{i_{\phi_i}}\phi_i=-q_i'C_{\gamma_i}(\gamma_i)_{,\phi}-\chi_i-P_i\dfrac{t_i^2}{2}\,.\end{cases}\tag{6.73}$$

6.5 Uncoupling between handling and ride

The $(6+2n)$ equations of motion obtained in the previous section constitute a set of linear second-order differential equations, even if the order of such set is only $(9+4n)$ as three of the unknown, namely x^*, y^* and ψ, are present only with their derivatives V, v and $\dot{\psi}$.

However, a detailed examination of such equations shows clearly that, if the speed V of the vehicle, which in the linearized model can be confused with its component u along x^* axis, is a known function of time, the equations form two completely uncoupled sets of $(3+n)$ equations each.

The first set contains only the generalized coordinates y^*, ψ, ϕ and ϕ_i: As a consequence it deals with the lateral behaviour of the vehicle, or, as is usually said, its handling.

The second set contains the generalized coordinates x^*, Z, θ and Z_i, dealing with the "suspension motion" of the vehicle, or better with its ride behaviour. This set can be further uncoupled by separating the first equation, that regarding x^* coordinate, i.e. dealing with the longitudinal dynamics of the vehicle, and the following $(2+n)$ containing coordinates Z, θ and Z_i which allow the study of the ride comfort in a proper sense.

This uncoupling is an interesting result, even if it is strictly linked with a number of assumptions and, as a consequence, becomes inapplicable if one of them is dropped. The first assumption is the existence of a plane of symmetry, the xz plane. Usually the lack of inertial symmetry of the structure and the differences between the characteristics of the individual springs and shock absorbers located at opposite sides of the vehicle are small enough to be neglected. However, it can happen that the payload of the vehicle is located in an unsymmetrical way, leading to a position of the centre of mass outside the symmetry plane and to nonvanishing moments of inertia J_{xy} and J_{yz}.

A second assumption is that of a perfect linearity of the behaviour of the springs and shock absorbers. The linearity of the elastic behaviour of springs and tires is an acceptable assumption in the motion about any equilibrium position, provided that its amplitude is small enough. On the contrary, the nonlinearity of the shock absorbers can be a factor which cannot be neglected even in the motion "in the small" if their force-velocity characteristic is unsymmetrical, as in the jounce and rebound movements they act with different damping coefficient even if the amplitude of the motion tends to zero. This issue will be dealt with in detail in Sec. 6.14.

A third assumption regards angles β, α_i, θ_0, θ, ϕ and ϕ_i, which must be small enough to allow the linearization of their trigonometric functions. This assumption holds only for small displacements from the equilibrium position and depends also on the characteristics of the vehicle: The harder the suspensions, the more extended is the range in which the uncoupling assumption holds. However, in general the mentioned angles are small enough in all normal driving conditions, except for vehicles with two wheels which can work with large roll angles. On the contrary, the linearization of the tire behaviour is not strictly required for uncoupling: Even if nonlinear laws $F_y(\alpha)$, $F_y(\gamma)$, $M_z(\alpha)$, etc. are introduced into the equations of motion, the two sets of equations for handling and ride would remain uncoupled, although nonlinear. This last statement is important, as the linear model for the behaviour of the tires holds only for values of angles α and γ which are far smaller than those for which the trigonometric functions can be linearized.

Some assumptions have been made on the modelling of the suspensions, which are better suited for solid axles than for independent suspensions. While the unavoidable kinematic errors cannot be accounted for in this way, it will be shown that this does not affect the uncoupling. On the contrary, the interaction between cornering forces and loads in x and z direction on the tires should actually couple all equations. If the same approximated approach seen for rigid vehicles is also adopted in the present case, however, it is possible to resort to uncoupled equations.

The uncoupled model, even if represents only a first approximation, is important for two reasons. Firstly, it shed some light on the actual behaviour of road vehicles and gives a theoretical foundation to the practice of using separate approximate models for the study of handling and ride characteristics of road vehicles. Secondly, simple linearized models, which allow closed form solutions, are well suited for optimization and parametric studies.

Clearly if detailed numerical simulations are performed there is no need to uncouple the equations and comprehensive, detailed nonlinear models can be used; the limit in this case can well be the unavailability of good estimates of the numerical values of many parameters which must be entered into the equations.

6.6 Handling of a vehicle on elastic suspensions

The explicit formulation of the mathematical model for the handling of a vehicle with two axles is then

$$[M]_1\{\ddot{q}\}_1 + [C]_1\{\dot{q}\}_1 + [K]_1\{q\}_1 = \{F\}_1 \,, \tag{6.74}$$

where

$$\{q\}_1 = \left[y^* \; \psi \; \phi \; \phi_1 \; \phi_2 \right]^T$$

$$[M]_1 = \begin{bmatrix} m & 0 & -m_S h & -m_1 q_{1_0} & -m_2 q_{2_0} \\ & J_S & -J_{xz} & -J^*_{xz_1} & -J^*_{xz_2} \\ & & J_x & 0 & 0 \\ & & & J^*_{x_1} & 0 \\ \text{symm.} & & & & J^*_{xz_2} \end{bmatrix}$$

$$[C]_1 = \begin{bmatrix} -\frac{Y_\beta}{V} & mV - Y_\psi & 0 & 0 & 0 \\ -\frac{N_\beta}{V} & -N_\psi & 0 & 0 & 0 \\ -\frac{L_\beta}{V} & -m_S h V & -L_\phi & -L_{\phi_1} & -L_{\phi_2} \\ -\frac{L_{1_\beta}}{V} & -m_1 q_{1_0} V - L_{1_\psi} & -L_{1_\phi} & -L_{1_{\phi_1}} & 0 \\ -\frac{L_{2_\beta}}{V} & -m_1 q_{2_0} V & L_{2_\psi} & -L_{1_\phi} & 0 & -L_{2_{\phi_2}} \end{bmatrix}$$

$$[K]_1 = \begin{bmatrix} 0 & 0 & -Y_\phi & -Y_{\phi_1} & -Y_{\phi_2} \\ 0 & 0 & m_S h V - N_\phi & m_1 q_{1_0} V - N_{\phi_2} & m_2 q_{2_0} V - N_{\phi_2} \\ 0 & 0 & -L_\phi & -L_{\phi_1} & -L_{\phi_2} \\ 0 & 0 & -L_{1_\phi} & -L_{1_{\phi_1}} & 0 \\ 0 & 0 & -L_{2_\phi} & 0 & -L_{1_{\phi_1}} \end{bmatrix}$$

$$\{F\}_1 = \delta \left[Y_\delta \; N_\delta \; 0 \; L_{1_\delta} \; L_{2_\delta} \right]^T + \left[F_{y_e} \; M_{z_e} \; 0 \; 0 \; 0 \right]^T .$$

As already stated, coordinates y^* and ψ are present only with their derivatives: The order of the set of differential equations is then 8 instead of 10. If a state-space approach is used (Eq. (5.68)), by introducing the state variables $p = \dot{\phi}$, $p_1 = \dot{\phi}_1$, $p_2 = \dot{\phi}_2$ and $r = \dot{\psi}$, the relevant vectors and matrices are:

-State vector

$$\{z\} = \left[v \; r \; p \; p_1 \; p_2 \; \phi \; \phi_1 \; \phi_2 \right]^T , \tag{6.75}$$

-dynamic matrix

$$[A] = \begin{bmatrix} -[M]_1^{-1}[C]_1 & -[M]_1^{-1}[K]_1^* \\ \begin{bmatrix} 0 & 0 & 1 & 0 & 0 \\ 0 & 0 & 0 & 1 & 0 \\ 0 & 0 & 0 & 0 & 1 \end{bmatrix} & \begin{bmatrix} 0 & 0 & 0 \\ 0 & 0 & 0 \\ 0 & 0 & 0 \end{bmatrix} \end{bmatrix} , \tag{6.76}$$

where $[K]_1^*$ is matrix $[K]_1$ in which the first two columns have been neglected.

-Input gain matrix

$$[B] = \begin{bmatrix} -[M]_1^{-1} \begin{bmatrix} Y_\delta & 1 & 0 \\ N_\delta & 0 & 1 \\ L_\delta & 0 & 0 \\ L_{1_\delta} & 0 & 0 \\ L_{2_\delta} & 0 & 0 \end{bmatrix} \\ [0]_{3\times 3} \end{bmatrix} , \tag{6.77}$$

-input vector

$$\{z\} = \begin{bmatrix} \delta \; F_{y_e} \; M_{z_e} \end{bmatrix}^T . \tag{6.78}$$

This approach can be used for the study of the stability of the vehicle or for computing the response of the vehicle to the various inputs, exactly in the same way seen for the rigid vehicle models. This model is only marginally more complex if numerical solution are searched.

Even if the complexity of the model is not an actual problem it is interesting to perform a simplification which allows one to reduce its size without sacrificing its applicability to real problems. As the stiffness of the tires in z direction is much higher than that of the suspensions, their compliance becomes important only in high frequency motions, much higher than the frequencies involved in the handling of the vehicle. As a consequence, if the compliance of the tire is neglected, which amounts to state that ϕ_1 and ϕ_2 and their derivatives are vanishingly small, the model reduces to a set of three equations (four first order equations in the state-space approach) which retains most of the features of the complete, five equations set. Note that again this statement is more correct in the case of solid axles than in that of independent suspensions: In the latter case it amounts to confuse distance d_0 between the centres of mass of the suspensions with the track t.

Since products like $P_i\phi_i$, which do not vanish when ϕ_i tends to zero and P_i to infinity, are present in the terms in N_{ϕ_1} and N_{ϕ_2} in the second equation, it is impossible to simply neglect all terms containing the roll angles of the axles. The load transfer between the wheels of the same axle can be evaluated as seen for the rigid vehicle, leading to a simple evaluation of the product $P_i\phi_i = 2\Delta Z_i/t_i$. By writing an equilibrium equation of the i-th axle about the roll centre RC$_i$ and neglecting its moment of inertia, it follows

$$P_i\phi_i = \frac{t_i^2}{2}\left[-\phi\chi_i - \dot{\phi}\Gamma_i - F_{y_i}\left(z_{i_0} + q_{i_0}\right)\right] . \tag{6.79}$$

As the stability derivatives N_i can be modified accordingly, introducing the variables $\beta = v/V$ and $r = \dot{\psi}$, the three equations of motion become

$$\begin{cases} mV\left(\dot{\beta}+r\right) - m_S h\ddot{\phi} = \left(Y_\beta - m\dot{V}\right)\beta + Y_{\dot{\psi}}r + Y_\phi\phi + Y_\delta\delta + F_{y_e} \\ J_z\dot{r} - J_{xz}\ddot{\phi} - m_S h\dot{V}\phi = N_\beta\beta + N_{\dot{\psi}}r + N_\phi\phi + N_\delta\delta + M_{z_e} \\ J_x\ddot{\phi} - m_S hV\left(\dot{\beta}+r\right) - J_{xz}\dot{r} = \left(L_\beta + m_S h\dot{V}\right)\beta + L_{\dot{\phi}}\dot{\phi} + L_\phi\phi , \end{cases} \tag{6.80}$$

where

$$\begin{cases} N_\beta = \sum_{\forall i}[-x_{w_i}C_i + (M_{z_i}),_\alpha] + \frac{1}{2}\rho V_a^2 Sl(C'_{M_z}),_\beta + \\ \quad +(f_0 + KV^2)\sum_{\forall i} C_i(z_{i_0} + q_{i_0}) \\ N_{\dot{\psi}} = \frac{1}{V}\left\{\sum_{\forall i}\left[x_{w_i}^2 C_i + x_{w_i}(M_{z_i}),_\alpha\right] + (f_0 + KV^2)\sum_{\forall i} x_{w_i}C_i(z_{i_0} + q_{i_0})\right\} \\ N_\phi = \sum_{\forall i}\left[x_{w_i}C_i(\delta_i),_\phi + x_{w_i}C_{\gamma_i}(\gamma_i),_\phi - (M_{z_i}),_\alpha(\delta_i),_\phi\right] - \frac{1}{2}\rho V_a^2 ShC_x + \\ \quad +(f_0 + KV^2)\sum_{\forall i}[C_i(z_{i_0} + q_{i_0})(\delta_i),_\phi + (z_{i_0} + q_{i_0})(Fy_i),_\gamma(\gamma_i),_\phi - \chi_i] \\ N_{\dot{\psi}} = -(f_0 + KV^2)\sum_{\forall i}\Gamma_i \\ N_\delta = \sum_{\forall i}[x_{w_i}K_i'C_i + x_{w_i}K_i'F_{xi_t} - (M_{z_i}),_\alpha] + \\ \quad +(f_0 + KV^2)\sum_{\forall i} C_i(z_{i_0} + q_{i_0}) . \end{cases} \tag{6.81}$$

Note that if the terms in $\dot{V}$ are dropped, the same set of equations frequently described in the literature[4] is obtained. There is however a difference: This model is here obtained from the complete model of the vehicle with elastic suspensions through uncoupling and controlled simplifications, while often is described as a model obtained through a number of more or less arbitrary assumptions.

---

[4]See for example W. Steeds, *Mechanics of Road Vehicles*, ILIFFE & Sons, London, 1960.

The study of either the stability or of the response to a steering input or to an external force or moment is straightforward and follows closely the same lines seen for the rigid vehicle. Note that here the presence of an equation containing the first and second derivative of a generalized coordinate ϕ together with the coordinate itself can induce an oscillatory behaviour. If the roll oscillations are strongly coupled with those of the other variables of the motion β and r, as it can be caused by roll steer, the overall behaviour can become strongly oscillatory and the dynamic stability can be decreased.

The steady-state response of the vehicle is easily obtained from the equation

$$\begin{cases} \dfrac{mV^2}{R} = Y_\beta \beta + Y_{\dot\psi} \dfrac{V}{R} + Y_\phi \phi + Y_\delta \delta + F_{y_e} \\ 0 = N_\beta \beta + N_{\dot\psi} \frac{V}{R} + N_\phi \phi + N_\delta \delta + M_{z_e} \\ -\dfrac{m_S h V^2}{R} = L_\beta \beta + L_\phi \phi \ , \end{cases} \tag{6.82}$$

where the steady-state curvature of the trajectory $1/R = r/V$ has been explicitly introduced.

By solving equations (6.82) in $1/R$ and neglecting external forces, the trajectory curvature gain $1/R\delta$ is readily obtained

$$\frac{1}{R\delta} = \frac{AY_\delta - BN_\delta}{VD} \ , \tag{6.83}$$

where

$$\begin{array}{ll} A = N_\beta L_\phi - N_\phi L_\beta & B = Y_\beta L_\phi - Y_\phi L_\beta \\ C = Y_\beta N_\phi - Y_\phi N_\beta & D = V(mA - m_S hC) + BN_{\dot\psi} - AY_{\dot\psi} \ . \end{array}$$

6.7 Ride comfort

The first equation of the second set of five differential equations, the one related to the longitudinal dynamics, is weakly coupled with the others and can be written in the form

$$\dot V = \frac{J_s \ddot\theta + \sum_{\forall i} F_{xi} + \frac{1}{2}\rho V_a^2 S C_x}{m} \ . \tag{6.84}$$

By introducing Eq. (6.84) into the other ones, the following set of four equations describing the suspension motions of a vehicle with two axles is obtained

$$[M]_2\{\ddot{q}\}_2 + [C]_2\{\dot{q}\}_2 + [K]_2\{q\}_2 = \{F\}_2 \, , \tag{6.85}$$

where

$$\{q\}_2 = \left[Z \; \theta \; Z_1 \; Z_2 \right]^T$$

$$[M]_2 = \begin{bmatrix} m_S & -m_S c & 0 & 0 \\ & J_y - \dfrac{J_s^2}{m} & 0 & 0 \\ & & m_1 & 0 \\ & \text{symm.} & & m_2 \end{bmatrix}$$

$$[C]_2 = \begin{bmatrix} Z_{\dot{z}} & Z_{\dot{\theta}} & Z_{\dot{z}_1} & Z_{\dot{z}_2} \\ & M_{\dot{\theta}} & M_{\dot{z}_1} & M_{\dot{z}_2} \\ & & Z_{1_{\dot{z}_1}} & 0 \\ & \text{symm.} & & Z_{2_{\dot{z}_2}} \end{bmatrix}$$

$$[K]_2 = \begin{bmatrix} Z_z & Z_\theta & Z_{z_1} & Z_{z_2} \\ M_z & M_\theta & M_{z_1} & M_{z_2} \\ & & Z_{1_{z_1}} & 0 \\ & \text{symm.} & & Z_{2_{z_2}} \end{bmatrix}$$

$$\{F\}_2 = \begin{Bmatrix} \dfrac{1}{2}\rho V_a^2 Sl(C_z)_{,\theta}\theta_0 + \displaystyle\sum_{\forall i}(x_i)_{,z}F_{xi} \\ \displaystyle\sum_{\forall i} F_{xi}\left(q_{i_0} + z_{i_C} - (x_i)_{,z}x_{s_i} - \dfrac{J_s}{m}\right) + \\ +\frac{1}{2}\rho V_a^2 S\left[-\left(h - \frac{J_s}{m}\right)C_z + l(C'_{M_y})_{,\theta}\theta_0\right] \\ 0 \\ 0 \end{Bmatrix}$$

and

$$\begin{cases} Z_{\dot{z}} = \sum_{\forall i} c_i & Z_{\dot{z}_i} = -c_i \\ Z_z = \sum_{\forall i} K_i & Z_{z_i} = -K_i \\ Z_{\dot{\theta}} = -\sum_{\forall i} c_i x_{d_i} & Z_\theta = -\sum_{\forall i} K_i x_{s_i} - \frac{1}{2}\rho V_a^2 Sl(C_z)_{,\theta} \\ M_{\dot{\theta}} = \sum_{\forall i} c_i x_{d_i}^2 & M_\theta = \sum_{\forall i} K_i x_{s_i}^2 - m_S g h - \frac{1}{2}\rho V_a^2 Sl(C_{M_y})_{,\theta} \\ M_{\dot{z}_i} = c_i x_{d_i} & M_{z_i} = K_i x_{s_i} \\ Z_{i_{\dot{z}_i}} = c_i + 2c_{t_i} & Z_{i_{z_i}} = K_i + 2P_i \\ M_z = -\sum_{\forall i} K_i x_{s_i} & \end{cases}$$

The expression for the generalized forces $\{F\}_2$ has been obtained by assuming that the reference position is the static equilibrium position, at standstill with no forces F_{xi}. In that condition all generalized coordinates are equal to zero, as they are defined as the displacements from that reference situation and then

$$\begin{array}{l} \sum_{\forall i} K_i L_i - m_S g = 0 \\ \sum_{\forall i} K_i x_{d_i} L_i - m_S g(c + h\theta_0) = 0 \\ K_i L_i - 2P_i L_{t_i} + m_i g = 0. \end{array}$$

Note that the stiffness matrix is symmetrical except for the terms in position 12 and 21: In Z_θ a term of aerodynamic origin is present, the change of lift due to pitching motions, which is absent in M_z, where it would have the meaning of a change of pitching moment due to vertical motions, which is clearly zero.

The equations have been written with reference to coordinate Z, i.e. with reference to the height on the road of point H in Fig. 6.20 and the result is inertial coupling. For the study of the suspensions motions it is more straightforward to refer to the height on the road of point H', in order to have a set of inertially uncoupled equation. By introducing the coordinate $z_S = Z - c(\theta + \theta_0)$ of point H' into the equations, the mass matrix becomes

$$[M]_3 = \begin{bmatrix} m_S & 0 & 0 & 0 \\ & J_y^* & 0 & 0 \\ & & m_1 & 0 \\ & symm. & & m_2 \end{bmatrix}, \tag{6.86}$$

where $J_y^* = J_y - c^2 m_S + J_s^2/m$. The damping and stiffness matrices remain unchanged, with the only difference of substituting distances x_i of the

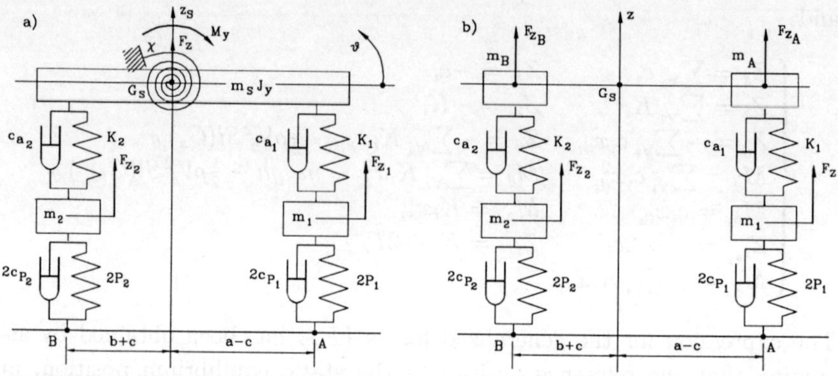

Fig. 6.23 (a) Model with four degrees of freedom for the study of ride comfort; (b) model in which the sprung mass is simulated by two separate masses. Lengths a and b are the same as for the rigid vehicle, $a = x_1$ and $b = -x_2$. Note that the longitudinal positions of the springs and shock absorbers are assumed to be coincident ($x_i = x_{s_i} = x_{d_i}$).

wheels, springs and shock absorbers with $x_i - c$. All matrices, except the stiffness matrix, remain symmetrical.

If the aerodynamic term which causes the lack of symmetry of the stiffness matrix is neglected, which introduces a small error owing to its smallness, the system can be sketched as in Fig. 6.23a. The vehicle is so modelled as a beam with elastic and damped supports which are connected to the ground through the unsprung masses.

The quasi-static equilibrium attitude of the vehicle, which is different from the reference position as it takes into account both the longitudinal forces on the tires and the aerodynamic forces, can be immediately obtained from the steady-state solution of Eq. (6.85). Note that even if the acceleration of the vehicle does not appear explicitly in the equations, it is accounted for through forces F_{xi}.

The dynamic response of the vehicle to the motion on uneven road is easily computed by assuming that points A and B move in a vertical direction with laws $h_A(t) = h(Vt)$ and $h_B(t) = h(Vt + l)$, where $h(x)$ is a function expressing the road profile. This amounts to excite the two masses m_1 and m_2 with two forces equal to $2[P_1 h_A(t) + c_{P_1} \dot{h}_A(t)]$ and $2[P_2 h_B(t) + c_{P_2} \dot{h}_B(t)]$ respectively.

6.8 Quarter-car models

Consider the model of Fig. 6.23a and assume that the torsional spring with stiffness χ, which is used to model the dependence of the aerodynamic pitching moment and the moment due to weight with the pitch angle can be neglected.

Define the dynamic index I_d of the sprung mass as

$$I_d = \frac{r_y^2}{(b+c)(a-c)} , \tag{6.87}$$

where r_y is the radius of gyration of the sprung mass about y-axis. If the dynamic index is equal to unity, i.e. if

$$J_y = m_S(b+c)(a-c) ,$$

the front and rear suspensions are located at conjugate centres of percussion of the sprung mass and the model of Fig. 6.23a reduces to the one of Fig. 6.23b. The suspension motions of the front and rear parts of the vehicle uncouple and the ride behaviour can be studied using two separate models with two degrees of freedom. This property can be demonstrated (see Section 6.9) both for the case in which the compliance of the tires is neglected and for the complete model.

Although this condition is never exactly met, the model in which the front and the rear parts of the vehicle uncouple is widely used for the study of the dynamic behaviour of the suspensions. Each one of the two parts in which the vehicle has been split is what is usually defined a "quarter car model", even if in this case it would be better defined a "half-car model"[5]. Three of the possible quarter car models are shown in Fig. 6.24.

The first model has a single degree of freedom. The tires are considered rigid bodies and the only mass considered is the sprung mass. This model holds well for motions taking place at low frequency, in the range of the natural frequency of the sprung mass (in most cases, up to 3–5 Hz).

The second model has two degrees of freedom. The tires are considered as massless springs and both the unsprung and the sprung masses are considered. This model holds well for frequencies up to the natural frequency of the unsprung mass and slightly over (in most cases, up to 30–50 Hz).

[5]Usually the quarter car model is thought as a single independent suspension with the part of sprung mass insisting on it.

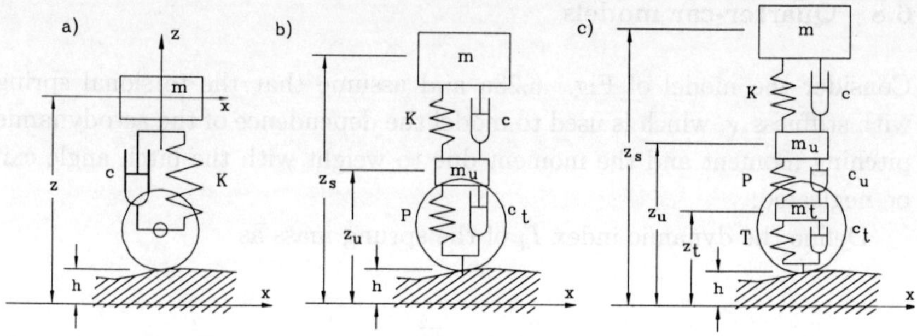

Fig. 6.24 Quarter-car models with one (a), two (b) and three (c) degrees of freedom.

The third model has three degrees of freedom. The tires are modelled as a spring-mass-damper system, which represent their dynamic characteristics in the lowest mode. This model allows to study motions taking place at frequencies in excess to the first natural frequency of the tires (up to 120–150 Hz).

If higher frequencies must be accounted for, it is possible to introduce a higher number of modes of the tires by inserting other masses. These models, which are essentially based on the modal analysis of the suspension-tire system, are clearly approximated as a tire can be considered a fairly highly damped system and is usually also nonlinear.

6.8.1 *Quarter-car with a single degree of freedom*

Consider the simplest quarter-car model, that shown in Fig. 6.24a. It has been used to demonstrate that the shock absorber must be a linear, symmetrical viscous damper[6].

The equation of motion of the system is very simple. Using the symbols shown in the figure it is

$$m\ddot{z} + c\dot{z} + Kz = c\dot{h} + Kh \;, \tag{6.88}$$

where z is the displacement from the static equilibrium position.

The frequency response of the quarter car is shown in Fig. 6.25a; it can be obtained simply by stating a harmonic input of the type $h = h_0 e^{i\lambda t}$.

[6]Bourcier De Carbon C.: *Théorie mathématiques et réalisation pratique de la suspension amortie des vehicules terrestres*, Proceedings SIA Conference, Paris, 1950.

The output is itself harmonic and can be expressed as $z = z_0 e^{i\lambda t}$, where both amplitudes h_0 and z_0 are complex numbers to account for the different phasing of response and excitation. The amplification factor, i.e. the ratio between the absolute values of the amplitudes of the response and the excitation and the phase of the first with respect to the second can be easily shown to be

$$\begin{cases} \dfrac{|z_0|}{|h_0|} = \sqrt{\dfrac{K^2 + c^2\lambda^2}{\left(K - m\lambda^2\right)^2 + c^2\lambda^2}} \\ \Phi = \arctan\left(\dfrac{-cm\lambda^3}{K\left(K - m\lambda^2\right) + c^2\lambda^2}\right) . \end{cases} \tag{6.89}$$

More than the frequency response expressing the ratio between the amplitudes of response and excitation, what matters in motor vehicle suspensions is the ratio between the amplitudes of the acceleration of the sprung mass and that of the displacement of the supporting point. As in harmonic motion the amplitude of the acceleration is equal to the amplitude of the displacement multiplied by the square of the frequency, $|(\ddot{z})_0|/|h_0|$ is equal to $\lambda^2|z_0|/|h_0|$. Both frequency responses are plotted in Fig. 6.25 for different values of the damping of the shock absorber, together with the phase Φ. The responses are plotted as functions of the nondimensional frequency $\lambda^* = \lambda\sqrt{m/K}$.

All curves pass through point A, located at a frequency equal to $\sqrt{2K/m}$. Since to obtain a good riding comfort the acceleration of the sprung mass must be kept to a minimum, a reasonable way to optimize the suspension is to choose a value of the damping of the shock absorber leading to a relative maximum, or at least point of stationarity, at point A on the curve related to the acceleration. By differentiating the expression of $\lambda^2|z_0|/|h_0|$ with respect to λ and equating the derivative to zero at point A the following value of the damping is obtained

$$c_{opt} = \sqrt{\frac{Km}{2}} = c_{cr}\frac{1}{2\sqrt{2}} , \tag{6.90}$$

where $c_{cr} = 2\sqrt{Km}$ is the critical damping of the suspension.

Although this way of optimizing the suspension can be easily criticized, as the comfort of a suspension is a far more complex concept that the simple reduction of the vertical acceleration (the so-called "jerk", i.e. the

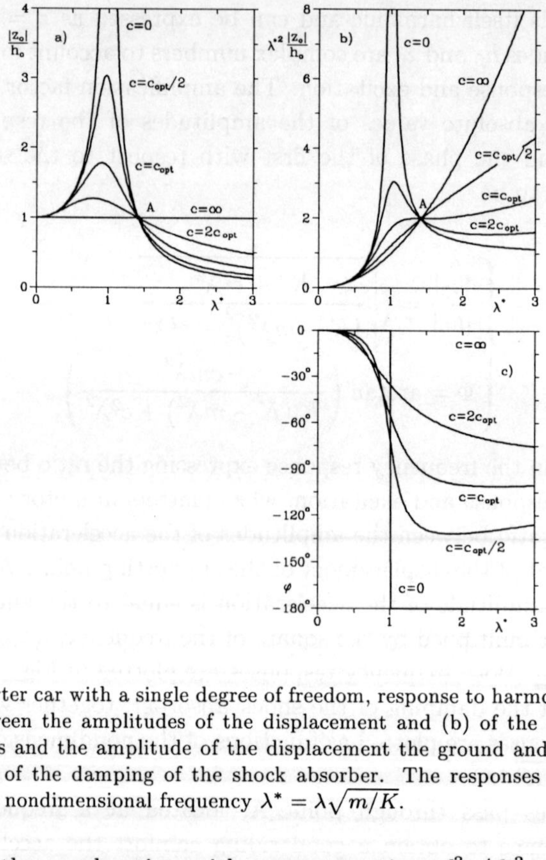

Fig. 6.25 Quarter car with a single degree of freedom, response to harmonic excitation. (a) Ratios between the amplitudes of the displacement and (b) of the acceleration of the sprung mass and the amplitude of the displacement the ground and (c) phase, for different values of the damping of the shock absorber. The responses are plotted as functions of the nondimensional frequency $\lambda^* = \lambda\sqrt{m/K}$.

derivative of the acceleration with respect to time d^3z/dt^3 plays also an important role) it already gives important indications. Note that the force the tire exerts on the ground is

$$F_z = c\left(\dot{z} - \dot{h}\right) + K(z - h) - m\ddot{z}\ , \tag{6.91}$$

where F_z is the dynamic component of the force.

To minimize the vertical acceleration leads to the minimization of the dynamic component of the vertical load on the tire, which has a very negative influence on the ability of the tire to exert cornering forces. The condition leading to the optimum comfort seems then to coincide with that leading to the optimum handling performances.

Equation (6.90) allows one to choose the value of the damping c. For the value of the stiffness K there is no such optimization: To minimize both the acceleration and the dynamic component of the force K should be as low as possible, the only limit to the softness of the springs coming from available space considerations: The softer the springs the larger the oscillations of the sprung mass. This conclusion is also drawn from an oversimplified model and can apply only in part to the real system.

As a last consideration, the optimum damping expressed by Eq. (6.90) is lower than the critical damping. The quarter-car model is underdamped and can undergo free oscillations, although highly damped, as the damping ratio $\zeta = c/c_{cr}$ is quite high, namely $1/2\sqrt{2} \approx 0.354$.

A simple way of demonstrating the advantages of symmetrical dampers, i.e. dampers acting both in the jounce and rebound stroke with equal damping coefficient, is the following[7]. Consider a quarter car with a single degree of freedom, i.e. with a rigid wheel, passing at high speed on a bump or a pothole. If the time needed to cross the road irregularity is far shorter than the period of oscillation of the sprung mass, an impulsive model can be used to study the effects of the force the latter receives from the spring and the shock absorber. The horizontal velocity of the sprung mass is unchanged, while it acquires a vertical velocity w such that the vertical momentum mw is equal to the total impulse due to the forces F_s exerted by the spring and F_d exerted by the shock absorber from time t_1 in which the wheel enters the road irregularity to time t_2 in which it rolls away

$$w = \frac{1}{m}\left(\int_{t_1}^{t_2} F_s dt + \int_{t_1}^{t_2} F_d dt\right) . \tag{6.92}$$

Note that the force due to the spring is only the part which exceeds the static value which compensates for the weight. If the integrals on the right hand side of Eq. (6.92) vanish, the suspension absorbs completely the irregularity without any perturbation being transmitted to the sprung mass. The first of the two integrals is assumed to be far smaller than the second one as in a small amplitude, high frequency disturbance the force due to the spring, which is proportional to the displacement, is negligible compared with the force due to the shock absorber, which is proportional to the velocity. By neglecting the first integral and assuming that the damper is symmetrical, the expression for the vertical velocity becomes

---

7 Bourcier De Carbon C.: *Théorie mathématiques et réalisation pratique de la suspension amortie des vehicules terrestres*, Proceedings SIA Conference, Paris, 1950.

$$w = \frac{1}{m}\int_{t_1}^{t_2} -c\dot{h}dt = -\frac{c}{m}(h_2 - h_1)\,, \tag{6.93}$$

where $h(x)$ is the law expressing the road profile.

From Eq. (6.93) it is clear that if $h_2 = h_1$ the suspension is able to insulate perfectly the sprung mass from the road irregularity, and this is due to the fact that the damping coefficient in the rebound is equal to that in the jounce.

This result is however plagued by the oversimplification of the model and it is well known that, while it is correct that the shock absorber must act in both the up- and the down-stroke, the damping coefficients must be unequal for best performances. This is easily explained by noting that while the instant value of the force due to the shock absorber is larger than that due to the spring, it is not true that the same inequality holds for the integrals, particularly when the first one vanishes. Also, the road-wheel constraint is unilateral. The disturbance when passing a bump at high speed is higher than when passing over a hole and in the first case the force due to the spring acts upwards; in this case a damping coefficient higher in the downstroke gives a negative value of the second integral of Eq. (6.92) which can compensate for the positive value of the first one. Some approximations are also linked to the use of the impulsive model.

6.8.2 *Quarter-car with two degrees of freedom*

The following model is that shown in Fig. 6.24b. It is well suited for the study of the behaviour of vehicle suspensions in a frequency range which goes beyond the natural frequency of the unsprung mass.

With reference to Fig. 6.24b, the equation of motion of the model is

$$\begin{bmatrix} m_s & 0 \\ 0 & m_u \end{bmatrix}\begin{Bmatrix} \ddot{z}_s \\ \ddot{z}_u \end{Bmatrix} + \begin{bmatrix} c & -c \\ -c & c+c_t \end{bmatrix}\begin{Bmatrix} \dot{z}_s \\ \dot{z}_u \end{Bmatrix} + \\ + \begin{bmatrix} K & -K \\ -K & K+P \end{bmatrix}\begin{Bmatrix} z_s \\ z_u \end{Bmatrix} = \begin{Bmatrix} 0 \\ c_t\dot{h} + Ph \end{Bmatrix}\,, \tag{6.94}$$

where z_s and z_u are the displacements from the static equilibrium position and are referred to an inertial frame.

The response to a harmonic excitation is readily obtained in the same way as seen for the previous model. By neglecting the damping of the tire c_t, which is usually very small, it follows

$$\begin{cases} \dfrac{|z_{s_0}|}{|h_0|} = P\sqrt{\dfrac{K^2 + c^2\lambda^2}{f^2(\lambda) + c^2\lambda^2 g^2(\lambda)}} \\ \dfrac{|z_{n_0}|}{|h_0|} = P\sqrt{\dfrac{\left(K - m\lambda^2\right)^2 + c^2\lambda^2}{f^2(\lambda) + c^2\lambda^2 g^2(\lambda)}}\, , \end{cases} \tag{6.95}$$

where

$$\begin{cases} f(\lambda) = m_s m_u \lambda^4 - \left[P m_s + K(m_s + m_u)\right]\lambda^2 + KP \\ g(\lambda) = (m_s + m_u)\lambda^2 - P\, . \end{cases}$$

The dynamic component of the force exerted in z direction by the tire on the ground can be easily computed in the same way seen for the model with a single degree of freedom, obtaining

$$\frac{|F_{z_0}|}{|h_0|} = P\lambda^2\sqrt{\frac{\left[K(m_s + m_u) - m_s m_u \lambda^2\right]^2 + c^2(m_s + m_u)\lambda^2}{f^2(\lambda) + c^2\lambda^2 g^2(\lambda)}}\, . \tag{6.96}$$

The frequency responses related to both the sprung and the unsprung masses are plotted in Fig. 6.26a and b for a system with $P = 4K$ and $m_s = 10m_u$. The plots, shown using the nondimensional frequency $\lambda^* = \lambda\sqrt{m/K}$, include curves obtained with different values of damping c; all curves lie in the nonshaded region of the graph.

If $c = 0$ the natural frequencies are two and the peaks are infinitely high. Also for $c \to \infty$ the peak, corresponding to the natural frequency of the whole system, which is now rigid, over the spring simulating the tire, goes to infinity.

The frequency responses of Fig. 6.26a and b multiplied by ${\lambda^*}^2$ are shown in Fig. 6.26c and d; they give the nondimensional ratio between the accelerations of the two masses and the displacement of the supporting point. All curves pass through points O, A, B and C. Between O and A and between B and C the maximum acceleration of the sprung mass increases with decreasing damping, while between A and B and from C upwards it increases with the damping

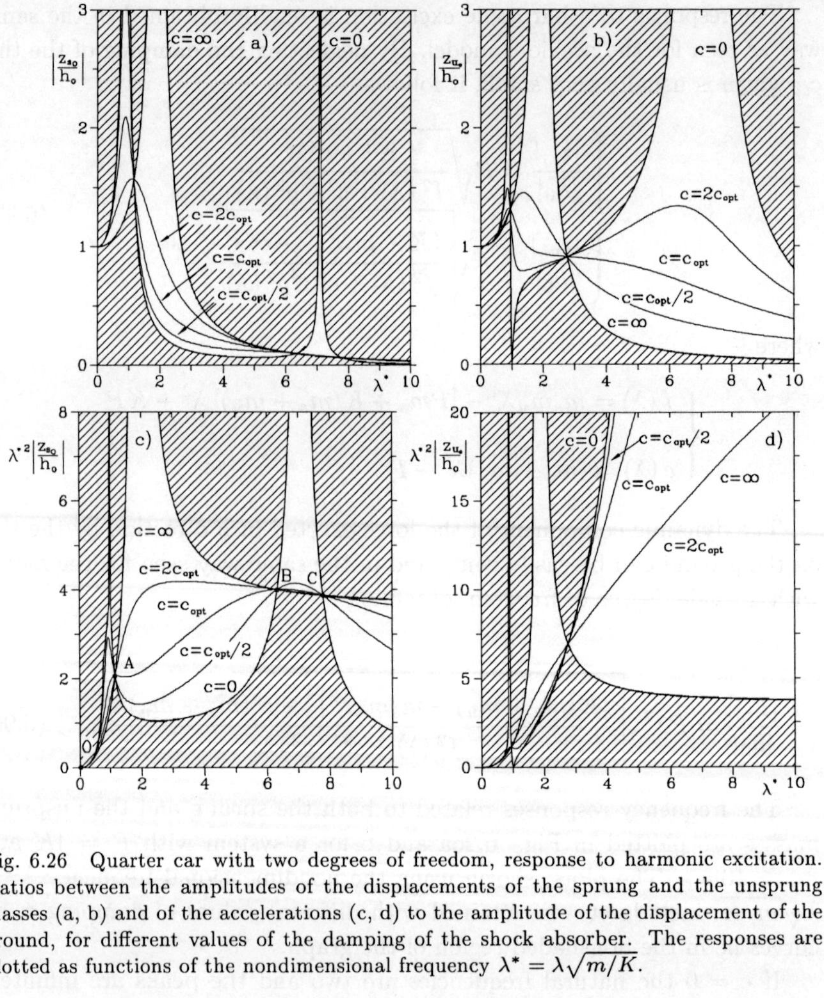

Fig. 6.26 Quarter car with two degrees of freedom, response to harmonic excitation. Ratios between the amplitudes of the displacements of the sprung and the unsprung masses (a, b) and of the accelerations (c, d) to the amplitude of the displacement of the ground, for different values of the damping of the shock absorber. The responses are plotted as functions of the nondimensional frequency $\lambda^* = \lambda\sqrt{m/K}$.

An optimum value of the damping can be found by trying to keep the acceleration as low as possible in a large field which goes up to the natural frequency of the unsprung mass, i.e. by looking for a curve which has a relative maximum or a stationary point in A. Operating as seen for the previous model the following value is obtained

$$c_{opt} = \sqrt{\frac{Km}{2}}\sqrt{\frac{P+2K}{P}}\,. \tag{6.97}$$

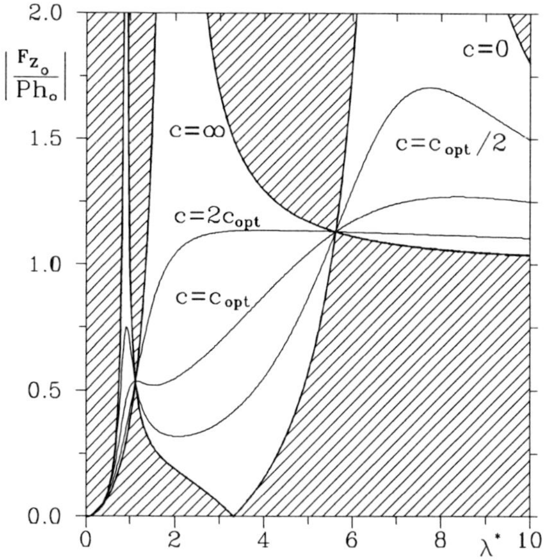

Fig. 6.27 Quarter car with two degrees of freedom, response to harmonic excitation. Ratio between the amplitude of the dynamic component of force F_z between tire and road and the displacement of the ground, made nondimensional by dividing it by the stiffness of the tire P, for different values of the damping of the shock absorber. The response is plotted as a function of the nondimensional frequency $\lambda^* = \lambda\sqrt{m/K}$.

As P is much larger than K, the value of $\sqrt{(P+2K)/P}$ is close to unity (in the case of Fig. 6.26 $\sqrt{(P+2K)/P} = 1.22$) and the optimum damping is only slightly larger than the one computed for the model with a single degree of freedom (Eq. (6.90)). From Fig. 6.26c it is clear that this value of the damping is effective in maintaining the acceleration low in wide frequency range.

The amplitude of the dynamic component of force F_z (Eq. (6.97)) is plotted in nondimensional form (divided by $P|h_0|$) as a function of the nondimensional frequency in Fig. 6.27. The value of the optimum damping expressed by Eq. (6.97) is effective in keeping as low as possible also the maximum value of the dynamic component of force F_z at least at low frequencies. At higher frequencies a slightly higher value of damping could be effective, even if it would result in a larger acceleration of the sprung mass.

The maximum value of the nondimensional amplitude of force F_z has been plotted as a function of ratio c/c_{opt} in Fig. 6.28a. When the damping

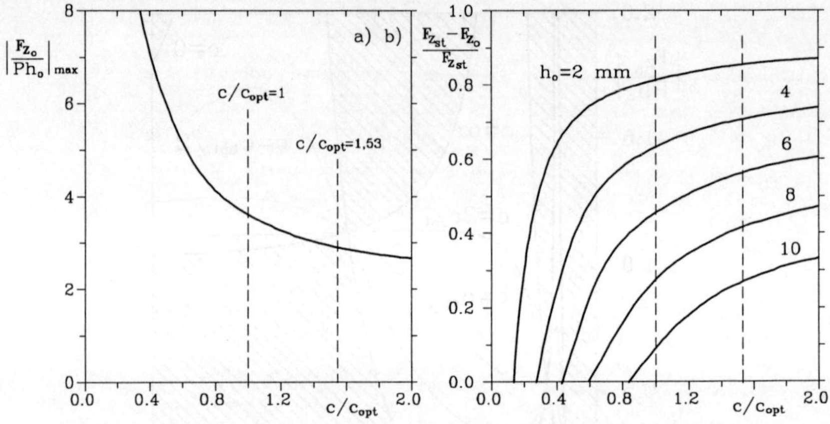

Fig. 6.28 (a) Maximum value of the amplitude of the dynamic component of force F_z in a frequency range between 0 and 30 $\sqrt{K/m}$ as a function of ratio c/c_{opt}. Same characteristics as the system studied in the previous figures. (b) Minimum value of the ground force (static force minus amplitude of the dynamic component) as a function of ratio c/c_{opt} for a quarter car model with parameters typical for a small car: $m_s = 238$ kg; $m_u = 38$kg; $K = 15.7$ kN/m; $P = 135$ kN/m; actual value of c/c_{opt} equal to 1.53.

goes beyond the optimum value computed above a certain decrease of the maximum amplitude of the force at high frequency is clearly obtained.

The minimum value of the force on the ground (computed as the static component minus the amplitude of the dynamic component) has been plotted as a function of ratio c/c_{opt} in Fig. 6.28b using data similar to those related to the front suspension of a small car. The curves refer to different amplitudes of the excitation h_0. If the damping is small enough the wheel can bounce on the road. Clearly when this occurs the present linear model is not applicable.

From the considerations seen above it is possible to draw the conclusion that the value of the damping coefficient expressed in Eq. (6.97) is optimal both from the viewpoint of the comfort and that of the handling, as it leads to low variations of the forces on the ground. A slightly higher damping can however improve slightly the handling performances. This conclusion, obtained from a highly simplified model, is not in good accordance with the experimental evidence, which states that the damping which optimizes riding comfort is lower than that optimizing handling. A common consideration on this issue is drawn from a plot obtained considering a random velocity input whose power spectrum is a white noise in a frequency range between 0.1 and 100 Hz. As it will be seen in Section 6.13.2, the power spectral density of the output of the system S_o is obtained from that of the input S_i by simply multiplying the latter by the square of the frequency response $H(\lambda)$

$$S_O = H^2(\lambda)S_i \ . \tag{6.98}$$

If the input is a white noise acting in a frequency range between λ_1 and λ_2, the r.m.s. (root mean square) value of the output is simply

$$O_{rms} = \sqrt{\int_{\lambda_1}^{\lambda_2} S_o d\lambda} = \sqrt{S_i}\sqrt{\int_{\lambda_1}^{\lambda_2} H^2(\lambda) d\lambda} \ . \tag{6.99}$$

The r.m.s. value of the acceleration is easily computed: As the output is the acceleration of the sprung mass and the input is the vertical velocity of the contact point, the frequency response is the first Eq. (6.95) multiplied by λ. Also the r.m.s. value of the dynamic component of force F_z is readily obtained, using the frequency response (6.96) divided by λ. An interesting result is the graph obtained by plotting the r.m.s. value of the acceleration versus that of the force for various values of the damping of the shock absorber (Fig. 6.29).

The conditions which lead to the optimum comfort (in the sense of minimum acceleration) and to the optimum handling (in the sense of minimum force variations) are readily identified: The first is obtained with a damping which is lower than the optimum damping defined above while the second for a value which is higher. While the adequateness of the use of a white noise velocity input will be discussed later (See Section 6.13.2), this result is in better accordance with the experimental results than the previous one.

As already stated, the presence of the tires has a negligible effect on the frequency response at low frequency, while at higher frequencies their stiffness must be accounted for. A comparison between the results obtained using the quarter car models with one and two degrees of freedom is shown in Fig. 6.30.

6.8.3 *Quarter-car model with dynamic vibration absorber*

A dynamic vibration absorber consists essentially of a mass connected to the system through a spring and possibly a damper (Fig. 6.31a). If properly tuned it can reduce the amplitude substantially at one of the resonances of

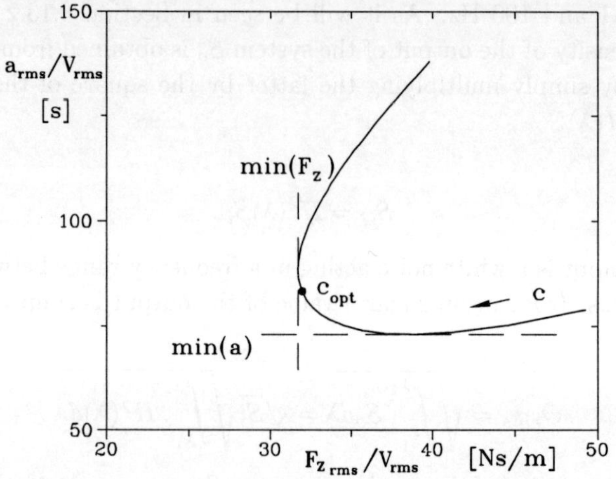

Fig. 6.29 Ratio of the r.m.s. value of the acceleration of the sprung mass and the r.m.s. value of the input velocity versus the ratio between the r.m.s. value of the dynamic component of force F_z and the r.m.s. value of the input velocity for various values of the damping c of the shock absorber. White noise velocity input in a range between 0.1 and 100 Hz. Quarter car model with two degrees of freedom with the same data as in Fig. 6.28b (m_s = 238 kg; m_u = 38kg; K = 15.7 kN/m; P =135 kN/m).

the original system but introduces one more resonance, whose peak amplitude is controlled by the value of its damping c_d.

The frequency response of the system of Fig. 6.31a is shown in Fig. 6.31c for 3 different values of the damping: If c_d is low, two resonance peaks are present while if c_d is high, there is only one peak. If the damping tends to zero the two peaks have an infinite height, while if it tends to infinity the system reduces to an undamped system with a single degree of freedom and thus has a single peak with infinite height. It is possible to demonstrate that the stiffness k_d which reduces the amplitude of the motion of mass m to a minimum is[8]

$$(K_d)_{opt} = K \frac{m m_d}{(m + m_d)^2} . \tag{6.100}$$

The value $(c_d)_{opt}$ of the damping which allows to obtain such minimum and the peak amplitude are respectively

[8] J.P. Den Hartog, *Mechanical vibrations*, McGraw Hill, New York, 1956.

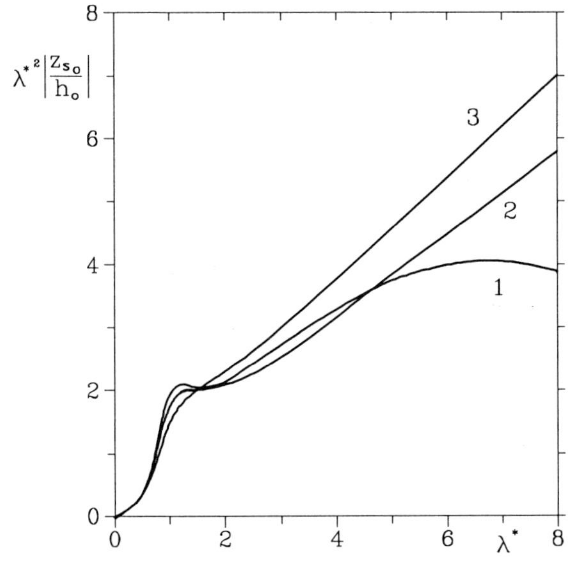

Fig. 6.30 Acceleration of the sprung mass as a function of the frequency for a unit displacement input. Comparison between the quarter car model with one and two degrees of freedom (in the latter case $P = 4K$, $m_s = 10m_u$). 1): 2 d.o.f.; 2): 1 d.o.f., damping defined by Eq. (6.97); 3): 1 d.o.f., damping defined by Eq. (6.90).

$$(c_d)_{opt} = \frac{m_d}{m+m_d}\sqrt{K\frac{3mm_d}{2(m+m_d)}}\,, \qquad \left|\frac{z_0}{h_0}\right|_{max} = \sqrt{1+\frac{2m}{m_d}}\,. \qquad (6.101)$$

Sometimes the vibration absorber can have no elastic element or can have a damper whose behaviour is better modelled by dry friction than by viscous damping.

Dynamic vibration absorbers are sometimes used in motor vehicle suspensions, as in the quarter car model of Fig. 6.31b, in which a standard shock absorber is also represented. The equation of motion of the system is simply

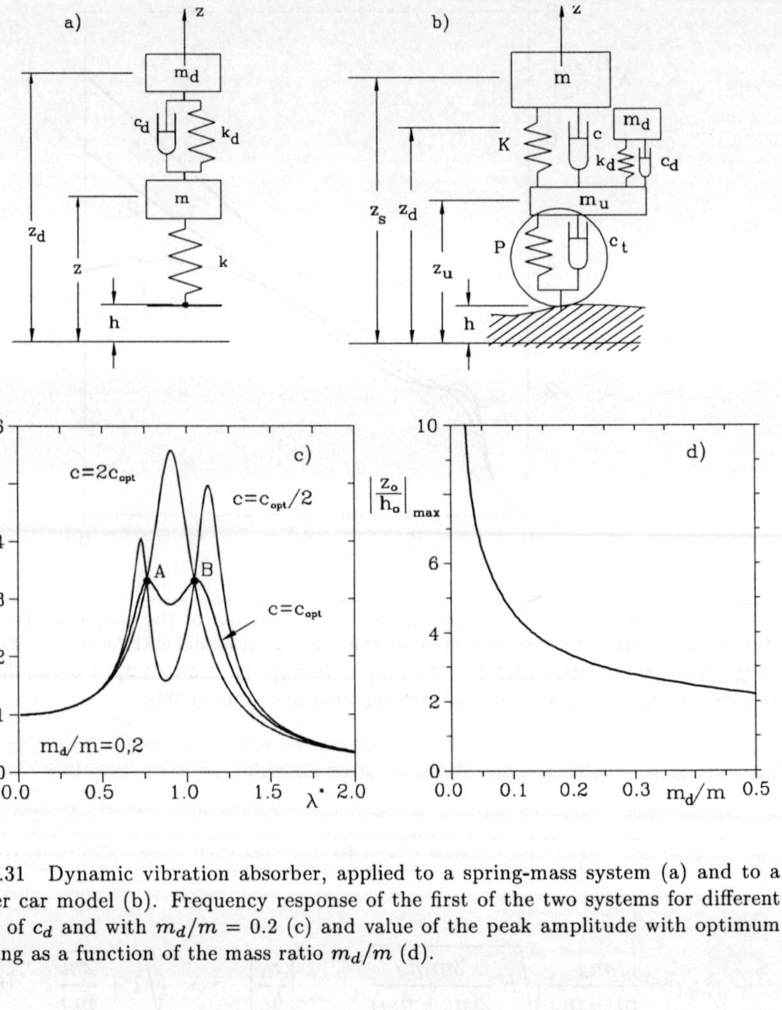

Fig. 6.31 Dynamic vibration absorber, applied to a spring-mass system (a) and to a quarter car model (b). Frequency response of the first of the two systems for different values of c_d and with $m_d/m = 0.2$ (c) and value of the peak amplitude with optimum damping as a function of the mass ratio m_d/m (d).

$$\begin{bmatrix} m_s & 0 & 0 \\ 0 & m_d & 0 \\ 0 & 0 & m_u \end{bmatrix} \begin{Bmatrix} \ddot{z}_s \\ \ddot{z}_d \\ \ddot{z}_u \end{Bmatrix} + \begin{bmatrix} c & 0 & -c \\ 0 & c_d & -c_d \\ -c & -c_d & c + c_d + c_t \end{bmatrix} \begin{Bmatrix} \dot{z}_s \\ \dot{z}_d \\ \dot{z}_u \end{Bmatrix} + \\ + \begin{bmatrix} K & 0 & -K \\ 0 & K_d & -K_d \\ -K & -K_d & K + K_d + P \end{bmatrix} \begin{Bmatrix} z_s \\ z_d \\ z_u \end{Bmatrix} = \begin{Bmatrix} 0 \\ 0 \\ c_t \dot{h} + Ph \end{Bmatrix} . \tag{6.102}$$

The various frequency responses can be obtained immediately in the same way as for the other quarter car models, the only difference being

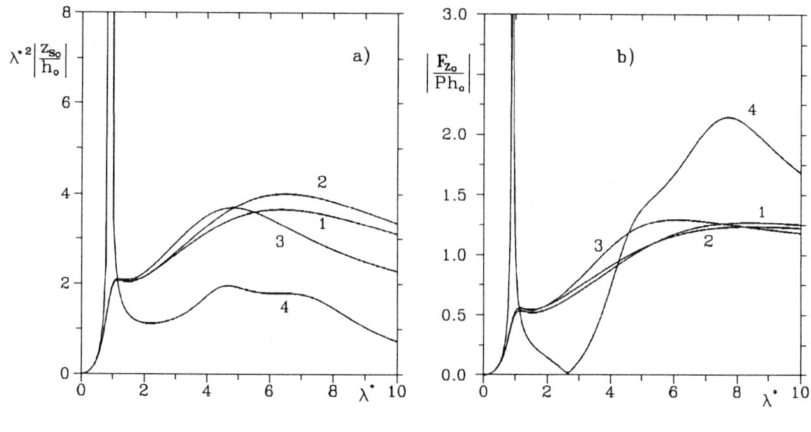

Fig. 6.32 Quarter car model with dynamic vibration absorber: Nondimensional amplitude of the acceleration (a) and of the dynamic component of the force F_z (b) as functions of the nondimensional frequency λ^* ($P = 4K$; $m_s = 10m_u$; $c = \sqrt{6mK/8}$). Line 1: Quarter car without vibration absorber; line 2: $m_d = 0.05m_s$, $\sqrt{k_d/m_d} = 0.89$, $c_d = 2\sqrt{k_d m_d}$; line 3: $m_d = 0.05m$, $\sqrt{k_d/m_d} = 7.09$, $c_d = 2\sqrt{k_d m_d}$; line 4: $m_d = 0.05m$, $c = 0$, $\sqrt{k_s/m_s} = 4.8$, $c_s = 0.8\sqrt{k_s m_s}$.

that in the present case the number of parameters is higher and the study is more complex. Some results are reported in Fig. 6.32. Since no attempt to optimize the suspension has been performed, the figure has only a qualitative interest.

Curve 1 deals with a conventional quarter car model as studied in the previous section. Curves 2 and 3 refer to a system in which a vibration absorber is applied to the unsprung mass, tuned on the first and the second natural frequency ($\sqrt{k_d/m_d} = 0.89$ and $\sqrt{k_d/m_d} = 7.09$). The mass of the vibration absorber is 1/20 of the sprung mass and damping c_d is $2\sqrt{k_d m_d}$. To add a vibration absorber to a conventional suspension changes little its performance, both in terms of acceleration of the sprung mass and of forces on the ground. But the interest in vibration absorbers lies in the possibility of using it instead of the conventional shock absorbers, like in the case shown by curve 4. In the case shown, in which the values of the parameters were obtained by a trial and error procedure without a true optimization, the acceleration of the sprung mass is quite low in the whole frequency range, except for a strong resonance peak at low frequency. The height of the peak is however limited, as clear for a damped system, and in practice is further limited by the other forms of damping which are present in the actual system, like that due to the tire. If the stiffness of the springs

K is low the peak occurs at very low frequency, where its importance can be marginal, and the advantages of the vibration absorber, mainly linked with the lower cost and complexity of the system due to the possibility of avoiding an element mounted between the body and the wheel, add to very good suspension performances.

Dynamic vibration absorbers were used instead of conventional shock absorbers with advantage on several low cost small cars with very soft suspensions; they can however be added to the conventional layout in luxury cars to further increase ride comfort.

6.9 Bounce and pitch motions

Consider the model of Fig. 6.23a. Its equation of motion is Eq. (6.85), with the modifications included into Eq. (6.85) to refer to point H′ instead of point H. By neglecting the very small aerodynamic terms and the term in $m_S g h$ in the equation for pitching motions, the equation of motion can be written in the very simple form

$$\begin{bmatrix} m_S & 0 & 0 & 0 \\ 0 & J_y^* & 0 & 0 \\ 0 & 0 & m_1 & 0 \\ 0 & 0 & 0 & m_2 \end{bmatrix} \begin{Bmatrix} \ddot{Z}_s \\ \ddot{\theta} \\ \ddot{Z}_1 \\ \ddot{Z}_2 \end{Bmatrix} + \\ + \begin{bmatrix} c_1 + c_2 & -d_1 c_1 + d_2 c_2 & -c_1 & -c_2 \\ & d_1^2 c_1 + d_2^2 c_2 & d_1 c_1 & d_2 c_2 \\ & & c_1 + 2c_{t_1} & 0 \\ \text{symm.} & & & c_2 + 2c_{t_2} \end{bmatrix} \begin{Bmatrix} \dot{Z}_s \\ \dot{\theta} \\ \dot{Z}_1 \\ \dot{Z}_2 \end{Bmatrix} + \\ + \begin{bmatrix} K_1 + K_2 & -d_1 K_1 + d_2 K_2 & -K_1 & -K_2 \\ & d_1^2 K_1 + d_2^2 K_2 & d_1 K_1 & d_2 K_2 \\ & & K_1 + 2P_1 & 0 \\ \text{symm.} & & & K_2 + 2P_2 \end{bmatrix} \begin{Bmatrix} Z_s \\ \theta \\ Z_1 \\ Z_2 \end{Bmatrix} = \\ = 2 \begin{Bmatrix} 0 \\ 0 \\ c_{t_1} \dot{h}_A + P_1 h_A \\ c_{t_2} \dot{h}_B + P_2 h_B \end{Bmatrix}, \tag{6.103}$$

where

$$d_i = |x_i - c|$$

and no distinction has been made between the position of the centre of the wheel, of the spring and the shock absorber.

The excitation vector has been written considering only the forcing functions due to the vertical motions of points A and B, neglecting longitudinal forces at the road-wheel interface. If a coupling between vertical and horizontal motions of the suspensions is present, the effects of the inertia of the wheels and driveline on the ride comfort are also neglected.

To demonstrate that when the dynamic index has a unit value the system uncouples into two independent quarter-cars, use coordinates Z_A and Z_B instead of Z_s and θ. The coordinate transformation can be expressed as

$$\begin{Bmatrix} Z_s \\ \theta \\ Z_1 \\ Z_2 \end{Bmatrix} = \frac{1}{l}\begin{bmatrix} d_2 & 0 & d_1 & 0 \\ -1 & 0 & 1 & 0 \\ 0 & l & 0 & 0 \\ 0 & 0 & 0 & l \end{bmatrix}\begin{Bmatrix} Z_A \\ Z_1 \\ Z_B \\ Z_2 \end{Bmatrix} . \tag{6.104}$$

The inertia matrix with the new coordinates can be obtained as

$$[M'] = [T]^T[M][T],$$

where $[T]$ is the transformation matrix defined by Eq. (6.104). All other matrices are obtained in the same way. Equation (6.103) becomes

$$\begin{bmatrix} m_S\dfrac{d_2^2+r^2}{l^2} & 0 & m_S\dfrac{d_1 d_2 - r^2}{l^2} & 0 \\ & m_1 & 0 & 0 \\ & & m_S\dfrac{d_1^2+r^2}{l^2} & 0 \\ \text{symm.} & & & m_2 \end{bmatrix}\begin{Bmatrix} \ddot{Z}_A \\ \ddot{Z}_1 \\ \ddot{Z}_B \\ \ddot{Z}_2 \end{Bmatrix} + \\ + \begin{bmatrix} c_1 & -c_1 & 0 & 0 \\ & c_1 + 2c_{t_1} & 0 & 0 \\ & & c_2 & -c_2 \\ \text{symm.} & & & c_2 + 2c_{t_2} \end{bmatrix}\begin{Bmatrix} \dot{Z}_A \\ \dot{Z}_1 \\ \dot{Z}_B \\ \dot{Z}_2 \end{Bmatrix} + \\ + \begin{bmatrix} K_1 & -K_1 & 0 & 0 \\ & K_1 + 2P_1 & 0 & 0 \\ & & K_2 & -K_2 \\ \text{symm.} & & & K_2 + 2P_2 \end{bmatrix}\begin{Bmatrix} Z_A \\ Z_1 \\ Z_B \\ Z_2 \end{Bmatrix} = 2\begin{Bmatrix} 0 \\ c_{t_1}\dot{h}_A + P_1 h_A \\ 0 \\ c_{t_2}\dot{h}_B + P_2 h_B \end{Bmatrix} . \tag{6.105}$$

It is thus clear that if $r^2 = d_1 d_2$ the first two equations uncouple from the other two, yielding two equations of motion of independent quarter cars with sprung masses $m_S d_2/l$ and $m_S d_1/l$ respectively. If the dynamic index

$$I_d = \frac{r^2}{d_1 d_2}$$

is equal to one, the free oscillations of the sprung mass are then rotations about the points of attachment of the suspensions: It is not possible to identify a bounce and a pitch mode but the free motion is better described in terms of front oscillation and rear oscillation.

In the study of the low frequency modes of the sprung mass the tires can be considered as rigid. A model in which the sprung mass is a beam directly suspended on the ground by the suspension springs and dampers can be used. Its equation of motion is simply

$$\begin{bmatrix} m_S & 0 \\ 0 & J_y^* \end{bmatrix} \begin{Bmatrix} \ddot{Z}_s \\ \ddot{\theta} \end{Bmatrix} + \begin{bmatrix} c_1 + c_2 & -d_1 c_1 + d_2 c_2 \\ -d_1 c_1 + d_2 c_2 & d_1^2 c_1 + d_2^2 c_2 \end{bmatrix} \begin{Bmatrix} \dot{Z}_s \\ \dot{\theta} \end{Bmatrix} +$$

$$+ \begin{bmatrix} K_1 + K_2 & -d_1 K_1 + d_2 K_2 \\ -d_1 K_1 + d_2 K_2 & d_1^2 K_1 + d_2^2 K_2 \end{bmatrix} \begin{Bmatrix} Z_s \\ \theta \end{Bmatrix} = \qquad (6.106)$$

$$= \frac{1}{l} \begin{Bmatrix} d_2 c_1 \dot{h}_A + d_1 c_2 \dot{h}_B + d_2 K_1 h_A + d_1 K_2 h_B \\ -c_1 \dot{h}_A + c_2 \dot{h}_B - K_1 h_A + K_2 h_B \end{Bmatrix} .$$

The modes of oscillation of the undamped system can be obtained by studying the homogeneous equation without the damping terms. It is thus clear that if $d_1 K_1 = d_2 K_2$ the bounce mode uncouples from the pitch mode: The first is a vertical translation of the sprung mass while the second one is a rotation about its centre of mass.

In general such condition is not met and both modes involve translation of the centre of mass and pitch rotation, i.e. both modes are rotations about two centres none of them coinciding with the centre of mass. If the dynamic index is equal to unity they coincide with the suspension points. Otherwise the positions of the centres depend on both the value of $-d_1 K_1 + d_2 K_2$ and on the dynamic index. In general the two points are located one in front and the other behind the centre of mass and one lies inside and one outside the wheelbase. The mode whose centre of rotation lies outside the wheelbase is mainly translational and is considered as a bounce mode, while the other being mainly rotational is considered as a pitch mode.

The natural frequencies λ of the undamped system can be obtained from the equation

$$\lambda^4 - \lambda^2 \frac{K_1(r^2 + d_1^2) + K_2(r^2 + d_2^2)}{m_S r^2} + K_1 K_2 \frac{l^2}{m_S^2 r^2} = 0 \,, \qquad (6.107)$$

where again the term in $m_S g h$ has been neglected.

If bounce and pitch uncouple $d_1 K_1 = d_2 K_2$, the natural frequencies are

$$\begin{cases} \lambda_1 = \sqrt{\dfrac{l K_1}{d_2 m_S}} & \text{bounce} \\ \lambda_2 = \sqrt{\dfrac{l d_1 K_1}{r^2 m_S}} = \lambda_1 \sqrt{\dfrac{d_1 d_2}{r^2}} & \text{pitch.} \end{cases} \qquad (6.108)$$

Note that from Eq. (6.108) it follows that when the dynamic index has a unit value the two natural frequencies coincide. This solves an apparent inconsistency: If $d_1 K_1 = d_2 K_2$ the centres of rotations are one in the centre of mass (pitch mode) and one at infinity (bounce mode) while when the dynamic index is equal to one the rotation centres are at the suspension points. When both conditions apply at the same time the natural frequencies coincide, and when two coincident eigenvalues are present any linear combination of the eigenvectors is itself an eigenvector. This means that, when dealing with a rigid beam, any point of the beam can be considered as rotation centre.

The bounce and pitch dynamics of the suspended mass are strictly related to each other. Some empirical criteria for the choice of the relevant parameters are here reported: They date back from the 1930s and were introduced by Maurice Olley[9].

– The vertical stiffness of the front suspension must be about 30% lower than that of the rear suspension;

– the pitch and bounce frequencies must be close to each other; the bounce frequency should be less than 1.2 times the pitch frequency;

– neither frequency should be greater than 1.3 Hz;

– the roll frequency should be approximately equal to the bounce and pitch frequency.

The first rule states that the natural frequency of the rear suspension is higher than that of the front one, at least if the weight distribution is not such that the rear wheels are far more loaded than the front ones. The importance of having a lower natural frequency of the front suspension can be explained by observing that any road input reaches the front suspension

[9] T. D. Gillespie, *Fundamentals of vehicle dynamics*, SAE, Warrendale, 1992.

and only after a certain time the rear one. If the natural frequency of the latter is higher, when the vehicle rides over a bump the rear part "catches up" rapidly the motion of the front part and after the first oscillation the body of the vehicle moves in bounce rather than in pitch, which is considered good for ride comfort. Then the rear part should lead the motion, but in the meantime the damping has caused the amplitude to decrease.

The second rule is easily fulfilled by modern cars. The problem here can be that of having the pitch frequency much higher than the bounce one, as it can occur when the dynamic index is smaller than unity (vehicle with long wheelbase and small front/rear overhang). Generally speaking, a dynamic index near unity is considered as a desirable condition for good ride properties, while a complete bounce-ride uncoupling as occurs when $d_1K_1 = d_2K_2$ is considered a nuisance. Coupling between bounce and pitching is good as it tends to avoid strong pitch oscillations.

A low value of natural frequencies leads to soft suspensions and large travels. Here the different traditions of the manufactures and of the particular markets to which the car is addressed can make a great difference.

If the value of the pitch natural frequency is too high, compared with that of bounce motions, ride comfort can be affected. To control the pitch and bounce natural frequency independently, without changing the wheel positions and the inertial properties of the body, the suspensions can be interconnected. If the front and rear wheels are connected by a spring which opposes to pitching motions in a way which is similar to antiroll bars used in rolling motions, the pitching frequency can be raised without increasing bouncing frequency and the damping of pitching motion is decreased. This is however the opposite of what is usually needed.

Various types of mechanical, hydraulic or pneumatic interconnections can be used, the latter particularly when air or hydraulic springs are used. A mechanical solution is shown in Fig. 6.33a: The vehicle is based on trailing arm suspensions, with springs located longitudinally under the sprung mass. The springs are connected to a further element, itself elastically connected to the vehicle body. The system is functionally similar (even if simpler) to the model shown in Fig. 6.33b, in which the intermediate element is a beam, hinged to the vehicle body and connected to the unsprung masses through springs. In the model the tires are considered as rigid bodies and have not been included.

If the beam and springs with stiffness χ_1 and χ_2 were not included, the equation of motion would have been Eq. (6.106), without the damping matrix if damping is neglected as in the figure. If the inertia of the beam is

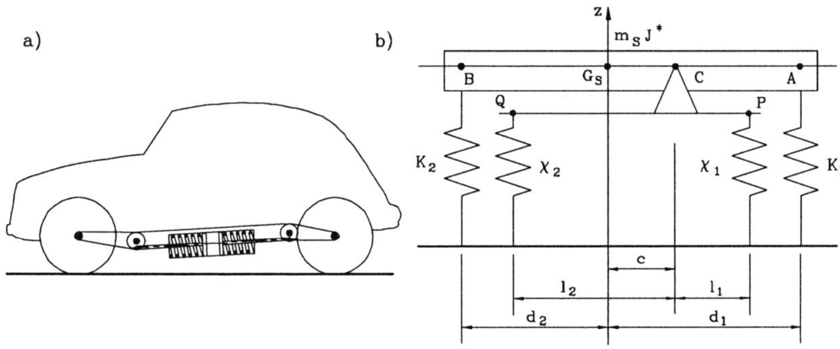

Fig. 6.33 Longitudinal interconnection of the suspensions. (a) Sketch of an application; (b) model in which the interconnection is implemented using a beam hinged to the sprung mass. The tires are considered as rigid bodies and not included into the model.

neglected, no further degree of freedom is needed, since the position of the beam is determined by the displacement z and the rotation θ of the sprung mass. The stiffness matrix can be obtained by simply adding the potential energy of springs χ_1 and χ_2 and performing the relevant derivatives.

The positions of points P and Q are simply

$$\begin{cases} z_P = z + c\theta - l_1\gamma \\ z_Q = z + c\theta + l_2\gamma \, , \end{cases} \tag{6.109}$$

where γ is the angle between line PQ and the horizontal and all relevant angles have been assumed to be small.

The potential energy due to the two added springs is

$$\begin{aligned} 2\mathcal{U} = \chi_1 z_P^2 + \chi_2 z_Q^2 = (\chi_1 + \chi_2)\left(z^2 + c^2\theta^2 + 2c\theta z\right) + \\ + \gamma^2(l_1^2\chi_1 + l_2^2\chi_2) + 2\gamma(z + c\theta)(l_1\chi_1 - l_2\chi_2) \, . \end{aligned} \tag{6.110}$$

The value of γ can be easily computed by stating that

$$\frac{\partial \mathcal{U}}{\partial \gamma} = 0 \, ,$$

which yields

$$\gamma = -(z + c\theta)\frac{l_1\chi_1 - l_2\chi_2}{l_1^2\chi_1 + l_2^2\chi_2} \, . \tag{6.111}$$

The final expression for the potential energy is then

$$\mathcal{U} = \frac{1}{2}(z + c\theta)^2 \frac{\chi_1 \chi_2 (l_1 + l_2)^2}{l_1^2 \chi_1 + l_2^2 \chi_2} . \tag{6.112}$$

By performing the relevant derivatives, the stiffness matrix becomes

$$[K] = \begin{bmatrix} K_1 + K_2 + \chi & -d_1 K_1 + d_2 K_2 + \chi c \\ -d_1 K_1 + d_2 K_2 + \chi c & d_1^2 K_1 + d_2^2 K_2 + \chi c^2 \end{bmatrix} , \tag{6.113}$$

where

$$\chi = \frac{\chi_1 \chi_2 (l_1 + l_2)^2}{l_1^2 \chi_1 + l_2^2 \chi_2} .$$

From Eq. (6.113) it is clear that the terms due to the interconnection between front and rear suspensions affect in a different way the various elements of the stiffness matrix and allow to modify independently the values of the bounce and pitch natural frequencies, possibly lowering the latter without affecting the former.

The study of the free bounce and pitch oscillations does not however supply a complete picture of the riding behaviour of the vehicle. The excitations applied to the front and rear wheels are not independent and the rear wheels are excited by the same forcing function as the front ones but with a delay $\tau = l/V$ equal to the time needed to travel a distance equal to the wheelbase.

If the wavelength is very long, i.e. the frequency of the forcing function is small, the excitation of the front and rear wheels occurs almost in phase, with the result of exciting very little pitching modes. A road input with a wavelength equal to the wheelbase or to its whole submultiples will excite in phase the two wheels and then will excite bouncing but not pitching modes. An input with a wavelength equal to twice the wheelbase (or an odd submultiple of it) will excite the two wheels at 180° phasing, producing pitching but not bouncing. The last statement is true only if the centre of mass is at mid-wheelbase; qualitatively however holds even if pitch and bounce motions are not uncoupled and *wheelbase filtering*, as this phenomenon is referred to, is present being immaterial the type of suspensions and the free response of the system.

Wheelbase filtering introduces a dependence of the response of the system from the speed. As an example, if the wheelbase is 2 m and the speed is 20 m/s, the time delay τ is 0.1 s. Maximum pitching and vanishing bouncing response occur with wavelengths of twice the wheelbase and its odd submultiples, 4, 4/3, 4/5, m. At 20 m/s the corresponding frequencies at which the bouncing response is cancelled are 5, 15, 25, Hz. Note that the lowest frequency at which wheelbase filtering occurs is quite high if compared with the bounce natural frequency of the sprung mass: The response of the vehicle is similar to that of the quarter car model, with some frequencies at which the response is filtered out in the high frequency range. The higher the speed the higher is the frequency range in which wheelbase filtering occurs.

In the same example maximum bouncing and cancellation of pitching response occur with wavelengths equal to the wheelbase and its submultiples, 2, 1, 0.5, ... m. At 20 m/s the corresponding frequencies at which the pitching response is cancelled are 10, 20, 30, ... Hz, even higher than those for which bounce is cancelled. However, no pitching excitation occurs at very low frequency, as stated above, and generally speaking very little pitching occurs in motorway driving.

In the case of trucks the longer wheelbase and the lower speeds, together with higher spring stiffness can change this picture: Here wheelbase filtering can lead to high pitching and low bouncing response. This is further aggravated by the fact that in tall vehicles pitching oscillations are felt, at locations above the centre of gravity, as horizontal oscillations which can be a nuisance to riding comfort.

The subjective feeling of riding comfort is also affected by the position in which the passengers are located; while near the centre of gravity pitching oscillations are little felt, far from it they detract to comfort to a greater extent.

The coupling between horizontal and vertical motions due to the suspension geometry can severely reduce comfort. For small oscillations about the equilibrium position it is possible to identify a pitch centre is a way similar to what has been done for the roll centre. Its position allows to study the coupling between vertical and longitudinal dynamics and particularly that between pitching and longitudinal oscillations.

6.10 Roll motions

As already stated, roll motions are coupled with handling and not with the ride comfort motion. However it is also true that rolling can affect strongly the subjective feeling of riding comfort.

If the element J_{xy} of the inertia matrix and the aerodynamic terms included into L_β are neglected together with distances h and q_{0_1}, roll motions can be studied through the equation

$$\begin{bmatrix} J_x & 0 & 0 \\ 0 & J_{x_1} & 0 \\ 0 & 0 & J_{x_1} \end{bmatrix} \begin{Bmatrix} \ddot{\phi} \\ \ddot{\phi}_1 \\ \ddot{\phi}_2 \end{Bmatrix} + \begin{bmatrix} \Gamma_1 + \Gamma_2 & -\Gamma_1 & \Gamma_2 \\ -\Gamma_1 & \Gamma_1 + \Gamma_{t_1} & 0 \\ -\Gamma_2 & 0 & \Gamma_2 + \Gamma_{t_2} \end{bmatrix} \begin{Bmatrix} \dot{\phi} \\ \dot{\phi}_1 \\ \dot{\phi}_2 \end{Bmatrix} +$$
$$+ \begin{bmatrix} \chi_1 + \chi_2 & -\chi_1 & \chi_2 \\ -\chi_1 & \chi_1 + \chi_{t_1} & 0 \\ -\chi_2 & 0 & \chi_2 + \chi_{t_2} \end{bmatrix} \begin{Bmatrix} \phi \\ \phi_1 \\ \phi_2 \end{Bmatrix} = 2 \begin{Bmatrix} 0 \\ \Gamma_{t_1}\dot{\alpha}_{t_1} + \chi_{t_1}\alpha_{t_1} \\ \Gamma_{t_2}\dot{\alpha}_{t_2} + \chi_{t_2}\alpha_{t_2} \end{Bmatrix}, \tag{6.114}$$

where χ_i, χ_{t_i}, Γ_i, Γ_{t_i} are respectively the torsional stiffness and damping of the suspensions and the tires. The excitation is given by the transversal slope of the road α_{t_1} and α_{t_2} at the front and rear suspensions.

Equation (6.114) can be solved numerically and allows the computation of the natural frequencies of roll oscillations. However, a further simplification allows to obtain some interesting informations: If the moments of inertia of the unsprung masses are neglected and the excitations at the front and rear wheels are assumed to be equal, the two unsprung masses can be considered as a single body and the equation of motion reduces to

$$\begin{bmatrix} J_x & 0 \\ 0 & 0 \end{bmatrix} \begin{Bmatrix} \ddot{\phi} \\ \ddot{\phi}_u \end{Bmatrix} + \begin{bmatrix} \Gamma & -\Gamma_u \\ -\Gamma_u & \Gamma \end{bmatrix} \begin{Bmatrix} \dot{\phi} \\ \dot{\phi}_u \end{Bmatrix} +$$
$$+ \begin{bmatrix} \chi & -\chi \\ -\chi & \chi + \chi_t \end{bmatrix} \begin{Bmatrix} \phi \\ \phi_u \end{Bmatrix} = 2 \begin{Bmatrix} 0 \\ \chi_t \alpha_t \end{Bmatrix}, \tag{6.115}$$

where

$$\chi = \chi_1 + \chi_2\,, \ \chi_t = \chi_{t_1} + \chi_{t_2}\,, \ \Gamma = \Gamma_1 + \Gamma_i\,,$$

ϕ_u is the rotation of the unsprung mass, which is modelled as a single body, and the damping of the tires has been neglected.

The equation of motion is formally identical to that of the quarter car model with two degrees of freedom, except for the fact that here the mass of the unsprung body is vanishingly small.

The optimum value of the damping can be obtained from Eq. (6.97)

$$\Gamma_{opt} = \sqrt{\frac{\chi}{J_x}}\sqrt{\frac{2\chi + \chi_t}{\chi_t}} \,. \tag{6.116}$$

This condition is usually not met, particularly when antiroll bars are present. The torsional damping of the suspensions is supplied by the same shock absorbers which are usually designed to optimize bounce behaviour and the damping supplied in rolling is generally lower than optimum. If antiroll bars are present, the increase of stiffness is not accompanied by an increase of damping, which cause, together with an increase of the natural frequency, also a decrease of the damping ratio. The stiffer the suspensions are in torsion, the more underdamped is rolling behaviour, if the stiffness increase has been obtained through antiroll bars. While they reduce rolling in steady state conditions, they can actually increase rolling in dynamic conditions and, in particular, they can cause a strong overshot in the response to a step input as it occurs when a rolling moment is applied suddenly, by a steering input, a wind gust or other causes. High rolling in dynamic conditions can well be a cause of rollover.

6.11 Models of deformable vehicles

The assumption that the vehicle body can be considered as a rigid body is clearly an approximation which can be, in some cases, quite rough. This is particularly the case for industrial vehicles and for some passenger vehicles, as open cars, whose stiffness is lower than usual.

If the body of the vehicle is a deformable body, the position of any point P in the reference frame xyz can be written as $[x+u_x, y+u_y, z+u_z]^T$, where $[x, y, z]^T$ is its position in the undeflected configuration and $[u_x, u_y, u_z]^T$ is its displacement due to the compliance of the body. Its position in the inertial frame XYZ is thus

$$\begin{aligned} \left\{ \begin{array}{c} X \\ Y \end{array} \right\}_P &= \left\{ \begin{array}{c} X \\ Y \end{array} \right\} + \{s_{xy}\} \\ Z_P &= Z + Z_0 + s_z \,, \end{aligned} \tag{6.117}$$

where

$$\{s_{xy}\} = [R_1^*] \left\{ [R_{11}] \left\{ \begin{array}{c} x + u_x \\ y + u_y \end{array} \right\} + [R_{12}] (z + u_z) \right\}$$

$$s_z = [R_{21}]\left\{\begin{array}{c} x+u_x \\ y+u_y \end{array}\right\} + [R_{22}]\,(z+u_z)\,.$$

The deformation of the sprung mass can be expressed in terms of its modal coordinates, i.e. as a linear combination of the eigenfunctions of the undamped system. Note that this remain true even if the sprung mass is a damped and even for nonlinear systems. Since the sprung mass has a plane of symmetry, its modes can be subdivided into symmetrical and skew-symmetrical modes, designated by subscripts *s* and *a* in the following equations.

The displacements $[u_x, u_y, u_z]^T$ can thus be expressed as

$$\left\{\begin{array}{c} u_x \\ u_y \\ u_z \end{array}\right\} = [Q_s(x,y,z)]\{\eta_s(t)\} + [Q_a(x,y,z)]\{\eta_a(t)\}\,, \tag{6.118}$$

where $[Q_i(x,y,z)]$ are matrices containing the eigenfunctions while $\{\eta_i(t)\}$ are vectors containing the modal coordinates. Equation (6.118) would be exact only if an infinity of eigenfunctions and of modal coordinates were considered; however a very good approximation is usually obtained taking into account a very small number of modes, particularly if the system is linear and lightly damped. Note that the modes considered in the equation are those of the free structure; the rigid-body modes however need not be considered as they are already included in the rigid-body analysis already performed.

Instead of using the eigenfunctions, a set of arbitrary functions of the space coordinates can be used, as it is common in the assumed modes methods for structural analysis; in this case however the number of coordinates needed to obtain a good approximation is higher and depends on the choice of the arbitrary functions; moreover the mass and stiffness matrices are not diagonal as it occurs with the eigenfunctions.

Let A and B be two points located in symmetrical positions with respect to xz plane. It follows that $u_{x_A} = u_{x_B}$, $u_{y_A} = -u_{y_B}$ and $u_{z_A} = u_{z_B}$ for symmetrical modes and $u_{x_A} = -u_{x_B}$, $u_{y_A} = u_{y_B}$ and $u_{z_A} = -u_{z_B}$ for skew-symmetrical ones. As a consequence of symmetry, some relevant integrals extended to the whole unsprung mass can be written in a much simplified form

$$
\begin{aligned}
&\int_m \left\{ \begin{array}{c} u_x \\ u_y \\ u_z \end{array} \right\} dm = \left\{ \begin{array}{c} [\mathcal{A}] \\ [0] \\ [\mathcal{C}] \end{array} \right\} \{\eta_s\} + \left\{ \begin{array}{c} [0] \\ [\mathcal{B}] \\ [0] \end{array} \right\} \{\eta_a\} \\
&\int_m x \left\{ \begin{array}{c} u_x \\ u_y \\ u_z \end{array} \right\} dm = \left\{ \begin{array}{c} [0] \\ [0] \\ [\mathcal{F}] \end{array} \right\} \{\eta_s\} + \left\{ \begin{array}{c} [0] \\ [\mathcal{D}] \\ [0] \end{array} \right\} \{\eta_a\} \\
&\int_m y \left\{ \begin{array}{c} u_x \\ u_y \\ u_z \end{array} \right\} dm = \left\{ \begin{array}{c} [\mathcal{E}] \\ [0] \\ [\mathcal{G}] \end{array} \right\} \{\eta_a\} \\
&\int_m \left(u_x^2 + u_y^2 + u_z^2\right) dm = \{\eta_s\}^T [\overline{M}_s] \{\eta_s\} + \{\eta_a\}^T [\overline{M}_a] \{\eta_a\} \, ,
\end{aligned}
\tag{6.119}
$$

where the diagonal matrices $[\overline{M}_s]$ and $[\overline{M}_a]$ are the modal mass matrices for symmetrical and skew-symmetrical modes.

The kinetic energy of a portion dm of mass located at point P is

$$
\begin{aligned}
d\mathcal{T} = \tfrac{1}{2} dm \Bigg\{ &\left\{ \begin{array}{c} \dot{X} \\ \dot{Y} \end{array} \right\}^T \left\{ \begin{array}{c} \dot{X} \\ \dot{Y} \end{array} \right\} + \{s_{xy}\}^T [\dot{R}_1^*]^T [\dot{R}_1^*] \{s_{xy}\} + \\
&+ \{\dot{s}_{xy}\}^T [R_1^*]^T [R_1^*] \{\dot{s}_{xy}\} + 2 \left\{ \begin{array}{c} \dot{X} \\ \dot{Y} \end{array} \right\}^T [\dot{R}_1^*] \{s_{xy}\} + 2 \{\dot{s}_{xy}\}^T [R_1^*] \{\dot{s}_{xy}\} + \\
&+ 2 \{s_{xy}\}^T [\dot{R}_1^*]^T [R_1^*] \{\dot{s}_{xy}\} + \dot{Z}^2 + \dot{s}_z^2 + 2 \dot{Z} \dot{s}_z \Bigg\} \, .
\end{aligned}
\tag{6.120}
$$

As usual in the linearized approach, all terms containing the products of three or more "small quantities" (pitch θ and roll ϕ angles and their derivatives, velocities v and $\dot{Z}$, and the yaw velocity $\dot{\psi}$) can be dropped. By introducing the moments of inertia of the sprung mass about axes xyz centred in H (Fig. 6.20), the expression of the kinetic energy of the sprung mass reduces to

$$\begin{aligned}\mathcal{T} = &\tfrac{1}{2}m_S\left(\dot{X}^2+\dot{Y}^2+\dot{Z}^2\right)+\tfrac{1}{2}J_{S_x}\dot{\phi}^2+\tfrac{1}{2}J_{S_y}\dot{\theta}^2+\tfrac{1}{2}J_{S_z}\dot{\psi}^2-J_{S_{xz}}\dot{\psi}\dot{\phi}+\\ &-m_S c\dot{Z}\dot{\theta}+\tfrac{1}{2}\{\dot{\eta}_s\}^T[\overline{M}_s]\{\dot{\eta}_s\}+\tfrac{1}{2}\{\dot{\eta}_a\}^T[\overline{M}_a]\{\dot{\eta}_a\}+\dot{\psi}\left[-[\mathcal{D}]+[\mathcal{E}]\right]\{\dot{\eta}_a\}+\\ &-\dot{\theta}[\mathcal{F}]\{\dot{\eta}_s\}+\dot{\phi}[\mathcal{G}]\{\dot{\eta}_a\}+\dot{Z}[\mathcal{C}]\{\dot{\eta}_s\}+V\Big[-(\theta+\theta_0)\,\dot{\theta}m_S c+\dot{\theta}m_S h+\\ &+\dot{\psi}\phi m_S h+[\mathcal{A}]\{\dot{\eta}_s\}+\dot{\theta}+[\mathcal{A}]\{\eta_s\}+(\theta+\theta_0)\,[\mathcal{C}]\{\dot{\eta}_s\}-\dot{\psi}[\mathcal{B}]\{\eta_a\}\Big]+\\ &+v\Big(-\dot{\phi}m_S h+\dot{\psi}m_S c+[\mathcal{B}]\{\dot{\eta}_a\}\Big)\,.\end{aligned} \tag{6.121}$$

The kinetic energy of the sprung masses can be expressed in the same way seen for the model of Section 6.4.1. When adding all terms of the various parts of the vehicle, the terms of the type of $(\theta+\theta_0)\,\dot{\theta}m_S c$ and $\dot{\psi}\phi m_S c$ cancel each other as point H is located on z-axis passing for the centre of mass of the whole vehicle.

The gravitational potential energy is still expressed by Eq. (6.45) to which the term

$$-(\theta+\theta_0)\,[\mathcal{A}]\{\eta_s\}+\phi[\mathcal{B}]\{\eta_a\}+[\mathcal{C}]\{\eta_s\}$$

is added. The deformation potential energy of the springs of the i-th suspension is still expressed by Eq. (6.46) to which the terms linked with the deformation modes are added. Assuming that the points of attachment of the springs of the left (right) i-th suspension are x_i, $\pm y_i$ and z_i, it follows

$$\mathcal{U}_{s_i}=\frac{1}{2}K_i\left(Z_i-Z+x_{s_i}\theta-\left[Q_{z_{s_i}}\right]\{\eta_s\}+L_i\right)^2+\frac{1}{2}\chi_i\left(\phi-\phi_i-\left[Q_{z_{a_i}}\{\eta_a\}\right]\right)^2, \tag{6.122}$$

where $\left[Q_{z_{s_i}}\right]$ and $\left[Q_{z_{s_i}}\right]$ are the parts of the matrices of the eigenfunctions for symmetrical and skew-symmetrical modes linked to u_z computed at the point of coordinates x_i, y_i, z_i.

The deformation potential energy of the tires of the i-th suspension is expressed by Eq. (6.47) without any change. The potential energy due to the deformation of the sprung mass is obviously

$$\mathcal{U}_d=\frac{1}{2}\{\eta_s\}^T[\overline{K}_s]\{\eta_s\}+\frac{1}{2}\{\eta_a\}^T[\overline{K}_a]\{\eta_a\}\,, \tag{6.123}$$

where $[\overline{K}_s]$ and $[\overline{K}_s]$ are the modal stiffness matrices.

In a similar way, the Rayleigh dissipation function of the shock absorbers of the i-th suspension is given by Eq. (6.49) suitably modified

$$\mathcal{F}_{s_i} = \frac{1}{2}c_i\left(\dot{Z}_i - \dot{Z} + x_{s_i}\dot{\theta} - \left[Q_{z_{s_i}}\right]\{\dot{\eta}_s\}\right)^2 + \frac{1}{2}\Gamma_i\left(\dot{\phi} - \dot{\phi}_i - \left[Q_{z_{a_i}}\right]\{\dot{\eta}_a\}\right)^2 , \tag{6.124}$$

where $\left[Q_{z_{s_i}}\right]$ and $\left[Q_{z_{a_i}}\right]$ are the same as seen above, but referred to the attachment points of the shock absorbers.

The Rayleigh function for the tires is unchanged (Eq. (6.50)) while that of the deformation modes of the sprung mass is

$$\mathcal{F}_d = \frac{1}{2}\{\dot{\eta}_s\}^T[\overline{C}_s]\{\dot{\eta}_s\} + \frac{1}{2}\{\dot{\eta}_a\}^T[\overline{C}_a]\{\dot{\eta}_a\} , \tag{6.125}$$

where $[\overline{C}_s]$ and $[\overline{C}_a]$ are the modal damping matrices. Note that the last expression is only approximated, as the damping matrices are not diagonal and can couple the modes to a greater extent. Other approximations are due to the way in which the presence of the suspensions has been taken into account, particularly in the case of independent suspensions, but these approximation are the same introduced in all the linearized models seen above.

If the virtual work due to external forces is computed neglecting the displacements due to deformation modes, the expressions of the generalized forces are the same as for the vehicle with rigid elements. This assumption is well suited for the present linearized model in which the exact kinematics of the suspensions is neglected.

A close inspection of the expressions of the potential and kinetic energy and of the dissipation function shows that if the forward velocity V is considered as a known function, the equations of motion split into two groups, exactly in the same way as for the case in which the deformations of the sprung mass are neglected.

A first set of equations deals with the generalized coordinates y^*, ψ, ϕ, ϕ_i and $\{\eta_a\}$. If the vehicle has n axles and m_a skew-symmetrical modes are considered, their number is $3 + n + m_a$. The differential set of equation for the handling behaviour of the vehicle is thus

$$\begin{bmatrix} [M]_1 & [M]_{a1}^T \\ [M]_{a1} & [\overline{M}_a] \end{bmatrix} \begin{Bmatrix} \{\ddot{q}\}_1 \\ \{\ddot{\eta}_a\} \end{Bmatrix} + \begin{bmatrix} [C]_1 & [C]_{1a} \\ [C]_{a1} & [C]_{aa} \end{bmatrix} \begin{Bmatrix} \{\dot{q}\}_1 \\ \{\dot{\eta}_a\} \end{Bmatrix} + \qquad (6.126)$$

$$+ \begin{bmatrix} [K]_1 & [K]_{1a} \\ [K]_{a1} & [K]_{aa} \end{bmatrix} \begin{Bmatrix} \{q\}_1 \\ \{\eta_a\} \end{Bmatrix} = \begin{Bmatrix} \{F\}_1 \\ \{0\} \end{Bmatrix},$$

where $\{q\}_1$, $[M]_1$, $[C]_1$, $[K]_1$ and $\{F\}_1$ are the same matrices and vectors included into Eq. (6.74) and

$$\begin{aligned}
[M]_{a1} &= \left[[\mathcal{B}]^T \; -[\mathcal{D}]^T + [\mathcal{E}]^T \; [\mathcal{G}]^T \; [0] \; [0] \right] \\
[C]_{1a} &= \left[[0] \; [0] \; -\textstyle\sum_{\forall i} \Gamma_i \left[Q_{z_{a_i}}\right] \; \Gamma_1 \left[Q_{z_{a_1}}\right] \; \Gamma_2 \left[Q_{z_{a_2}}\right] \right]^T \\
[C]_{a1} &= \left[[0] \; V[\mathcal{B}]^T \; -\textstyle\sum_{\forall i} \Gamma_i \left[Q_{z_{a_i}}\right]^T \; \Gamma_1 \left[Q_{z_{a_1}}\right]^T \; \Gamma_2 \left[Q_{z_{a_2}}\right]^T \right]^T \\
[C]_{aa} &= [\overline{C}_a] + \sum_{\forall i} \Gamma_i \left[Q_{z_{a_i}}\right]^T \left[Q_{z_{a_i}}\right] \\
[K]_{1a} &= \left[[0] \; -\dot{V}[\mathcal{B}] \; g[\mathcal{B}] - \textstyle\sum_{\forall i} \chi_i \left[Q_{z_{a_i}}\right] \; \chi_1 \left[Q_{z_{a_1}}\right] \; \chi_2 \left[Q_{z_{a_2}}\right] \right]^T \\
[K]_{a1} &= \left[[0] \; [0] \; g[\mathcal{B}]^T - \textstyle\sum_{\forall i} \chi_i \left[Q_{z_{a_i}}\right]^T \; \chi_1 \left[Q_{z_{a_1}}\right]^T \; \chi_2 \left[Q_{z_{a_2}}\right]^T \right]^T \\
[K]_{aa} &= [\overline{K}_a] + \sum_{\forall i} \chi_i \left[Q_{z_{a_i}}\right]^T \left[Q_{z_{a_i}}\right] .
\end{aligned}$$

The last expressions refer to a two-axle vehicle but can easily be generalized.

As in the previous models, coordinates y^* and ψ are present only with their derivatives: The order of the set of differential equations is then $4 + 2n + 2m_a$.

The second set of equations deals with the generalized coordinates x^*, Z, θ, Z_i and $\{\eta_s\}$. If m_s symmetrical modes are considered, their number is $3 + n + m_s$.

Again the first equation, that describing the longitudinal dynamics of the vehicle, is weakly coupled with the others. It can be studied separately and its expression is still Eq. (6.84), to which a term $[\mathcal{A}]\{\ddot{\eta}_s\}/m$ is added to the left hand side.

The set of $2 + n + m_s$ equations which remains after separating the first one describes the suspension motions of the vehicle, or as it is usually referred to, its ride behaviour. It is

$$\begin{bmatrix} [M]_2 & [M]_{s2}^T \\ [M]_{s2} & [\overline{M}_s] \end{bmatrix} \left\{ \begin{array}{c} \{\ddot{q}\}_2 \\ \{\ddot{\eta}_s\} \end{array} \right\} + \begin{bmatrix} [C]_2 & [C]_{s2}^T \\ [C]_{s2} & [C]_{ss} \end{bmatrix} \left\{ \begin{array}{c} \{\dot{q}\}_2 \\ \{\dot{\eta}_s\} \end{array} \right\} +$$

$$+ \begin{bmatrix} [K]_2 & [K]_{s2}^T \\ [K]_{s2} & [K]_{ss} \end{bmatrix} \left\{ \begin{array}{c} \{q\}_2 \\ \{\eta_s\} \end{array} \right\} = \left\{ \begin{array}{c} \{F\}_s \\ \{0\} \end{array} \right\} , \tag{6.127}$$

where $\{q\}_2$, $[M]_2$, $[C]_2$, $[K]_2$ and $\{F\}_2$ are the same matrices and vectors included into Eq. (6.85) and

$$\begin{aligned}
&[M]_{s2} = \left[[\mathcal{C}]^T \; -[\mathcal{F}]^T \; [0] \; [0] \right] \\
&[C]_{s2} = \left[\sum\nolimits_{\forall i} c_i \left[Q_{z_{s_i}}\right]^T \; - \sum\nolimits_{\forall i} c_i x_i \left[Q_{z_{s_i}}\right]^T \; -c_1 \left[Q_{z_{s_1}}\right]^T \; c_2 \left[Q_{z_{s_2}}\right]^T \right]^T \\
&[C]_{ss} = [\overline{C}_s] + \sum_{\forall i} c_i \left[Q_{z_{s_i}}\right]^T \left[Q_{z_{s_i}}\right] \\
&[K]_{s2} = \left[\sum\nolimits_{\forall i} K_i \left[Q_{z_{s_i}}\right] \dot{V} [\mathcal{C}]^T - g[\mathcal{A}]^T - \sum\nolimits_{\forall i} K_i x_i \left[Q_{z_{s_i}}\right]^T \right. \\
&\qquad \left. -K_1 \left[Q_{z_{s_1}}\right]^T \; -K_2 \left[Q_{z_{s_2}}\right]^T \right]^T \\
&[K]_{ss} = [\overline{K}_a] + \sum_{\forall i} K_i \left[Q_{z_{s_i}}\right]^T \left[Q_{z_{s_i}}\right] \\
&\{F\}_s = - \left(\dot{V}[\mathcal{C}] - g[\mathcal{A}] - \sum_{\forall i} K_i x_i \left[Q_{z_{s_i}}\right] \right) \theta_0 - g[\mathcal{C}] + \sum_{\forall i} K_i L_i \left[Q_{z_{s_i}}\right] .
\end{aligned}$$

All matrices are symmetrical, except for the stiffness matrix due to the lack of symmetry of $[K]_2$ caused by the usual aerodynamic terms. Note that the aerodynamic forces have been considered as independent from deformation modes; if this assumption were dropped, the equations would have been different but the uncoupling would not have been affected.

Symmetrical and skew-symmetrical deformation modes thus play a very different role in the dynamic behaviour of the vehicle. The first ones, like bending modes in xz plane, can affect riding comfort but have no importance in the study of handling. The most important skew-symmetrical modes are those related to torsional deformations and it is well known that their influence on handling is very important, particularly in sport cars and above all in Formula 1 racers. Also transversal bending can have a similar effect.

The considerations here seen can be extended to other types of vehicles: In an articulated vehicle for instance the motion about the hinge can be considered as a sort of deformation mode of the sprung mass. Motions due to the articulation which are skew-symmetrical, as the rotation about a vertical axis (like angle θ in the model of Section 5.10) or a horizontal axis (like the roll rotation of the trailer) are coupled with the handling behaviour. Symmetrical motions, as pitching rotations of the trailer, affect comfort.

Similarly, also the deformation of the unsprung masses can be included without changing the overall conclusions: This is important as configurations based on compliant unsprung masses are more and more common.

But the uncoupling is even more a general feature. The above considerations can be applied to vehicles with two wheels, with the only exceptions that the roll angle can easily take values which are beyond the range in which the linearization of trigonometric functions apply and that the lateral movements of the driver, aimed to displace the centre of mass and to produce unsymmetrical aerodynamic forces, destroy the symmetry on which the uncoupling is based.

No particular assumption on the nature of the forces which support the vehicle has been done. The same uncoupling holds also for vehicles supported by hydrostatic, aerostatic or aerodynamic forces. In the first case the assumption of existence of a roll axis of the suspension is substituted by that of existence of a roll axis fixed to the hull in its undeflected configuration. The small roll oscillations are thus demonstrated to be uncoupled with pitch and bounce motions, which are coupled with each other. In the case of aircraft, roll and yaw oscillations are known to be coupled (dutch roll) while bounce and pitch oscillations are coupled to each other. Even the presence of aerodynamic forces due to the deformations of the structure does not change the overall picture, provided that they can be assumed to depend linearly on the modal coordinates $\{\eta_s\}$ and $\{\eta_a\}$.

6.12 Gyroscopic moments and other second-order effects

In the previous sections some minor effects which can alter to a lesser extent the dynamic behaviour of the vehicle have been neglected. One of them is the gyroscopic effect of the wheels and of other rotating parts. If the polar moment of inertia of a wheel and its angular velocity, directed as axis y_i of the i-th unsprung mass are respectively J_p and ω, any angular velocity of the vehicle about axes x_i and z_i will cause a gyroscopic moment expressed, in x_i, y_i, z_i frame as

$$M_g = \omega J_p \begin{Bmatrix} \dot{\psi} \\ 0 \\ -\dot{\phi}_i \end{Bmatrix} = J_p \frac{V}{R_0} \begin{Bmatrix} \dot{\psi} \\ 0 \\ -\dot{\phi}_i \end{Bmatrix} . \tag{6.128}$$

Gyroscopic moments can be easily included into the models seen in the present and the previous chapters simply by including the angular velocity of the rotating elements into the expression of the kinetic energy before using Lagrange equations. The linearized expression of the angular velocity of a wheel rotating about axis y_i with speed ω is (see Eq. (6.20))

$$\begin{cases} p = \dot{\phi}_i - \theta\dot{\psi} \\ q = \dot{\theta} + \phi_i\dot{\psi} + \omega \\ r = \dot{\psi} - \phi_i\dot{\theta} \ . \end{cases} \tag{6.129}$$

If the moment of inertia of the wheel about a diametral axis is J_t, its rotational kinetic energy is

$$\mathcal{T} = \frac{1}{2}\left[J_t\left(\dot{\phi}_i - \theta\dot{\psi}\right)^2 + J_p\left(\dot{\theta} + \phi_i\dot{\psi} + \omega\right)^2 + J_t\left(\dot{\psi} - \phi_i\dot{\theta}\right)^2\right] . \tag{6.130}$$

The terms which do not contain ω are those already included in the previous expressions of the kinetic energy for the non rotating element $\mathcal{T}_{nr}$. It then follows

$$\mathcal{T} = \mathcal{T}_{nr} + \frac{1}{2}J_p\left[\omega^2 + 2\omega\left(\dot{\theta} + \phi_i\dot{\psi}\right)\right] . \tag{6.131}$$

By performing all relevant derivatives of the kinetic energy, it follows immediately that, if the angular velocity ω is kept constant, no new term enters the pitch equation, i.e that related to the degree of freedom θ. On the left hand side of the equation related to roll, i.e. to the degree of freedom ϕ_i a term $-\omega J_p\dot{\psi}$, which comes from the derivative $\partial\mathcal{T}/\partial\phi_i$ must be added. If it is moved to the right hand side of the equation it yields the component along x-axis of the gyroscopic moment M_g (Eq. (6.128)). Finally, the left hand side of the equation related to yaw, i.e. to the degree of freedom ψ contains the term $\omega J_p\dot{\phi}_i$, which comes from the derivative with respect to time of $\partial\mathcal{T}/\partial\dot{\psi}$. If it is moved to the right hand side of the equation the component along z-axis of the gyroscopic moment M_g is found.

In case of steering wheels the rotation axis can be different from axis y_i. Gyroscopic moments can be computed following the same procedure, simply by introducing the components of the angular velocity ω along axes x and y, if the kingpin axis is directed along z-axis. All three components of the angular velocity can be present if the wheel has a camber angle or if the kingpin axle is inclined.

In steady-state motion gyroscopic moments are generally small. Neglecting camber, $\dot{\psi} = V/R$ and $\dot{\phi}_i = 0$: The component M_{g_z} vanishes while

$$M_{g_x} = J_p \frac{V^2}{RR_0} . \tag{6.132}$$

Gyroscopic moment is thus proportional to the centrifugal acceleration V^2/R, which is limited. If J_p and R_o are equal to 6 kg m^2 and 0.5 m, values related to an industrial vehicle, for instance, and the centrifugal acceleration is of 5 m/s^2, a gyroscopic moment of 60 Nm is obtained. If the track of the axle is 1.4 m, the load transfer due to the gyroscopic moment of the two wheels is of about 43 N.

Gyroscopic moments of the wheels can however be more important in nonstationary conditions and, above all, can affect to a larger extent the dynamics of the steering system: Their effect on the free-controls dynamics can thus be important. In rigid-axle suspensions of steering axles, very strong reactions on the steering wheel due to gyroscopic moments of the wheels can be present when travelling on uneven road. They can cause severe discomfort and make driving difficult.

Gyroscopic moments due to the engine or other rotating elements of the vehicle are usually less important, except in particular cases, as that of electric railway engines in which they can cause an increase of wear of the rim of the wheels. As a last consideration, a rotating mass with axis parallel to z axis can couple roll and pitch motions, making uncoupling between handling and comfort impossible even if the assumptions of small displacements, linearity and symmetry hold. This can even be exploited, as in the case of the flywheel stabilizers in ships, where the coupling allows to use the larger pitch moment of inertia to limit the roll oscillations. Also a mass rotating about x-axis has a coupling effect between pitch and yaw, while a mass rotating about y-axis couples roll and yaw, which are already coupled in the handling behaviour.

Other second-order dynamic effects can be of some importance. In nonstationary motion, for example, the angular acceleration of rotating masses can produce inertia torques of non-negligible entity.

Other effects, neglected in the previous models, can be present. As an example, the transmission of the driving torque to the wheels can introduce reaction torques which, being exerted between the various parts of the vehicle, do not affect the global dynamics, at least as a first approximation. They can however modify the configuration of the vehicle and so affect the forces the vehicle receives from the road or, less likely, the air. In a vehicle with longitudinal engine and a rear live axle which includes the differential, the driving torque causes a relative roll displacement between the body and the driving axle. The small roll angle of the body can induce roll steer on the other axles and affect the global behaviour of the vehicle. These effects are usually neglected, owing to their small importance, but to include them into the model is straightforward.

A larger effect can have, as seen in Sec. 6.1.3 when dealing with anti-dive and anti-squat design, the reaction torques exerted on the suspensions when they are not transferred directly to the vehicle body by the links but load the suspension spring and cause lift of dive of a given axle. For this effect to be present the torque must be applied to the sprung mass. For driving torques this occurs only in the case of live axles, while in case of braking torques the brakes are almost always located on the sprung masses, with the only exception of the little used layout in which they are near the differential gears in De Dion axles or independent suspensions of driving wheels. Then the suspension layout must allow a vertical movement when a torque is applied to it, i.e. the derivative $\partial z/\partial M_y$ must be different from zero, an example being that of trailing arm suspensions.

6.13 Excitation sources

6.13.1 *Internal excitation*

The sources of vibration on board of a vehicle are essentially three: The wheels, the driveline and the engine. All contain rotating parts and as a consequence a first cause of dynamic excitation is unbalance. A rotor is perfectly balanced when its rotation axis coincides with one of its principal axes of inertia; however this condition can only be met approximately and balancing tolerances must be stated for any rotating object[2]. As a consequence of the residual unbalance a rotating object exerts on its supports a

[2]G. Genta, *Vibration of structures and machines*, Springer, New York, 1995.

force whose frequency is equal to the rotational speed ω and its amplitude is proportional to its square ω^2. As the engine, the driveline and the wheels rotate at different speeds, the excitations due to them are characterized by different frequencies.

Apart from the excitation due to unbalance, there are other effects which are peculiar of each element. Wheels can have geometrical and structural irregularities. The outer shape of the tires cannot be exactly circular and is characterized by a runout (eccentricity) which has the same effect as mass unbalance, exciting vibrations with a frequency equal to the rotational speed, plus other harmonic components which excite higher harmonics. An ovalization of the shape excites a vibration with frequency equal to 2ω, a triangular shape with frequency 3ω, etc. The very presence of the tread excites higher frequencies, which usually lie in the acoustic range; to avoid a strong excitation with a period equal to the time of passage of the single tread element, the pattern of the tread is usually made irregular, with randomly spaced elements.

The same occurs for variations of stiffness, which induce dynamic forces with frequencies equal to rotational speed and its multiples. As various harmonics exist at different extent in different tires the spectrum of the dynamic force exerted by the tire on the unsprung mass is typical of each tire. As it is common in dynamics of machinery, such typical spectrum is referred to as the mechanical signature of the tire.

When the wheel is called to exert longitudinal and transversal forces, the irregularities, both geometrical and structural, introduce also dynamic components in these directions. However, the tire-wheel assembly is a complex mechanical element with given elastic and damping properties, which can filter out some of the frequencies produced at the road-tire interface. High frequencies are mostly filtered out by the tire itself, before being further filtered by the suspension, and are felt onboard mainly as noise, as airborne vibration.

The excitation due to the driveline unbalance is usually transferred to the vehicle body through its soft mountings. The transmission is however made of flexible elements and, particularly at high frequency, they can have resonances. A long driveshaft can have its own critical speeds, and in the case of a two-span shaft with a central joint as is common in the front-engine, rear-drive layouts a critical speed, corresponding to a mode in which the two span behave as rigid bodies on a compliant central support, is usually located within the working range. If the balancing of the central joint is poor, strong vibrations when crossing this critical speed are present.

When Hooke's joints are present, torque pulsations occur when the input and output shaft are at an angle and can be a major problem. In modern front wheel drive cars the joints near the wheels are of the constant-speed type to avoid the vibrations caused by them, but care must be taken to design the driveline layout to avoid excitations from joints.

The engine is a major source of vibration and noise for the presence of unbalance of the rotating parts, of the inertial forces due to reciprocating elements and of the time variation of the driving torque. These vibrations have a fundamental frequency which, in four-stroke engines, is equal to half the rotational speed but a large number of harmonics are usually present. A reciprocating engine usually has a number of torsional critical speeds and its dynamics is quite complicated. It has been the object of many studies and many books have been written on the subject[3].

The design of the engine suspension system is a complex issue, and active elements have been proposed and are in the designing stage. The elimination of the sources of vibrations, e.g. using dampers on the crankshaft or counterbalance shafts spinning at twice the rotational speed, the proper insulation of the engine from the vehicle structure by using adequate soft mountings and dampers and the noise insulation of the passenger compartment are all very useful provisions for increasing ride comfort.

The engine suspension can be used as a sort of dynamic vibration absorber. The engine mass, the compliance and the damping of its support constitute a damped vibration absorber which can be tuned on the main wheel hop resonance, about 12–15 Hz, to control main vertical shake vibration due to wheel excitation.

To increase acoustical comfort active noise cancellation is very promising. It is already applied in aeronautics and the first automotive applications both in the form of active engine mufflers and of passenger compartment noise control are due to appear soon. With the introduction of active noise control more advanced goals than pure noise suppression can be achieved. As an example, experience in rail transportation field has shown that a complete noise suppression is not considered satisfactory by most passengers, as it decreases privacy allowing to listen to what others are saying, for instance. A completely noiseless machine may seem unnatural and can be in some cases even dangerous; perhaps the ultimate goal is that of achieving a noise that the users find "pleasant".

Similarly absolute vibration suppression can be undesirable as vibration

[3]See, for instance, W. Thompson, *Fundamental of automobile engines balancing*, Mech. Eng. Publ. Ltd., 1978.

conveys many useful information to the driver and can give symptoms of anomalies which are occurring. Also here the goal seems more to supply a vibrational environment which the user finds satisfactory than to suppress completely all vibrational input.

6.13.2 *Road excitation*

The excitation due to motion on uneven road is important for the riding comfort, for the ability of the tires to exert forces in x and y direction, as it causes a variable normal load F_z, and for the stressing of the structural elements. Such excitation cannot be studied with a deterministic approach and the methods used for random vibrations must be applied.

A number of studies have been devoted to characterize experimentally the road profiles and to interpret statistically the results. From experimental measurements of the road profile a law $h(x)$ can be defined and its power spectral density can be obtained through harmonic analysis. Note that the profile is a function of space and not of time and the frequency referred to space $\overline{\lambda}$ is expressed in rad/m or cycles/m and not in rad/s or in Hz. The power spectral density $\overline{S}$ of law $h(x)$ is thus expressed in $\text{m}^2/(\text{rad/m})$ or in $\text{m}^2/(\text{cycles/m})$.

Some examples of power spectral density $\overline{S}$ for tarmac, concrete roads and pavè roads are plotted as functions of $\overline{\lambda}$ in Fig. 6.34. Also the curves proposed by I.S.O. to classify road profiles are shown[4].

The law $\overline{S}(\overline{\lambda})$ can be expressed by a straight line on a logarithmic plot, i.e. by the law $\overline{S} = c\overline{\lambda}^{-n}$, where n is a nondimensional constant while the dimensions of c depend on n (if $n = 2$, c is expressed in $\text{m}^2(\text{cycles/m})$, for instance).

I.S.O. standards suggest $n = 2$ for road ondulations, i.e. for disturbances with a wavelength greater than 6 m, and $n = 1.37$ for irregularities, with a wavelength smaller than the mentioned value.

If the vehicle travels with velocity V it is possible to transform the law $h(x)$ into a law $h(t)$ and compute a frequency λ and a power spectral density S (measured in $\text{m}^2/(\text{rad/s})$ or m^2/Hz) referred to time from $\overline{\lambda}$ and $\overline{S}$ defined with respect to space

[4]B.S.I. Proposal for Generalized Road Inputs to Vehicles, ISO/TC 108/WG9 Document 5, 1972.

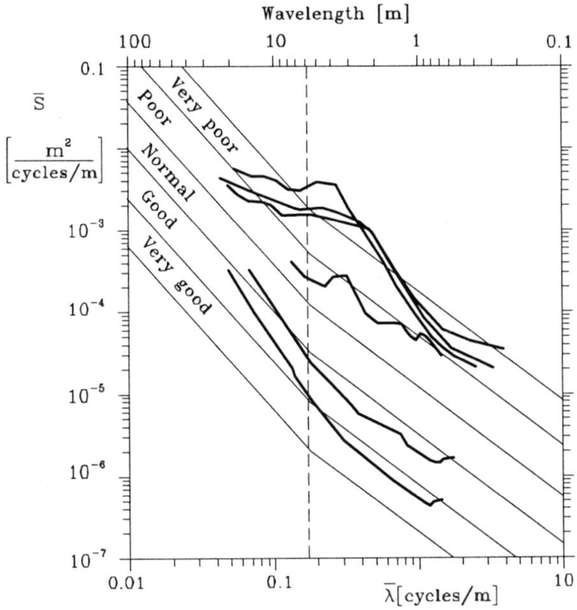

Fig. 6.34 Power spectral density of some road profiles and ISO classification.

$$\begin{cases} \lambda = V\overline{\lambda} \\ S = \dfrac{\overline{S}}{V} . \end{cases} \tag{6.133}$$

The dependence of S from λ is thus

$$S = cV^{n-1}\lambda^{-n} . \tag{6.134}$$

Note that if $n = 2$, as is suggested for road ondulations, the power spectral density of the displacement is proportional to λ^{-2} and then the power spectral density of the vertical velocity is constant: Road excitation is then equivalent to a white noise in terms of vertical velocity of the contact point. This justifies the use of an input of this type for the plot of Fig. 6.29. The power spectral density of the vertical acceleration is proportional to λ^2, showing that high frequency disturbances are strongly felt as acceleration. This is even more true for the ISO standard road profile, which at high frequency shows a lower decrease (power -1.37 instead of -2).

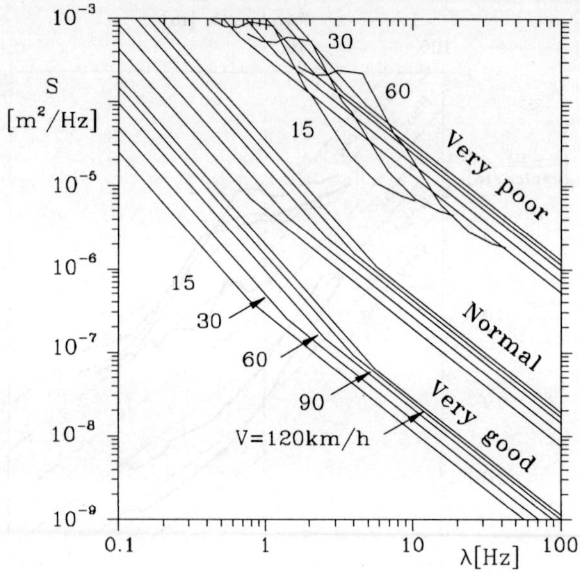

Fig. 6.35 Power spectral density of the displacement $h(t)$ as a function of the frequency λ at various speeds for a pavè road (from G.H. Tidbury, *Advances in Automobile Engineering*, part III, Pergamon Press, London, 1965). Also the curves "very good", "normal" and "very poor" roads following ISO standard are reported.

The law $S(\lambda)$ at various speeds for a pavè road and for roads defined as "very good", "normal" and "very poor" following ISO standard are plotted in Fig. 6.35.

Once that the power spectral density $S(\lambda)$ of the excitation (namely of function $h(t)$) and the frequency response $H(\lambda)$ of the vehicle are known, the power spectral density of the response $S_r(\lambda)$ is easily computed as

$$S_r(\lambda) = H^2(\lambda)S(\lambda) . \tag{6.135}$$

If the frequency response $H(\lambda)$ is the ratio between the amplitude of the acceleration of the sprung mass and that of the displacement of the contact point, the power spectral density $S_r(\lambda)$ refers directly to the acceleration of the sprung mass. The root mean square value of the acceleration, for instance, computed with reference to a given frequency range, is simply

$$a_{rms} = \sqrt{\int_{\lambda_1}^{\lambda_2} S_a(\lambda)d\lambda} \,. \tag{6.136}$$

The response, in terms of acceleration of the sprung and the unsprung masses of a quarter car model moving at various speeds on two different roads is shown in Fig. 6.36.

The figures on the left are obtained for the optimum value of the damping of the shock absorber while those on the right for a value equal to half of that. The figure shows that the response is very strong, particularly for what the unsprung mass is concerned on the pavè road, which constitutes a very severe test and a sort of limiting case, and that a decrease of the damping coefficient leads to a strong increase of the response.

In the case of a vehicle, road irregularities excite bounce, pitch and roll motions; for bounce and pitch excitation wheelbase filtering must be accounted for. Consider the model expressed by Eq. (6.106) and introduce in the vector of the forcing functions the expressions

$$\begin{cases} h_A = h_o \sin(\lambda t) \\ h_B = h_o \sin[\lambda(t+\tau)] = h_o[\sin(\lambda t)\cos(\lambda\tau) + \cos(\lambda t)\sin(\lambda\tau)] \,, \end{cases} \tag{6.137}$$

where $\tau = l/V$ is the time needed to travel a distance equal to the wheelbase.

The expression for the excitation vector is then

$$\frac{h_0}{l}\left\{\begin{array}{l} \Big[d_2K_1 + d_1K_2\cos(\lambda\tau) - d_1c_2\lambda\sin(\lambda\tau)\Big]\sin(\lambda t) + \\ \qquad + \Big[d_2c_1\lambda + d_1c_2\lambda\cos(\lambda\tau) + d_1K_2\lambda\sin(\lambda\tau)\Big]\cos(\lambda t) \\ \Big[-K_1 + K_2\cos(\lambda\tau) - c_2\lambda\sin(\lambda\tau)\Big]\sin(\lambda t) + \\ \qquad + \Big[-c_1\lambda + c_2\lambda\cos(\lambda\tau) + K_2\lambda\sin(\lambda\tau)\Big]\cos(\lambda t) \end{array}\right\} . \tag{6.138}$$

The amplitude of the excitation which is included in the first equation, that for bounce motion, is easily computed by adding the squares of the in-phase and in-quadrature components of the first element of the excitation vector and computing the square root. The result is

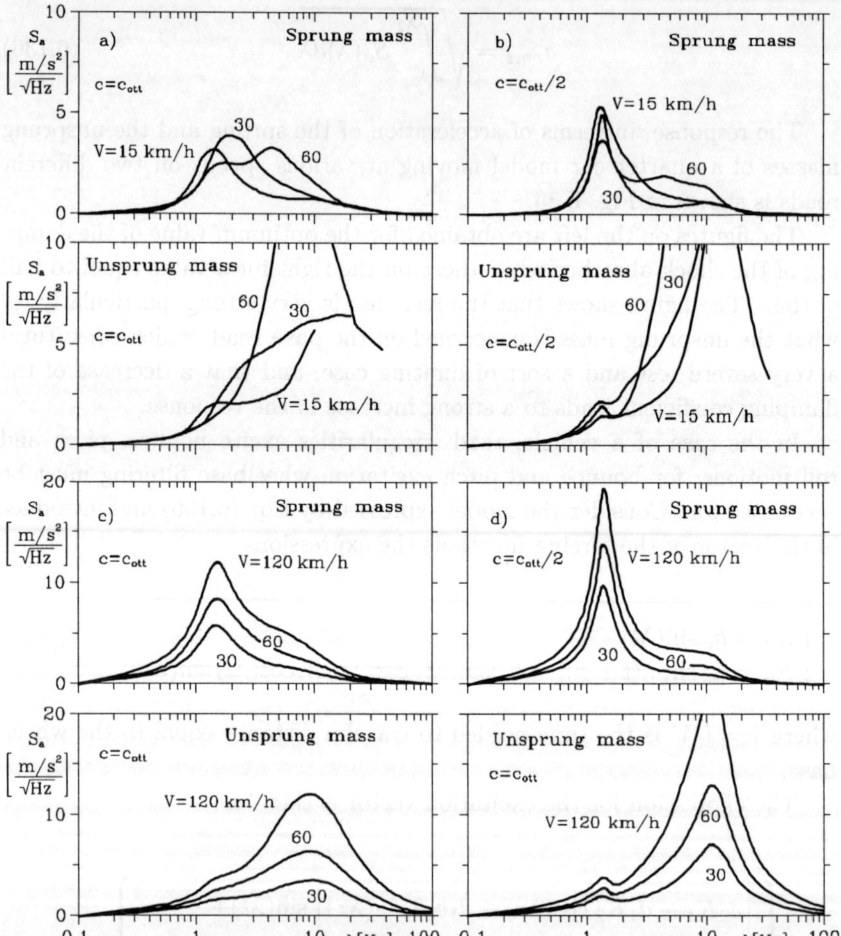

Fig. 6.36 Response of a quarter car model travelling at various speeds on a pavè road (a, b) and on a road defined as normal following ISO standard (c, d). Note the different scales and different speed ranges. The response is reported in the form of the spectrum of the amplitude of the acceleration (i.e. the square root of the power spectral density). The pavè road is a stretch of the MIRA test track and the quarter car model has the following data: m_S = 230 kg, m_u = 20 kg, K = 20.4 kN/m, P = 86 kN/m.

$$\frac{h_0}{l}\left(d_2^2K_1^2 + d_1^2K_2^2 + d_1^2c_2^2\lambda^2 + d_2^2c_1^2\lambda^2 + 2d_1d_2(K_1K_2 + c_1c_2\lambda^2)\cos(\lambda\tau) + \right.$$
$$\left. +2d_1d_2\lambda(c_1K_2 - c_2K_1)\sin(\lambda\tau)\right)^{1/2} . \tag{6.139}$$

In a similar way, the amplitude of the excitation in the second equation, that for pitch motion, is

$$\frac{h_0}{l}\left(K_1^2 + K_2^2 + c_2^2\lambda^2 + c_1^2\lambda^2 - 2(K_1K_2 + c_1c_2\lambda^2)\cos(\lambda\tau) + \right. \\ \left. -2\lambda(c_1K_2 - c_2K_1)\sin(\lambda\tau)\right)^{1/2} . \tag{6.140}$$

In order to obtain a first approximation evaluation of wheelbase filtering, assume that $d_2K_1 = d_1K_2$ and $d_1c_2 = d_2c_1$. The amplitude of bounce excitation is

$$\frac{h_0}{l}\sqrt{d_2^2K_1^2 + d_1^2c_2^2\lambda^2}\,\sqrt{2[1+\cos(\lambda\tau)]}\,. \tag{6.141}$$

Similarly, if $K_1 = K_2$ and $c_2 = c_1$, the amplitude of pitch excitation is

$$\frac{h_0}{l}\sqrt{K_1^2 + c_2^2\lambda^2}\,\sqrt{2[1-\cos(\lambda\tau)]}\,. \tag{6.142}$$

Wheelbase filtering has then the effect of multiplying the frequency response of the vehicle by $\sqrt{2[1+\cos(\lambda\tau)]}$ for bounce and $\sqrt{2[1-\cos(\lambda\tau)]}$ for pitch. Such functions are reported as functions of the frequency together with the frequency response of the quarter car model of Fig. 6.36 in Fig. 6.37. The values of the speed and the wheelbase used for the figure are respectively of 30 m/s and 2.16 m.

6.13.3 *Effects of vibration on man*

The optimization of the suspension has been performed with the aim of reducing to a minimum the vertical accelerations of the sprung mass. This would be correct if the discomfort were caused directly by the acceleration, being immaterial the frequency at which the acceleration is applied.

The ability of the human body to withstand vibration and the related discomfort has been the object of countless studies and several standards on the subject have been stated. Only the plot of Fig. 6.38, taken from ISO 2631 standard, will be reported here.

The standard states the r.m.s. value of the acceleration which causes, in a given time, a reduction of the physical efficiency. The exposure limits can be obtained by multiplying by 2 the values reported in the figure, while

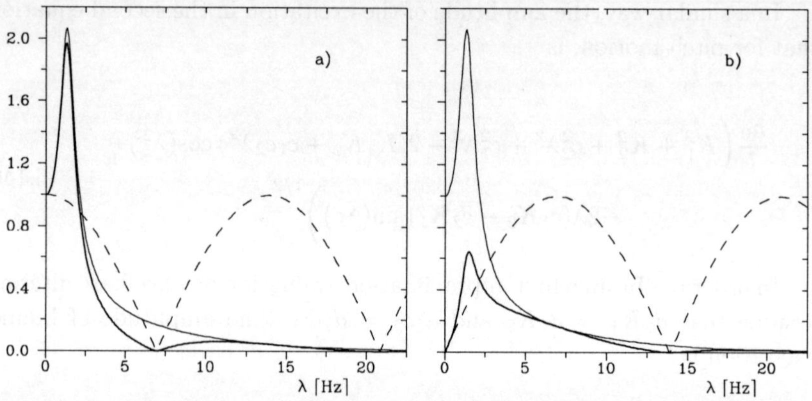

Fig. 6.37 Wheelbase filtering. (a): Function $\sqrt{2[1+\cos(\lambda\tau)]}$ (dotted line), inertance of the quarter car model of Fig. 6.36 and product of the two; (b): Same as in (a), but for function $\sqrt{2[1-\cos(\lambda\tau)]}$. $V = 30$ m/s; $l = 2.16$ m.

the "reduced comfort boundary" is obtained by dividing the same values by 3.15 (i.e., by decreasing the r.m.s. value by 10 dB). From the plot it is clear that the frequency field in which man is more affected by vibration lies between 4 and 8 Hz.

Frequencies lower than 1 Hz produce sensations which can be assimilated to motion sickness. They depend on many parameters other than acceleration and vary from individuals. Above 80 Hz the effect of vibration depends very much on the part of the body which is interested and on the skin conditions, as local vibrations become the governing factor, to give general guidelines. There are also resonance fields at which some parts of the body vibrate with particular large amplitudes. As an example, the thorax-abdomen system has a resonant frequency at about 3 –6 Hz, although all resonant frequency values are much dependent on individual characteristics. The head-neck-shoulder system has a resonant frequency at about 20–30 Hz, and many other organs have more or less pronounced resonances at other frequencies (e.g., the eyeball at 60–90 Hz, the lower jaw-skull system at 100–220 Hz, etc.).

At low frequency the curves are, on a logarithmic plane, straight lines sloping downwards while at high frequency they are lines parallel to the bisector of the first quadrant: At low frequency the derivative of the acceleration with respect to time, i.e. the jerk d^3z/dt^3, has the same importance as the acceleration, while at high frequency the amplitude of the velocity is

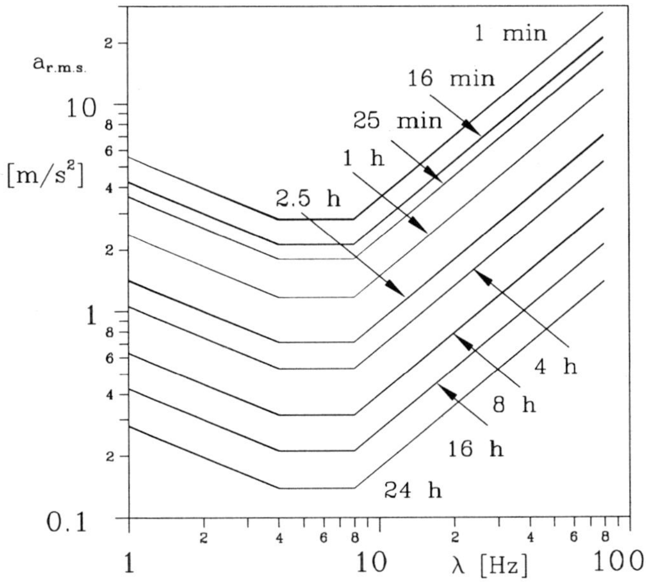

Fig. 6.38 rms value of the vertical acceleration causing reduced physical efficiency to a sitting subject as a function of the frequency. The curves for different exposure times have been reported (ISO 2631 standard).

constant. This suggests that what actually causes discomfort is either the vertical jerk or the acceleration or the velocity depending on the frequency range and not the acceleration alone.

6.14 Concluding remarks on ride comfort

The linearized study of suspension motions, based mostly on quarter-car models, shows that the value of the damping of the shock absorbers which optimizes comfort is the same as that reducing to a minimum the dynamic component of the force on the ground and hence optimizing handling. However, some results obtained considering the root mean square value of the acceleration and of the dynamic component of the force show that, even using a simplified linearized model, the value optimizing comfort is lower than that optimizing handling.

The last statement is also confirmed by other considerations. Firstly, to optimize handling the reduction of the force is not the only goal and the displacement of the sprung mass with respect to the unsprung ones is also important. Every type of suspension has some deviations from a perfect kinematic guide and thus causes the wheels to be set in a position which is different from the nominal one (e.g., changes of the camber angles, roll steer etc.); this affects negatively the handling characteristics of the vehicle. The larger the displacement of the sprung mass the worse is this problem.

Operating in the same way as for minimizing the acceleration, it can be shown that the value of the damping which minimizes the displacement is

$$c = \sqrt{\frac{m(P+K)(P+2K)}{2P}}\,, \tag{6.143}$$

which is higher than the optimum value computed above.

Note that this also suggests the increase of the stiffness of the suspensions and goes against the criterion of "the softer the better" which derives from the consideration of the vertical acceleration alone.

Another point is linked to roll oscillations. The damping of the shock absorbers is usually chosen considering mainly bounce; this causes rolling motions to be in most cases excessively underdamped. When antiroll bars are used the situation becomes worse: By increasing the roll stiffness without increasing the corresponding damping they cause a more marked underdamped behaviour and a decrease of the dynamic stability of roll motions. This not only increases the amplitude of rolling motions, while lowering the roll angle in steady state conditions, and the dynamic load transfer but also makes the rollover of the vehicle in dynamic conditions easier.

The increase of the damping of the shock absorbers above the value defined above as optimum is effective in reducing these effects, which affect more the handling characteristics than comfort.

On the contrary, the importance of reducing the jerk to increase comfort goes in the opposite direction. The value of damping which minimizes jerk is lower than that which minimizes the acceleration leading to a better comfort when the damping is decreased.

The effect of the stiffness of the springs on comfort is in a way contradictory: On one hand, as already stated, the need of reducing the acceleration suggests to reduce the stiffness as much as possible, but this would lead to very low natural frequencies which can in turn cause motion sickness and similar effects.

As already stated the behaviour of the shock absorbers is usually far from being linear. Moreover, the force they exert is not only a function of the velocity but also of the displacement and of other parameters, like the temperature. Usually this last consideration is neglected and the force F is considered as a function of the vertical velocity alone ($F = F(w)$) but this function is almost always nonsymmetric, with a higher damping capacity in the rebound stroke than in the jounce one. If the force is proportional to the velocity, the law $F(w)$ is at best made by two straight lines through the origin. The linear model seen above would thus seem not to be applicable even to the case of small oscillations.

Any law $F(w)$ can be thought as the sum of an odd function and an even function. It is possible to demonstrate[5] that the even function causes a displacement of the centre of the oscillations in dynamic conditions from the static equilibrium position but has little effect on the response of the system. If the even part of the law $F(w)$ is neglected, the characteristic of the shock absorber can be linearized in the origin and an equivalent linear damping can be used for the study of the small oscillations of the system: This explains why linearized models can be used with some confidence even in a case in which the effect of the nonlinearities would seem to be important in all working conditions.

Dry friction in leaf springs introduces hysteresis and an apparent increase of stiffness in low amplitude motion. A qualitative force-deflection characteristics of a leaf spring is shown in Fig. 6.39: The hysteresis cycle is well visible. Note that the overall elastic behaviour is practically linear, with a hysteresis cycle occurring about the straight line representing the average stiffness. If small amplitude oscillations occur about the equilibrium position the apparent stiffness is strongly dependent on the amplitude, with a value tending to infinity when the amplitude tends to zero. This behaviour is typical of dry friction and causes the spring to lock when very small movements are required. The stiffness for the small oscillations which are typical of ride behaviour can be far larger than the overall stiffness of the spring.

The presence of dry friction makes linear models not applicable, or at least makes their results quite inaccurate and causes a deterioration of the ride qualities of the suspension.

Other nonlinearities can be introduced by nonlinear springs, which are sometimes used for industrial vehicles in order to avoid large variations of the natural frequencies with the load. Also air springs are widely used on industrial vehicles, and their characteristic is strongly nonlinear. However,

[5] G.Genta, P.Campanile, *An Approximated Approach to the Study of Motor Vehicle Suspensions with Nonlinear Shock Absorbers, Meccanica*, Vol. 24, 1989, pp. 47-57.

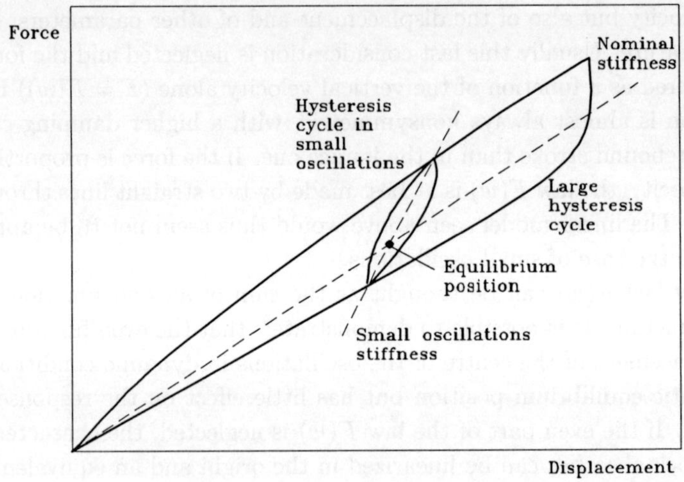

Fig. 6.39 Load-deflection characteristics of a leaf spring exhibiting hysteretic behaviour. The hysteresis cycle for small displacements about the equilibrium position is shown.

the nonlinearities of the behaviour of the springs can be dealt with in the same way as those which are due to the kinematics of the suspensions and, in the motion about any equilibrium condition, a linearized study holds with very good approximation.

6.15 Simple model for the handling of motorcycles

A simple handling model for two wheeled vehicles is shown in the present section. When modelling a vehicle with two wheels it is impossible to neglect roll motion and also the steer angle must be considered; as a consequence a model with a minimum of five degrees of freedom is required. The model can then be linearized if the stability in straight running is considered or to simulate manoeuvres with low lateral acceleration, but conditions in which the roll angle reaches values high enough to compel to resort to a nonlinear model occur very frequently, not only in competition driving, but also in normal driving conditions.

Further difficulties are linked to the facts that a two wheeled vehicle is intrinsically unstable and the driver also has to perform as a stabilizer for the capsize mode and that the mass of the driver, who controls the vehicle not only acting on the steering but also displacing its body, can be a substantial fraction of the total mass. To model the vehicle without modelling also the driver can lead to poor results but, as usual, to build a model for the human behaviour is a very difficult task.

Several mathematical models for the behaviour of the motorcycles have been built. Almost all of them rely on rough simplification of the actual system, as assuming that the contact points of the wheels lies in the plane of symmetry of the vehicle, and on a more or less large linearization. In the following sections a model developed in A. Vietti, *Modellistica dei motoveicoli da competizione*, Degree thesis, Dept. of Mechanics, Politecnico di Torino, May 1996, will be followed.

6.15.1 *Geometrical definitions and degrees of freedom*

A sketch of the vehicle model is shown in Fig. 6.40; it is made by two rigid bodies, one including the vehicle frame, the rear wheel and the driver (mass m_2, centre of mass G_2) and the second including the steering system and the front wheel (mass m_1, centre of mass G_1). They are connected to each other through a cylindrical hinge. A reference frame $Hx_2y_2z_2$ is rigidly connected to the first one, with origin at point H defined in the same way as for the model of the vehicle on elastic suspensions (Fig. 6.20). Its position is defined by the yaw and roll angles ψ and ϕ defined as in Fig. 6.19, with the only difference that here the pitch angle θ vanishes as the presence of the suspensions is not accounted for.

Frame $x_1y_1z_1$ is rigidly connected to the steering system, with axis z_1 coinciding with the axis of the hinge and axis x_1 passing through the centre of mass G_1. The steering angle δ is the angle between axis x_1 and plane x_2z_2.

The generalized coordinates are the coordinates X and Y of point H in the inertial reference frame XYZ and the yaw, roll and steering angles ψ, ϕ and δ.

Operating in the same way as for the previous models, the relevant equations will be obtained through Lagrange equations. The kinetic and potential energies, the Rayleigh dissipation function and the generalized forces must then be obtained.

6.15.2 *Kinetic and potential energy, Rayleigh function*

The positions of points G_1 and G_2 are

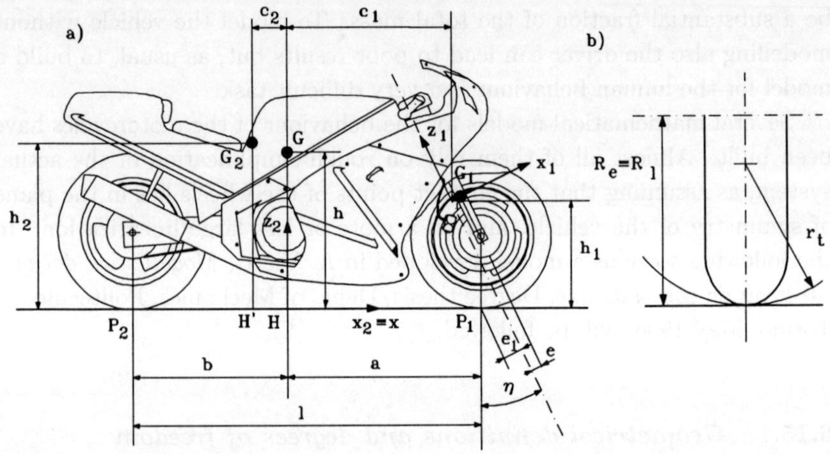

Fig. 6.40 Model for a two wheeled vehicle. (a): Reference frames and main geometrical definitions; (b) sketch of the transversal profile of the tires; definition of the transversal curvature radius r_t.

$$\overline{(G_2 - O)} = \begin{Bmatrix} X \\ Y \\ 0 \end{Bmatrix} + [R_1] \left(\begin{Bmatrix} 0 \\ -r_t \tan(\phi) \\ r_t \end{Bmatrix} + [R_1] \begin{Bmatrix} -c_2 \\ d \\ h_2 - r_t \end{Bmatrix} \right), \tag{6.144}$$

$$\overline{(G_1 - O)} = \begin{Bmatrix} X \\ Y \\ 0 \end{Bmatrix} + [R_1] \left[\begin{Bmatrix} 0 \\ -r_t \tan(\phi) \\ r_t \end{Bmatrix} + \right.$$

$$\left. +[R_2] \left(\begin{Bmatrix} -c_2 \\ d \\ h_2 - r_t \end{Bmatrix} + [R_\eta][R_3] \begin{Bmatrix} e \\ 0 \\ 0 \end{Bmatrix} \right) \right], \tag{6.145}$$

where d is the lateral displacement of point G_2 from the midplane, due to the lateral displacement of the driver, and r_t is the radius of the transversal profile of the tire. The latter has been introduced to take into account the fact that in modern motor cycles the tires, which are very stiff, have a transversal profile of the type sketched in Fig. 6.40b. r_t is the radius of curvature of the part of the tire in contact with the ground, assumed to be constant about the midplane of the wheel for all the arc of contact, up to the position of maximum roll. If two different tires are used, r_t is the average of the radii; this approximation can be rough, but is the only assumption which prevents the roll axis from becoming skew with the plane of symmetry during the roll motion. The assumption of rigid tires for the study of handling behaviour is consistent with the uncoupling of handling and ride; moreover, the equivalent rolling radius of the tire R_e is assumed to coincide with the loaded radius R_l.

Matrices $[R_i]$ are simply

$$[R_1] = \begin{bmatrix} \cos(\psi) & -\sin(\psi) & 0 \\ \sin(\psi) & \cos(\psi) & 0 \\ 0 & 0 & 1 \end{bmatrix} , \qquad [R_2] = \begin{bmatrix} 1 & 0 & 0 \\ 0 & \cos(\phi) & -\sin(\phi) \\ 0 & \sin(\phi) & \cos(\phi) \end{bmatrix} ,$$

$$[R_3] = \begin{bmatrix} \cos(\delta) & -\sin(\delta) & 0 \\ \sin(\delta) & \cos(\delta) & 0 \\ 0 & 0 & 1 \end{bmatrix} , \qquad [R_\eta] = \begin{bmatrix} \cos(\eta) & 0 & -\sin(\eta) \\ 0 & 1 & 0 \\ \sin(\eta) & 0 & \cos(\eta) \end{bmatrix} .$$

The expression of the translational kinetic energy is then

$$\mathcal{T}_t = \frac{1}{2} m_2 \overline{(\dot{\mathrm{G}_1 - \mathrm{O}})}^T \overline{(\dot{\mathrm{G}_1 - \mathrm{O}})} + \frac{1}{2} m_1 \overline{(\dot{\mathrm{G}_2 - \mathrm{O}})}^T \overline{(\dot{\mathrm{G}_2 - \mathrm{O}})} . \tag{6.146}$$

In a similar way the angular velocities of the two rigid bodies are

$$\{\omega_1\} = \begin{Bmatrix} 0 \\ 0 \\ \dot{\delta} \end{Bmatrix} + [R_3]^T [R_\eta]^T \begin{Bmatrix} \dot{\phi} \\ 0 \\ 0 \end{Bmatrix} + [R_3]^T [R_\eta]^T [R_2]^T \begin{Bmatrix} 0 \\ 0 \\ \dot{\psi} \end{Bmatrix} , \tag{6.147}$$

$$\{\omega_2\} = \begin{Bmatrix} \dot{\phi} \\ 0 \\ 0 \end{Bmatrix} + [R_2]^T \begin{Bmatrix} 0 \\ 0 \\ \dot{\psi} \end{Bmatrix} . \tag{6.148}$$

To include into the model the gyroscopic moments due to the wheels, their kinetic energy due to rotation must be included into the expression of the rotational kinetic energy of the vehicle. The total angular velocity of the wheels can thus be expressed as

$$\{\omega_{w_i}\} = \{\omega_i\} + \begin{Bmatrix} 0 \\ V/R_{e_i} \\ 0 \end{Bmatrix} , \tag{6.149}$$

with i=1, 2 for the front and rear wheels respectively. To assume that the rolling radius is constant is an approximation even in the case of rigid wheels due to the toroidal profile of the tires.

The rotational kinetic energy is thus

$$\mathcal{T}_r = \frac{1}{2}\{\omega_1\}^T[J_1]\{\omega_1\} + \frac{1}{2}\{\omega_2\}^T[J_2]\{\omega_2\} + \sum_{i=1}^{2}\frac{1}{2}\{\omega_{w_i}\}^T[J_{w_i}]\{\omega_{w_i}\} . \tag{6.150}$$

As the wheels are gyroscopic solids, the kinetic energy due to rotation about their transversal axes can be included into the rotational kinetic energy of the frame and the steering system simply by including the transversal moment of inertia of the wheels J_{t_i} into that of the main bodies. The inertia matrices $[J_{w_i}]$ are thus simply

$$[J_{w_i}] = \begin{bmatrix} 0 & 0 & 0 \\ 0 & J_{p_i} & 0 \\ 0 & 0 & 0 \end{bmatrix} .$$

The steering system is symmetrical with respect to xz plane. Its inertia matrix $[J_1]$ is

$$[J_1] = \begin{bmatrix} J^*_{1_x} & 0 & J^*_{1_{xz}} \\ 0 & J^*_{1_y} & 0 \\ J^*_{1_{xz}} & 0 & J^*_{1_z} \end{bmatrix} .$$

Note that the moments of inertia $J^*_{1_{ij}}$ are baricentric and referred to a frame with z^*_1 axis parallel to the hinge axis z_1 (see Fig. 6.40).

The inertia matrix of the frame of the vehicle, which includes the driver, $[J_2]$ would have the same structure if the driver were located symmetrically with respect of xz plane, i.e. if distance d were vanishingly small. However to consider in a correct way the position of the driver would lead to much complicated analytical formulations, which are at any rate approximated as it is impossible to model in a correct way the geometry of a moving human body. As an example, the upper part of the body can be considered as a rigid body hinged at the seat location, but this is also a rough approximation.

In the present model the lateral displacement of the driver is accounted for simply by adding to the moments of inertia $J_{2_{ij}}$ computed assuming that the driver is symmetrically located, the effect of a concentrated mass equal to the mass m_d of the driver, located at its centre of mass and laterally displaced of d

$$[J_2] = \begin{bmatrix} J_{2_x} + \dfrac{m_2^2 - m_2 m_d}{m_d} d^2 & 0 & J_{2_{xz}} \\ & J_{2_y} & \dfrac{m_2^2 - m_2 m_d}{m_d} \Delta h d \\ \text{symm.} & & J_{2_z} + \dfrac{m_2^2 - m_2 m_d}{m_d} d^2 \end{bmatrix},$$

where Δh is the vertical distance between G_2 and the centre of mass of the vehicle frame.

The gravitational potential energy is simply

$$\mathcal{U} = m_1 g \begin{bmatrix} 0 & 0 & 1 \end{bmatrix} (\overline{G_1 - O}) + m_2 g \begin{bmatrix} 0 & 0 & 1 \end{bmatrix} (\overline{G_2 - O}) . \tag{6.151}$$

The Rayleigh dissipation function due to the steering damper is

$$\mathcal{F} = \frac{1}{2} c_\delta \dot{\delta}^2 . \tag{6.152}$$

6.15.3 *Generalized forces*

The expressions of the sideslip angles of the vehicle and of the wheels are close to the ones seen for the models of vehicles with four or more wheels, with the added simplification that in the present case the contact points of the wheels are in, or very close to, the symmetry plane and that there is a single wheel per axle. Also the sideslip angles of the wheels are in the case of vehicles with two wheels usually very small and the linearization of their trigonometric functions is even more justified than in the previous cases. A major difference is however the fact that the inclination of the steering axle η is usually not negligible.

With simple computations it is possible to verify that the linearized expressions for the sideslip angles and the camber angles are

$$\begin{cases} \alpha_1 = \dfrac{v}{V} + \dfrac{a}{V}\dot{\psi} - \dfrac{e_1}{V}\dot{\delta} - \delta\cos(\eta) \\ \\ \alpha_2 = \dfrac{v}{V} - \dfrac{b}{V}\dot{\psi} \end{cases} \tag{6.153}$$

$$\begin{cases} \gamma_1 = \phi - \delta\sin(\eta) \\ \gamma_2 = \phi\,. \end{cases} \tag{6.154}$$

Generalized forces due to wheel-road contact and to aerodynamic actions are obtained in the same way as seen for the previous models: Virtual displacements of the application points are first written and then the virtual work is computed and differentiated with respect to the virtual displacements. This procedure is however quite intricate and was performed through a symbolic computer code.

Note that in the case of motorcycles, and particularly for high performance racing machines, aerodynamic forces, mainly drag, due to the driver are quite important. Depending on the driving stile, the driver stands up with his torso to act as an aerodynamic brake and displaces laterally the point of application of the drag to create a yawing moment. In particular, he opens the knee on side towards the centre of the curve not only to lightly touch the ground with the knee protection, which is useful to feel with precision the roll angle and, as some drivers insist, to improve equilibrium, but mainly to produce an aerodynamic yawing moment which tends to turn the vehicle towards the inside of the curve. This reduces the understeering behaviour (or increase the oversteering one) of the vehicle.

In the present model aerodynamic forces are applied to the centre of mass G_2. The dependence of forces and moments on the generalized coordinates and on the control variable d is considered linear

$$\begin{cases} F_{x_{aer}} = -\frac{1}{2}\rho V_r^2 S C_x \\ F_{y_{aer}} = \frac{1}{2}\rho V_r^2 S \left[(C_y)_{,\beta}\beta + (C_y)_{,\delta}\delta\right] \\ F_{z_{aer}} = \frac{1}{2}\rho V_r^2 S C_z \\ M_{x_{aer}} = \frac{1}{2}\rho V_r^2 Sl \left[(C_{M_x})_{,\beta}\beta + (C_{M_x})_{,\delta}\delta + (C_{M_x})_{,d}d\right] \\ M_{y_{aer}} = \frac{1}{2}\rho V_r^2 Sl C_{M_y} \\ M_{z_{aer}} = \frac{1}{2}\rho V_r^2 Sl \left[(C_{M_z})_{,\beta}\beta + (C_{M_z})_{,\delta}\delta + (C_{M_z})_{,d}d\right] . \end{cases} \tag{6.155}$$

Note that the variations of drag, lift, side force and pitching moment due to the driver movements have been neglected.

6.15.4 *Linearized model*

Even the linearized model is complicated enough to require the use of a symbolic computer code. When linearizing the expressions of the normal forces due to the tires, the terms containing the variables of motion disappear and the expressions reduce to those seen for the vehicle with four wheels, with the difference that now the axle and the wheel coincide.

As was the case for a four-wheeled vehicle, if the speed V is assumed as a known function of time and the equations are written in a reference frame centred at point H and with x axis coinciding with x_2 axis, the first equation of motion uncouples from the others. It reduces to the obvious expression

$$m\dot{V} = F_{x_1} + F_{x_2} - \frac{1}{2}\rho V_r^2 S C_x ,$$

where m is the total mass of the vehicle.

The other four equations can be written in the form

$$[M]\{\ddot{x}\} + [C]\{\dot{x}\} + [K]\{x\} = \{F_c\} + \{F_e\} , \tag{6.156}$$

where the generalized forces $\{F_c\}$ due to the driver are kept separate from external forces $\{F_e\}$.

The vector containing the generalized coordinates is

$$\{x\} = \left[y\ \psi\ \phi\ \delta\right]^T . \tag{6.157}$$

The mass matrix is

$$[M] = \begin{bmatrix} m & 0 & -mh & m_1 e \\ & J_z & J_{xz} & m_1 e c_1 + J_{1_z}\cos(\eta) + J_{1_{xz}}\sin(\eta) \\ & & J_x & -m_1 e h_1 - J_{1_z}\sin(\eta) + J_{1_{xz}}\cos(\eta) \\ \text{symm.} & & & J_{1_x} \end{bmatrix}, \tag{6.158}$$

where J_{ij} are the moments of inertia of the whole vehicle and $J_{1_{ij}}$ are those of the steering system referred to frame $x_1y_1z_1$ which is not baricentric.

The elements of matrix $[C]$ are

$$\begin{aligned}
C_{11} &= \frac{1}{V}\left[C_1 + C_2 - \frac{1}{2}\rho V_r^2 S (C_y)_{,\beta}\right] \\
C_{12} &= mV + \frac{1}{V}(C_1 a - C_2 b) \\
C_{13} &= C_{33} = 0 \\
C_{14} &= -\frac{1}{V} C_1 e_1 \\
C_{21} &= -\frac{1}{V}\left[-C_1 a + C_2 b + (C_{M_{z_1}})_{,\alpha} + (C_{M_{z_2}})_{,\alpha} + \frac{1}{2}\rho V_r^2 S\Big(l(C_{M_z})_{,\beta} - c_2 (C_y)_{,\beta}\Big)\right]
\end{aligned}$$

$$\begin{aligned}
C_{22} &= \frac{1}{V}\left[C_1 a^2 + C_2 b^2 - a(C_{M_{z_1}})_{,\alpha} + b(C_{M_{z_2}})_{,\alpha}\right] \\
C_{23} &= V\left(\frac{J_{p_1}}{R_{0_1}} + \frac{J_{p_2}}{R_{0_2}}\right) \\
C_{24} &= -\frac{e_1}{V}\left[C_1 a - (C_{M_{z_1}})_{,\alpha}\right] - V\frac{J_{p_1}}{R_{0_1}}\sin(\eta) \\
C_{31} &= \frac{1}{V}\left[\frac{1}{2}\rho V_r^2 S\Big(-l(C_{M_x})_{,\beta} + h_2 (C_y)_{,\beta}\Big)\right] \\
C_{32} &= -V\left(mh + \frac{J_{p_1}}{R_{0_1}} + \frac{J_{p_2}}{R_{0_2}}\right)
\end{aligned} \tag{6.159}$$

$$
\begin{aligned}
C_{34} &= -V\cos(\eta)\frac{J_{p_1}}{R_{0_1}}\\
C_{41} &= -\frac{1}{V}\left[C_1 e_1 + (C_{M_{z_1}})_{,\alpha}\cos(\eta)\right]\\
C_{42} &= V\left(m_1 e + \frac{J_{p_1}}{R_{0_1}}\sin(\eta)\right) - \frac{a}{V}\left[C_1 e_1 + (C_{M_{z_1}})_{,\alpha}\cos(\eta)\right]\\
C_{43} &= V\cos(\eta)\frac{J_{p_1}}{R_{0_1}}\\
C_{44} &= \frac{e_1}{V}\left[C_1 e_1 + (C_{M_{z_1}})_{,\alpha}\cos(\eta)\right] + c_\delta\ .
\end{aligned}
$$

Note that the damping matrix is not symmetrical, with some skew symmetrical gyroscopic terms easily identified.

Similarly, the elements of matrix $[K]$ are

$$
\begin{aligned}
K_{11} &= K_{12} = K_{21} = K_{22} = K_{31} = K_{32} = K_{41} = K_{42} = 0\\
K_{13} &= -(F_{y_1})_{,\gamma} - (F_{y_2})_{,\gamma} + \frac{1}{2}\rho V_r^2 S C_z\\
K_{14} &= -(C_1 + F_{x_1})\cos(\eta) + (F_{y_1})_{,\gamma}\sin(\eta) - \frac{1}{2}\rho V_r^2 S (C_y)_{,\delta}\\
K_{23} &= mh\dot{V} - a(F_{y_1})_{,\gamma} + b(F_{y_2})_{,\gamma} - F_{x_1} r_{p_1} - F_{x_2} r_{p_2} +\\
&\quad + \tfrac{1}{2}\rho V_r^2 S\left(h_2 C_x - c_2 C_z + l C_{M_y}\right)
\end{aligned}
$$

$$
\begin{aligned}
K_{24} = m_1 e\dot{V} &- \left[a C_1 - (C_{M_{z_1}})_{,\alpha} + F_{x_1}\left(a + \frac{e_1}{\cos(\eta)}\right)\right]\cos(\eta)\\
&+ a(F_{y_1})_{,\gamma}\sin(\eta) - \tfrac{1}{2}\rho V_r^2 S\left(l (C_{M_z})_{,\delta} - c_2 (C_y)_{,\delta}\right)
\end{aligned}
\tag{6.160}
$$

$$
\begin{aligned}
K_{33} &= mg(h - r_t) - \frac{1}{2}\rho V_r^2 S r_t C_z\\
K_{34} &= m_1 g e - \frac{1}{2}\rho V_r^2 S\left(h_2 (C_y)_{,\delta} - l (C_{M_x})_{,\delta}\right)\\
K_{43} &= m_1 g e - F_{x_1} r_{p_1}\cos(\eta) + e_1 (F_{y_1})_{,\gamma}\\
K_{44} &= -\left[m_1 g e + e_1 (F_{y_1})_{,\gamma}\right]\sin(\eta) + \left[C_1 e_1 + (C_{M_{z_1}})_{,\alpha}\cos(\eta) - m_1 e\dot{V}\right]\cos(\eta)\ .
\end{aligned}
$$

As usual, the first two columns of the stiffness matrix vanish, as the two coordinates y and ψ do not appear directly in the equations of motion. The order of the linear set of equations is then 6 and not 8, yielding a state space model of six first order differential equations.

The vector containing the control forces is

$$\{F_c\} = \begin{Bmatrix} 0 \\ m_2 d\dot{V} + \frac{1}{2}\rho V_r^2 S d\left(C_x + l(C_{M_z})_{,d}\right) \\ -m_2 g d + \frac{1}{2}\rho V_r^2 S d\left(C_z + l(C_{M_x})_{,d}\right) \\ M_s \end{Bmatrix}, \qquad (6.161)$$

where M_s is the moment applied by the driver to the handlebar.

The vector of the external forces is

$$\{F_e\} = \left[F_{y_e}\ M_{z_e}\ -hF_{y_e}\ eF_{y_{e_1}}\right]^T, \qquad (6.162)$$

where $F_{y_{e_1}}$ is the part of the total lateral force F_{y_e} which is applied to the centre of mass G_1.

The state space model of the open loop system is thus

$$\{\dot{z}\} = [A]\{z\} + [B_c]\{u_c\} + [B_e]\{u_e\}, \qquad (6.163)$$

where

$$\{z\} = \left[v\ r\ p\ v_\delta\ \phi\ \delta\right]^T,$$

p and v_δ being the derivatives of ϕ and δ with respect to time;

$$[A] = \begin{bmatrix} -[M]^{-1}[C] & -[M]^{-1}[K^*] \\ [0]_{2\times 2}\ [I]_{2\times 2} & [0]_{2\times 2} \end{bmatrix},$$

where $[K^*]$ is taken from matrix $[K]$ such that the first two columns have been cancelled,

$$[B_c] = \begin{bmatrix} [M]^{-1}\begin{Bmatrix} 0 & 0 \\ 0 & m_2\dot{V} + \frac{1}{2}\rho V_r^2 S\left(C_x + l(C_{M_z})_{,d}\right) \\ 0 & -m_2 g + \frac{1}{2}\rho V_r^2 S\left(C_z + l(C_{M_x})_{,d}\right) \\ 1 & 0 \end{Bmatrix} \\ [0]_{2\times 2} \end{bmatrix},$$

$$[B_e] = \begin{bmatrix} [M]^{-1} \begin{Bmatrix} 1 & 0 & 0 \\ 0 & 1 & 0 \\ -h & 0 & 0 \\ 0 & 0 & e \end{Bmatrix} \\ [0]_{2\times 2} \end{bmatrix} ,$$

$$\{u_c\} = \begin{Bmatrix} M_s \\ d \end{Bmatrix} , \qquad \{u_e\} = \begin{Bmatrix} F_{y_e} \\ M_{z_e} \\ F_{y_{e1}} \end{Bmatrix} .$$

6.15.5 *Open loop stability*

The stability with free controls can be studied simply by searching the eigenvalues of the dynamic matrix [A]. The eigenproblem usually yields two real eigenvalues and two complex conjugated pairs. Of the real solutions, one is negative and has practically no importance in the behaviour of the system and the other is positive and hence unstable. It corresponds to the capsize mode and must be stabilized by the driver or by some control device. This eigenvalue decreases with increasing speed, as gyroscopic moments of the wheels reduce the velocity at which the motorcycle falls on its side.

The two complex conjugate pairs are related to the so-called weave and wobble modes: The first one is mainly a yaw oscillation of the whole vehicle but involves also the roll and steering degrees of freedom while the other one is mainly an oscillation of the steering system about its axis. The weave frequency is lower than the wobble frequency, and is usually more affected by the speed. While the first mode is usually damped, the latter can become unstable, particularly at high speed. To stabilize wobble motion it is possible to introduce a steering damper, which has been introduced into the present mathematical model. The damper has the effect of reducing wobble instability, but can affect negatively weave and too large a damping can trigger weave instability: The value of coefficient c_δ must be chosen with care and the present mathematical model can supply useful guidelines.

Example 6.1

Study the open loop stability of the motorcycle of Appendix 1.5. Plot the eigenvalues as functions of the speed and the roots locus, assuming that the damping coefficient of the steering damper is either 0 or 8 Nms/rad.

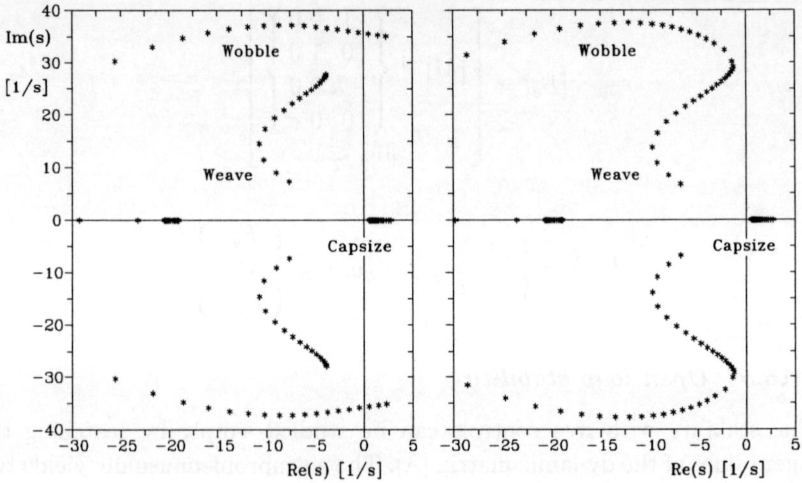

Fig. 6.41 Roots loci for the motorcycle of Appendix 1.5. (a): $c_\delta = 0$; (b) $c_\delta = 8$ Nms/rad.

The roots loci and the eigenvalue plots are reported in Figures 6.41, 6.42 and 6.43. In the first case, $c_\delta = 0$, the wobble mode becomes unstable slightly over 210 km/h while in the second one stability is insured in the whole speed range. With $c_\delta = 8$ Nms/rad frequencies and decay rates for weave and wobble modes tend to the same values at high speed.

6.15.6 *Steady-state response*

Equation (6.156) can be used to compute the steady state response by simply assuming that v, r, ϕ and δ, the velocity V and the control variables are all constant. In this case however there is an added difficulty: The control variables are two, the lateral displacement of the driver causing the centre of mass to displace laterally of d and the moment applied to the handlebar M_s.

A simple way of dealing with the steady-state response can be that of fixing a value of d and either of r or δ and solving the equations in v, ϕ, M_s and either δ or r. Perhaps the more straightforward approach is the first one, as in steady state conditions to assume the value of r means assuming a value of the radius of the trajectory R. The problem can thus be stated in the following terms: Given a circular trajectory, a lateral displacement of

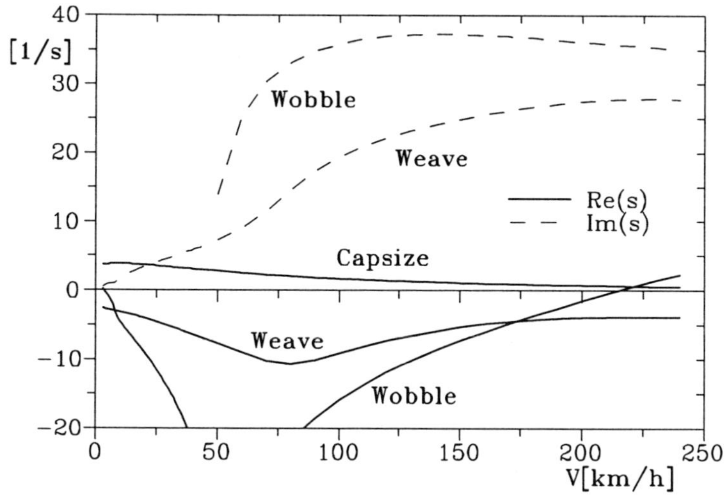

Fig. 6.42 Plot of the eigenvalues versus speed for the case of Fig. 6.41a: $c_\delta = 0$.

the driver, a value of the speed and possibly a given set of external forces, compute the sideslip, roll and steering angles ($\beta = v/V$, ϕ and δ) and the steering moment M_s. The relevant equation is

$$\begin{bmatrix} C_{11} & K_{13} & K_{14} & 0 \\ C_{21} & K_{23} & K_{24} & 0 \\ C_{31} & K_{33} & K_{34} & 0 \\ C_{41} & K_{43} & K_{44} & -1 \end{bmatrix} \begin{Bmatrix} V\beta \\ \phi \\ \delta \\ M_s \end{Bmatrix} = r \begin{Bmatrix} -C_{12} \\ -C_{22} \\ -C_{32} \\ -C_{42} \end{Bmatrix} +$$

$$+d \begin{Bmatrix} 0 \\ \frac{1}{2}\rho V_r^2 S\Big(C_x + l(C_{M_z})_{,d}\Big) \\ -m_2 g + \frac{1}{2}\rho V_r^2 S\Big(C_z + l(C_{M_x})_{,d}\Big) \\ 0 \end{Bmatrix} + \begin{Bmatrix} F_{y_e} \\ M_{z_e} \\ -hF_{y_e} \\ eF_{y_{e_1}} \end{Bmatrix} . \tag{6.164}$$

By solving Eq. (6.164) it is possible to compute the trajectory curvature gain $1/R\delta = r/V\delta$, the sideslip angle gain $\beta/\delta = v/V\delta$, the roll angle gain ϕ/δ and the steering moment gain M_s/δ. Owing to the linearity of the model, if the lateral displacement d and the external forces are assumed to be equal to zero, the mentioned gains are a function of the velocity only,

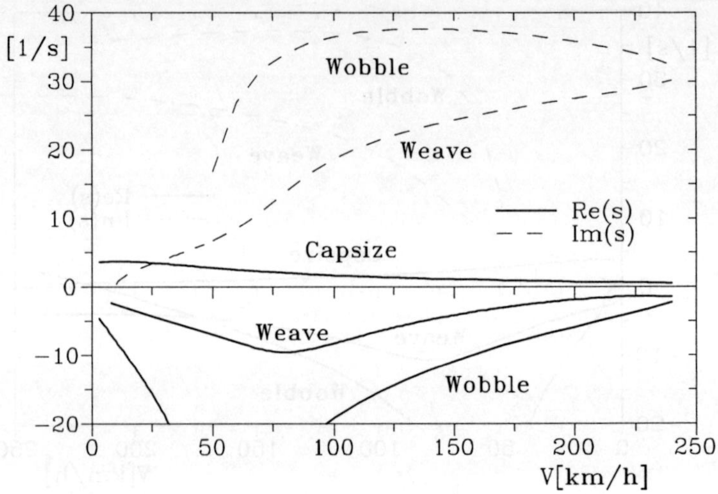

Fig. 6.43 Plot of the eigenvalues versus speed for the case of Fig. 6.41b: $c_\delta = 8$ Nms/rad.

i.e. do not depend on the radius of the trajectory. Equation (6.164) can thus be assumed to hold also for very high speeds, if a large enough radius R is considered.

Once the trajectory curvature gain has been obtained, the usual definitions of understeer, neutral steer and oversteer can be applied.

The present model is linearized and hence holds only if the relevant angles are small enough to allow the linearization of their trigonometric functions. The limit is in this case mainly due to the roll angle ϕ which can easily go beyond the value of about 20° which can be considered a limit for linearization.

As a last consideration, the computation of the steady state response assumes that the motion is possible, which implies that the driver stabilizes the capsize mode, which is unstable in open-loop operation.

Example 6.2

Compute the steady state steering response of the motorcycle of Appendix 1.5 assuming that the driver is not displaced from the symmetry plane and no external force is present. Compute the values of β, δ, ϕ, M_s on a curve with a radius of 200 m at a speed of 80 km/h.

Compute then the response as a function of speed when

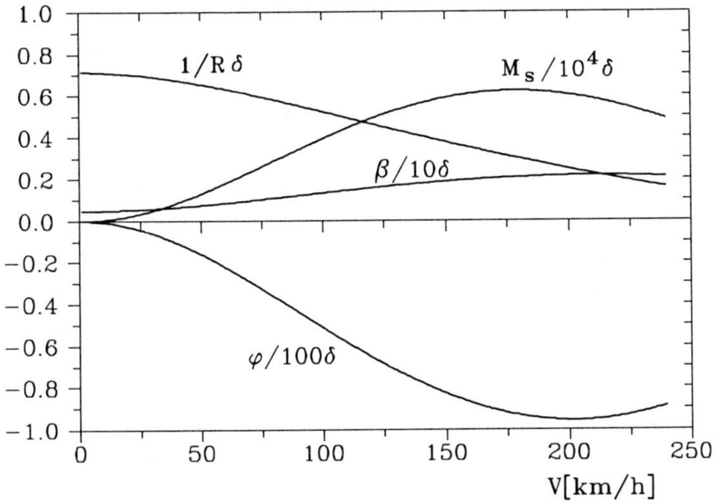

Fig. 6.44 Values of the trajectory curvature gain $1/R\delta = r/V\delta$, the sideslip angle gain $\beta/\delta = v/V\delta$, the roll angle gain ϕ/δ and the steering moment gain M_s/δ for the motorcycle of Appendix 1.5 versus speed V.

travelling on a curve with a radius of 200 m with the centre of mass of the driver-frame complex displaced of 50 mm from the symmetry plane. Repeat the computation of the values of β, δ, ϕ, M_s on a curve with a radius of 200 m at a speed of 80 km/h.

The various gains are plotted versus the speed in Fig. 6.44. The nondimensional curves hold for any value of the radius R: They are plotted for speeds up to 249 km/h, even if on a curve with a radius of 200 m the limit for linearization occurs at speeds slightly in excess of 80 km/h.

At 80 km/h their values are $1/R\delta = 0.58$ 1/m; $\beta/\delta = 1.08$, $\phi/\delta = -36.5$ and $M_s/\delta = 2858$ Nm. On a curve of 200 m radius the steering angle is $\delta = 0.0086$ rad $= 0.49$ °. As the wheelbase is 1.3159 m, the kinematic value is $\delta_c = 0.00658$ rad $= 0.38$ °. The vehicle is then understeer.

The other values are $\beta = 0.0093$ rad $= 0.53$ °, $\phi = -0.306$ rad $= -17.6°$ and $M_s = 24.7$ Nm. Note that the value of the roll angle is quite close to the limit for linearization.

The same gains but for a lateral displacement of the centre

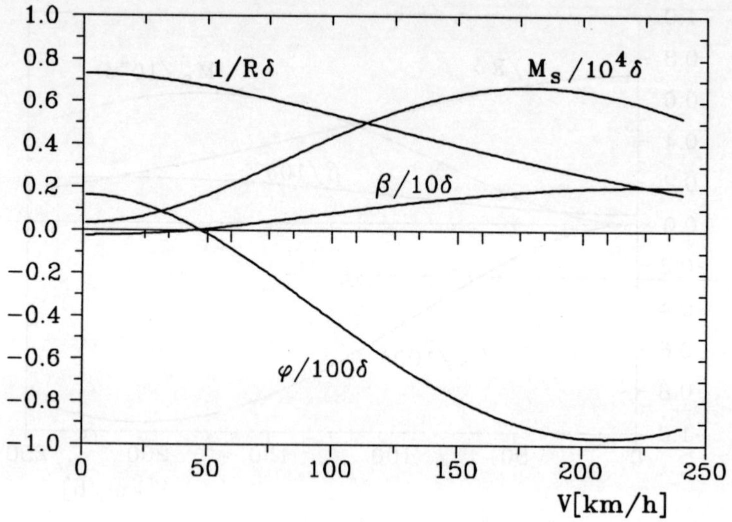

Fig. 6.45 Values of the trajectory curvature gain $1/R\delta = r/V\delta$, the sideslip angle gain $\beta/\delta = v/V\delta$, the roll angle gain ϕ/δ and the steering moment gain M_s/δ for the motorcycle of Appendix 1.5 versus speed V. Lateral displacement of the centre of mass $d = 50$ mm; radius of the trajectory $R = 200$ m.

of mass of 50 mm are plotted versus speed in Fig. 6.45. In this case the curves hold only for a radius of 200 m and lose any validity at a speed at which on such a radius the roll angle gets too large for the linearization to hold.

At 80 km/h their values are $1/R\delta = 0.598$ 1/m; $\beta/\delta = 0.509$, $\phi/\delta = -24.5$ and $M_s/\delta = 3231$ Nm. On a curve of 200 m radius the steering angle is so $\delta = 0.0084$ rad $= 0.48$ °, $\beta = 0.0043$ rad $= 0.24$ °, $\phi = -0.2049$ rad $= -11.7°$ and $M_s = 27.0$ Nm. From the plot it is clear that the displacement of the centre of mass makes the vehicle only slightly less oversteer but has a major effect on the sideslip angle, which at low speed can even become negative, and on the roll angle: Both are largely reduced. The steering moment is not much affected by the lateral displacement.

6.15.7 *Nonstationary motion*

Since any two-wheeled vehicle is unstable, a driver model acting at least as a capsize stabilizer must be introduced into the model to study the nonstationary motion. As already stated in the previous chapter, the task of building a mathematical model of human behaviour is a difficult one and even results obtained using very complex models are not completely satisfactory.

In the present model there is an added complexity due to the fact that the driver has been assumed to control the vehicle by both applying the moment M_s to the handlebar and displacing laterally the centre of mass of the frame-driver complex of distance d. The observation of the behaviour of racing drivers has however suggested that these two control actions are generally used with different aims: The steering torque is generally applied to supply a quick corrective action to perturbations, with frequencies even as high as 2 Hz (low frequency and even constant components of the moment are also present, but this has little to do with the control of the unstable capsize mode), while the lateral displacement is usually a low frequency action, used mostly at the initial stage of the preparation of a curve. Higher frequencies can be present while handling a "chicane" or a slalom test, but even here the frequencies involved are seldom as high as 1 Hz.

A possible strategy is then the following: The lateral displacement can be considered as an open loop feedforward action, applied when preparing a manoeuvre while the actual feedback control able to stabilize the vehicle is provided by the application of moment M_s to the handlebar.

The driver can thus be assumed to be a linear PID tracking system, whose time domain model is

$$M_s(t+\tau) = K\left[e(t) + T_L\frac{de(t)}{dt} + \frac{1}{T_s}\int_0^t e(u)du\right] , \tag{6.165}$$

where the error $e(t)$ is defined as the difference $\phi_r(t) - \phi(t)$ between the reference value of the roll angle, i.e. the value the driver wants to apply to perform the manoeuvre, and the actual value. The values of the time delay τ, of the gain K, of the predictive time T_L and of the integral gain $1/T_s$ are not easy to assess. An approach which has been used is that of writing the whole driver-vehicle model in the Laplace domain and to use a pole placement approach.

Once the driver parameters have been stated, there is no difficulty in integrating Eqs. (6.163) and (6.165) with a given law $\phi_r(t)$ and given initial conditions. An interesting manoeuvre is for example a step or a ramp input of the reference roll angle ϕ_r starting from a state in which all state variable have vanishingly small values and d has a stated value. The manoeuvre corresponds to the insertion of the vehicle on a steady state curve and the results give an immediate feeling of the manoeuvrability of the vehicle. The values of the steering moment M_s allow to draw some conclusion on the effort required to steer the vehicle. Simulations of the manoeuvre performed when straightening the vehicle after a curve, of a slalom manoeuvre or of the response to an external force or moment are equally straightforward and yield interesting results.

Note that in actual driving these manoeuvres are usually accompanied by changes in the speed. There is no difficulty in taking this into account when performing a numerical simulation, but the changes of the normal forces on the ground and then of the characteristics of the tires must be accounted for. Also, the relevant matrices change during the manoeuvre and this increases the amount of computational work which must be performed.

Finally, the changes of speed induce pitching and this in turn can cause some changes in the geometrical characteristics of the vehicle: This effect cannot be accounted for in a model which is based on the rigid-body assumption.

6.15.8 *Nonlinear approach*

The model described above can be very easily transformed into a semi-linearized model, simply by substituting the linearized expressions of the forces due to the tires with nonlinear ones, as those referred to as the magic formula. The changes to the model are very simple but this approach is questionable: In normal driving and even in races the sideslip angles of the wheels are quite small while the nonlinearity problems are in this case due to the roll angle, which in semilinearized models is considered small. The limitations of semilinearized models are thus almost the same as those of linear models, with a greater complexity and the impossibility to perform a stability analysis.

A complete nonlinear model in which the roll angle can assume any value, up to the values of $50°-60°$ which are common in race driving is required. All 5 degrees of freedom, including displacement in x direction, are coupled and the equations are far more complicated but can be obtained from the expressions of the kinetic and potential energies and the virtual work of forces through a symbolic computer code. The equations can then be integrated numerically, to perform more accurate simulations of the behaviour of the vehicle or, better, of the vehicle-driver system.

Another important result obtainable from the nonlinear model is the study of the stability in the small about any steady-state condition. The values of the state variables in steady state motion are first computed and then the model is linearized about this condition. The solution of the eigenproblem allows one to study the stability in the small and can yield results which are different from those obtained by considering the vehicle in the upright condition on which the linearized model seen in the previous sections is based.

As a concluding remark, the numerical simulation of two-wheeled vehicles is quite complicated and has received far less attention than that of four-wheeled ones. However, even if in some cases it yields results which are more qualitative than quantitative, it allows one to obtain a good insight of some phenomena which are difficult to study with the usual experimental and empirical approach and supplies interesting comparative informations on different solutions and setups.

Chapter 7

Road accidents

Road transportation, like any human activity, has a number of risks and the number of road accidents is quite high in all countries. Their cost, both in terms of human lives and economic losses is quite high and the goal of increasing the safety of motor vehicles is generally considered one of the technical and social priorities. Actually, the efforts of the past years have succeeded in reducing the number and the consequences of road accidents, both in relation to the number of vehicles and in absolute terms.

The actions taken are both technical and legal and involve many disciplines. Vehicle dynamics is involved both in the stage of the design of the vehicle, as handling characteristics are essential in avoiding the occurrence of accidents and sometimes to reduce their consequences, and in the reconstruction of their occurring. While the previous chapters deal with the motion of a single vehicle in normal driving conditions, the present one is devoted to the study of the motion of the vehicle on the road during an accident and to the interaction between the vehicles which are involved in it.

It must be noted that very often the reconstruction of an accident is a difficult task as the expert, who acts usually as a consultant of the court or of one of the persons involved, has only a partial knowledge of the situation. The only known data are in many cases the positions of the vehicles after the accident plus some marks which may remain on the road, and sometimes even them are uncertain and affected by large errors. A road accident is an event occurring in a very short time and usually in an unexpected way and has a large psychological impact on witnesses and protagonists. This, together with the limited technical culture of the persons involved and the economical interests at stake can make the reports of the witnesses very difficult to interpret and to weigh correctly, particularly when they lead to conflicting reconstructions.

The traditional methods used in the reconstruction of the accidents are based on very rough approximations, which are however justified by the uncertainties and sometimes the quick variations of the parameters of the problem. Recently however the use of more elaborated numerical models, implemented on computers, allowed to reach a better accuracy.

7.1 Vehicle collision: Impulsive model

One of the problems which must be solved more frequently in the reconstruction of accidents is the study of collisions between vehicles or between a vehicle and an obstacle. The simplest model deals with the collision as an impulsive phenomenon, i.e., assumes that the time during which the vehicles remain in contact (typically of the order of 0.1 s) is vanishingly short and that the forces they exchange are infinitely large. Mathematically the forces can be represented by a Dirac impulse function and the study is based on the momentum theorem, which states that the variation of the momenta of the vehicles is equal to the impulse of the forces they exchange.

Since the forces the vehicles exchange during the collision are larger by orders of magnitude than the other forces acting on them, this approach is quite correct, even if it has the disadvantage of not allowing to study what happens during the impact but only how the motion changes between instant t_1 preceding the collision and instant t_2 following it. Actually, it is meaningless to ask what happens during a phenomenon whose duration is zero by definition.

7.1.1 *Central head-on collision*

The simplest case is that of a head-on collision in which the velocities of the centres of gravity of the vehicles lie along the same straight line (Fig. 7.1a). Actually, if the velocities $\dot{x}_A$ and $\dot{x}_B$ have the same sign a rear collision occurs while a head-on collision is characterized by opposite signs of the velocities.

The relative velocity

$$V_R = \dot{x}_B - \dot{x}_A \tag{7.1}$$

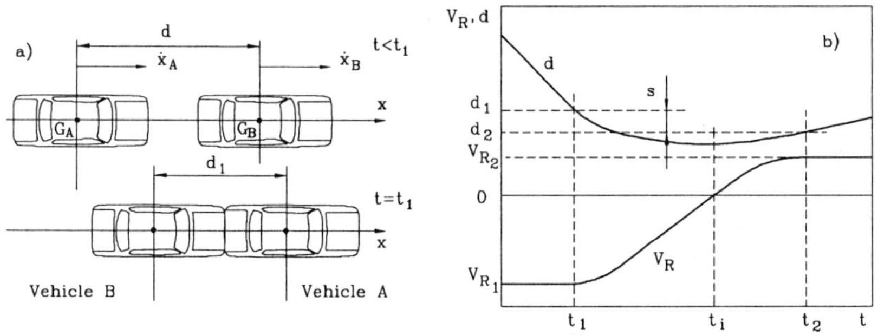

Fig. 7.1 Central head-on collision. (a) Situations before the collision and at time t_1; (b) relative velocity and distance between the centres of gravity as functions of time.

must be negative, otherwise the vehicles do not approach each other.

During the collision the absolute value of the relative velocity decreases between times t_1 and time t_i, when the centres of gravity are at their minimum distance. In the rebound phase, between times t_i and t_2, the relative velocity becomes positive (Fig. 7.1b). Also the distance between the centres of gravity is plotted in the same figure. The residual crush s of the vehicles is the difference between d_1 and d_2.

As already said, however, the impulsive model does not allow to study what happens between t_1 and t_2. The conservation of the momentum allows to write the relationship

$$m_A \dot{x}_{A_1} + m_B \dot{x}_{B_1} = m_A \dot{x}_{A_2} + m_B \dot{x}_{B_2} \ . \tag{7.2}$$

Since the collision is generally inelastic, the kinetic energy is not conserved

$$\frac{1}{2} m_A \dot{x}_{A_1}^2 + \frac{1}{2} m_B \dot{x}_{B_1}^2 \geq \frac{1}{2} m_A \dot{x}_{A_2}^2 + \frac{1}{2} m_B \dot{x}_{B_2}^2 \ ; \tag{7.3}$$

the two sides being equal in the case of a perfect elastic collision.

The position and the velocity of the centre of mass of the system made by the two vehicles are simply

$$x_G = \frac{m_A x_A + m_B x_B}{m_A + m_B} \ , \tag{7.4}$$

$$\dot{x}_G = \frac{m_A \dot{x}_A + m_B \dot{x}_B}{m_A + m_B} . \tag{7.5}$$

A consequence of the conservation of the momentum is the conservation of the velocity of the centre of mass of the system: $\dot{x}_{G_1} = \dot{x}_{G_2}$.

The velocities of the vehicles can then be expressed as functions of the velocity of the centre of mass, which remains constant, and of the relative velocity, function of time

$$\dot{x}_A = \dot{x}_G - V_R \frac{m_B}{m_A + m_B} , \qquad \dot{x}_B = \dot{x}_G + V_R \frac{m_A}{m_A + m_B} . \tag{7.6}$$

The kinetic energy of the system can thus be written in the form

$$\mathcal{T} = \frac{1}{2} \left[\dot{x}_G^2 (m_A + m_B) + V_R^2 \frac{m_A m_B}{m_A + m_B} \right] . \tag{7.7}$$

As the velocity of the centre of mass is constant, the maximum energy dissipation occurs when the relative velocity vanishes. It is therefore possible to state a lower and an upper bound to the kinetic energy after the collision

$$\frac{1}{2} \dot{x}_G^2 (m_A + m_B) \leq \mathcal{T} \leq \frac{1}{2} \left[\dot{x}_G^2 (m_A + m_B) + V_R^2 \frac{m_A m_B}{m_A + m_B} \right] . \tag{7.8}$$

The minimum kinetic energy (expression on the left) occurs when the collision is perfectly inelastic and the two vehicles remain attached to each other. In the case of a perfectly elastic collision, the absolute value of the relative velocity after the impact is equal to that at time t_1: $V_{R_2} = -V_{R_1}$.

A restitution coefficient e^* is usually defined as[1]

$$e^* = -\frac{V_{R_2}}{V_{R_1}} . \tag{7.9}$$

In case of a perfectly elastic collision $e^* = 1$, while $e^* = 0$ for inelastic impacts. In all real impacts $0 < e^* < 1$.

The energy dissipated during the collision is then

---

[1]Symbol e^* will be used for the restitution coefficient instead of e to avoid confusion with the base of natural logarithms.

$$\mathcal{T}_2 - \mathcal{T}_1 = \frac{1}{2}\frac{m_A m_B}{m_A + m_B}\left(V_{R_1}^2 - V_{R_2}^2\right) = \frac{1}{2}\frac{m_A m_B}{m_A + m_B} V_{R_1}^2 \left(1 - e^{*^2}\right) . \quad (7.10)$$

In the first part of the collision, from time t_1 to time t_i, the relative velocity decreases to zero and the kinetic energy reduces to a minimum, given by the expression on the left hand side of Eq. (7.8). The energy related to velocity V_{R_1} transforms partly in elastic deformation energy and part of it is dissipated as heat. In the second part of the collision, from time t_i to time t_2, a fraction of this energy is transformed back to kinetic energy. The ratio between the energy restituted in the second phase and that subtracted from the kinetic energy in the first one is e^{*^2}.

In the case of motor vehicle collisions the value of e^* is low, typically in the range of 0.05–0.2 in case of impacts with large permanent deformations. However it depends on the relative velocity and can be higher in low speed collisions, tending to unity when no permanent deformations are left.

The velocities of the vehicles after the impact are

$$\begin{cases} \dot{x}_{A_2} = \dot{x}_{A_1} + m_B V_{R_1} \dfrac{1+e^*}{m_A + m_B} \\ \\ \dot{x}_{B_2} = \dot{x}_{B_1} - m_A V_{R_1} \dfrac{1+e^*}{m_A + m_B} . \end{cases} \quad (7.11)$$

7.1.2 *Oblique collision*

Very seldom the situation is that described in Fig. 7.1; usually the velocity vectors of each one of the two vehicles do not pass through the centre of mass of the other as in Fig. 7.2. If the collision is considered as an impulsive phenomenon, the conservation of the momentum still holds

$$\begin{cases} m_A \dot{x}_{A_1} + m_B \dot{x}_{B_1} = m_A \dot{x}_{A_2} + m_B \dot{x}_{B_2} \\ m_A \dot{y}_{A_1} + m_B \dot{y}_{B_1} = m_A \dot{y}_{A_2} + m_B \dot{y}_{B_2} \end{cases} \quad (7.12)$$

but does not allow to solve the problem of finding the situation after the collision from that which precedes it. Also the assessment of a coefficient of restitution does not allow to obtain a solution, since the conditions to be stated are two and not one (Eq. (7.12) contains two conditions and 4 unknowns).

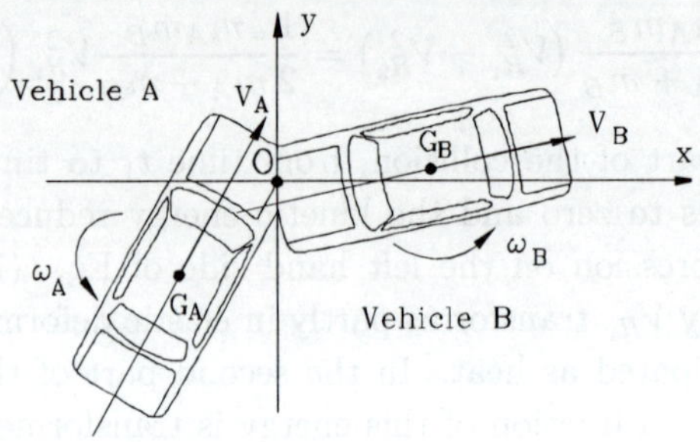

Fig. 7.2 Oblique collision.

Assume that the contact surface at the moment of maximum deformation is flat and state a reference frame Oxy centred at point O in which the resultant of the contact forces is applied and with x-axis perpendicular to the contact area. Note that the components of velocities V_i of the vehicles are plotted in the figure in the direction of the positive axes: Generally speaking some of them are negative in any practical case. The vehicles can also have an angular velocity ω; if not at time t_1 for sure at time t_2.

The impulsive model implies that the duration of the impact is zero: It can thus be thought as the collision between two rigid bodies, predeformed as shown in the figure.

The velocity of point O, considered as belonging alternatively to vehicle A and vehicle B, is

$$\vec{V}_{A_O} = \left\{ \begin{array}{c} \dot{x}_{G_A} - \omega_A y_A \\ \dot{y}_{G_A} + \omega_A x_A \end{array} \right\} \qquad \vec{V}_{B_O} = \left\{ \begin{array}{c} \dot{x}_{G_B} - \omega_B y_B \\ \dot{y}_{G_B} + \omega_B x_B \end{array} \right\}, \tag{7.13}$$

where x_A, y_A, etc. are the coordinates of the centres of gravity in the Oxy frame. Note that in the figure both x_A and y_A are negative.

The relative velocity of vehicle B with respect to vehicle A at point O is

$$\vec{V}_R = \vec{V}_{B_O} - \vec{V}_{A_O} = \left\{ \begin{array}{c} \dot{x}_{G_B} - \dot{x}_{G_A} - \omega_B y_B + \omega_A y_A \\ \dot{y}_{G_B} - \dot{y}_{G_A} + \omega_B x_B - \omega_A x_A \end{array} \right\}. \tag{7.14}$$

The x component of the relative velocity, hereafter indicated as $V_{R_\perp}$ is the velocity at which the two surfaces approach each other, or better, move away for each other, as a positive value means that the distance between the surfaces increases. The y component $V_{R_\parallel}$ is the velocity at which the surfaces slide over each other and can be either positive or negative.

Like in the case of the central impact, it is possible to define a restitution coefficient

$$e^* = -\frac{V_{R_{\perp 2}}}{V_{R_{\perp 1}}} \ .$$

Each vehicle receives an impulse from the other which is equal to its change of momentum. Indicating with $\vec{I}$ the impulse received by vehicle A from vehicle B (the impulse received by vehicle B from vehicle A is then $-\vec{I}$), the momentum theorem applied to the two vehicles can be written in the form

$$\begin{cases} I_x = m_A\left(\dot{x}_{A_2} - \dot{x}_{A_1}\right) \\ I_y = m_A\left(\dot{y}_{A_2} - \dot{y}_{A_1}\right) \end{cases} \qquad \begin{cases} -I_x = m_B\left(\dot{x}_{B_2} - \dot{x}_{B_1}\right) \\ -I_y = m_B\left(\dot{y}_{B_2} - \dot{y}_{B_1}\right) \ . \end{cases} \tag{7.15}$$

Similarly the conservation of the angular momentum can be written in the form

$$\begin{cases} I_x y_A - I_y x_A = J_A\left(\omega_{A_2} - \omega_{A_1}\right) \\ -I_x y_B + I_y x_B = J_B\left(\omega_{B_2} - \omega_{B_1}\right) \ . \end{cases} \tag{7.16}$$

A single relationship is still needed. It is possible to relate the components of the impulse in a direction perpendicular to the impact surface I_x with the tangential component I_y with a simple relationship

$$I_y = \lambda I_x \ , \tag{7.17}$$

where λ is a sort of friction coefficient in the zone in which the two vehicles are in contact. As the vehicles usually interlock with each other, its value can be far higher than that of an actual coefficient of friction. Moreover, its value changes in time and only an average value is required. Its sign is the same as that of ratio $V_{R_\perp}/V_{R_\parallel}$.

By introducing Eqs. (7.15) and (7.16) into Eq. (7.14), it follows

$$V_{R_{\perp 2}} = V_{R_{\perp 1}} - I_x\left(a - \lambda b\right) \ , \tag{7.18}$$

where

$$\begin{cases} a = \dfrac{1}{m_A} + \dfrac{1}{m_B} + \dfrac{y_A^2}{J_A} + \dfrac{y_B^2}{J_B} \\ \\ b = \dfrac{x_A y_A}{J_A} + \dfrac{x_B y_B}{J_B} \,. \end{cases}$$

The value of the x component of the impulse is then

$$I_x = V_{R_{\perp 1}} \frac{1+e^*}{a - \lambda b} \,. \tag{7.19}$$

The direct problem, i.e. that of finding the conditions after the collision (time t_2) from those at time t_1 is easily solved. Once that the components $V_{R_{\perp 1}}$ and $V_{R_{\parallel 1}}$ are computed, the sign of λ is known and the impulse can be computed from Eqs. (7.19) and (7.17). The velocities after the collisions are thus computed through Eqs. (7.15) and (7.16).

For the inverse problem, i.e. that of finding the conditions before the collision (time t_1) from those at time t_2, Eq. (7.19) can be modified as

$$I_x = -V_{R_{\perp 2}} \frac{1+e^*}{e^* \, (a - \lambda b)} \tag{7.20}$$

and a procedure similar to the previous one can be followed.

The problem is the assessment of the values of coefficients λ and e^* and of the positions of the centres of mass of the vehicles with respect to point O, since the position of the latter cannot usually be evaluated with precision. It is possible to repeat the computation with different values of the uncertain parameters in order to define zones in the parameter space in which the solutions must lie and to rule out possible reconstructions of the accident. Some solutions can be ruled out by remembering that $a - \lambda b$ must be positive, as both I_x and V_{i_1} are negative, if vehicle A is in the half plane with negative x.

7.1.3 *Collision against a fixed obstacle*

A particular case is that of the collision against a fixed obstacle (Fig. 7.3). The equations seen above still apply, provided that $1/m_B = 1/J_B = \dot{x}_B = \dot{y}_B = 0$, i.e. the fixed obstacle is considered as a vehicle with infinite mass travelling at zero speed.

It then follows

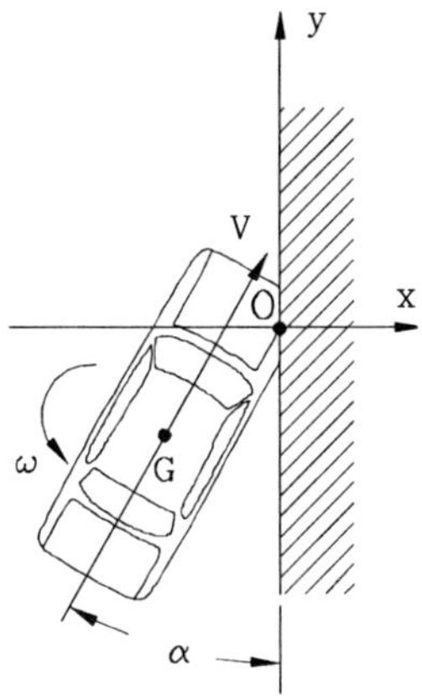

Fig. 7.3 Collision against a fixed obstacle.

$$\vec{V}_R = \left\{ \begin{array}{c} -V\cos(\alpha) - \omega y_G \\ -V\sin(\alpha) + \omega x_G \end{array} \right\} \tag{7.21}$$

$$I = \frac{mr^2(1+e^*)[-V_1\cos(\alpha) - \omega_1 y_G]}{r^2 + y_G(y_G - \lambda x_G)}\,, \tag{7.22}$$

where $r = \sqrt{J/M}$ is the radius of gyration of the vehicle.

If the angular velocity ω_1 is negligible, the expressions for the velocity after the impact are

$$\left\{ \begin{array}{l} \dot{x}_2 = \dot{x}_1 \dfrac{y_G(y_G - \lambda x_G) - r^2 e^*}{y_G(y_G - \lambda x_G) + r^2} \\ \dot{y}_2 = \dot{y}_1 - \dot{x}_1 \dfrac{\lambda r^2(1+e^*)}{y_G(y_G - \lambda x_G) + r^2} \\ \omega_2 = \dot{x}_1 \dfrac{(y_G - \lambda x_G)(1+e^*)}{y_G(y_G - \lambda x_G) + r^2}\,. \end{array} \right. \tag{7.23}$$

If $\dot{x}_2 > 0$ or $\omega_2 > 0$ the vehicle has a second collision with the obstacle with the rear part of the body or, in practice, it undergoes a deformation which displaces rearward the point of contact, with a change of the values of x_G and y_G: It is impossible that in an actual case at the end of the collision $\dot{x}_2 > 0$. In case of a perfectly inelastic impact the vehicle slides finally along the obstacle and the deformations are such that $y_G - \lambda x_G \approx 0$;

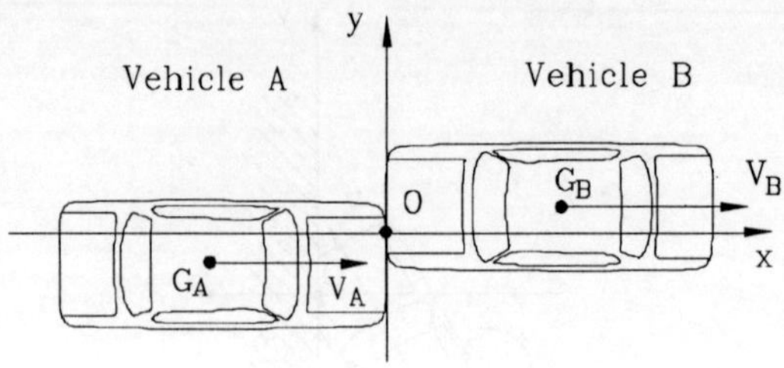

Fig. 7.4 Non-central head-on collision.

$\dot{x}_2 = \omega_2 = 0$ and $\dot{y}_2 = \dot{y}_1 - \lambda\dot{x}_1$. If the obstacle is a flat surface, λ is the coefficient of friction.

7.1.4 *Non-central head-on collision*

Consider a head-on collision in which the velocities of the centres of gravity of the vehicles do not lie along the same straight line (Fig. 7.4).

If the vehicles have no angular velocity before the collision, it follows that $\dot{y}_{A_1} = \dot{y}_{B_1} = \omega_A = \omega_B = 0$ and $\dot{x}_{A_1} = V_A$; $\dot{x}_{B_1} = V_B$. The components of the relative velocity are then $V_{R_{\perp_1}} = V_B - V_A$ and $V_{R_{\parallel_1}} = 0$ and, as the slip velocity is nil, λ can be assumed to vanish.

The expressions for the velocities after the impact are

$$\begin{cases} \dot{x}_{A_2} = V_A + (V_B - V_A)\,\dfrac{1+e^*}{am_A} \\ \dot{x}_{B_2} = V_B - (V_B - V_A)\,\dfrac{1+e^*}{am_B} \\ \omega_{A_2} = (V_B - V_A)\,\dfrac{y_A(1+e^*)}{aJ_A} \\ \omega_{B_2} = -\,(V_B - V_A)\,\dfrac{y_B(1+e^*)}{aJ_B} \end{cases} . \tag{7.24}$$

7.1.5 *Side collision*

Consider a lateral collision in which the velocities of the centres of gravity of the vehicles are perpendicular to each other (Fig. 7.5).

If before the collision the angular velocities of the vehicles are vanishingly small, it follows that $\dot{y}_{A_1} = \dot{x}_{B_1} = \omega_A = \omega_B = 0$ and $\dot{x}_{A_1} = V_A$; $\dot{y}_{B_1} = V_B$. It follows that $V_{R_{\perp_1}} = -V_A$ and $V_{R_{\parallel_1}} = V_B$ and then

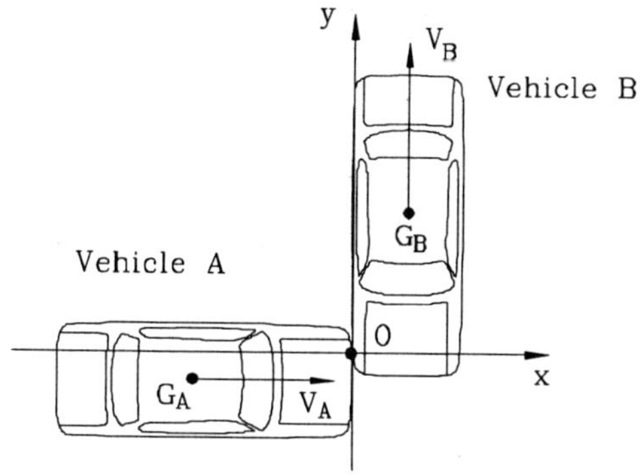

Fig. 7.5 Lateral collision.

$$I_x = -V_A \frac{1+e^*}{a-\lambda b}, \tag{7.25}$$

where λ is negative if both V_A and V_B are positive. The expressions for the velocities after the impact are

$$\begin{cases} \dot{x}_{A_2} = V_A \left[1 - \dfrac{1+e^*}{m_A(a-\lambda b)}\right] \\ \dot{y}_{A_2} = -V_A \lambda \dfrac{1+e^*}{m_A(a-\lambda b)} \\ \omega_{A_2} = -\left(y_A - \lambda x_A\right) \dfrac{(1+e^*)}{J_A(a-\lambda b)} \end{cases}$$
$$\begin{cases} \dot{x}_{B_2} = V_A \dfrac{1+e^*}{m_B(a-\lambda b)} \\ \dot{y}_{B_2} = V_B + V_A \lambda \dfrac{1+e^*}{m_A(a-\lambda b)} \\ \omega_{B_2} = \left(y_B - \lambda x_B\right) \dfrac{(1+e^*)}{J_A(a-\lambda b)} \end{cases} \tag{7.26}$$

7.1.6 *Simplified approach*

Often the problem is however simpler. If the directions of the velocity vectors of the vehicles before and after the impact can be estimated independently, there is no need to assume the values of e^* and λ. With reference to Fig. 7.6, Eq. (7.12) can be written as

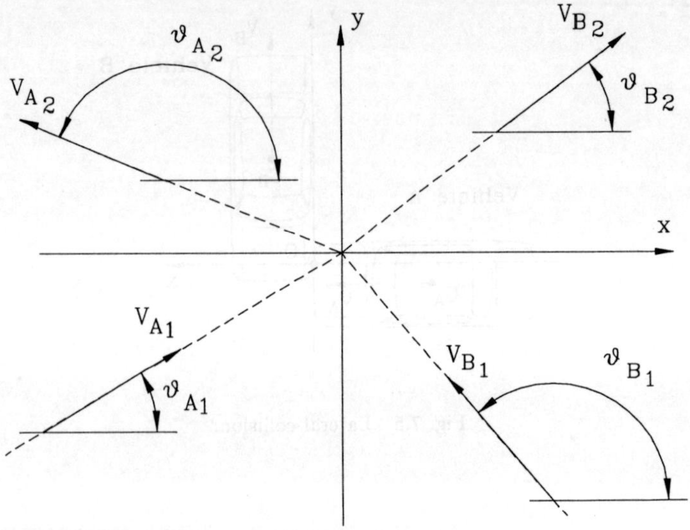

Fig. 7.6 Velocities at times t_1 and t_2.

$$\begin{cases} m_A V_{A_1}\cos(\theta_{A_1}) + m_B V_{B_1}\cos(\theta_{B_1}) = m_A V_{A_2}\cos(\theta_{A_2}) + m_B V_{B_2}\cos(\theta_{B_2}) \\ m_A V_{A_1}\sin(\theta_{A_1}) + m_B V_{B_1}\sin(\theta_{B_1}) = m_A V_{A_2}\sin(\theta_{A_2}) + m_B V_{B_2}\sin(\theta_{B_2}) \,. \end{cases} \tag{7.27}$$

If all angles θ_i at times t_1 and t_2 are known or can be assumed, both the direct or the inverse problem are easily solved. Once that the velocities have been computed it is possible to obtain the angular velocities, the components of the impulse, e^* and λ.

In the case of the inverse problem, usually the velocity after the impact is obtained from the distance travelled by the two vehicles between time t_2 and the instant t_3 in which all motion is extinguished. Often the wheels are assumed to be blocked, which is correct only if the deformations due to the collision are large enough, or if the sideslip angles are large enough to cause the wheels to slide on the ground. If the vehicle slides for a distance d before coming to a stop without hitting any other obstacle, the velocity V_2 can be computed by equating the kinetic energy of the vehicle after the collision with the energy dissipated by friction $mgfd$

$$V_2 = \sqrt{2gdf} , \tag{7.28}$$

where f is the coefficient of friction between the tires and the road.

If at time t_3 the vehicle hits an obstacle with a speed V_3 after sliding for a distance d, it follows

$$V_2 = \sqrt{2gdf + V_3^2} . \tag{7.29}$$

Velocity V_3 can be assessed only from the damage suffered by the vehicle during the secondary collision, which is difficult and affected by large uncertainties and errors. The larger the value of the friction coefficient f or of the distance d the less the errors in the estimate of V_3 affect the results.

Another source of uncertainties is the evaluation of the value of f, which is affected not only by the road conditions and the possibility that some of the wheels are rolling instead of slipping but also by the actual motion of the vehicle.

The motion of the vehicle on the road after the collision is actually not a simple translatory motion but a combination of translation and rotation. As already stated, the angular velocity of the vehicle after the collision can be computed, even if the approximations can be higher as the position of the point of application of the shock load can only be guessed. At a first glance it would seem that the rotation $\psi_3 - \psi_2$ of the vehicle during the time from t_2 to t_3 could be computed as done for the translational motion

$$\psi_3 - \psi_2 = \frac{J_z \omega_2^2}{2mg\overline{d}f} , \tag{7.30}$$

where $\overline{d}$ is the average distance between the centre of gravity and the centres of the contact areas of the wheels.

This is not correct as the coefficient of friction cannot be applied to the translational and rotational motions separately. However, as it will be shown in Section 7.3.1, it is possible to approximate the correct results by studying separately the two motions with two "equivalent" friction coefficients, which are both smaller in magnitude than the actual coefficient f.

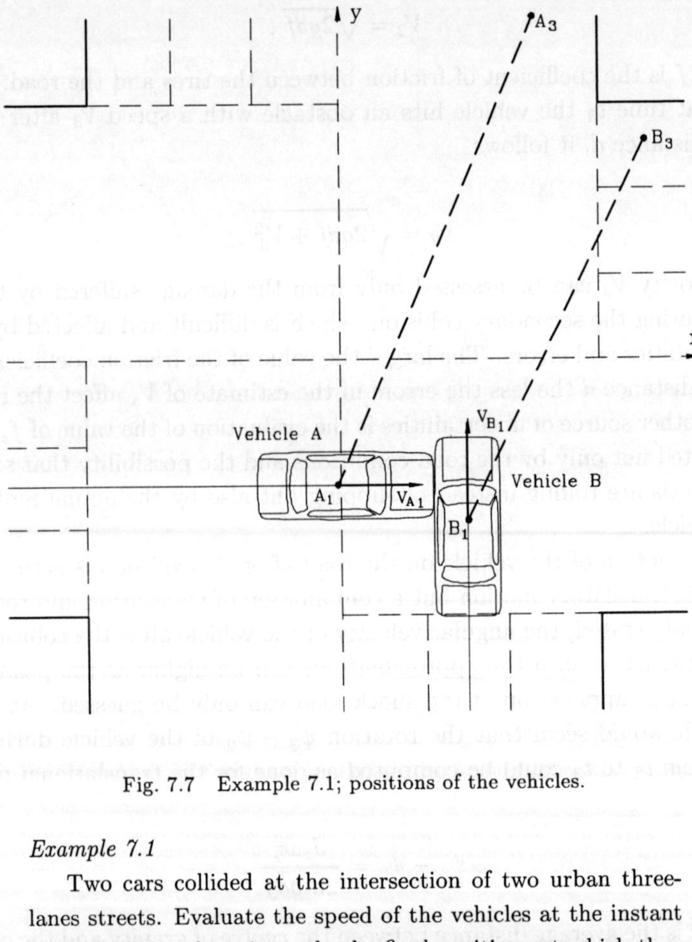

Fig. 7.7 Example 7.1; positions of the vehicles.

Example 7.1

Two cars collided at the intersection of two urban three-lanes streets. Evaluate the speed of the vehicles at the instant of the collision, knowing only the final positions at which they stopped and their directions at time t_1. Vehicle A stopped without any secondary collision while vehicle B ended against a wall. From the deformations caused by the secondary impact, a value $V_3 \approx 11$ m/s ≈ 40 km/h is assumed.

With reference to the xy frame shown in Fig. 7.7, the co-ordinates of the centres of mass of the vehicles at time t_3 are $x_{A_3} = 9.2$ m, $y_{A_3} = 12.8$ m, $x_{B_3} = 13.2$ m and $y_{B_3} = 12.0$ m. The masses of the vehicles are $m_A = 850$ kg and $m_B = 870$ kg.

Since the exact position of the vehicles at time t_1 is unknown, each one of them will be assumed to travel in the centre of the right, centre and left lane. As a result, nine possible positions of the impact point will be considered. With simple geometrical computations, the values of y_{A_1} and y_{B_1} as functions of the position of vehicle A and those of x_{A_1} and x_{B_1} as functions of the position of vehicle B are

	Vehicle A		Vehicle B	
Lane	y_{A_1}	y_{B_1}	x_{A_1}	x_{B_1}
Right	-9.8	10.2	6.0	8.6
centre	-6.3	-6.7	2.6	5.2
Left	-2.8	-3.2	-0.8	1.8

It is straightforward to compute the distances d_A and d_B for the nine cases under study. Vehicle A stops without hitting any obstacle. As its velocity was almost perpendicular to its longitudinal axis and some rotation occurred, owing to the fact that the front wheels were blocked as a consequence of the impact, a fairly high value of the friction coefficient can be assumed, namely $f_A = 0.45$.

Vehicle B had only one wheel locked and moved in a direction less inclined with respect to its longitudinal axis; an average value of the friction coefficient $f_B = 0.30$ is then assumed. In all nine cases the values of V_{A_2} and V_{B_2} can be easily obtained

$$V_{A_2}=\sqrt{2gd_Af_A}\ , \qquad V_{B_2}=\sqrt{2gd_Bf_B + V_{B_3}^2}\ .$$

Angles θ_{A_1} and θ_{B_1} are respectively $\theta_{A_1} = 0$ and $\theta_{B_1} = 90°$. Angles θ_{A_2} and θ_{B_2} can be computed as

$$\theta_{A_2}=\arcsin\left(\frac{y_{A_3}-y_{A_1}}{d_A}\right)\ , \qquad \theta_{B_2}=\arcsin\left(\frac{y_{B_3}-y_{B_1}}{d_B}\right)\ .$$

By solving Eq. (7.27) in V_{A_1} and V_{B_1}, the following values of the velocities are obtained:

Velocity V_{A_1} (in km/h)

		Vehicle A		
	Lane	Right	centre	Left
	Left	49.3	53.3	59.0
Vehicle B	centre	34.9	38.5	42.8
	Left	19.1	21.2	24.1

Velocity V_{B_1} (in km/h)

		Vehicle A		
	Lane	Right	centre	Left
	Left	99.7	91.4	81.7
Vehicle B	centre	103.3	95.7	86.8
	Left	105.5	98.6	90.7

It is very unlikely that vehicle B was on the left lane, as this would yield too low a value for the velocity of vehicle A. However its position has very little effect on the value of the velocity of vehicle B, which is in this case the most important parameter to be determined. Some uncertainty on this result still remains, as values between 81.7 and 105.5 km/h are possible.

This approach which can be defined as traditional, has still a wide uncertainty margin, mainly linked to the evaluation of the friction coefficients and of the velocities after the impact, in case a vehicle stops hitting an obstacle. Its simplicity allows one to perform several times the computations and obtain upper and lower bounds of the results. The values of the initial velocities can then be used to perform more accurate numerical simulations.

Example 7.2

Two cars collided at the intersection of two urban streets (Fig. 7.8). Both the collision point and the final positions at which they stopped (both vehicles stopped without hitting any obstacle) are known. Compute the speeds at which the two vehicles reached the intersection and the positions taken by the vehicles after the impact.

The inertial properties of the vehicles, the coordinates of their centre of mass and their yaw angles at times t_2 and t_3 with reference to the frame xy shown in the figure are $m_A = 1130$ kg, $m_B = 890$ kg, $J_A = 1780$ kg m^2, $J_B = 1400$ kg m^2, $x_{A_1} = 0.3$ m, $y_{A_1} = 0.6$ m, $x_{B_1} = 0.7$ m, $y_{B_1} = 1.4$ m, $x_{A_3} = -4.4$

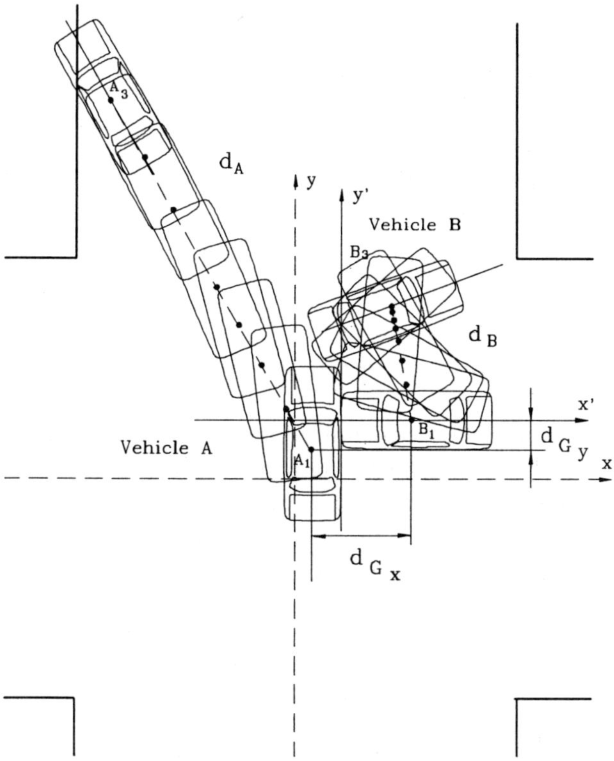

Fig. 7.8 Example 7.2; positions of the vehicles at times t_1 and t_3 and reconstruction of the accident; computed positions at times 0.05, 0.1, 0.15, 0.2, 0.3 and 0.5 s from time t_2.

m, $y_{A_3} = 8.5$ m, $x_{B_3} = 1.7$ m, $y_{B_3} = 4.9$ m, $\psi_{A_1} = 90°$, $\psi_{B_1} = 180°$, $\psi_{A_3} = 120°$ and $\psi_{B_3} = 20°$.

The positions of the centres of mass of the vehicle with reference to frame $x'y'$ centred on the impact zone are $x'_{G_A} = -0.825$ m, $y'_{G_A} = -0.80$ m, $x'_{G_B} = 1.575$ m and $y'_{G_B} = 0$.

For the computation of the velocities after the impact a value $f_A = 0.15$ was assumed for the first vehicle (after the impact it travelled with little sideslip) and $f_B = 0.50$ was assumed for the second one. As the distances travelled after the collision are $d_A = 9.123$ m and $d_B = 3.640$ m, the velocities at time t_2, computed through Eq. (7.28), are $V_{A_2} = 18.7$ m/s $=$ 67.3 km/h and $V_{B_2} = 5.98$ m/s $= 21.5$ km/h.

The velocity vectors are then

$$\vec{V}_{A_2} = \begin{Bmatrix} -9.65 \\ 16.22 \end{Bmatrix} \text{ m/s} , \qquad \vec{V}_{B_2} = \begin{Bmatrix} -1.64 \\ 5.75 \end{Bmatrix} \text{ m/s} .$$

The velocities before the collision are easily obtained from Eq. (7.27)

$$\vec{V}_{A_1} = \begin{Bmatrix} 0 \\ 20.75 \end{Bmatrix} \text{ m/s} , \qquad \vec{V}_{B_1} = \begin{Bmatrix} -13.89 \\ 0 \end{Bmatrix} \text{ m/s} .$$

The two cars were then travelling at 74.7 km/h and 50.0 km/h.

This result was obtained using a much simplified model. A more detailed approach can be used to confirm the results, using Eqs. (7.25) and following, with the only difference that in this case vehicle B hits the other one. As $V_{R_{\perp_1}} = V_{B_1} = 13.89$ m/s and $V_{R_{\parallel_1}} = -V_{A_1} = -20.75$ m/s, it follows

$$I_x = V_B \frac{1+e^*}{a-\lambda b} = -\frac{13,890(1+e^*)}{2.368-0.371\lambda} .$$

The components of the impulse can be easily computed from the velocities before and after the impact: $I_x = -10,905$ Ns and $I_y = -5,119$ Ns. The value $\lambda = 0.469$ is easily obtained, which is high but realistic owing to the possibility of interlocking between the vehicles.

By equating the two values of I_x the value $e^* = 0.723$ for the restitution coefficient is obtained. It is also quite high, but is again justified from the low permanent deformations found on the vehicles.

The angular velocities after the impact are then obtained

$$\omega_{A_2} = \frac{I_x y'_A - I_y x'_A}{J_A} = 2.52 \text{ rad/s} ,$$

$$\omega_{B_2} = \frac{-I_x y'_B + I_y x'_B}{J_B} = -5.75 \text{ rad/s} .$$

To compute the rotations of the vehicles during the accident Eq. (7.30) can be used. The value $f = 0.7$ can be used for both vehicles (see Sec. 7.3.1) and the geometrical parameters are $\overline{d}_A = 1.45$ m and $\overline{d}_B = 1.35$ m. The results are $|\psi_{A_3} - \psi_{A_2}| = 0.502$ rad $= 29°$, $|\psi_{B_3} - \psi_{B_2}| = 2.813$ rad $= 161°$, which are close to the ones measured on the road. The positions of the vehicle at times between t_2 and t_3 are reported in Fig. 7.8.

7.2 Vehicle collision: Advanced model

7.2.1 *Head on collision against a fixed obstacle*

The model studied in the previous sections was based on the assumption that the collision is an impulsive phenomenon and consequently it was impossible to assess what happens during the impact. The shock has however a duration, which is short but finite, and it is possible to study how the displacement, velocity and acceleration change between t_1 and t_2.

Consider at first a head-on collision against a fixed obstacle, like that occurring during a crash test. The force that the vehicle receives from the obstacle has a time history of the type shown in Fig. 7.9a. The curve is not smooth as the compliance of the front part of the vehicle changes strongly during the crushing process, owing to geometric nonlinearities, buckling and other phenomena. The experimental law can however be approximated by a smooth curve $F(t)$ while retaining the most important features of the actual behaviour of the vehicle (dashed curve in the figure).

The force is linked to the acceleration of the vehicle in a complex way, as its configuration changes in time and each point of the vehicle has its own acceleration. However, with the exception of the front part which undergoes the crushing process, the vehicle can be considered as a rigid body and the position of the centre of mass can be considered as fixed to the undeformed part of the vehicle. The acceleration can thus be obtained directly from the force $F(t)$; actually in a crash test the acceleration is usually measured by accelerometers located on the vehicle and the force is obtained from it.

From the acceleration it is straightforward to compute the velocity and the crushing of the vehicle.

A law which can approximate the acceleration is[2]

---

[2]R.H. Macmillan, *Dynamic of Vehicle Collisions*, Inderscience Enterprises, Jersey 1983.

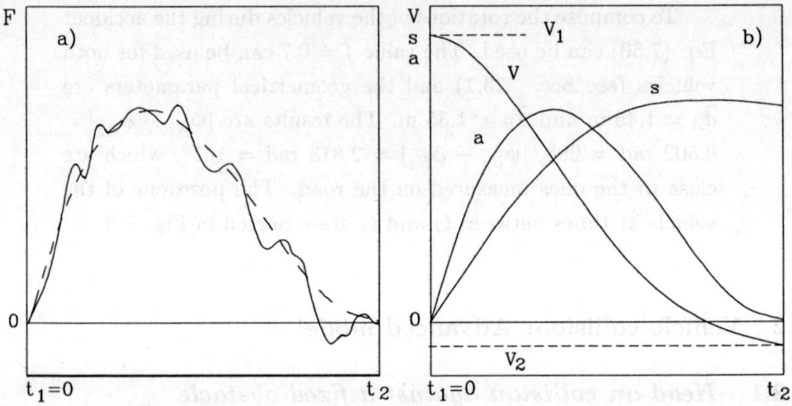

Fig. 7.9 (a) Force the vehicle receives from the obstacle during a crash test as a function of time. Experimental curve and mathematical empirical law. (b) Time histories $V(t)$, $a(t)$ and $s(t)$ obtained from the empirical law $F(t)$.

$$a = \tau(1-\tau)^{\beta}\frac{cV_1}{t_2}\,, \tag{7.31}$$

where t_2 is the duration of the impact, $\tau = t/t_2$ is the nondimensional time $(0 \le \tau \le 1)$ and c and β are nondimensional constants.

Such law has vanishing derivative (jerk equal to zero) at the end of the collision ($t = t_2$) while the jerk at the beginning of the collision ($t = 0$) is different from zero; both these features comply with the intuitive physical interpretation of the phenomenon.

The velocity can be obtained integrating Eq. (7.31)

$$V = -cV_1\left[\frac{(1-\tau)^{\beta+1}}{\beta+1} - \frac{(1-\tau)^{\beta+2}}{\beta+2}\right] + K\,. \tag{7.32}$$

The constant of integration K can be computed since at time $t = 0$ the speed is $V = V_1$

$$K = V_1\left[1 + \frac{c}{(\beta+1)(\beta+2)}\right]. \tag{7.33}$$

By remembering the definition of the restitution coefficient $e^* = -V_2/V_1$, it is possible to compute the value of constant c. By computing the velocities at times t_2 and t_1, i.e. for $\tau = 0$ and $\tau = 1$, and equating their ratio to $-e^*$, it follows that

$$c = -(1 + e^*)(\beta + 1)(\beta + 2) . \tag{7.34}$$

The final expression of the velocity is thus

$$V = V_1 \left\{ (1 + e^*) \left[1 + \tau(\beta + 1) \right] (1 - \tau)^{\beta+1} - e^* \right\} . \tag{7.35}$$

A further integration gives the distance travelled s

$$s = V_1 t_2 \left\{ -(1 + e^*) \left[(1 - \tau)^{\beta+2} - \frac{\beta + 1}{\beta + 3} (1 - \tau)^{\beta+3} \right] - e^* \tau + K_1 \right\} . \tag{7.36}$$

If at time $t = 0$ the distance is assumed to be nil, s also has the meaning of the crushing of the front part of the vehicle. This statement allows to compute the value of the integration constant K_1

$$K_1 = \frac{2(1 + e^*)}{\beta + 3} . \tag{7.37}$$

The final expression of the displacement is thus

$$s = V_1 t_2 \left(\frac{1 + e^*}{\beta + 3} \left\{ 2 - (1 - \tau)^{\beta+2} \left[2 + (\beta + 1)\tau \right] \right\} - e^* \tau \right) . \tag{7.38}$$

At time $\tau = 1$, the displacement gives directly the residual crushing of the vehicle s_2

$$s_2 = V_1 t_2 \left[\frac{2(1 + e^*)}{\beta + 3} - e^* \right] . \tag{7.39}$$

In case of an elastic collision, the residual crushing must vanish: If $e^* = 1$ then $s_2 = 0$. From this statement a first relationship between β and e^* can be stated: $\beta = 1$ if $e^* = 1$.

Parameters β and e^* which characterize the impact depend on many factors. Firstly on the structural characteristics of the vehicle but also on the type of impact and, in the case of head-on collision against a fixed obstacle, on the impact velocity V_1 and time t_2.

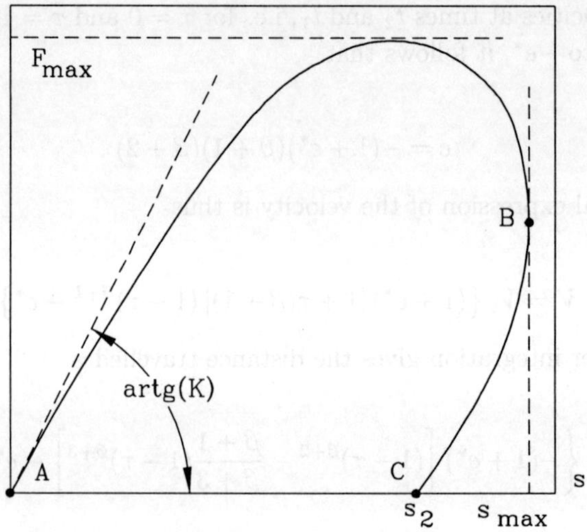

Fig. 7.10 Force received by the vehicle during a head-on collision against the obstace as a function of the crushing s.

A first characteristic of the vehicle is its stiffness at the instant it enters in contact with the obstacle, namely its *crushing modulus* $K = (dF/ds)_{\tau=0}$ (Fig. 7.10). With simple kinematic considerations it follows

$$K = m\left(\frac{da}{ds}\right)_{\tau=0} = m\left(\frac{da}{dt}\right)_{\tau=0}\left(\frac{ds}{dt}\right)_{\tau=0}^{-1} = \frac{m}{V_1}\left(\frac{da}{dt}\right)_{\tau=0} . \qquad (7.40)$$

By introducing the expression (7.31) of the acceleration into Eq. (7.40), it follows

$$K = m\frac{(1+e^*)(\beta+1)(\beta+2)}{t_2^2} . \qquad (7.41)$$

The value of K can be obtained from crash tests; values between 1 and 2 MN/m being found in the literature.

The crushing process is far from being linear. A qualitative plot of force F against the displacement s is reported in Fig. 7.10: The inelastic behaviour of the vehicle and the hysteresis cycle are clearly shown. The area below the line from point A to point B is the energy absorbed by the vehicle from time t_1 to the instant t_i at which the maximum displacement

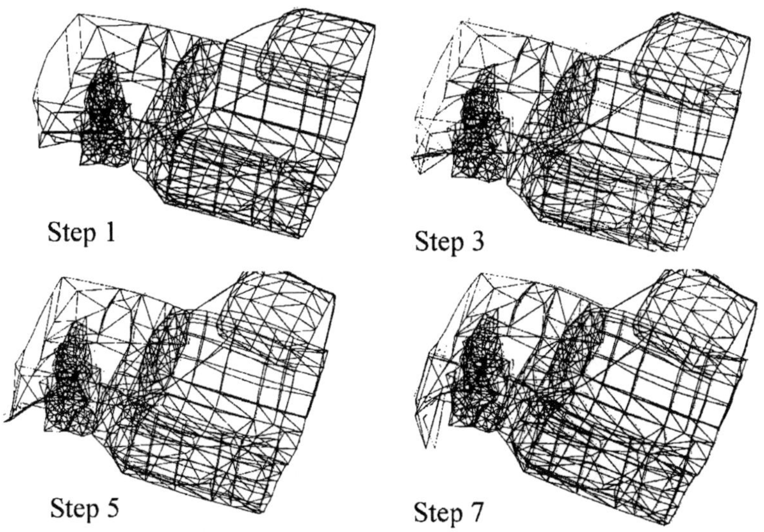

Fig. 7.11 Numerical simulation of a crash test. Deformed shape of the vehicle computed at four different times through numerical integration in time.

is reached. In the case of collision against a fixed rigid obstacle it is equal to the kinetic energy of the vehicle.

The area below the line from point B to point C is the energy which is transformed back as kinetic energy during the rebound phase from time t_i to t_2. If $e^* = 0$ such area vanishes and $s_2 = s_{max}$. If on the contrary $e^* = 1$ line BC is superimposed to line AB and the area of the hysteresis cycle vanishes.

The plot is usually obtained from a crash test, both materially performed on an actual vehicle or simulated by computer. The numerical simulation is a very complex task, owing to the very complex geometry of the front part of the vehicle and of the nonlinearities of all types involved and requires large and fast computers. Some results obtained through the finite element method, which as present is the only one which allows to tackle problems of this complexity, are reported in Fig. 7.11. The computation has been performed through step by step integration in time of the equations of motion; the deformed mesh at four different times has been reported in the figure.

Often the law $F(s)$, obtained through the empirical law $F(t)$ defined above, is assumed to be independent of the deformation rate ds/dt. Such

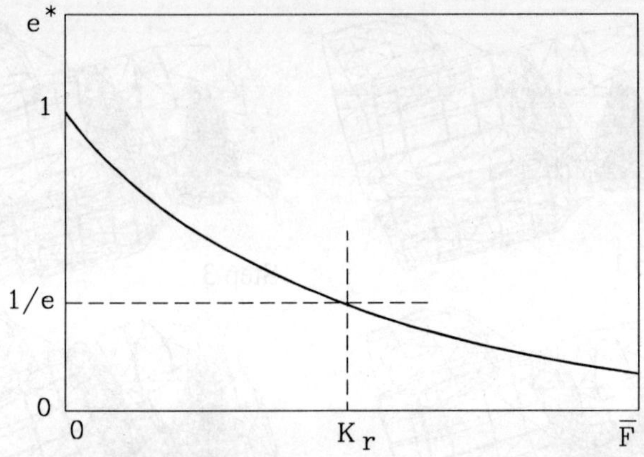

Fig. 7.12 Law $e^*(\overline{F})$ approximated by Eq. (7.44).

assumption has no theoretical background and is justified by the fact that it is usually not strongly contradicted by the experimental evidence, at least if the deformation rate is not too large.

The mean value of the force received by the vehicle

$$\overline{F} = \frac{1}{t_2}\int_0^{t_2} F dt \tag{7.42}$$

can be computed easily by remembering that the total impulse received by the vehicle is equal to the change of the momentum

$$\overline{F} = m\frac{V_2 - V_1}{t_2} = -m\frac{V_1}{t_2}(1 + e^*) . \tag{7.43}$$

If the mean value of force F is small, the permanent deformations are usually negligible and the impact is close to be elastic, i.e. $e^* \approx 1$. With increasing $\overline{F}$ the collision becomes more and more inelastic, i.e. e^* decreases with increasing $\overline{F}$. It is possible to approximate the dependence of the restitution coefficient e^* on the average force $\overline{F}$ with an exponential law (Fig. 7.12)

$$e^* = e^{-\overline{F}/K_r} , \tag{7.44}$$

where K_r is the value of $-\overline{F}$ at which the coefficient of restitution takes the value $e^* = 1/e = 0.368$. It is referred to as the *impact resistance modulus*. Also K_r can be obtained from a crash test. From Eqs. (7.41) and (7.44), it follows that

$$K_r = m\frac{V_1(1+e^*)}{t_2 \ln(1/e^*)} , \tag{7.45}$$

which allows to compute K_r from quantities which can be measured or computed. Values between 40 and 100 kN have been reported for K_r in the literature.

As seen above, constant β depends on e^*: It takes a unit value when $e^* = 1$ and increases when e^* decreases. Assume that, at least in the case of strongly inelastic impacts, i.e. near the condition $e^* = 0$, the law $\beta(e^*)$ can be approximated as

$$\beta = \beta_0 - (\beta_0 - 1)e^* , \tag{7.46}$$

which, although assumed for $e^* \approx 0$, gives the correct result also for $e^* = 1$.

As $\beta > 1$, also β_0 is always larger than unity. Also β_0, which is nondimensional, can be considered as a characteristic of the vehicle, and will be referred to as *structural index*. A large value of β_0 characterizes vehicles with a very stiff front part, while a compliant front section is typical of vehicles with low β_0. Its values are usually in the vicinity of 2.

Parameters K, K_r and β_0 characterize completely a vehicle for what the head-on collision against a rigid obstacle is concerned. Once they have been measured, it is possible to compute the laws $F(t)$, $a(t)$ and $s(t)$ from the collision conditions, namely the velocity V_1 and the mass of the vehicle m.

The values reported in Tables 7.1 and 7.2 have been obtained from crash tests published by manufacturers [3]. The values in the first table are not exactly comparable with the others as they have been computed from tests very well characterized, while for the others not all the parameters were known and the values of β_0 and s_2 had to be assumed.

To solve the direct problem, i.e. to obtain the conditions after the collision, mainly s_2 and V_2 from V_1, it is possible to obtain t_2 from Eq. (7.45) and to introduce it into Eq. (7.41), obtaining

[3] R.H. Macmillan, *Dynamic of Vehicle Collisions*, Inderscience Enterprises, Jersey 1983.

Table 7.1 Parameters obtained from crash tests on some European passenger vehicles.

	BMC Mini	BMC 1100	Ford Anglia	BMC 1800
m [kg]	720	950	1000	1250
V_1[m/s]	14	11.5	14	14
e^*	0.10	0.08	0.5	0.11
t_2[ms]	92	102	103	97
β_0	2.35	2.79	2.89	2.16
K [MN/m]	1.27	1.67	1.81	1.80
K_r [kN]	52.3	45.8	47.6	90.7

Table 7.2 Parameters obtained from crash tests on some American passenger vehicles.

	Sub compact	Compact	Interm.	Standard 71/72	Standard 73/74
m [kg]	1135	1545	1820	2045	2045
V_1[m/s]	12	12	10.5	10	10.3
e^*	0.01	0.01	0.20	0.20	0.20
t_2[ms]	102	102	151	147	143
β_0	2.50	2.50	2.00	2.00	2.00
K [MN/m]	1.60	2.17	1.03	1.09	1.16
K_r [kN]	28.6	38.8	96.7	93.6	108.8

$$KmV_1^2 = K_r^2\left[\ln\left(\frac{1}{e^*}\right)\right]\frac{(\beta+1)(\beta+2)}{1+e^*}\,. \tag{7.47}$$

Substituting Eq. (7.46) into Eq. (7.47) the latter yields

$$\frac{KmV_1^2}{K_r^2} = \left[\ln\left(\frac{1}{e^*}\right)\right]\frac{(\beta_0+1)(\beta_0+2) - e^*(\beta_0-1)(2\beta_0+3) + e^{*^2}(\beta_0-1)^2}{1+e^*}\,, \tag{7.48}$$

which can be solved numerically in e^* and then allows to solve completely the direct problem.

The inverse problem, i.e. to obtain the conditions before the collision, mainly V_1, from the crushing s_2, is also easily solved. From Eqs. (7.37), (7.41) and (7.45) it follows that

$$s_2\frac{K}{K_r} = \ln\left(\frac{1}{e^*}\right)\frac{(\beta+1)(\beta+2)\left[2(1+e^*)-(\beta+3)e^*\right]}{\beta+3}\,. \tag{7.49}$$

Substituting Eq. (7.46) into Eq. (7.49) the latter yields

$$s_2\frac{K}{K_r} = \ln\left(\frac{1}{e^*}\right)\left[{e^*}^2(\beta_0-1)^2 - e^*(2\beta_0^2+\beta_0-3)+3\beta_0+\beta_0+2\right]$$

$$\times\frac{e^*(e^*-1)(\beta_0-1)+2}{-e^*(\beta_0-1)+\beta_0+3}\,, \tag{7.50}$$

which can be solved numerically in e^* and then allows to solve the inverse problem.

Example 7.3

Consider a car with $\beta_0 = 2$, $K = 1.2$ MN/m, $K_r = 65$ kN and $m = 1000$ kg impacting an obstacle at 20 m/s. Compute the parameters of the impact and the laws $F(t)$, $a(t)$ $V(t)$ and $s(t)$.

The numerical solution of Eq. (7.48) yields $e^* = 0.0416$ and then $\beta = 1.958$, $t_2 = 0.1$ s. The laws $F(t)$, $a(t)$ $V(t)$ and $s(t)$ for this case are plotted in Fig. 7.13; the residual crush is $s_2 =$ 757 mm.

7.2.2 *Head-on collision between vehicles*

The head-on collision between vehicles can be studied using the same model seen in the previous section, if the contact surface is assumed to remain planar and if the characteristics of the impact, and particularly the law $F(s)$, are independent of the deformation rate. Each vehicle can thus be assumed to impact against a moving massless obstacle (Fig. 7.13a). Each vehicle is modelled as a point mass, provided with a nonlinear spring, whose characteristics (K, K_r and β_0) are the same as seen for the collision against a fixed obstacle.

The deformation rates

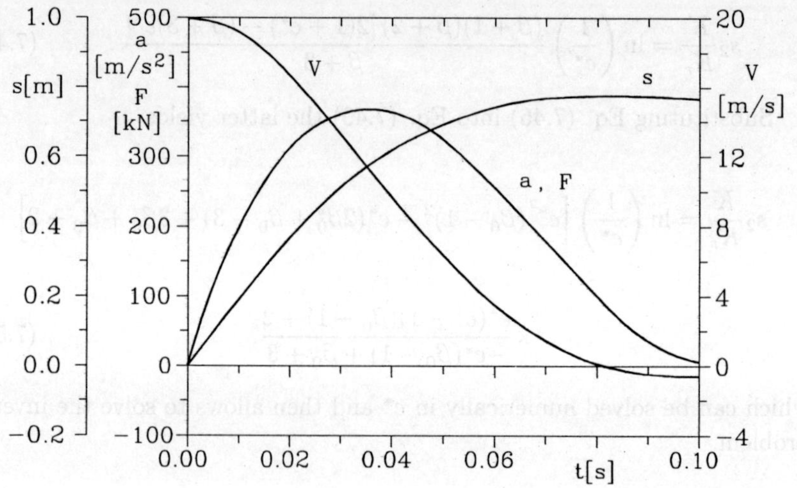

Fig. 7.13 Laws $F(t)$, $a(t)$, $V(t)$ and $s(t)$ for Example 7.3.

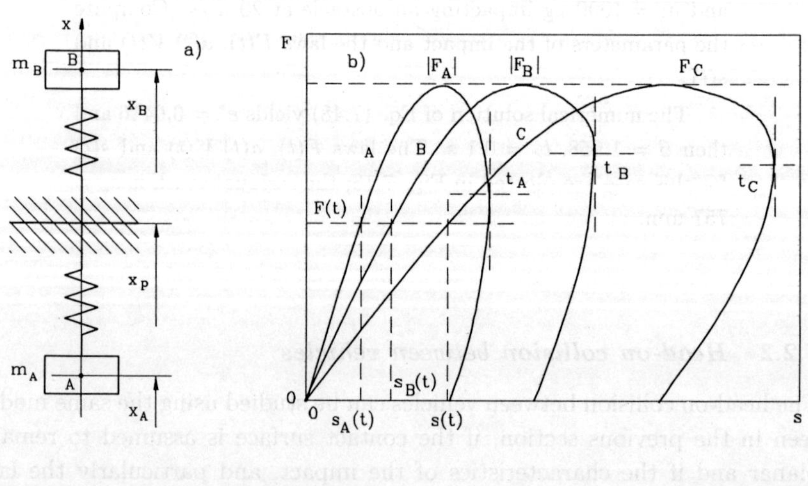

Fig. 7.14 Head-on collision between vehicles. (a) model; (b) forces as functions of the crush of the vehicles.

$$\begin{cases} \dot{s}_A = \dot{x}_b - \dot{x}_A \\ \dot{s}_B = -\dot{x}_b + \dot{x}_B \end{cases} \tag{7.51}$$

are positive when the springs are compressed.

The relative velocity V_R is then

$$V_R = \dot{x}_B - \dot{x}_A = -(\dot{s}_A + \dot{s}_B) \, . \tag{7.52}$$

If the curves $F(s)$ for the two vehicles are known, a plot of the type shown in Fig. 7.14b can be drawn. As the total force acting on the virtual obstacle must vanish, for each value of the time $|F_A| = |F_B|$. The intersections of the curves $|F_A(s_A)|$ and $|F_B(s_B)|$ with any line F = constant give thus the deformations s_A and s_B in the same instant. A third curve in which the force is plotted as a function of the total deformation $s = s_A + s_B$, i.e. of the change of the distance between the centres of mass, can be plotted.

The maximum values of the deformations need not be reached at the same time: In the figure the maximum deformation is reached at time t_A and t_B for the two vehicles, while the distance between the centres of mass is minimum at time t_C.

As the forces acting on the vehicles are simply $F_A = m_A \ddot{x}_A$ and $F_B = m_B \ddot{x}_B$, it follows that

$$F_A = -F_B = \dot{V}_R \frac{m_A m_B}{m_A + m_B} \, . \tag{7.53}$$

If the derivative of the relative velocity has a time history which is of the same type of that assumed for the acceleration in Eq. (7.31),

$$\dot{V}_R = -\frac{c V_{R_1}}{t_2} \tau (1 - \tau)^{\beta} \, , \tag{7.54}$$

where, as usual, $\tau = t/t_2$, the same model used for the collision against an obstacle holds for the present case as well.

Instead of m, K, V and s the relevant equations now contain

$$m_C = \frac{m_A m_B}{m_A + m_B} \, , \qquad K_C = \frac{K_A K_B}{K_A + K_B} \, , \qquad V_R \, , \qquad s_C = s_A + s_B \, .$$

Assume that the characteristics of the vehicles and the residual deformation of one of them, say s_{A_2}, are known. As the curve $F(s)$ has been assumed to be the same which characterized the impact against the barrier, from s_{A_2} it is possible to compute directly e^*_A from Eq. (7.50) and then the velocity V'_{A_1}[4] which causes the same residual deformation s_{A_2} in an impact with a rigid obstacle.

The maximum value F_{max_A} received by the first vehicle is

$$F_{max_A} = K_{r_A} \beta_A^{\beta_A} \ln\left(\frac{1}{e_A^*}\right) \frac{(\beta_A + 1)(\beta_A + 2)}{(\beta_A + 1)^{\beta_A + 1}} . \tag{7.55}$$

The absolute value of such force is equal to the maximum force acting on vehicle B. It is thus possible to write an equation identical to Eq. (7.55) with the characteristics of the second vehicle. As

$$\beta_B = \beta_{0_B} - (\beta_{0_B} - 1)e_B^* ,$$

it is possible to obtain e_B^* and then t_{B_2}, V_{B_1} and s_{B_2}.

The two equivalent collisions against a rigid obstacle are thus completely characterized. As the curves $F(s)$ are the same in the actual collision and the equivalent ones, the energy dissipated in the former is equal to the one dissipated in the latter

$$\Delta E = \Delta E_A + \Delta E_B = \frac{1}{2} m_A V_{A_1}^2 \left(1 - e_A^{*^2}\right) + \frac{1}{2} m_B V_{B_1}^2 \left(1 - e_B^{*^2}\right) . \tag{7.56}$$

It is thus possible to consider the actual collision, described by curve C in Fig. 7.14b, by using the same equations seen in the previous chapter with the mentioned substitutions. The maximum force, already expressed by Eq. (7.55), and the energy dissipated take the values

$$F_{max} = \frac{s_{C_2} K_C \beta_C^{\beta_C} (\beta_C + 3)}{(\beta_C + 1)^{\beta_C + 1} [2 - e_C^*(\beta_C + 1)]} \tag{7.57}$$

$$\Delta E = \frac{1}{2} m_C V_{C_1}^2 \left(1 - e_C^{*^2}\right) . \tag{7.58}$$

Equations (7.57) and (7.58) contain 3 unknowns β_C, e_C^* and V_R (or V_{C_1}, which is the same). A third equation can be obtained considering that force F_{max} can also be expressed as

$$F_{max} = K_C V_R t_2 \frac{\beta_C^{\beta_C}}{(\beta_C + 1)^{\beta_C + 1}} . \tag{7.59}$$

[4] V' is the velocity in the equivalent collision against a rigid obstacle.

As

$$t_2^2 = \frac{m_C}{K_C}(1 + e_C^*)(\beta_C + 1)(\beta_C + 2) , \tag{7.60}$$

computing V_R from Eq. (7.58) and substituting the value so obtained into Eqs. (7.59) and (7.60), it follows that

$$F_{max}^2 = K_C \frac{2\Delta E \beta^{2\beta}(\beta + 2)}{(1 - e^*)(\beta + 1)^{2\beta+1}} . \tag{7.61}$$

The last equation can be solved in e^*, obtaining

$$e^* = 1 - K_C \frac{2\Delta E \beta^{2\beta}(\beta + 2)}{F_{max}^2(\beta + 1)^{2\beta+1}} . \tag{7.62}$$

Finally, introducing Eq. (7.62) into Eq. (7.57) the equation

$$F_{max} = s_C K_C \frac{\beta^\beta(\beta + 3)}{(\beta + 1)^{\beta+1}} \left[1 - \beta + \frac{2\Delta E K_C}{F_{max}^2}(\beta + 2)\left(\frac{\beta}{\beta + 1}\right)^{2\beta} \right]^{-1} \tag{7.63}$$

is obtained, which can be easily solved numerically in β. It is thus possible to obtain the values of e^* and V_R which solve the problem.

The inverse problem, consisting in obtaining the parameters characterizing the collision once the relative velocity V_R is known is more difficult as it must be solved in an iterative way. A value of the final crushing s_2^* of one of the two vehicles can be assumed and from it the relative velocity V_R^* can be computed as seen above. A new value of the residual crushing, for example obtained as $s_2^{**} = s_2^* V_R / V_R^*$, can then be computed and a new relative velocity, which is closer to the correct one, is obtained. The procedure should converge quickly to the required result. The velocities of the vehicle after the collisions can then be obtained without further problems.

Example 7.4

Consider the head-on collision between two cars whose characteristics are known from crash tests, e.g. the vehicles of the first and fourth columns of Tab. 7.1. The characteristics are then m_A = 720 kg, m_B = 1250 kg, β_{0_A} = 2.35, β_{0_B} = 2.16, K_A = 1.27 MN/m, K_B = 1.80 MN/m, K_{r_A} = 52.3 kN and K_{r_B}

= 90.7 kN. The residual crush of the first vehicle is $s_{A_2} = 400$ mm. Compute the parameters of the impact and the relative velocity at time t_1.

The parameters which characterize the collision are $m_C = 456.85$ kg and $K_C = 0.745$ MN/m. The coefficient of restitution for the first vehicle is easily computed from the residual crush by solving numerically Eq. (7.50): $e_A^* = 0.105$. The values of β_A, t_{A_2} and V_{A_1} are then $\beta_A = 2.208$, $t_{A_2} = 0.092$ s and $V_{A_1} = 13.63$ m/s.

Equations (7.55), (7.46), (7.41) and (7.45) yield $F_{max} = 218$ kN, $e_B^* = 0.252$, $\beta_B = 1.87$, $t_{B_2} = 0.098$ s, $V_{B_1} = 7.85$ m/s and $s_{B_2} = 202$ mm. The energy dissipated and the total crush are then $\Delta E = 102$ kJ and $s_C = 602$ mm.

Solving Eq. (7.63) in β and using Eqs. (7.58) and (7.62) the final results are obtained: $\beta = 2.037$, $e^* = 0.159$ and $V_R = 21.43$ m/s.

Example 7.5

Consider the same vehicles of the previous examples colliding head-on with velocities $V_{A_1} = 26.7$ m/s $= 96$ km/h and $V_{B_1} = -13.1$ m/s $= -47$ km/h. Compute the velocities after the impact.

The relative velocity is $V_R = 39.8$ m/s. By assuming a residual crush of the first vehicle $s_{A_2} = 400$ mm in the previous example a relative velocity $V_R = 21.43$ m/s has been obtained. Correcting linearly the assumed residual crush, a new trial value $s_{A_2} = 743$ mm is obtained, which yields a relative velocity $V_R = 36.16$ which is already close to the correct values.

With two further iterations a crush $s_{A_2} = 823$ mm is obtained, together with $e^* = 0.048$, $F_{max} = 378$ kN and $\Delta E = 360$ kJ.

As a final results the velocities after the collision $V_{A_2} = 0.234$ m/s $= 0.84$ km/h and $V_{B_2} = 2.144$ m/s $= 7.72$ km/h are obtained.

7.2.3 *Oblique collision between vehicles*

A first problem in the study of oblique collisions is the evaluation of the characteristics of the vehicle: If it is already difficult to obtain the values of K, K_r and β_0 for head-on collisions, it is almost impossible to obtain them

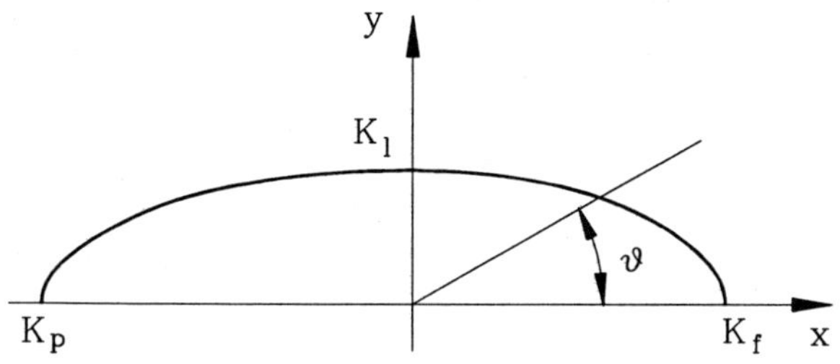

Fig. 7.15 Polar diagram $K(\theta)$ approximated by two arcs of ellipse.

for a generic impact direction. If it is possible to find the relevant values for side or rear impacts, the dependence of the characteristics from angle θ (Fig. 7.15) can be approximated by two arcs of ellipse, one for the front and one for the rear.

Following the notation of the figure, function $K(\theta)$ can be expressed as

$$K = \begin{cases} \dfrac{K_f K_l}{\sqrt{K_f \sin^2(\theta) + K_l \cos^2(\theta)}} & \text{for } 0 \leq \theta \leq 90^\circ \\ \dfrac{K_p K_l}{\sqrt{K_p \sin^2(\theta) + K_l \cos^2(\theta)}} & \text{for } 90^\circ \leq \theta \leq 180^\circ . \end{cases} \tag{7.64}$$

Similar relationships can be written to approximate the other characteristics of the vehicle.

If the collision, albeit oblique, were central, i.e. the two velocities were aligned, and the two angular velocities were equal to zero (Fig. 7.16a), the procedure seen above for the head-on collision would still hold, provided that the correct characteristics of the vehicles were used.

In case of non-central oblique collisions the vehicles are subject also to angular accelerations about the yaw axis z. An approximated but very simple way to take this into account is that of substituting the mass of the vehicle against which each vehicle collides with an effective mass, lower than the actual one. It is questionable whether it is worthwhile to attempt to refine any further this model, since the assumptions on which the present models are based and the uncertainties on the data do not allow at any rate to obtain high precision.

Consider the situation of Fig. 7.16b: Vehicle A hits the front part of vehicle B. Neglecting the friction in the contact area ($\lambda = 0$), the x component of the momentum of the second vehicle does not change and

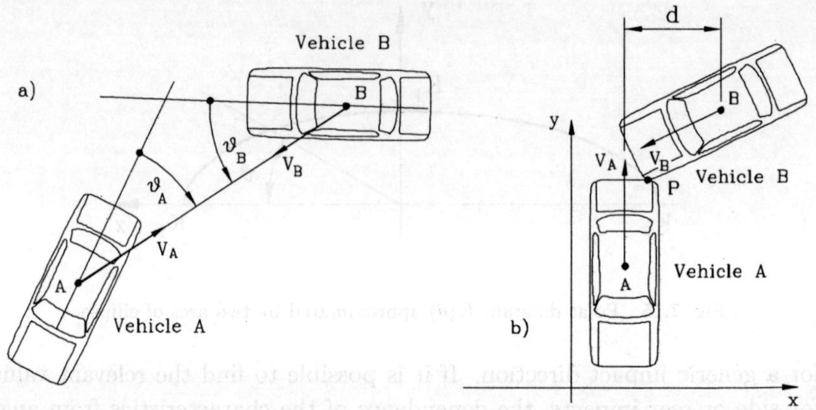

Fig. 7.16 (a) Central oblique collision. (b) Oblique collision in which vehicle A hits the front part of vehicle B.

the impulse has only its y component. Vehicle B undergoes an acceleration in y direction and an angular acceleration

$$\ddot{y}_{G_B} = \frac{F}{m_B}\,, \qquad \ddot{\theta}_B = \frac{Fd}{J_{z_B}}\,. \tag{7.65}$$

The acceleration of point P, thought as belonging to vehicle B, is then

$$\ddot{y}_{P_B} = \ddot{y}_{G_B} + d\ddot{\theta}_B = \frac{F}{m_B}\left(1 - \frac{d^2}{r_B^2}\right)\,, \tag{7.66}$$

where r_B is the radius of gyration. The acceleration of the contact point is thus equal to the acceleration of the centre of mass of a vehicle having a *reduced mass*

$$m_{r_B} = m_B \frac{r_B^2}{r_B^2 + d^2}\,.$$

If neither vehicle collides head-on with the other, reference must be made to the surface on collision, as seen in Section 7.1.2. If no allowance is taken for the friction between the vehicles, the reduced masses of the two vehicles can be computed with reference to the perpendicular to such surface (Fig. 7.17a). If friction is taken into account, reference must be made to a direction inclined of an angle arctan(λ) with respect to the perpendicular to the collision surface (Fig. 7.17b).

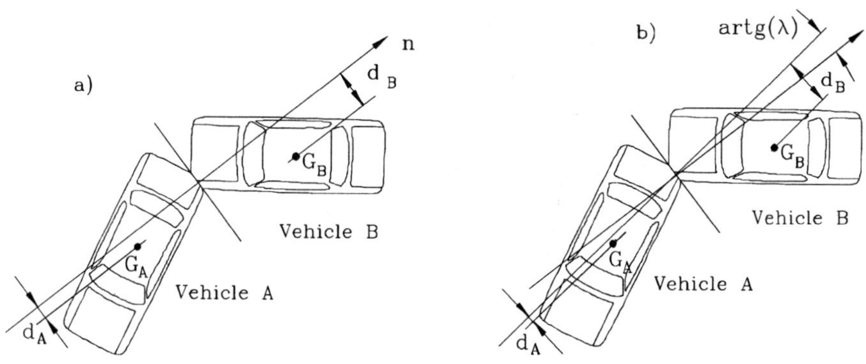

Fig. 7.17 Oblique collision; definition of distances d_A and d_B for the computation of the reduced mass. (a) No friction ($\lambda = 0$) and (b) $\lambda \neq 0$.

7.3 Motion after the collision

7.3.1 *Vehicle with locked wheels*

If after the collision the motion of the vehicle is simply translational the distance travelled can be easily computed as seen in Section 7.1.6. This however is seldom the case: After the impact a certain yaw velocity $\omega = \dot{\psi}$ is usually present and it is incorrect to take into account independently translational and rotational motion.

Consider a vehicle which, after the collision, moves with the wheels completely locked, as if the brakes were fully applied or the deformations of the body were large enough to prevent the wheels from rotating. With reference to Fig. 7.18a, the components of the velocity of the centre of contact P_i of the i-th wheel u_i and v_i in the directions of axes x and y fixed to the vehicle are

$$\begin{cases} u_i = V\cos(\beta) - \omega r_i \sin(\chi_i) \\ v_i = V\sin(\beta) + \omega r_i \cos(\chi_i) \end{cases}, \tag{7.67}$$

where r_i and χ_i are linked to the coordinates x_i and y_i of point P_i by the obvious relationships

$$r_i = \sqrt{x_i^2 + y_i^2}\,, \qquad \chi_i = \arctan\left(\frac{y_i}{x_i}\right).$$

The absolute value of force F_i exchanged by the i-th locked wheel is

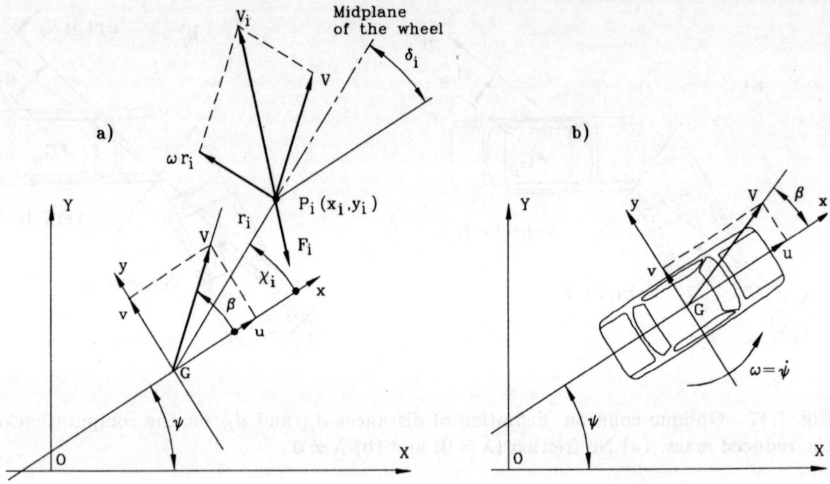

Fig. 7.18 (a) Velocity of the centre of contact of the i-th wheel; (b) inertial reference frame and variables of the motion.

simply $|F_i| = fZ_i$ and its direction is equal to that of the velocity V_i, with opposite sign. The components of force F_i are then

$$\begin{cases} F_{x_i} = -fZ_i \dfrac{u_i}{|V_i|} = fZ_i \dfrac{-V\cos(\beta) + \omega r_i \sin(\chi_i)}{|V_i|} \\ F_{y_i} = -fZ_i \dfrac{v_i}{|V_i|} = -fZ_i \dfrac{V\sin(\beta) + \omega r_i \cos(\chi_i)}{|V_i|} \end{cases}, \tag{7.68}$$

where

$$|V_i| = \sqrt{V^2 + \omega^2 r_i^2 + 2V\omega r_i \sin(\beta - \chi_i)} \ . \tag{7.69}$$

The moment of force F_i about the yaw axis z is

$$M_i = F_{y_i} r_i \cos(\chi_i) - F_{X_i} r_i \sin(\chi_i) = fZ_i r_i \frac{\omega r_i + V\sin(\beta - \chi_i)}{|V_i|} \ . \tag{7.70}$$

The trajectory can thus be easily computed by numerical integration of the equations of motion. No linearization is possible in this case, as the slip angle of the vehicle β can be very large. The equations of motion are Eq. (5.29), where the forces acting on the wheels are those expressed by Eq. (7.68) and then rotated in the inertial reference frame by multiplying them by a suitable rotation matrix. The model here used is essentially a

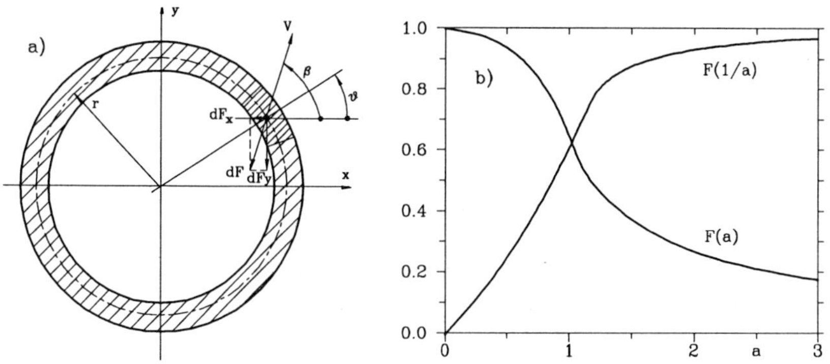

Fig. 7.19 (a) Simplified model for the study of the trajectory of a vehicle which moves with locked wheels. Circular contact area which is substituted for the actual contact area. (b) Functions $F(a)$ and $F(1/a)$ for a between 0 and 3.

three-degrees of freedom, rigid body model in which all forces except those due to tire-road interaction have been neglected. It would be at any rate very difficult to take into account aerodynamic forces in conditions in which angle β_a is large and rapidly varying.

At time $t = 0$, which coincides with time t_2 immediately after the collision, the position and the velocity of the vehicle and also angle $\beta = \arctan(\dot{x}'/\dot{y}' - \dot{\psi})$ are known and the numerical integration can be started.

Note that, owing to the impossibility of linearizing the equations of motion, it is impossible to work in terms of axles and the wheels must be considered one by one; in particular velocity V_i is different for the wheels of the same axle. It is difficult to take into account load transfer between the wheels of the same axle with a model based on the assumption of rigid body; in the present conditions the load transfer can be strongly influenced by roll rotations and a more complete model is required if this effect is to be taken into account. Forces F_{x_i} and F_{y_i} and moment M_i change continuously during the motion. Also the total forces and moments acting on the vehicle change, but to a lesser extent.

For a very simple approach, which can be used to study the motion of the vehicle without having to perform a numerical simulation, it is possible to substitute the actual contact area between the vehicle and the ground with a circle having a radius r equal to the average distance of the centres of the contact areas of the wheels and the centre of gravity (Fig. 7.19a).

The force and the moment exerted on an arc of amplitude $d\theta$ of such a circumference can be expressed by equations of the same type of equations

(7.68) and (7.70). Assuming that the vertical load mg exerted on the contact area is evenly distributed on the circumference, the components of the force and the yawing moment are

$$\begin{cases} dF_x = f\dfrac{mg}{2\pi}\dfrac{-V\cos(\beta)+\omega r\sin(\theta)}{\sqrt{V^2+\omega^2r^2+2V\omega r\sin(\beta-\theta)}}d\theta \\ dF_y = -f\dfrac{mg}{2\pi}\dfrac{V\sin(\beta)+\omega r\cos(\theta)}{\sqrt{V^2+\omega^2r^2+2V\omega r\sin(\beta-\theta)}}d\theta \\ dM = -f\dfrac{mgr}{2\pi}\dfrac{V\sin(\beta-\theta)+\omega r}{\sqrt{V^2+\omega^2r^2+2V\omega r\sin(\beta-\theta)}}d\theta\,. \end{cases} \tag{7.71}$$

Force dF can be decomposed along directions parallel and perpendicular to the velocity V. The first component, which is tangential to the trajectory, is

$$dF_\parallel = dF_x\cos(\beta)+dF_y\sin(\beta) = -f\frac{mg}{2\pi}\frac{V+\omega r\sin(\beta-\theta)}{\sqrt{V^2+\omega^2r^2+2V\omega r\sin(\beta-\theta)}}d\theta\,. \tag{7.72}$$

Its effect is that of reducing the speed of the vehicle. The second component, which acts in a direction perpendicular to trajectory and then acts to bend the path of the vehicle, is

$$dF_\perp = -dF_x\sin(\beta)+dF_y\cos(\beta) = -f\frac{mg}{2\pi}\frac{\omega r\cos(\beta-\theta)}{\sqrt{V^2+\omega^2r^2+2V\omega r\sin(\beta-\theta)}}d\theta\,. \tag{7.73}$$

By integrating the expression of the forces and moments, it follows

$$\begin{cases} F_\parallel = -f\dfrac{mg}{2\pi}\displaystyle\int_0^{2\pi}\dfrac{1+a\sin(\zeta)}{\sqrt{1+a^2+2a\sin(\zeta)}}d\zeta \\ F_\perp = -f\dfrac{mg}{2\pi}\displaystyle\int_0^{2\pi}\dfrac{a\cos(\zeta)}{\sqrt{1+a^2+2a\sin(\zeta)}}d\zeta = 0 \\ M = -f\dfrac{mgr}{2\pi}\displaystyle\int_0^{2\pi}\dfrac{1+\frac{1}{a}\sin(\zeta)}{\sqrt{1+\left(\frac{1}{a}\right)^2+\frac{2}{a}\sin(\zeta)}}d\zeta\,, \end{cases} \tag{7.74}$$

where $\zeta = \beta-\theta$ (as β does not depend on θ, $d\zeta = d\theta$) and the nondimensional parameter a is the ratio between the component of the velocity due to rotation and velocity V

$$a = \omega r/V \ .$$

The component of force F perpendicular to the trajectory is equal to zero: This means that the trajectory is straight, at least within the assumptions used in the present model. If a more accurate model were used the trajectory would bend, even if not much.

The integrals which appear in the expressions of the forces and the moments are functions of parameter a only. By introducing the function $F(a)$ defined as

$$F(a) = \frac{1}{2\pi}\int_0^{2\pi} \frac{1 + a\sin(\zeta)}{\sqrt{1 + a^2 + 2a\sin(\zeta)}} d\zeta \ . \tag{7.75}$$

The expressions of $F_{\parallel}$ and M are simply

$$\begin{cases} F_{\parallel} = -fmgF(a) \\ M = -fmgrF\left(\frac{1}{a}\right) \ . \end{cases} \tag{7.76}$$

Function $F(a)$ must be obtained numerically; its plot is reported in Fig. 7.19b and some values are reported in Table 7.3.

Table 7.3 Values of functions $F(a)$ and $F(1/a)$ for some values of a.

a	$F(a)$	$F(1/a)$	a	$F(a)$	$F(1/a)$
0	1	0	1,4	0,3860	0,8566
0,2	0,9899	0,1005	1,6	0,3306	0,8936
0,4	0,9587	0,2043	1,8	0.2900	0.9177
0,6	0,9028	0,3158	2,0	0,2587	0,9342
0,8	0,8125	0,4441	2,5	0,2043	0,9587
1,0	0,6366	0,6366	3,0	0,1691	0,9716
1,2	0,4685	0,7926	∞	0	1

The equations which govern the motion of the vehicle are two differential equations

$$\begin{cases} \dfrac{dV}{dt} = -fgF\left(\dfrac{r\omega}{V}\right) \\[2ex] \dfrac{d\omega}{dt} = -fgr\dfrac{m}{J_z}F\left(\dfrac{V}{r\omega}\right) \ . \end{cases} \tag{7.77}$$

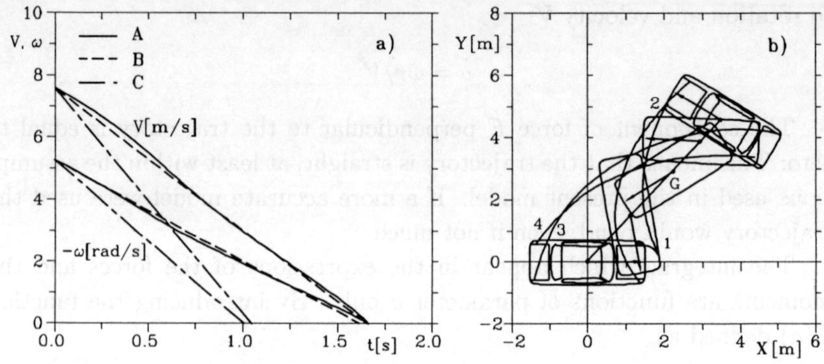

Fig. 7.20 Motion of a vehicle with locked wheels after a collision. (a) Time histories $V(t)$ and $\omega(t)$ computed through numerical integration of the equations of motion (curves A), numerical integration of Eq. (7.77) (B) and by considering separately translations and rotations (C). (b) Trajectory corrsponding to curve (A).

They must be integrated numerically, as parameter a changes continuously during motion and function $F(V/r\omega)$ cannot be expressed in closed form. Once that the laws $V(t)$ and $\omega(t)$ are known, the yaw angle and the position on the trajectory can be obtained by further integration

$$\psi = \int_0^t \omega(u)du \, , \qquad s = \int_0^t V(u)du \, . \tag{7.78}$$

Example 7.6

Consider a vehicle with $m = 1000$ kg, $J_z = 2000$ kg m^2, $a =$ 1.4 m, $b = 1.4$ m and $t = 1.08$ m. The vehicle, after the collision, moves with a speed $V = 7.36$ m/s, angular velocity $\omega = -5.19$ rad/s and angle β equal to 53°. Compute the trajectory and the positions taken by the vehicle until it stops.

The results obtained by integrating numerically the equations of motion are reported in Fig. 7.20a, curves A. A value of 0.7 for the friction coefficient has been assumed. The integration was performed using a time step $\Delta t = 0.01$ s. The trajectory is reported in Fig. 7.20b. The vehicle stops in a time of 1.70 s, after a displacement of 6.33 m and a rotation of 4.19 rad (240°).

Very similar results are obtained by integrating Eq. (7.77) (curves B). The values of function $F(a)$ used for the integration were taken from Table 7.3 and interpolated linearly. The time needed to extinguish the motion, the displacement and the rotation are respectively 1.62 s, 6.11 m and 4.06 rad (233°).

Wrong results would have been obtained by considering separately translatory and rotational motion (curves C). The time needed to extinguish the motion, the displacement and the rotation would have been respectively 1.01 s, 3.94 m and 2.62 rad (150°).

Laws $V(t)$ and $\omega(t)$ are almost linear. If this would occur exactly the value of a would remain constant during the motion and no numerical integration would be needed: A constant rate deceleration with the values of dV/dt and $d\omega/dt$ given by Eq. (7.77) would occur. Note that this is equivalent to study the motion as a translation and a rotation occurring separately, with "reduced" coefficients of friction equal to $fF(a)$ and $fF(1/a)$ respectively. In the example, immediately after the collision the value of a is 1.05, $F(a) = 0.62$ and $F(1/a) = 0.66$. By multiplying the friction coefficient by these values and considering separately the two motions, a time of 1.7 s for coming to a standstill is obtained. This value is very close to the one obtained through more complex models.

7.3.2 *Vehicle with free wheels*

In most cases the wheels of the vehicle, or at least some of them, remain free to rotate after the collision. There is little difficulty in integrating numerically the equations of motion, obviously written without any linearization. Equation (5.29) can be used, together with Eq. (5.39) yielding the sideslip angles of the wheels. A model of the tire which can be used for all values of α from 0 to 360° is needed. The "magic formula" (2.21) is at present probably the best choice, particularly for what the precision of the results is concerned.

The only difference from what is seen when dealing with the nonlinear model of the vehicle is the fact that some of the wheels may be locked. However, if the angular velocity of the vehicle is high, the sideslip angles remain at high values for a long time and the importance of modelling very accurately the behaviour of the tires at low sideslip angle is not great. In this situation a very simple model for the cornering force of the tire as the one shown in Fig. 7.21 can be used. With reference to the figure, the force

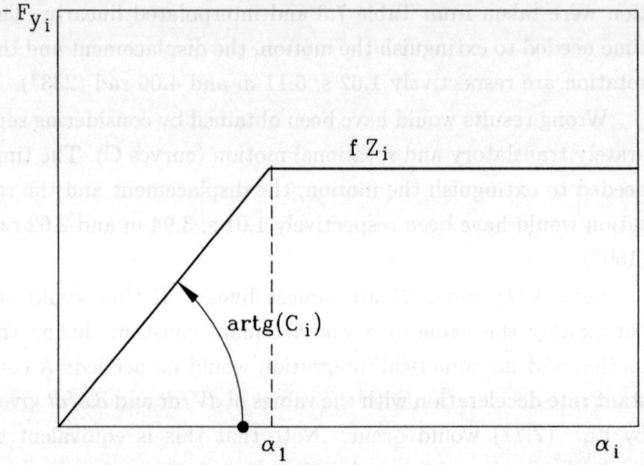

Fig. 7.21 Cornering force of the tire as a function of the sideslip angle. Simplified model in which the curve $F_y(\alpha)$ is approximated by two straight lines.

which the tire receives from the road is

$$\begin{cases} F_{x_i} = Z_i \dfrac{u_i}{|u_i|} \left[-f_r \cos(\delta_i) - f \sin(\delta_i)\right] \\ F_{y_i} = Z_i \dfrac{v_i}{|v_i|} \left[-f_r \sin(\delta_i) + f \cos(\delta_i)\right] \end{cases} \quad \text{if } \alpha > \alpha_1$$

$$\begin{cases} F_{x_i} = Z_i \dfrac{u_i}{|u_i|} \left[-f_r \cos(\delta_i) - f \dfrac{\alpha_i}{C_i} \sin(\delta_i)\right] \\ F_{y_i} = Z_i \dfrac{v_i}{|v_i|} \left[-f_r \sin(\delta_i) + f \dfrac{\alpha_i}{C_i} \cos(\delta_i)\right] \end{cases} \quad \text{if } \alpha \leq \alpha_1, \tag{7.79}$$

where f_r is the rolling coefficient.

The moment about z-axis is expressed by the first part of Eq. (7.70). The rolling drag is usually very small if compared with the other forces and could be neglected, but as the equation has at any rate to be integrated numerically there is no actual need to do so.

Example 7.7

Repeat the study of Example 7.6 assuming that the wheels are free and all steer angles δ_i are equal to zero.

By assuming a law $F_y(\alpha)$ of the type shown in Fig. 7.21 with $\alpha_1 = 8°$ and $f_r = 0.02$ the results shown in Fig. 7.22 are obtained through numerical integration. The computations

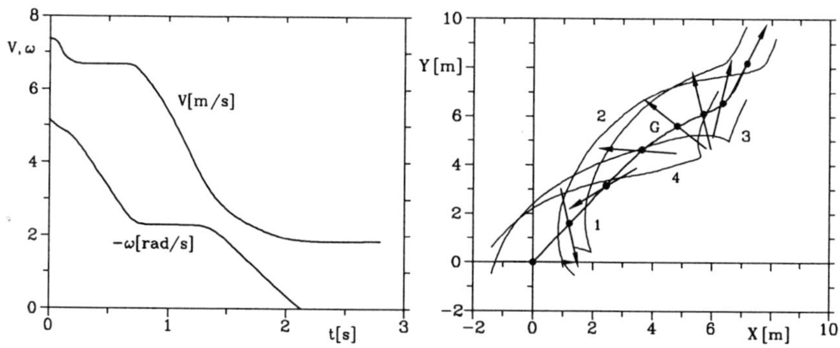

Fig. 7.22 Motion of a vehicle with free wheels after a collision. (a) Time histories $V(t)$ and $\omega(t)$ computed through numerical integration of the equations of motion and (b) trajectories of the centre of mass and of the points of contact of the wheels.

were repeated with different values of angle α_1, obtaining practically the same results in all cases. Also with $\alpha_1 = 0$ the laws $V(t)$ and $\omega(t)$ do not change, except for some oscillations due to numerical problems.

Note that the vehicle at the end of the simulation is aligned with its velocity and rolls freely in forward direction. In other cases the simulation can end with the vehicle rolling away in reverse.

7.4 Rollover

7.4.1 *Quasi-static rollover*

As was seen in Section 5.3, rollover of a rigid vehicle in static or quasi-static conditions is usually impossible, except in the case of vehicles with particularly narrow track or high centre of gravity: The conditions for slipping are reached well before those for rollover.

Rollover on a flat surface is controlled by parameter $t/2h_G$, which is sometimes referred to as rollover threshold: It constitutes the limit to the ratio between the lateral acceleration and the gravitational acceleration a_y/g. If the road has a transversal slope $i_t = \tan(\alpha_t)$ the threshold becomes $t/2h_G - i_t$. The threshold increases linearly with the slope if the external part of the curve is raised and decreases otherwise.

If $i_t = t/2h_G$ the vehicle overturns naturally without the application of

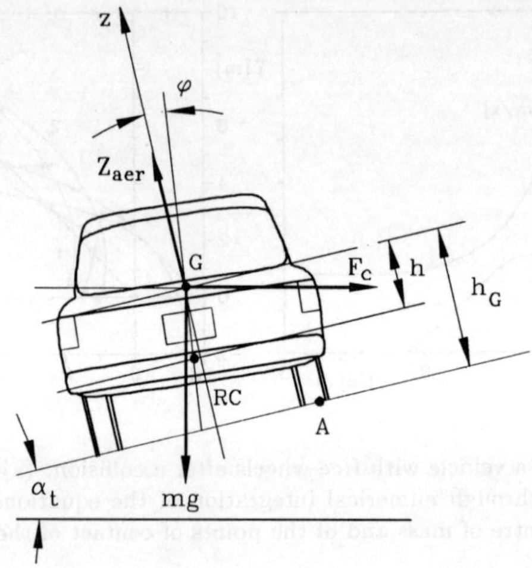

Fig. 7.23 Quasi-static rollover of a vehicle on elastic suspensions.

external forces, but this can occur only in case of off-road motion, as the adverse slope must be quite large.

To release the assumption of rigid vehicle the roll of the vehicle body must be accounted for. With reference to Fig. 7.24, the forces acting in the direction parallel and perpendicular to the road surface are

$$\begin{cases} F_{\parallel} = \dfrac{mV^2}{R}\cos(\alpha_t) - mg\sin(\alpha_t) \\ F_{\perp} = \dfrac{mV^2}{R}\sin(\alpha_t) + mg\cos(\alpha_t) - Z_{aer}\,. \end{cases} \tag{7.80}$$

If the inertia of the unsprung masses and the compliance of the tires are neglected and the roll angle is small enough, its value is

$$\phi \approx \frac{F_{\parallel}h}{K_t}\,, \tag{7.81}$$

where h is the height of the centre of mass over the roll axis and K_T is the torsional stiffness of the vehicle.

The vehicle rolls over if

$$\frac{F_{\|}}{F_{\perp}} > \frac{t - 2h\phi}{2h_G} \,. \tag{7.82}$$

Neglecting aerodynamic lift and assuming that the road surface is flat, the above equations can be solved in the lateral acceleration, yielding

$$\frac{V^2}{Rg} > \frac{t}{2h_G} \frac{1}{1 + \dfrac{mgh^2}{K_t h_G}} \,. \tag{7.83}$$

The fraction at right hand side is a factor smaller than one which expresses the reduction of resistance to rollover due to the presence of suspensions. The rollover threshold decreases with increasing roll compliance $1/K_t$ of the suspensions and distance h between the centre of gravity and the roll centre. A stiff suspension with high roll centre can substantially reduce rollover danger and bring the rollover conditions close to those characterizing the rigid vehicle.

The rollover threshold can be reduced, owing to roll compliance, of a few percent (typically of 5%) in passenger vehicles, with lower reductions in case of sports cars and greater ones in large luxury saloon cars.

The compliance of the tires increases slightly this effect, and some effects can be due to the exact geometry of the suspensions, particularly for what their lateral deflection is concerned, the lateral deflection of the tires and the inclination of the roll axis due to the differences between front and rear suspensions. To keep all these factors into account a detailed mathematical model of the vehicle must be built and solved on a computer.

Alternatively, the whole vehicle can be put on a "tilt-table", i.e. on a platform which can be inclined laterally, and the lateral slope can be increased until the load on the less loaded wheels reduces almost to zero, denouncing that the threshold of rollover has been reached. Note that the tilt-table arrangement exactly simulates rollover due to the lateral slope of the road, as can occur in off-road driving, but does not reproduce exactly conditions taking place on flat road. The difference can be small if the vehicle rolls over on a lateral slope which is not too steep, i.e. about 20° – 25°, while the errors build up when the lateral slope reaches values up to 45°, where errors of about 30% can be present. To avoid this problem, a "cable-pull" test can be devised, in which the lateral forces are applied directly to the vehicle at the location of the centre of mass by cables.

In spite of these effects, static rollover remains a rare occurrence for all vehicles but those with very high centre of mass and narrow track, as the available tire lateral forces are not high enough to prevent lateral slipping before rollover.

7.4.2 *Dynamic rollover*

On the contrary, dynamic conditions, linked to roll oscillations, can lead to rollover if the external rolling moment adds to an inertia torque due to roll accelerations.

If a step roll input is applied to the vehicle, the response is a damped roll oscillation about the static equilibrium position which would have been reached if the same input would be applied very slowly. Note that an input is applied slowly if its characteristic times are larger than the period of the lowest natural frequency involved; as the roll period is usually of the order of one second most roll inputs are faster than that and have the characteristics of a dynamic load. If the vehicle were critically damped in roll no oscillations would take place but the roll angle would increase monotonically, reaching asymptotically the static value.

As all vehicles are underdamped in roll, the roll angle overshoots and then oscillates; the lower the damping ratio of the system the higher the overshoot and thus the danger of rollover. The presence of antiroll bars makes things worse from this viewpoint. When antiroll bars are added to an existing suspension the roll stiffness is increased but usually roll damping is not changed, as the damping of shock absorbers is optimized for bounce and pitch behaviour. The effect is an increase of the natural frequency and of the underdamped behaviour of the system. While the static roll angle is decreased by the added stiffness both the overshoot and the natural frequency of roll oscillations are increased, the latter effect increasing roll inertia torques.

The model for roll oscillations shown in Section 6.10 can be used to study rollover and, in particular, roll resonances, but only up to the point when the load on one side reduces to zero. The restoring moment due to weight and the external rolling moment can be easily added to the first equation.

Consider a vehicle which undergoes roll oscillations (Fig. 7.24). The inertia torque $J_x\ddot{\phi}$ due to roll oscillations is sketched as a function of time in the figure. It is always lower than $mgt/2$, the restoring moment due to weight on level road, and hence no danger of rollover is present. However,

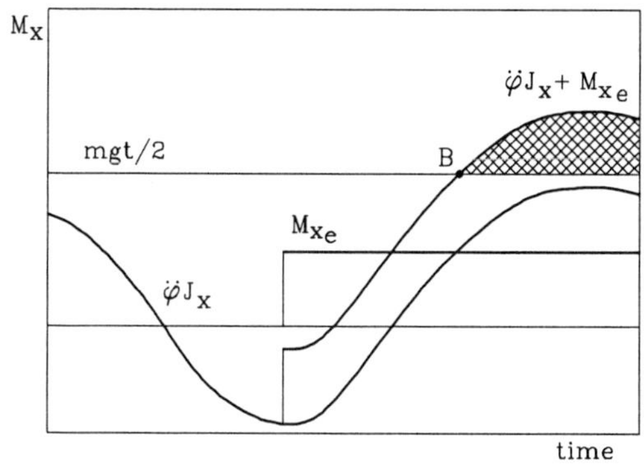

Fig. 7.24 Roll inertia torques causing rollover: In B the conditions for rollover are reached.

if an external force is applied which causes the rolling moment M_{x_e}, the conditions for incipient rollover occur in B.

From that point on the wheels at one side lose contact with the ground. The weight is still stabilizing, until a large enough roll rotation to bring weight force outside point A in Fig. 7.24 is completed, but the vehicle continues the roll over motion. Note that this very simple model is inadequate to predict what happens when some of the wheels are no more in contact with the ground and the roll angle gets large.

A driver reacting very quickly can prevent rollover even in these conditions by giving appropriate steering inputs, as testified by stunt drivers who can proceed with the car on two wheels, without either rolling over or rolling back on four wheels, but it is most unlikely that a normal driver in normal conditions performs stunts of this type and usually once rollover starts it goes on with its full consequences.

Roll dynamics is closely coupled with yaw dynamics and rollover is no exception. When giving a sinusoidal steering input the lateral tire forces of the rear wheels are delayed with respect to the ones of the front wheels; this delay is larger in long vehicles and the whole effect is even more pronounced in articulated ones. The combined roll-yaw dynamics with its resonances can facilitate rollover, particularly when selected frequencies are present in the excitation.

Tractor-trailer combinations are particularly subject to rollover due to sudden steering inputs: The roll of the trailer is delayed with respect to that of the tractor and can be large enough to cause the former to rollover.

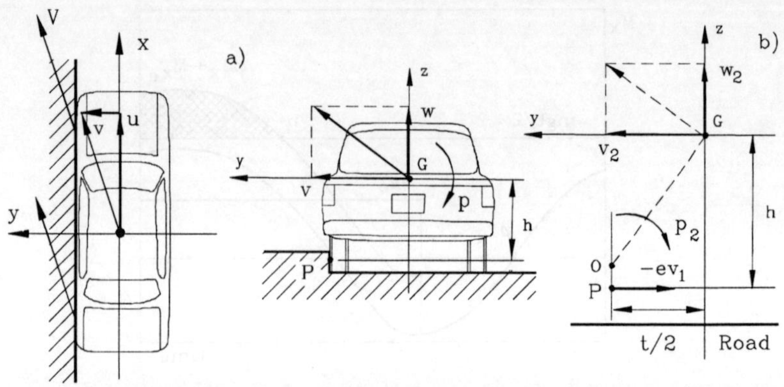

Fig. 7.25 Side impact against the curb. (a) Sketch; (b) determination of the centre of rotation at time t_2.

A hitch providing roll coupling between trailer and tractor helps in this case, as the second one collaborates in resisting the tendency to overturn of the former.

7.4.3 *Lateral collision with the curb*

Consider a vehicle modelled as a rigid body and assume that its velocity V is not contained in the plane of symmetry (Fig. 7.25). Both wheels at one side enter in contact at time t_1 with the curb. At time t_1 the components of the velocity are $u = u_1$, $v = v_1$, $w = p = q = r = 0$[5], i.e. the velocity of the vehicle is contained in a plane parallel to the road and the angular velocity is nil.

If friction between the vehicle and the curb is neglected, the component u of the velocity along x-axis remains constant and the components q and r of the angular velocity along y and z axes remain vanishingly small. The other components of the velocity after the impact, i.e. at time t_2, are $v = v_2$, $w = w_2$, and $p = p_2$.

The component of the velocity in direction perpendicular to the impact surface in P at time t_1 is $V_{\perp_1} = v_1$. Assuming a partially inelastic impact with coefficient of restitution e^*, at time t_2 the same velocity is $V_{\perp_2} = -e^* v_1$.

If the condition that the motion of point P between time t_1 and t_2

[5]The components ω_x, ω_y and ω_z, of the angular velocity are here indicated as p, q and r.

occurs in a plane parallel to the ground is added, the position of the centre of rotation at time t_2 can be easily found and a relationship linking w_2 and p_2 to v_1 can be obtained.

With reference to Fig. 7.25b it follows

$$\begin{cases} p_2 = -\dfrac{v_2 + e^* v_1}{h} \\ w_2 = -p_2 \dfrac{t}{2} = (v_2 + e^* v_1) \dfrac{t}{2h} \, . \end{cases} \tag{7.84}$$

The components of the impulse the vehicle receives during the impact can be related to the variations of the momentum and the angular momentum

$$\begin{cases} I_y = m(v_1 - v_2) \\ I_z = -m w_2 \\ I_y h - I_z \dfrac{t}{2} = -J_x p_2 \, . \end{cases} \tag{7.85}$$

By introducing Eq. (7.84) into Eq. (7.85) it follows

$$\begin{cases} v_2 = -v_1 \dfrac{A^2 - e^*(1 + B^2)}{1 + A^2 + B^2} \\ p_2 = -v_1 \dfrac{A^2(1 + e^*)}{h(1 + A^2 + B^2)} \, , \end{cases} \tag{7.86}$$

where

$$A = \frac{2h}{t} \, , \qquad B = \frac{2}{t}\sqrt{\frac{J_x}{m}} \, .$$

The rollover motion starts at time t_2. If the component of the velocity of point P in y direction vanishes rapidly owing to the friction on the road, rollover actually occurs if the centre of gravity crosses a line perpendicular to the road in P. This means that the vehicle rolls over only if its centre of mass moves upwards of a distance Δh equal to

$$\Delta h = \sqrt{\frac{t^2}{4} + h^2} - h = h \left(\sqrt{1 + \frac{1}{A^2}} - 1 \right) \, . \tag{7.87}$$

This can occur only if the kinetic energy associated to velocities v_2, w_2 and p_2 is at least equal to the potential energy $mg\Delta h$. With simple computations it can be shown that the condition for completing rollover is

$$v_2^2 + w_2^2 + \frac{J_x}{m} p_2^2 \geq 2hg \left(\sqrt{1 + \frac{1}{A^2}} - 1 \right) . \tag{7.88}$$

As

$$v_2^2 + w_2^2 = p_2^2 \left(\frac{t^2}{4} + h^2 \right) ,$$

the condition for rollover becomes

$$v_1^2 \frac{A^2(1+e^*)^2}{1 + A^2 + B^2} \geq 2hg \left(\sqrt{1 + \frac{1}{A^2}} - 1 \right) . \tag{7.89}$$

Consider for instance a vehicle with $A = 0.6$ and $B = 0.7$ hitting a curb in a perfectly elastic way ($e^* =1$), which is the most dangerous condition. From Eq. (7.89) the rollover condition is $v_1^2 \geq 2.42gh$.

Equation (7.89) can be written explicitly in terms of forward velocity V and impact angle θ

$$V^2 \sin^2(\theta) \geq 2hg \frac{1 + A^2 + B^2}{A^2(1+e^*)^2} \left(\sqrt{1 + \frac{1}{A^2}} - 1 \right) . \tag{7.90}$$

If $h = 0.36$ m and the impact angle is $\theta = 15°$, the mentioned vehicle will roll over for forward velocities greater than 11 m/s (40 km/h). Note that this result depends strongly on the value of e^*: The ratio between the velocities V needed for rollover when $e^* = 0$ and $e^* = 1$ is equal to $\sqrt{2}$. It is reasonable that in this case the restitution coefficient e^* is greater than in the case of impacts between vehicles, but in general it depends on the impact conditions.

It must also be noted that if the height of the curb is small, and consequently the value of h is large, the vehicle tends not to engage against the curb but to drive over it and the whole study is not applicable. The assumption of rigid body is also questionable: The impact usually occurs between the curb and the unsprung mass and the latter can undergo plastic and elastic deformations, accompanied by deformations of the tires.

7.4.4 *Effect of the transversal slope and the curvature of the road*

The model seen above is based on the assumptions that the road is flat and that the curb is straight. To keep into account the transversal slope of the road it is sufficient to assume that y-axis is inclined. The only difference is that of changing the expression of the potential energy due to the vertical displacement of the centre of mass of the vehicle, as a vertical line passing for point P is no longer perpendicular to the road.

If the curb is not straight the motion is more complex and it is not possible to study the motion in yz plane independently of that in x direction. However, if the curb follows a circular path with a radius far greater than the length of the vehicle and of the displacements in y direction involved in the rollover motions, it is possible to simplify the problem.

Consider the situation shown in Fig. 7.26a. The motion in yz plane can be studied with reference to the non-inertial xyz frame, which rotates about line CC′ with angular velocity $V\cos(\theta)/R$. Both centrifugal and Coriolis acceleration must be introduced in the study of the motion starting from time t_2.

Centrifugal acceleration is directed radially while Coriolis acceleration acts in x direction. Their values are

$$\begin{cases} a_c = \dfrac{u^2}{R'} = \dfrac{V^2}{R'}\cos^2(\theta) \\ \\ a_{cor} = -2\dfrac{V}{R'}\cos(\theta)\left[v_2\cos(\alpha_t) + w_2\sin(\alpha_t)\right] \, . \end{cases} \tag{7.91}$$

Only the first needs to be considered in the study of the motion in yz plane. It can be considered constant, neglecting the fact that R' and u change during the rollover motion. The same study seen for the previous case can be repeated by simply introducing the resultant of centrifugal and gravitational accelerations (here decomposed in directions parallel and perpendicular to the road surface)

$$\begin{cases} a_\perp = g\cos(\alpha_t) - a_c\sin(\alpha_t) \\ a_\parallel = g\sin(\alpha_t) + a_c\cos(\alpha_t) \end{cases} \tag{7.92}$$

for the gravitational acceleration (Fig. 7.26a,b). The vehicle will rollover if its centre of mass goes beyond line PQ, whose direction is that of the

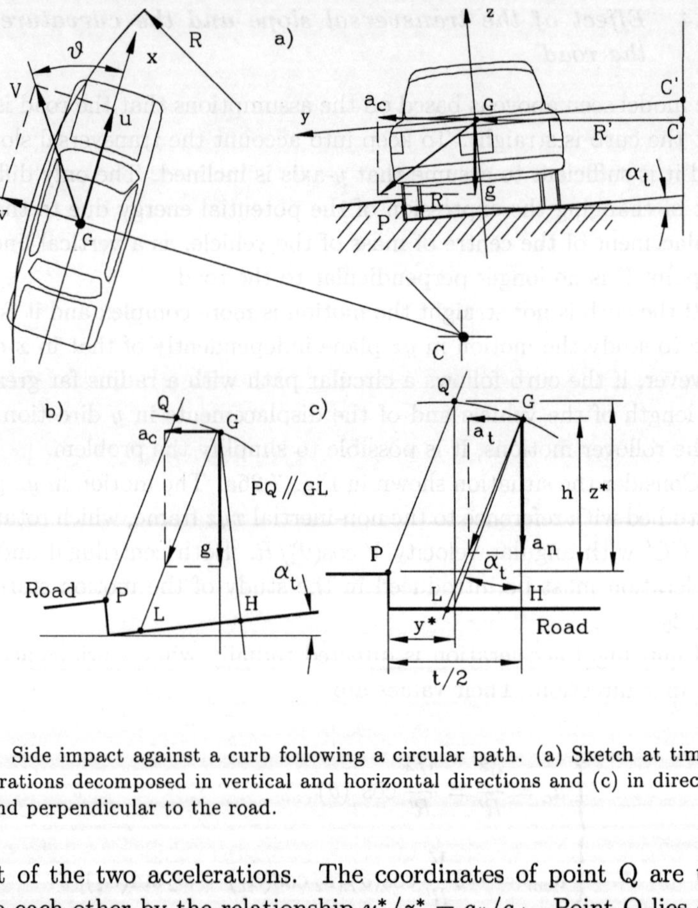

Fig. 7.26 Side impact against a curb following a circular path. (a) Sketch at time t_2; (b) accelerations decomposed in vertical and horizontal directions and (c) in directions parallel and perpendicular to the road.

resultant of the two accelerations. The coordinates of point Q are then linked to each other by the relationship $y^*/z^* = a_\parallel/a_\perp$. Point Q lies on a circle centred in P passing through point G. It then follows that

$$y^{*^2} + z^{*^2} = h^2 + \frac{t^2}{4} \tag{7.93}$$

By intersecting the circle with line PQ it follows that

$$y^* = h\frac{\sqrt{1+\frac{1}{A^2}}}{\sqrt{1+\left(\frac{a_\perp}{a_\parallel}\right)^2}}\,, z^* = h\frac{\sqrt{1+\frac{1}{A^2}}}{\sqrt{1+\left(\frac{a_\parallel}{a_\perp}\right)^2}}\,, \tag{7.94}$$

where $A = 2h/t$.

The total increase of potential energy in the motion from G to Q to be substituted for $mg\Delta h$ into the rollover condition is

$$\Delta\mathcal{U} = mha_{\perp}\left\{\sqrt{\left(1+\frac{1}{A^2}\right)\left[1+\left(\frac{a_{\perp}}{a_{\parallel}}\right)^2\right]}-1-\frac{a_{\parallel}}{Aa_{\perp}}\right\}. \tag{7.95}$$

The final form of the rollover condition is thus

$$v_1^2\frac{A^2(1+e^*)^2}{1+A^2+B^2} \geq 2ha_{\perp}\left\{\sqrt{\left(1+\frac{1}{A^2}\right)\left[1+\left(\frac{a_{\perp}}{a_{\parallel}}\right)^2\right]}-1-\frac{a_{\parallel}}{Aa_{\perp}}\right\}. \tag{7.96}$$

Clearly if $a_{\perp}/a_{\parallel} = A$ the vehicle rolls over even for $v_1 = 0$: Points G and Q coincide and the system is statically on the verge of instability. If the curb is along a straight line ratio $a_{\parallel}/a_{\perp}$ is equal to $\tan(\alpha_t) = i_t$, transversal slope of the road.

If on the contrary the road is flat but curved, $a_{\parallel}/a_{\perp} = a_c/g = V^2\cos^2(\theta)/Rg$.

The nondimensional velocity $v_1/\sqrt{a_{\perp}h}$ needed for rollover is plotted against ratio $a_{\parallel}/a_{\perp}$ for various values of e^* in Fig. 7.27. The plot has been obtained for $A = 0.6$ and $B = 0.7$.

7.5 Motion of transported objects during the impact

7.5.1 *Free objects*

The study of the motion of the objects onboard vehicles during a collision is important both for the study of the effects on transported persons and to devise suitable restraint systems for persons and packaging for objects.

If the object is simply supported on a surface parallel to the road, when the vehicle stops abruptly owing to a collision, which in the following will be considered as a central impact against a fixed object, it continues its motion until it collides with the front wall of the load compartment. This first part of the collision process is easily studied under the assumption that the mass of the object is negligible with respect to that of the vehicle and the motion of the former does not influence that of the latter.

Consider the situation sketched in Fig. 7.28, in which the elastic and

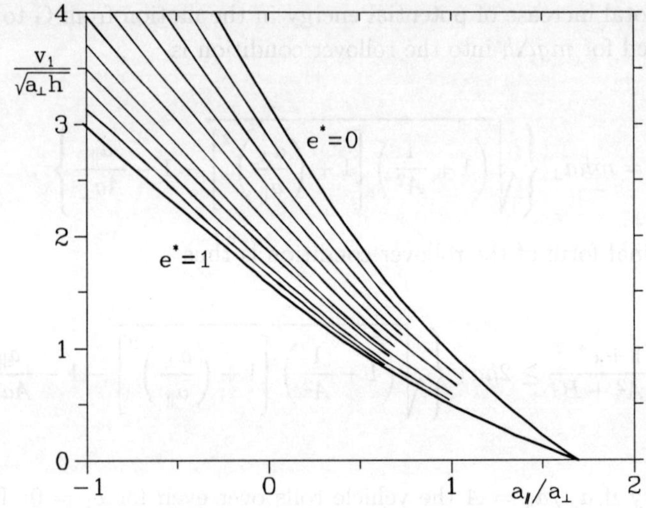

Fig. 7.27 Nondimensional velocity $v_1/\sqrt{a_\perp h}$ needed for rollover as a function of $a_\parallel/a_\perp$ for various values of e^*. Vehicle with $A = 0.6$ and $B = 0.7$.

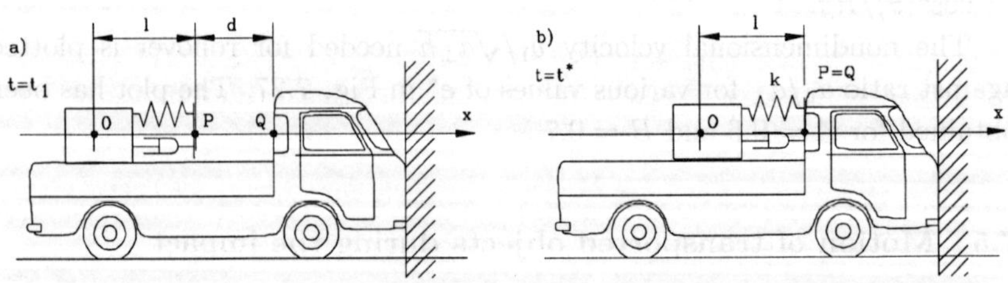

Fig. 7.28 Motion of the object O free to move on a vehicle which hits a rigid barrier. The elastic and damping properties of the object are modelled by a spring-damper system. (a) Situation at time t_1. (b) Situation at time t^*, when the object hits the front wall of the load compartment.

damping properties of the transported object are modelled by a spring-damper system. The motion of the vehicle is assumed to follow the model described in Section 7.2.1 and the motion of the object will be studied with reference to the xz frame centred in the position of point P, belonging to the front part of the object, at time t_1, in which the vehicle hits the obstacle.

Neglecting friction, the motion of point P at the instant following the collision is described by the law

$$x_P = V_1 t = V_1 t_2 \tau \ , \tag{7.97}$$

where V_1 is the velocity of the vehicle before the collision and the nondimensional time $\tau = t/t_2$ is defined with respect to the duration of the collision t_2.

Assuming the model of Section 7.2.1 to describe the motion of the vehicle, the x coordinate of point Q is

$$x_Q = d + V_1 t_2 \left(\frac{1+e^*}{\beta+3} \left\{ 2 - \left[2 + (\beta+1)\tau \right] (1-\tau)^{\beta+2} \right\} - e^* \tau \right) . \tag{7.98}$$

The object hits the front wall of the load compartment at time t^*, when the distance between points P and Q vanishes, i.e. when

$$\tau^* - \frac{2 - \left[2 + (\beta+1)\tau^* \right] (1-\tau^*)^{\beta+2}}{\beta+3} = \frac{d}{V_1 t_2 (1+e^*)} . \tag{7.99}$$

Equation (7.99) can be easily solved in τ^*; however this solution holds only if $\tau^* \leq 1$, i.e. if the objects collides before time t_2. If this condition does not hold the secondary collision inside the vehicle occurs when the latter has started its motion backward after rebounding from the obstacle and Eq. (7.98) does not hold any more. The condition for which Eq. (7.99) holds, i.e. for which $\tau^* \leq 1$, is simply

$$d \leq V_1 t_2 (1+e^*) \frac{\beta+1}{\beta+3} . \tag{7.100}$$

If condition (7.100) is satisfied, the relative velocity is easily computed by considering that at time τ^* the velocities of the object and of the vehicle are both known

$$V_R = V_1(1+e^*) \left\{ 1 - \left[1 + (\beta+1)\tau^* \right] (1-\tau^*)^{\beta+1} \right\} . \tag{7.101}$$

If on the contrary condition (7.100) is not satisfied, the computation is even simpler. If the rebound velocity is considered constant $V_2 = -e^* V_1$ after time t_2, the relative velocity is simply $V_R = V_1(1+e^*)$. The time at which the secondary collision takes place can be assumed to be $t^* = t_2 + t_3$, where t_3 is the time needed to travel at the velocity V_R defined above for a distance d_2 which still separates the object and the vehicle at time t_2. d_2

can be computed stating $\tau = 1$ into Eq. (7.98), i.e. by Eq. (7.100) with '='. Operating in this way, it follows that

$$t_3 = \frac{d}{V_1(1+e^*)} - t_2\frac{\beta+1}{\beta+3} \tag{7.102}$$

The motion of the object after it hits the wall of the load compartment can be performed in different ways. It is possible to resort to a semiempirical time history, as it was done for the vehicle colliding against an obstacle, or to model the mechanical properties of the object and to use that model to obtain the equation of motion.

By assuming that the object and the wall are rigid bodies with interposed a linear spring-damper system, as sketched in Fig. 7.28, the equation of motion of point O for $t > t^*$ is

$$m\ddot{x}_O + c(\dot{x}_O - \dot{x}_Q) + k(x_O + l - x_Q) = 0\,, \tag{7.103}$$

with the initial conditions $x_O = x_Q - l$ and $\dot{x}_O = V_1$ for $t = t^*$.

By introducing the nondimensional coordinate $\chi = (x_O + l - x_Q)/V_1 t_2$ and the nondimensional time τ, the equation of motion reduces to

$$\frac{d^2\chi}{d\tau^2} + \frac{ct_2}{m}\frac{d\chi}{d\tau} + \frac{kt_2^2}{m}\chi = \frac{d^2\chi_Q}{d\tau^2}\,, \tag{7.104}$$

with the initial conditions

$$\chi = 0 \quad \text{and} \quad \frac{d\chi}{d\tau} = 1 - \left(\frac{d\chi_Q}{d\tau}\right)_{\tau=\tau^*} \quad \text{for} \quad t = t^*.$$

The expressions of $d^2\chi_Q/d\tau^2$ and of $d\chi_Q/d\tau$ can be easily computed from t^* to t_2 ($0 \le \tau \le 1$) from Eqs. (7.31) and (7.35) while for $t > t_2$ ($\tau > 1$) they are simply $d^2\chi_Q/d\tau^2 = 0$ and $d\chi_Q/d\tau = -e^*$.

If the secondary collision occurs before time t_2 the equation of motion of the object is

$$\frac{d^2\chi}{d\tau^2} + \frac{ct_2}{m}\frac{d\chi}{d\tau} + \frac{kt_2^2}{m}\chi = \begin{cases} (\beta+1)(\beta+2)(1+e^*)\tau(1-\tau)^\beta & \text{if } \tau^* \le \tau \le 1 \\ 0 & \text{if } \tau \ge 1, \end{cases} \tag{7.105}$$

with the initial conditions $\chi = 0$ and

$$\frac{d\chi}{d\tau} = (1 + e^*) \left\{1 - [1 + (\beta + 1)\tau^*] (1 - \tau)^{\beta+1}\right\}$$

for $t = t^*$.

If the secondary shock occurs during the rebound phase, after the vehicle has lost contact from the obstacle, the equation of motion is the second of Eqs. (7.105) and the initial condition on $d\chi/d\tau$ reduces to $d\chi/d\tau = (1+e^*)$.

These equations of motion hold only up to the time in which the object rebounds back losing contact with the wall of the load compartment. This instant can be easily computed by looking for the time at which the acceleration of the object vanishes, as the force acting between points P and Q reduces to zero.

The acceleration can be computed as

$$\begin{cases} \dfrac{d^2x_O}{d\tau^2} = \dfrac{V_1}{t_2}\left[\dfrac{d^2\chi}{d\tau^2} - (\beta + 1)(\beta + 2)(1 + e^*)\tau(1 - \tau)^{\beta}\right] & \text{if } \tau^* \leq \tau \leq 1 \\ \dfrac{d^2x_O}{d\tau^2} = \dfrac{V_1}{t_2}\dfrac{d^2\chi}{d\tau^2}i & \text{if } \tau \geq 1. \end{cases} \tag{7.106}$$

This expression of the acceleration is one of the most interesting results of this study, since the aim of the elastic system with stiffness k and damping c is that of allowing the object to survive the shock of the collision, which means to reduce its acceleration within allowable limits.

Another important result is the value of the maximum displacement χ, i.e. the maximum compression of the spring; it cannot be higher than a given allowable limit, beyond which the elastic system is crushed or, at least, shows nonlinear characteristics with increasing stiffness. In practice, to limit the acceleration the spring must be soft but the decrease of the value of k is limited by the available space as it causes the maximum travel to increase.

The whole process is governed by five nondimensional parameters: β and e^*, linked to the way in which the vehicle collides with the obstacle, ratio d/V_1t_2, linked to the clearance between the object and the wall, kt_2^2/m and ct_2/m, related to the elastic and damping characteristics of the object.

Parameter kt_2^2/m can be written in the form

$$\frac{kt_2^2}{m} = (\lambda_n t_2)^2 = 4\pi^2 \left(\frac{t_2}{T_n}\right)^2 , \tag{7.107}$$

where λ_n and T_n are the circular frequency and the period of the undamped free oscillations of a spring-mass system with mass m and stiffness k. If this parameter has a value equal to π^2 the half-period of the undamped oscillations is equal to t_2.

Parameter ct_2/m is linked to kt_2^2/m and to the damping ratio ζ by the relationship

$$\frac{ct_2}{m} = \zeta\sqrt{2}\sqrt{\frac{kt_2^2}{m}} . \tag{7.108}$$

The absolute value of the acceleration of an object onboard the vehicle studied in Fig. 7.13 hitting a fixed obstacle at 20 m/s is plotted against time in Fig. 7.29. The value of kt_2^2/m has been assumed to be equal to π^2; the various curves correspond to different values of ct_2/m. In the case studied, the lowest maximum acceleration is obtained for $ct_2/m = \pi/\sqrt{2}$, i.e. for $\zeta = 0.5$. Note that this value of the damping ratio coincides with the "optimum value" defined in Section 6.8.1 for the quarter car model with a single degree of freedom. As the two models are different, such instance cannot be generalized and the values of ζ which minimizes the acceleration must be obtained for each case, as it can be verified by plotting the same figure with different values of the parameters.

The maximum values of the acceleration and of the displacement are plotted versus parameter kt_2^2/m in Fig. 7.30. The plot has been obtained with the same values of β and e^* used for Fig. 7.29 and assuming a damping ratio equal to 0.5. From the figure it is clear that the distance d must be kept to a minimum to have low values of the acceleration. From the maximum allowable value of the displacement χ it is possible to obtain the minimum value of k and then the value of the peak acceleration occurring during the impact. The results shown in Figs. 7.29 and 7.30 were obtained by numerical integration in time of the equation of motion.

The simple models here shown allow simple and straightforward qualitative assessments, with a limited number of parameters involved. There is however no difficulty in building more complete models which allow to obtain results yielding good quantitative predictions of the actual behaviour of the system.

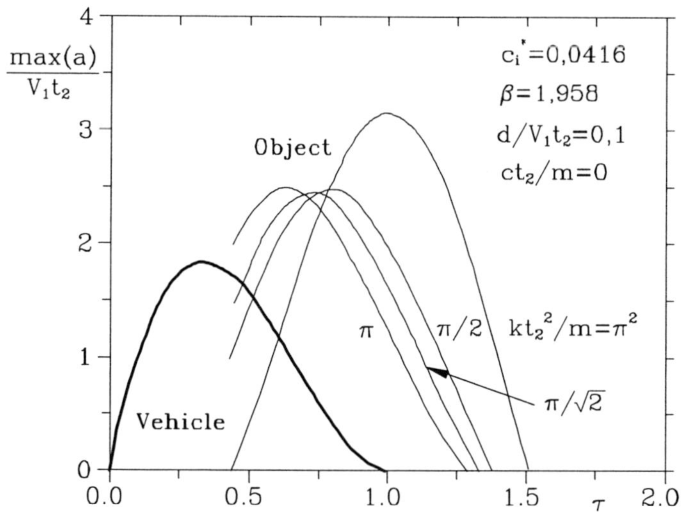

Fig. 7.29 Time history of the acceleration for the vehicle studied in Fig. 7.13 hitting a fixed obstacle at 20 m/s and for an object carried by it.

7.5.2 *Constrained objects*

The behaviour of an object constrained onboard a vehicle is qualitatively not different from the one seen for free objects. In the case of constrained objects the distance d is smaller, being the clearance of the restraining device, and can even be equal to zero.

The motion of a constrained object can be studied in four distinct phases (Fig. 7.31).

The first phase begins in the instant the vehicle enters in contact with the obstacle starting its deceleration and ends at time t^* when the object contacts the restraining system. Usually, if the clearance is not too large, time t^* occurs before time t_2.

The second phase spans between time t^* and time t_2. The collision of the vehicle against the obstacle is completed and the object is retained by the constraining device. At the end the vehicle rebounds freely back.

The third phase is between time t_2 and time t_3. The object rebounds backward and, at the end, loses contact with the restraining system. The vehicle moves backward, with a speed which is low if the collision is strongly inelastic.

The fourth phase goes from time t_3 to time t_4 in which the object,

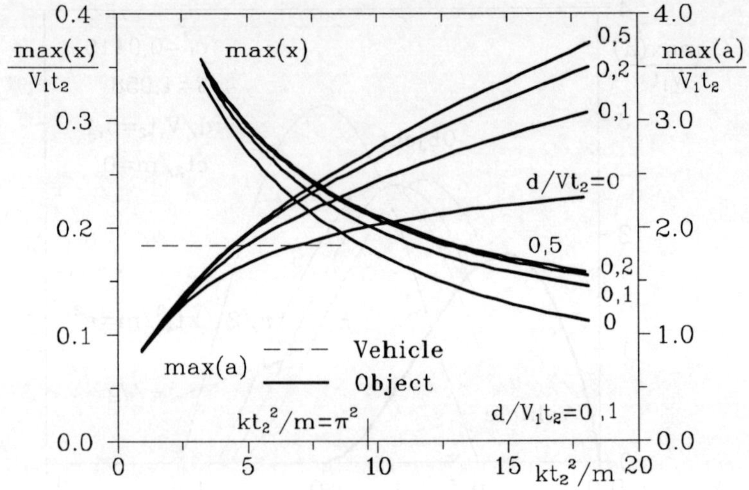

Fig. 7.30 Peak values of the acceleration and of the displacement as a function of the stiffness of the spring for various values of the clearance d. Same vehicle as in Fig. 7.29; damping equal to half of the critical damping (ζ =0.5).

moving backward at a speed higher than that of the vehicle, hits the back wall of the load compartment. The study can then proceed, as at this point there can be a further rebound, which can be particularly dangerous.

A case of particular importance which can be studied, at least as a first approximation, with the present technique, is that of the motion of a person wearing a seat belt. In this case the collision with the seat at time t_4 can be the most dangerous event.

To increase the adequacy of the model to supply also quantitative information it is possible to take into account the nonlinearities which are always present in actual cases. A nonlinear law stating the dependence of the force provided by the restraint as a function of the displacement x and of the velocity $\dot{x}$ can be introduced without adding much to the complexity of the numerical simulation.

The restraining system should have a perfectly plastic behaviour to reduce to a minimum the acceleration peak, as in this case the energy dissipation is maximum for a given value of applied force and displacement (Fig. 7.32a). A more realistic behaviour is an elasto-plastic law (Fig. 7.32b) while actual systems show a more complex force-displacement characteristic (Fig. 7.32c).

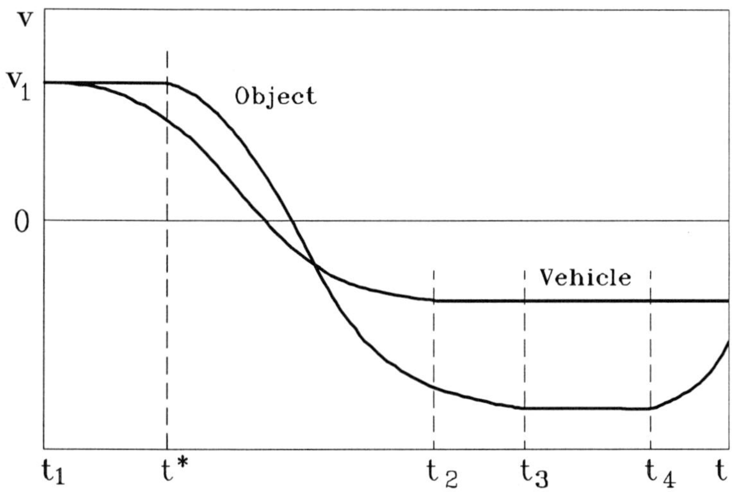

Fig. 7.31 Time history of velocity of the motion of the vehicle and of an object constrained on board.

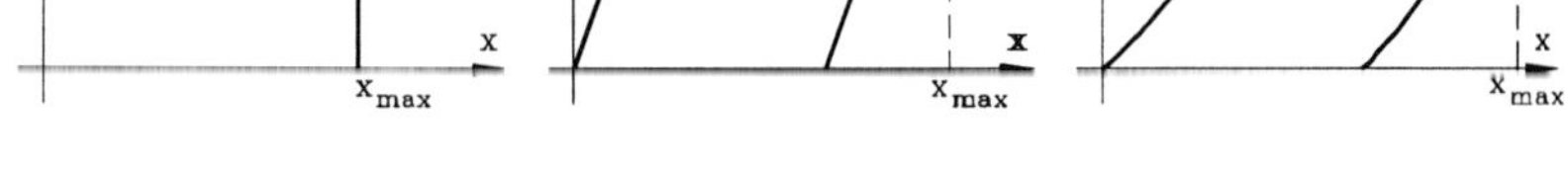

Fig. 7.32 Force exerted by the restraining system as a function of the deformation x; the energy dissipated is the area under the curve. (a) Perfectly plastic, (b) elasto-plastic and (c) actual behaviour.

Appendix A

This appendix contains fairly complete data of five vehicles of different types: A small city car, a saloon car, a fast sport car, an articulated truck and a racing motorcycle. They are used throughout the text in the examples and can be used by the reader to repeat the computations shown for different kinds of motor vehicles.

Although not being exactly any actual vehicle, their characteristics are typical. The reader is encouraged to repeat the examples shown in the various chapters for the different vehicles, varying some data like those linked with the load conditions or some design choices (e.g. front wheel drive versus rear wheel drive, etc.).

A.1 Example 1: small front-wheel drive car

The vehicle is a typical five-seat hatchback car with a 1 liter spark ignition engine (Fig. A.1). The map of the engine is also reported.

The main geometrical data and inertial properties of the vehicle, wheels, engine, transmission are

length = 3640 mm	width = 1560 mm	height = 1410 mm
$m = 830$ kg	$t_1 = 1284$ mm	$t_2 = 1277$ mm
$J_x = 290$ kg m^2	$J_y = 1094$ kg m^2	$J_z = 1210$ kg m^2
$J_{xz} = -84$ kg m^2		
J_w(each) = 0.4 kg m^2	$J_e = 0.085$ kg m^2	$J_t = 0.05$ kg m^2

$P_{max} = 38.3$ kW at 5200 rpm; $T_{max} = 87$ Nm at 3000 rpm.
Data for rolling coefficient and for the "magic formula" for lateral forces

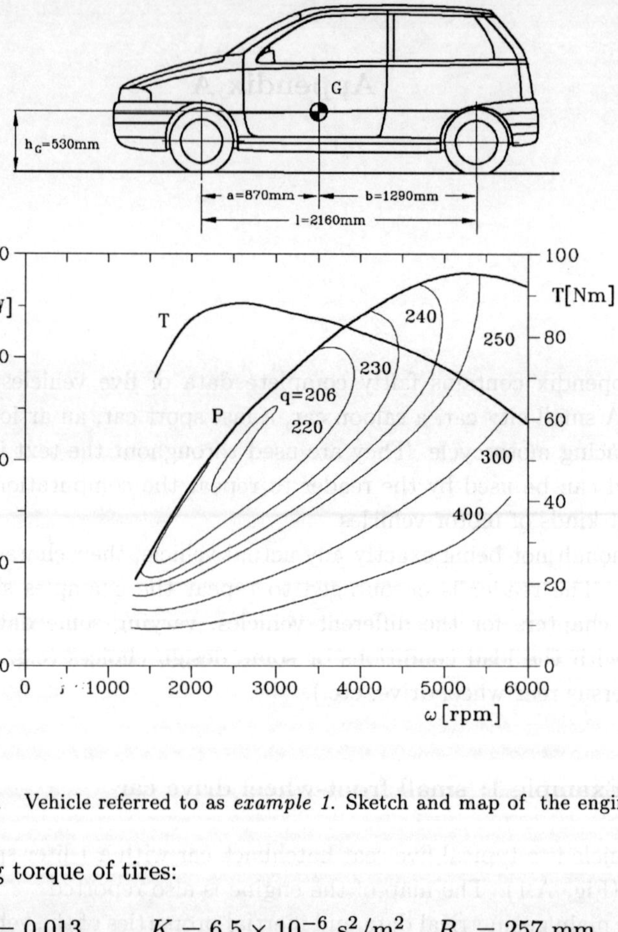

Fig. A.1 Vehicle referred to as *example 1.* Sketch and map of the engine.

and aligning torque of tires:

$f_0 = 0.013$	$K = 6.5 \times 10^{-6}$ s^2/m^2	$R_e = 257$ mm
$a_0 = 1.3000$	$a_1 = -53.3100$	$a_2 = 1190.0000$
$a_3 = 588.6000$	$a_4 = 2.5212$	$a_5 = 0$
$a_6 = -0.5178$	$a_7 = 1.0000$	$a_8 = 0$
$a_9 = 0$	$a_{10} = 0$	$a_{111} = 0$
$a_{112} = 0$	$a_{12} = 0$	$a_{13} = 0$
$c_0 = 2.4000$	$c_1 = -4.4037$	$c_2 = -1.3600$
$c_3 = -4.0978$	$c_4 = -3.2800$	$c_5 = 0.2446$
$c_6 = 0$	$c_7 = -0.07918$	$c_8 = 0$
$c_9 = 1.0000$	$c_{10} = 0$	$c_{11} = 0$
$c_{12} = 0$	$c_{13} = 0$	$c_{14} = 0$
$c_{15} = 0$	$c_{16} = 0$	$c_{17} = 0$

The coefficients for longitudinal force (Eq. (2.15)) are

$$A = 1.12 \quad C = 0.625 \quad D = 1$$
$$n = 0.6 \quad k = 46 \quad d = 5$$

Aerodynamic data:

$$S = 1.7\ \text{m}^2 \quad C_x = 0.32 \quad C_z = -0.21$$
$$C_{M_y} = -0.09 \quad (C_y)_{,\beta} = -2.2\ 1/\text{rad} \quad (C_{M_z})_{,\beta} = -0.6\ 1/\text{rad}$$

Transmission (front wheel drive):

$$\tau_I = 65/14 \quad \tau_{II} = 63/23 \quad \tau_{III} = 34/21$$
$$\tau_{IV} = 21/22 \quad \tau_f = 1/.2884 \quad \eta_t = 0.91$$

A.2 Example 2: large front-wheel drive saloon car

The vehicle is a typical five-seat European saloon car with a 2 liter spark ignition supercharged engine (Fig. A.2).

The main geometrical data and inertial properties of the vehicle, wheels, engine, transmission are

$$\text{length} = 4690\ \text{mm} \quad \text{width} = 1830\ \text{mm} \quad \text{height} = 1450\ \text{mm}$$
$$m = 1150\ \text{kg} \quad t_1 = 1490\ \text{mm} \quad t_2 = 1482\ \text{mm}$$
$$J_x = 530\ \text{kg m}^2 \quad J_y = 1630\ \text{kg m}^2 \quad J_z = 1850\ \text{kg m}^2$$
$$J_{xz} = -120\ \text{kg m}^2$$
$$J_w(\text{each}) = 0.6\ \text{kg m}^2 \quad J_e = 0.19\ \text{kg m}^2 \quad J_t = 0.07\ \text{kg m}^2$$

P_{max} = 122 kW at 5500 rpm; T_{max} = 255.1 Nm at 2500 rpm (with overboost 284.4 Nm at 2750 rpm).

Data for rolling coefficient and for the "magic formula" for lateral forces, longitudinal forces and aligning torque of tires:

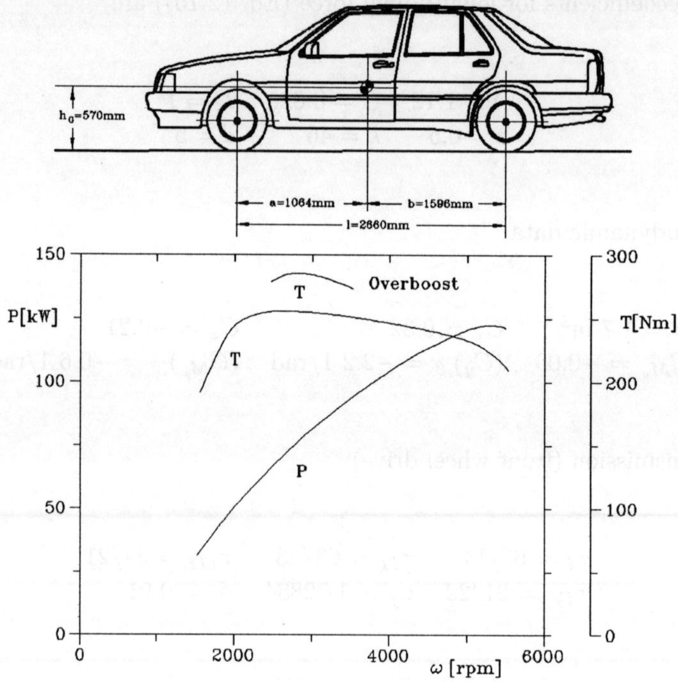

Fig. A.2 Vehicle referred to as *example 2.* Sketch and engine power and torque curves.

$f_0 = 0.013$	$K = 6.5 \times 10^{-6}\ \mathrm{s^2/m^2}$	$R_e = 287$ mm
$a_0 = 1.6929$	$a_1 = -55.2084$	$a_2 = 1271.2800$
$a_3 = 1601.8000$	$a_4 = 6.4946$	$a_5 = 4.7966 \times 10^{-3}$
$a_6 = -0.3875$	$a_7 = 1.0000$	$a_8 = -4.5399 \times 10^{-2}$
$a_9 = 4.2832 \times 10^{-3}$	$a_{10} = 8.6536 \times 10^{-2}$	$a_{111} = -7.9730$
$a_{112} = -0.2231$	$a_{12} = 7.6680$	$a_13 = 45.8764$
$b_0 = 1.6500$	$b_1 = -7.6118$	$b_2 = 1122.6$
$b_3 = -7.3600 \times 10^{-3}$	$b_4 = 144.8200$	$b_5 = -7.6614 \times 10^{-2}$
$b_6 = -3.8600 \times 10^{-3}$	$b_7 = 8.5055 \times 10^{-2}$	$b_8 = 7.5719 \times 10^{-2}$
$b_9 = 2.3655 \times 10^{-2}$	$b_{10} = 2.3655 \times 10^{-2}$	
$c_0 = 2.2264$	$c_1 = -3.0428$	$c_2 = -9.2284$
$c_3 = 0.500088$	$c_4 = -5.56696$	$c_5 = -0.25964$
$c_6 = -1.29724 \times 10^{-3}$	$c_7 = -0.358348$	$c_8 = 3.74476$
$c_9 = -15.1566$	$c_{10} = 2.1156 \times 10^{-3}$	$c_{11} = 3.4600 \times 10^{-4}$
$c_{12} = 9.13952 \times 10^{-3}$	$c_{13} = -0.244556$	$c_{14} = 0.100695$
$c_{15} = -1.3980$	$c_{16} = 0.44441$	$c_{17} = -0.998344$

Aerodynamic data:

$S = 2.06$ m^2	$C_x = 0.36$	$C_z = -0.12$
$C_{M_y} = -0.05$	$(C_y)_{,\beta} = -1.8$ 1/rad	$(C_{M_z})_{,\beta} = -0.5$ 1/rad

Transmission (front wheel drive):

$\tau_I = 1/3.750$	$\tau_{II} = 1/2.235$	$\tau_{III} = 1/1.518$
$\tau_{IV} = 1/1.132$	$\tau_V = 1/0.028$	$\tau_f = 20/59$
$\eta_t = 0.91$		

A.3 Example 3: large rear-wheel drive sports car

The vehicle is a mid-engine two-seater sports car with a 3.5 liter spark ignition supercharged engine (Fig. A.3).

The main geometrical data and inertial properties of the vehicle, wheels, engine, transmission are

length = 4250 mm	width = 1900 mm	height = 1160 mm
$m = 1480$ kg	$t_1 = 1502$ mm	$t_2 = 1578$ mm
$J_x = 590$ kg m^2	$J_y = 1730$ kg m^2	$J_z = 1950$ kg m^2
$J_{xz} = -50$ kg m^2	J_{w_1}(each) $= 0.7$ kg m^2	J_{w_2}(each) $= 0.7$ kg m^2
$J_e = 0.7$ kg m^2	$J_t = 0.08$ kg m^2	

$P_{max} = 235$ kW at 7200 rpm; $T_{max} = 324$ Nm at 5000 rpm. The power and torque curves of the engine are reported in Fig. A.3.

Data for rolling coefficient and for the "magic formula" for lateral forces, longitudinal forces and aligning torque of tires:

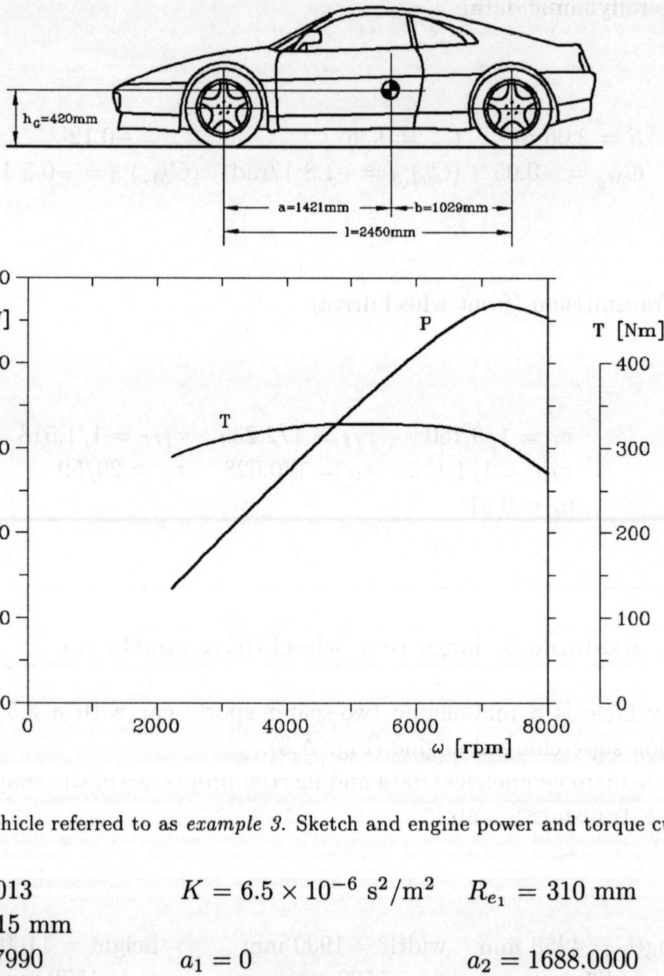

Fig. A.3 Vehicle referred to as *example 3*. Sketch and engine power and torque curves.

$f_0 = 0.013$	$K = 6.5 \times 10^{-6}\ \mathrm{s^2/m^2}$	$R_{e_1} = 310$ mm
$R_{e_2} = 315$ mm		
$a_0 = 1.7990$	$a_1 = 0$	$a_2 = 1688.0000$
$a_3 = 4140.0000$	$a_4 = 6.0260$	$a_5 = 0$
$a_6 = -0.3589$	$a_7 = 1.0000$	$a_8 = 0$
$a_9 = -6.1110 \times 10^{-3}$	$a_{10} = -3.2240 \times 10^{-2}$	$a_{111} = 0$
$a_{112} = 0$	$a_{12} = 0$	$a_1 3 = 0$
$b_0 = 1.65$	$b_1 = 0$	$b_2 = 1688$
$b_3 = 0$	$b_4 = 229$	$b_5 = 0$
$b_6 = 0$	$b_7 = 0$	$b_8 = -10$
$b_9 = 0$	$b_{10} = 0$	
$c_0 = 2.0680$	$c_1 = -6.4900$	$c_2 = -21.850$
$c_3 = 0.4160$	$c_4 = -21.3100$	$c_5 = 2.9420 \times 10^{-2}$
$c_6 = 0$	$c_7 = -1.1970$	$c_8 = 5.2280$

$$c_9 = -14.8400 \qquad c_{10} = 0 \qquad c_{11} = 0$$
$$c_{12} = -3.7360 \times 10^{-3} \qquad c_{13} = 3.8910 \times 10^{-2} \qquad c_{14} = 0$$
$$c_{15} = 0 \qquad c_{16} = 0.6390 \qquad c_{17} = 1.6930$$

Aerodynamic data:

$$S = 1.824 \text{ m}^2 \qquad C_x = 0.335 \qquad C_z = -0.34$$
$$C_{M_y} = 0 \qquad (C_y)_{,\beta} = -2.3 \text{ 1/rad} \qquad (C_N)_{,\beta} = -0.3 \text{ 1/rad}$$

Transmission (rear wheel drive) (the value of τ_f is inclusive of the reduction gears located between engine and gearbox):

$$\tau_I = 1/3.214 \qquad \tau_{II} = 1/2.105 \qquad \tau_{III} = 1/1.458$$
$$\tau_{IV} = 1/1.094 \qquad \tau_V = 1/0.861 \qquad \tau_f = 1/4.051$$
$$\eta_t = 0.87$$

A.4 Example 4: articulated truck

The vehicle is an articulated truck with a two-axle tractor and a three-axle trailer (Fig. A.4).

Tractor

The main geometrical data and inertial properties of the vehicle, wheels, engine, transmission are:

$$m = 7150 \text{ kg} \qquad t_1 = 2100 \text{ mm} \qquad t_2 = 1835 \text{ mm}$$
$$J_x = 4500 \text{ kg m}^2 \qquad J_y = 25800 \text{ kg m}^2 \qquad J_z = 27000 \text{ kg m}^2$$
$$J_{xz} = -3800 \text{ kg m}^2 \qquad J_w = 2.5 \text{ kg m}^2 \text{ (each)}$$
$$J_e = 2.55 \text{ kg m}^2 \qquad J_t = 1.1 \text{ kg m}^2$$

The vertical and torsional stiffnesses of the suspensions are:

$$K_1 = 2100 \text{ N/m} \qquad K_2 = 2800 \text{ N/m} \qquad K_{t_1} = 3790 \text{ Nm/rad}$$
$$K_{t_2} = 3790 \text{ Nm/rad}$$

Aerodynamic data:

$$S = 5.14 \text{ m}^2 \qquad C_x = 0.45 \qquad C_x = 0.65 \text{ (whole vehicle)}$$
$$C_z = 0 \qquad (C_y)_{,\beta} = -2.2 \text{ 1/rad} \qquad (C_{M_z})_{,\beta} = -1.5 \text{ 1/rad}$$

Transmission (rear-wheel drive):

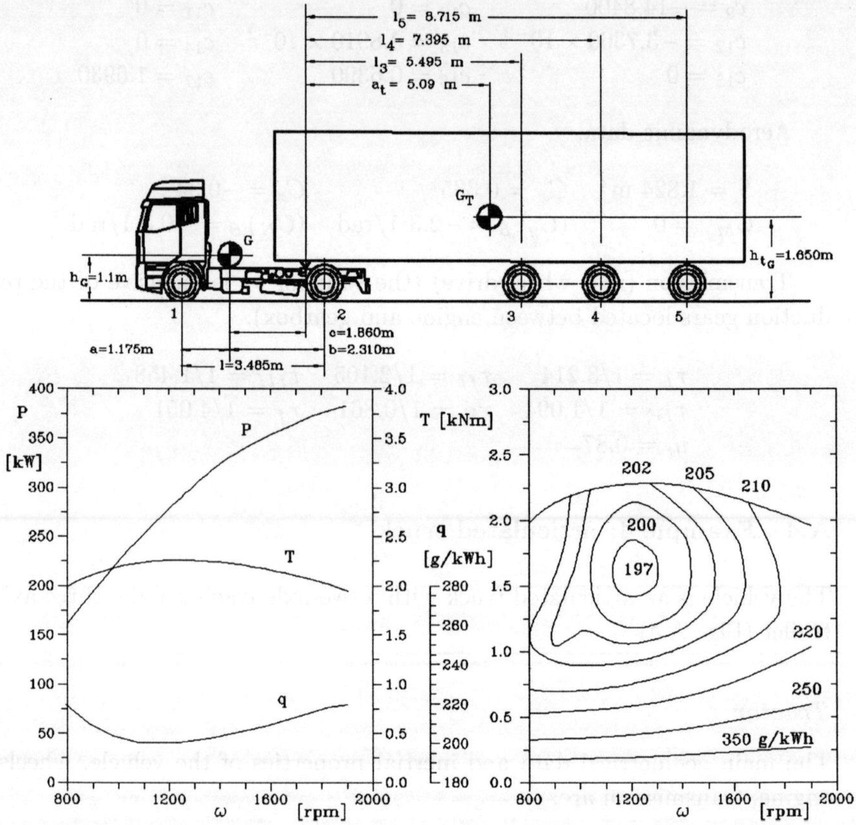

Fig. A.4 Vehicle referred to as *example 4*. Sketch and map of the engine.

$\tau_I = 1/12.5$ $\tau_{II} = 1/8.35$ $\tau_{III} = 1/6.12$
$\tau_{IV} = 1/4.56$ $\tau_V = 1/3.38$ $\tau_{VI} = 1/2.47$
$\tau_{VII} = 1/2.14$ $\tau_{VIII} = 1/1.81$ $\tau_{IX} = 1/1.57$
$\tau_X = 1/1.35$ $\tau_{XI} = 1/1.17$ $\tau_{XII} = 1.00$
$\tau_{XIII} = 1/0.87$ $\tau_f = 1/4.263$ $\eta_{tI} = 0.81$
$\eta_{tII} = 0.84$ $\eta_{tIII} = 0.84$ $\eta_{tIV} = 0.84$
$\eta_{tV} = 0.89$ $\eta_{tVI} = 0.87$ $\eta_{tVII} = 0.84$
$\eta_{tVIII} = 0.87$ $\eta_{tIX} = 0.84$ $\eta_{tX} = 0.87$
$\eta_{tXI} = 0.84$ $\eta_{tXII} = 0.93$ $\eta_{tXIII} = 0.89$

Trailer

Inertial properties of the trailer:

$m = 32000$ kg	$t_3 = 1835$ mm	$t_4 = 1835$ mm
$t_5 = 2100$ mm	$J_x = 30{,}000$ kg m^2	$J_z = 285{,}000$ kg m^2
$J_z = 285{,}000$ kg m^2		

Vertical and torsional stiffnesses of the suspensions:

$K_3 = 2150$ N/m	$K_4 = 2150$ N/m	$K_5 = 1380$ N/m
$K_{t_4} = 3200$ Nm/rad	$K_{t_2} = 3200$ Nm/rad	$K_{t_2} = 2800$ Nm/rad

Aerodynamic data:

$S = 7.5$ m^2	$C_x = 0.25$	$C_z = 0$
$(C_y)_{,\beta} = -2.35$ 1/rad	$(C_{M_z})_{,\beta} = -0.6$ 1/rad	referred to l_5

Tires

Axles 1 and 5 have single tires, axles 2, 3 and 4 have twin tires. Data for rolling coefficient and basic data for cornering forces and aligning torque:

$f_0 = 0.008$	$K = 0$	$R_e = 460$ mm
$a_3 = 5019.3$	$a_4 = 65.515$	
$c_3 = -0.6100$	$c_4 = -2.3400$	$c_5 = 0.02727$

A.5 Example 5: racing motorcycle

The vehicle is a racing motorcycle (Fig. A.5).

The main geometrical data and inertial properties of the vehicle (moment of inertia J_x is referred to an axis laying on the ground), wheels, engine, transmission are:

$m_1 = 20$ kg	$m_2 = 200$ kg	$m_d = 70$ kg (driver)
$J_x = 80$ kg m^2	$J_y = 110$ kg m^2	$J_z = 40$ kg m^2
$J_{xz} = 0$	$J_{1_z} = 2$ kg m^2	$J_{1_{xz}} = 1$kg m^2
$J_{p_1} = 0.4$	$J_{p_2} = 0.4$ kg m^2	$J_e = 0.08$ kg m^2
$l = 1315.9$ mm	$a = 641.7$ mm	$b = 674.2$ mm
$c_1 = 561.5$ mm	$c_2 = 54.2$ mm	$e = 50$ mm
$e_1 = 95$ mm	$h = 495.6$ mm	$h_1 = 432$ mm
$h_2 = 500$ mm	$\eta = 23\,^\circ$	$c_\delta = 8$ Nms/rad

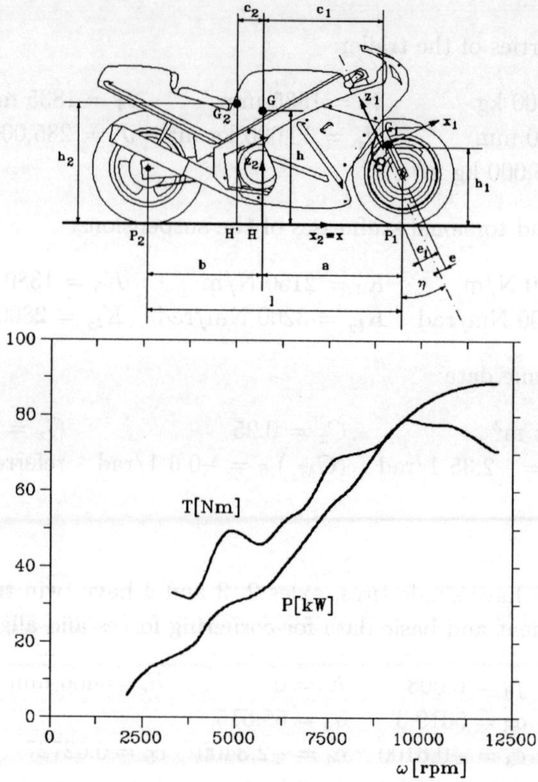

Fig. A.5 Vehicle referred to as *example 5.* Sketch and engine power and torque curves.

$P_{max} = 88.2$ kW at 11900 rpm; $T_{max} = 76.3$ Nm at 9750 rpm.
Wheels and tires, geometrical and linearized data:

$Re_1 = 300$ mm	$Re_2 = 300$ mm	$r_{t_1} = 70$ mm
$r_{t_2} = 110$ mm	$r_t = 90$ mm	
$f_0 = 0.01$	$K = 4 \times 10^{-6}$ s^2/m^2	$\mu_{x_{max}} = 1.1$
$(C_1),_Z = 27.27$ 1/rad	$(C_2),_Z = 30.0$ 1/rad	$(M_{z_1}),_{\alpha,Z} = 0.210$ m/rad
$(M_{z_2}),_{\alpha,Z} = 0.228$ m/rad	$(F_{y_1}),_{\gamma,Z} = -1.177$ 1/rad	$(F_{y_2}),_{\gamma,Z} = -1.367$ 1/rad

Data for the "magic formula" for lateral forces, longitudinal forces and aligning torque, front tire:

$a_0 = 1.5000$	$a_1 = -27.8700$	$a_2 = 1275.0000$
$a_3 = -1000.000$	$a_4 = 4.0000$	$a_5 = 0.015$
$a_6 = -0.3500$	$a_7 = -1.9950$	$a_8 = 0.058$
$a_9 = 0$	$a_{10} = 0$	$a_{11} = 5$
$a_{12} = 0$	$a_{13} = 0$	
$b_3 = 49.6$	$b_4 = 226$	$b_5 = 0.069$
$c_0 = 2.4000$	$c_1 = -2.7200$	$c_2 = -2.2800$
$c_3 = 1.8600$	$c_4 = 2.7300$	$c_5 = 0.1100$
$c_6 = 0.0300$	$c_7 = -0.07000$	$c_8 = 0.643$
$c_9 = -4.0400$	$c_{10} = -0.0700$	$c_{11} = -0.0150$
$c_{12} = 0$	$c_{13} = 0$	$c_{14} = -0.0660$
$c_{15} = 0.9450$	$c_{16} = 0$	$c_{17} = 0$

Data for the "magic formula" for lateral forces, longitudinal forces and aligning torque, rear tire:

$a_0 = 1.5000$	$a_1 = -27.8700$	$a_2 = 1275.0000$
$a_3 = -1100.000$	$a_4 = 4.0000$	$a_5 = 0.010$
$a_6 = -0.3500$	$a_7 = -1.9950$	$a_8 = 0.058$
$a_9 = 0$	$a_{10} = 0$	$a_{11} = 5$
$a_{12} = 0$	$a_{13} = 0$	
$b_3 = 49.6$	$b_4 = 226$	$b_5 = 0.069$
$c_0 = 2.4000$	$c_1 = -2.7200$	$c_2 = -2.2800$
$c_3 = 1.8600$	$c_4 = 2.7300$	$c_5 = 0.1100$
$c_6 = 0.0300$	$c_7 = -0.07000$	$c_8 = 0.643$
$c_9 = -4.0400$	$c_{10} = -0.0700$	$c_{11} = -0.0150$
$c_{12} = 0$	$c_{13} = 0$	$c_{14} = -0.0660$
$c_{15} = 0.9450$	$c_{16} = 0$	$c_{17} = 0$

Aerodynamic data:

$S = 1\ \mathrm{m}^2$	$C_x = 0.23$	$C_z = 0.10$
$C_{M_y} = 0$	$(C_y)_{,\beta} = 0.026$ 1/rad	$(C_{M_z})_{,d} = 0.065$ 1/rad

Transmission (the values of τ_i are inclusive of the reduction gear located between engine and gearbox):

$\tau_I = 1/4.91$	$\tau_{II} = 1/3.84$	$\tau_{III} = 1/3.22$
$\tau_{IV} = 1/2.81$	$\tau_V = 1/2.5$	$\tau_{VI} = 1/2.29$
$\tau_f = 1/3$	$\eta_t = 0.88$	

Bibliography

The information about motor vehicle dynamics is mostly found in papers published in journals or presented at conferences. Although relevant papers can be found in journals which deal with mechanical systems, dynamics or mechanical engineering in general (e.g. *Dynamics & Control*, Kluwer Academic Publ., Dordrecht, the Netherlands; *Journal of Guidance, Control and Dynamics*, AIAA, New York, USA; many journals of American Society of Mechanical Engineers (ASME) and other similar Societies), there are specialized journals addressing to Motor Vehicle Technology.

The Society of Automotive Engineers (SAE) publishes specialized journals, organizes conferences and promotes Motor Vehicle Technology in many other ways.

Among the specialized journals, the following can be listed (in alphabetical order)
ATA, Associazione Tecnica dell'Automobile, Torino, Italy,
ATZ, Automobiltechnische Zeitschrift, Berlin, Germany,
Automotive Engineer, SAE, New York, Warrendale (PA) USA,
International Journal of Vehicle Design, Inderscience, Geneva, Switzerland,
Vehicle System Dynamics, Swets & Zeitlinger, Lisse, the Netherlands.

Some books dealing with the aspects of motor vehicle dynamics mentioned in the present text are

(1) E. Koenig, R. Fachsenfield, *Aerodynamik des Kraftfahrzeuge*, Verlag der Motor, Rundshou-Umsha Verlag, Frankfurt A.M. 1951
(2) Bussien, *Automobiltechnisches Handbuch*, Technischer Verlag H Cam. Berlin, 1953
(3) M. Bencini, *Dinamica del veicolo*, Tamburini, Milano, 1956
(4) C. Deutsch, *Dynamique des vehicles routiers*, Organisme National de

Sécurité Routière
(5) W. Steeds, *Mechanics of Road Vehicles*, Iliffe & Sons, London, 1960
(6) G.H. Tidbury, *Advances in Automobile Engineering*, Pergamon Press, London, 1965
(7) F. Pernau, *Die entscheidenden Reifeneigen shaften*, Vortragstext, Eszter, 1967
(8) J.R. Ellis, *Vehicle Dynamics*, Business Books Ltd., London, 1969
(9) G. Pollone, *Il veicolo*, Levrotto & Bella, Torino, 1970
(10) H.C.A Van Eldik Thieme, H.B. Pacejka, *The tire as a vehicle component*, Vehicle Res. Lab., Delft University of Technology, 1971
(11) M. Mitschke, *Dinamik der Kraftfahzeuge*, Springer, Berlin, 1972
(12) A. Morelli, "Costruzioni automobilistiche", in *Enciclopedia dell'Ingegneria*, ISEDI, Milano, 1972
(13) A.J. Scibor Ryilski, *Road Vehicle Aerodynamics*, Pentech Press, London, 1975
(14) M.D. Artamonov, V.A. Ilarionov, M.M. Morin, *Motor Vehicles, Fundamentals and Design*, MIR, Moscow, 1976
(15) W.H. Hucho, "The Aerodynamic drag of cars. Current understanding, unresolved problems, and future prospects", in *Aerodynamic drag mechanism of bluff bodies and road vehicles*, Plenum Press, New York, 1978
(16) R. H. Macmillan, *Dynamics of Vehicle Collisions*, Inderscience Enterprises, Jersey, 1983
(17) E. Fiala, *Ingegneria automobilistica*, in Manuale di ingegneria meccanica, part 2, EST, Milan, 1985
(18) G.G. Lucas, *Road Vehicle Performance*, Gordon & Breach, London, 1986
(19) J.C. Dixon, *Tyres, Suspension and Handling*, Cambridge University Press, Cambridge, 1991
(20) T. D. Gillespie, *Fundamentals of Vehicle Dynamics*, Society of Automotive Engineers, Warrendale, 1992
(21) D. Bastow, G.P. Howard, *Car Suspension and Handling*, Pentech Press, London, and Society of Automotive Engineers, Warrendale, 1993
(22) G. Genta, *Meccanica dell'autoveicolo*, Levrotto & Bella, Torino, 1993
(23) W.F. Milliken, D.L. Milliken, *Race Car Vehicle Dynamics*, Society of Automotive Engineers, Warrendale, 1995

Index

Series on Advances in Mathematics for Applied Sciences

Aims and Scope

This Series reports on new developments in mathematical research relating to methods, qualitative and numerical analysis, mathematical modeling in the applied and the technological sciences. Contributions related to constitutive theories, fluid dynamics, kinetic and transport theories, solid mechanics, system theory and mathematical methods for the applications are welcomed.

This Series includes books, lecture notes, proceedings, collections of research papers. Monograph collections on specialized topics of current interest are particularly encouraged. Both the proceedings and monograph collections will generally be edited by a Guest editor.

High quality, novelty of the content and potential for the applications to modern problems in applied science will be the guidelines for the selection of the content of this series.

Instructions for Authors

Submission of proposals should be addressed to the editors-in-charge or to any member of the editorial board. In the latter, the authors should also notify the proposal to one of the editors-in-charge. Acceptance of books and lecture notes will generally be based on the description of the general content and scope of the book or lecture notes as well as on sample of the parts judged to be more significantly by the authors.

Acceptance of proceedings will be based on relevance of the topics and of the lecturers contributing to the volume.

Acceptance of monograph collections will be based on relevance of the subject and of the authors contributing to the volume.

Authors are urged, in order to avoid re-typing, not to begin the final preparation of the text until they received the publisher's guidelines. They will receive from World Scientific the instructions for preparing camera-ready manuscript.

SERIES ON ADVANCES IN MATHEMATICS FOR APPLIED SCIENCES

Published:

Vol. 1 Mathematical Topics in Nonlinear Kinetic Theory II: The Enskog Equation
by N. Bellomo et al.

Vol. 2 Discrete Models of Fluid Dynamics
ed. A. S. Alves

Vol. 3 Fluid Dynamics Applications of the Discrete Boltzmann Equation
by R. Monaco and L. Preziosi

Vol. 4 Waves and Stability in Continuous Media
ed. S. Rionero

Vol. 5 A Theory of Latticed Plates and Shells
by G. I. Pshenichnov

Vol. 6 Recent Advances in Combustion Modelling
ed. B. Larrouturou

Vol. 7 Linear Kinetic Theory and Particle Transport in Stochastic Mixtures
by G. C. Pomraning

Vol. 8 Local Stabilizability of Nonlienar Control Systems
by A. Bacciotti

Vol. 9 Nonlinear Kinetic Theory and Mathematical Aspects of Hyperbolic Systems
ed. V. C. Boffi

Vol. 10 Nonlinear Evolution Equations. Kinetic Approach
by N. B. Maslova

Vol. 11 Mathematical Problems Relating to the Navier-Stokes Equation
ed. G. P. Galdi

Vol. 12 Thermodynamics and Kinetic Theory
eds. W. Kosiński et al.

Vol. 13 Themomechanics of Phase Transitions in Classical Field Theory
by A. Romano

Vol. 14 Applications of Padé Approximation Theory in Fluid Dynamics
by A. Pozzi

Vol. 15 Advances in Mathematical Modelling of Composite Materials
ed. K. Z. Markov

Vol. 16 Advanced Mathematical Tools in Metrology
eds. P. Ciarlini et al.

SERIES ON ADVANCES IN MATHEMATICS FOR APPLIED SCIENCES

Vol. 17 The Fokker–Planck Equation for Stochastic Dynamical Systems and Its Explicit Steady State Solution
by C. Soize

Vol. 18 Calculus of Variation, Homogenization and Continuum Mechanics
eds. G. Bouchitté et al.

Vol. 19 A Concise Guide to Semigroups and Evolution Equations
by A. Belleni-Morante

Vol. 20 Global Controllability and Stabilization of Nonlinear Systems
by S. Nikitin

Vol. 21 High Accuracy Non-Centered Compact Difference Schemes for Fluid Dynamics Applications
by A. I. Tolstykh

Vol. 22 Advances in Kinetic Theory and Computing: Selected Papers
ed. B. Perthame

Vol. 23 Waves and Stability in Continuous Media
eds. S. Rionero and T. Ruggeri

Vol. 24 Impulsive Differential Equations with a Small Parameter
by D. Bainov and V. Covachev

Vol. 25 Mathematical Models and Methods of Localized Interaction Theory
by A. I. Bunimovich and A. V. Dubinskii

Vol. 26 Recent Advances in Elasticity, Viscoelasticity and Inelasticity
ed. K. R. Rajagopal

Vol. 27 Nonstandard Methods for Stochastic Fluid Mechanics
by M. Capinski and N. J. Cutland

Vol. 28 Impulsive Differential Equations: Asymptotic Properties of the Solutions
by D. Bainov and P. Simeonov

Vol. 29 The Method of Maximum Entropy
by H. Gzyl

Vol. 30 Lectures on Probability and Second Order Random Fields
by D. B. Hernández

Vol. 31 Parallel and Distributed Signal and Image Integration Problems
eds. R. N. Madan et al.

Vol. 32 On the Way to Understanding The Time Phenomenon: The Constructions of Time in Natural Science — Part 1. Interdisciplinary Time Studies
ed. A. P. Levich

SERIES ON ADVANCES IN MATHEMATICS FOR APPLIED SCIENCES

Vol. 33 Lecture Notes on the Mathematical Theory of the Boltzmann Equation
ed. N. Bellomo

Vol. 34 Singularly Perturbed Evolution Equations with Applications to Kinetic Theory
by J. R. Mika and J. Banasiak

Vol. 35 Mechanics of Mixtures
by K. R. Rajagopal and L. Tao

Vol. 36 Dynamical Mechanical Systems Under Random Impulses
by R. Iwankiewicz

Vol. 37 Oscillations in Planar Dynamic Systems
by R. E. Mickens

Vol. 38 Mathematical Problems in Elasticity
ed. R. Russo

Vol. 39 On the Way to Understanding the Time Phenomenon: The Constructions of Time in Natural Science — Part 2. The "Active" Properties of Time According to N. A. Kozyrev
ed. A. P. Levich

Vol. 40 Advanced Mathematical Tools in Metrology II
eds. P. Ciarlini et al.

Vol. 41 Mathematical Methods in Electromagnetism Linear Theory and Applications
by M. Cessenat

Vol. 42 Constrained Control Problems of Discrete Processes
by V. N. Phat

Vol. 43 Motor Vehicle Dynamics: Modeling and Simulation
by G. Genta

Vol. 44 Microscopic Theory of Condensation in Gases and Plasma
by A. L. Itkin and E. G. Kolesnichenko

Vol. 45 Advanced Mathematical Tools in Metrology III
eds. P. Cianlini et al.

Vol. 46 Mathematical Topics in Neutron Transport Theory — New Aspects
by M. Mokhtar-Kharroubi

Vol. 47 Theory of the Navier–Stokes Equations
eds. J. G. Heywood et al.

Vol. 48 Advances in Nonlinear Partial Differential Equations and Stochastics
eds. S. Kawashima and T. Yanagisawa

SERIES ON ADVANCES IN MATHEMATICS FOR APPLIED SCIENCES